Lecture Notes in Mathematics 1796

Springer
Berlin
Heidelberg
New York
Hong Kong
London
Milan
Paris
Tokyo

Jens M. Melenk

hp-Finite Element Methods for Singular Perturbations

Springer

Author

Jens M. Melenk

Max Planck Institute for Mathematics in the Sciences
Inselstr. 22
04103 Leipzig
Germany

e-mail: melenk@mis.mpg.de

Cataloging-in-Publication Data applied for

Die Deutsche Bibliothek - CIP-Einheitsaufnahme

Melenk, Jens M.:
hp-finite element methods for singular perturbations / Jens M. Melenk. -
Berlin ; Heidelberg ; New York ; Barcelona ; Hong Kong ; London ; Milan ;
Paris ; Tokyo : Springer, 2002
(Lecture notes in mathematics ; Vol. 1796)
ISBN 3-540-44201-4

Mathematics Subject Classification (2000): 65N30, 65N35, 58J37, 35J25

ISSN 0075-8434
ISBN 3-540-44201-4 Springer-Verlag Berlin Heidelberg New York

Springer-Verlag Berlin Heidelberg New York a member of BertelsmannSpringer
Science + Business Media GmbH

http://www.springer.de

Typesetting: Camera-ready TEX output by the author

SPIN: 10891013 41/3142/ du - 543210 - Printed on acid-free paper

Preface

Many partial differential equations arising in practice are parameter-dependent problems and are of singularly perturbed type for small values of this parameter. These include various plate and shell models for small thickness in solid mechanics, the convection-diffusion equation, Oseen equation, and Navier-Stokes equations in fluid flow problems where the fluid is assumed to have small viscosity, and finally equations arising in semi-conductor device modelling. Analysis of such equations by numerical methods such as the finite element method is an important task in today's computational practice. A significant design aspect of numerical methods for such parameter-dependent problems is robustness, that is, that the performance of the numerical method is independent of, or at least fairly insensitive to, the parameter. Numerous methods have been proposed and analyzed both theoretically and computationally for a variety of singularly perturbed problems—we merely refer at this point to the three recent monographs [97,99,108] and their extensive bibliographies.
Most numerical methods employed in the study of singularly perturbed problems are low order methods such as the classical h-version finite element method (FEM), where convergence is obtained by refining the mesh while keeping the approximation order fixed. In high order methods such as the hp-version of the finite element method (hp-FEM), mesh refinement can be combined with increasing the approximation order; for certain problem classes, this added flexibility of the hp-FEM allows it to achieve exponential rates of convergence.
The present book is devoted to a complete analysis of the hp-FEM for a class of singularly perturbed problems on curvilinear polygons. To the knowledge of the author this work represents the first *robust exponential convergence* result for a class of singularly perturbed problems under realistic assumptions on the input data, that is, piecewise analyticity of the coefficients of the differential equation and the geometry of the domain.
This work is at the intersection of several active research areas that have their own distinct approaches and techniques: numerical methods for singular perturbation problems, high order numerical methods for elliptic problems in non-smooth domains, regularity theory for singularly perturbed problems in terms of asymptotic expansions, and regularity theory for elliptic problems in curvilinear-polygons. Although, naturally, the present work draws on techniques employed in all of these fields, new tools and regularity results for the solutions had to be developed for a rigorous robust exponential convergence proof.

This book comprises research undertaken during my years at ETH Zürich. I take this opportunity to thank Prof. Dr. C. Schwab for many stimulating discussions on the topics of this book and for his support and encouragement over the years.

Leipzig, June 2002 *J.M. Melenk*

Preface

Contents

Notation

General notation

$\mathbb{N}$	The set of positive integers, $\{1, 2, \ldots, \}$.
$\mathbb{N}_0$	The set of non-negative integers $\mathbb{N} \cup \{0\}$.
$\mathbb{Z}$	The set of integers $\mathbb{N}_0 \cup -\mathbb{N}$.
$\mathbb{R}, \mathbb{R}^+, \mathbb{R}_0^+$	The real, the positive real, and the non-negative real numbers.
$\mathbb{C}$	The complex numbers.
$\mathbf{i}$	The imaginary unit with $\mathbf{i}^2 = -1$.
$\Gamma(\cdot)$	The Gamma function with $\Gamma(j+1) = j!$ for $j \in \mathbb{N}_0$.
δ_{ij}	The Kronecker symbol: $\delta_{ij} = 0$ for $i \neq j$ and $\delta_{ii} = 1$.
C, C', γ, γ', K, b	Generic constants independent of critical parameters such as ε, the differentiation order, the polynomial degree, etc. These constants may be different in different instances.
$\lfloor \cdot \rfloor$	$\lfloor x \rfloor = \max\{n \in \mathbb{Z} \mid n \leq x\}$.
$[\,\cdot\,]$	In Section 5.5: $[p] = \max\{1, p\}$ for $p \in \mathbb{Z}$ (see p. 198). In all other sections $[\,\cdot\,]$ denotes the jump operator.
$\subset\subset$	Compact embedding.
$\lvert K \rvert$	for $K \subset \mathbb{R}^n$ represents the Lebesgue measure (volume) of K.
$\mathbb{E}_\Omega$	The characteristic function of the set Ω.

Matrices

$\mathbb{M}^n$	The set of (real) $n \times n$ matrices.
$\mathbb{S}^n$	$\mathbb{S}^n \subset \mathbb{M}^n$ are the symmetric matrices.
$\mathbb{S}^n_>$	The set of symmetric positive definite matrices.
$:$	For matrices A, B, we set $A : B = \sum_{i,j} A_{ij} B_{ij}$.

Sets, balls, sectors, neighborhoods

$B_r(x)$	The ball of radius r around the point x.
B_R, B_R^+, B_R^-	Ball and half balls with radius R; see (5.5.1).
$\mathcal{U}_\kappa(K)$	The κ-neighborhood of the set K, i.e., $\cup_{x \in K} B_\kappa(x)$.
S	A generic sector, Definition 4.2.1, p. 146.
$S_R(\omega)$	A sector with opening angle ω, see (4.1.2).

$S_R^{0,\delta}(\omega)$, $S_R^{\omega,\delta}(\omega)$	Conical neighborhood of the lateral parts Γ_0, Γ_ω of the sector $S_R(\omega)$, see (5.4.30).
I_x, I_y	Intervals on $\mathbb{R}$; I_y is the form $I_y = [0,b]$ for a $b > 0$; see outset of Section 7.3.1.
S_X, $S_X(\delta)$	Complex neighborhoods of interval I_x, see (7.3.11).

Norms, differential operators, standard function spaces

$L^2(\Omega)$	The space of square integrable functions.
$H^k(\Omega)$	Sobolev space H^k of L^2-functions whose distributional derivatives of order up to k are also in L^2; cf. [1].
$H_0^1(\Omega)$	Sobolev space of H^1-functions with vanishing trace on $\partial\Omega$; cf. [1].
$H_{0,\varepsilon}^1(\Omega)$	The Sobolev space of H^1 functions with vanishing trace on $\partial\Omega$ equipped with the energy norm $\|\cdot\|_{L^2(\Omega)} + \varepsilon\|\nabla\cdot\|_{L^2(\Omega)}$; cf. p. 184.
$H^{1/2}(\Omega)$	The usual Sobolev space $H^{1/2}$; cf. [1].
$H_{00}^{1/2}(\Omega)$	The usual Sobolev space $H_{00}^{1/2}$; cf. [1].
$\|\cdot\|_\varepsilon$	Energy norm $\|u\|_\varepsilon \sim \|u\|_{L^2(\Omega)} + \varepsilon\|\nabla u\|_{L^2(\Omega)}$; cf. p. 3.
$\|\cdot\|_{\varepsilon,\alpha}$	Exponentially weighted energy norm, cf. p. 243.
$D^\alpha u$	For multi-indices $\alpha = (\alpha_1,\dots,\alpha_n) \in \mathbb{N}_0^n$ and (smooth) functions u defined on an open subset of $\mathbb{R}^n$: $D^\alpha u = \partial_{x_1}^{\alpha_1}\partial_{x_2}^{\alpha_2}\cdots\partial_{x_n}^{\alpha_n} u$.
$\nabla^p u(x)$	$\|\nabla^p u(x)\|^2 = \sum_{\alpha_1,\dots,\alpha_p=1}^{n} \|\partial_{\alpha_1}\partial_{\alpha_2}\cdots\partial_{\alpha_p} u(x)\|^2$, where, for tensor-valued functions $u = (u_i)_{i=1}^N$ and shorthand $\partial_\alpha u$ for $\partial_{x_\alpha} u$ $\|\partial_{\alpha_1}\partial_{\alpha_2}\cdots\partial_{\alpha_p} u(x)\|^2 = \sum_{i=1}^N \|\partial_{\alpha_1}\partial_{\alpha_2}\cdots\partial_{\alpha_p} u_i(x)\|^2$.
$\mathcal{A}(G)$	For domains $G \subset \mathbb{R}^n$ (or $\mathbb{C}^n$) $\mathcal{A}(G)$ denotes the set of functions analytic on G. For closed sets $\overline{G}$, $f \in \mathcal{A}(\overline{G})$ is understood to imply the existence of an open neighborhood of $\overline{G}$ on which f is analytic; see also (1.2.2).
$\mathcal{A}(G,\mathbb{R}^n)$	The set of vector-valued functions that are (componentwise) analytic on G.
$\mathcal{A}(G,\mathbb{S}_>^n)$	The set of functions from G to the symmetric positive definite matrices $\mathbb{S}_>^n$ that are (componentwise) analytic on G.
L_ε	The differential operator, (1.2.1).
$[\cdot]$	The jump operator across a curve. Only in Sec. 5.5: $[p] = \max\{1,p\}$ for $p \in \mathbb{Z}$; p. 198.
∂_{n_A}	The co-normal derivative operator $\mathbf{n}^T\nabla\cdot$.

Weight functions and weighted spaces

$\hat{\Phi}_{p,\beta,\varepsilon}$	Weight function in a sector, p. 147.
$\Phi_{p,\beta,\varepsilon}$	Weight function in a curvilinear polygon, p. 184.
$\hat{\Psi}_{p,\beta,\varepsilon,\alpha}$	Exponentially weighted weight function in a sector, p. 231.
$\Psi_{p,\beta,\varepsilon,\alpha}$	Exponentially weighted weight function in Ω; see (7.4.7).
$H^{m,l}_{\beta,\varepsilon}$	Weighted Sobolev space, p. 149.
$\mathcal{B}^{l}_{\beta,\varepsilon}$	Countably normed space, p. 149.
$H^{m,l}_{\beta,\varepsilon,\alpha}$	Exponentially weighted Sobolev space, p. 232.
$\mathcal{B}^{l}_{\beta,\varepsilon,\alpha}$	Exponentially weighted countably normed space, p. 232.
$\mathcal{E}$	Smallest characteristic length scale, p. 177.

Semi norms for controlling high order derivatives

$N_{R,p}(u)$, $N'_{R,p,q}(u)$, $N'^{,\pm}_{R,p,q}(u)$	Bounds on higher derivatives of u, p. 198.
$M_{R,p}(f)$	Bounds on higher derivatives of f, p. 202.
$M'_{R,p}(f)$, $\tilde{M}_{R,p}(f)$, $N'_{R,p}(u)$	Bounds on higher derivatives of f and u, p. 208.
$N'^{,\pm}_{R,p}(u)$, $H_{R,p}(u)$, $M'^{,\pm}_{R,p}(f)$	Bounds on higher derivatives of u and f, p. 217.

Description of the boundary and corner layer

ψ_j	Boundary fitted coordinates $(x,y) = \psi_j(\rho_j, \theta_j)$ in neighborhood of arc Γ_j, where ρ_j measures the distance of the point (x,y) to Γ_j; see Notation 2.3.3
A_j	Vertex of the curvilinear domain Ω, Section 1.2.
Γ_j	Analytic arc being part of the boundary of the curvilinear polygon Ω, Section 1.2.
S_j, S_j^+, S_j^-	Sectors near A_j for the definition of corner layer, (7.4.2), (7.4.3).
Ω_j, χ^{BL}, χ^{CL}	Subdomains of Ω and cut-off functions associated with arcs Γ_j and the vertices A_j; see Notation 2.3.3 and outset of Section 7.4.1.
Ω_j, χ_j^{BL}, χ_j^{CL}	Subdomains of Ω and cut-off functions associated with arcs Γ_j and the vertices A_j; see Notation 2.3.3.
s_κ, $\tilde{s}_{j,\kappa}$	Anisotropic and anisotropic stretching maps; see Notation 2.4.3.

Polynomials, approximation, and projections

I, S, T	The reference interval $I = (0,1)$, square $S = I \times I$, and triangle $T = \{(x,y) \mid 0 < x < 1, 0 < y < 1 - x\}$, p. 87.
$\underline{S}$, $\underline{T}$	The references square and triangle in Section 3.2.3; see (3.2.19).
$\mathcal{P}_p(T)$, $\mathcal{Q}_p(S)$, $\Pi_p(K)$	Spaces of polynomials, p. 87.
$P_p^{(\alpha,\beta)}$	Jacobi polynomials, see [124].
L_p, $\tilde{L}_p$	$L_p = P_p^{(0,0)}$ is the usual Legendre polynomial; $\tilde{L}_p(x) = L_p(2x-1)$.
$\psi_{p,q}$	Orthogonal polynomials on the triangle, (3.2.23).
$\mathcal{GL}$	Gauss-Lobatto points, p. 87.
i_p, j_p	1D and 2D Gauss-Lobatto interpolation operators, p. 87.
$i_{p,\Gamma}$	Gauss-Lobatto interpolation operator on an edge Γ, p. 88.
E	Polynomial extension operator from the boundary, p. 89.
Π_p^∞	Polynomial projector defined in Theorem 3.2.20, p. 103.
$\Pi_p^{1,\infty}$	Polynomial projector defined in Theorem 3.2.24, p. 108.
$\Pi_p^{L^2}$	The L^2 projector into the space $\mathcal{P}_p$.

Meshes and finite element approximation

$\mathcal{T}$	Triangulation, p. 39.
$S^p(\mathcal{T})$, $S_0^p(\mathcal{T})$	Spaces of piecewise mapped polynomials of degree p on the mesh $\mathcal{T}$, p. 113.
$\Pi_{p,\mathcal{T}}^\infty$	Elementwise application of Π_p^∞ on a mesh $\mathcal{T}$, p. 113.

1. Introduction

1.1 Introduction

1.2 Problem class and assumptions

This work presents numerical analysis and regularity results for singularly perturbed equations of the form (1.2.1). Such equations are ubiquitous, appearing, for example in convection-dominated fluid flow, in semi-conductor device modelling, and solid mechanics (where their analysis is crucial for an understanding of the layer structure of Reissner-Mindlin plate models, [9, 10]).
We consider the following class of singularly perturbed equations:

$$L_\varepsilon u_\varepsilon := -\varepsilon^2 \nabla \cdot (A(x)\nabla u_\varepsilon) + b(x) \cdot \nabla u_\varepsilon + c(x) u_\varepsilon = f \text{ on } \Omega, \quad (1.2.1a)$$

$$u_\varepsilon = g \text{ on } \partial\Omega. \quad (1.2.1b)$$

The bounded Lipschitz domain $\Omega \subset \mathbb{R}^2$ is assumed to be a *curvilinear polygon* as depicted in Fig. 1.2.1. The boundary $\partial\Omega$ is assumed to consist of finitely

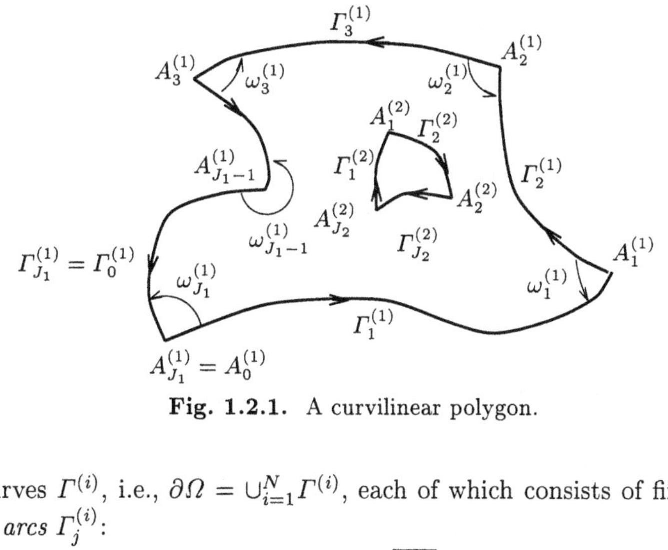

Fig. 1.2.1. A curvilinear polygon.

many curves $\Gamma^{(i)}$, i.e., $\partial\Omega = \cup_{i=1}^N \Gamma^{(i)}$, each of which consists of finitely many *analytic arcs* $\Gamma_j^{(i)}$:

$$\Gamma^{(i)} = \cup_{i=1}^{J_i} \overline{\Gamma_j^{(i)}}.$$

The arcs $\Gamma_j^{(i)}$ are parametrized by

$$\Gamma_j^{(i)} = \{(x_j^{(i)}(\theta), y_j^{(i)}(\theta) \mid \theta \in (0,1)\},$$

where the functions $x_J^{(i)}$, $y_j^{(i)}$ are analytic on a neighborhood of the interval $[0,1]$. We assume that

$$\left|\frac{d}{d\theta}x_j^{(i)}\right|^2 + \left|\frac{d}{d\theta}y_j^{(i)}\right|^2 > 0 \quad \text{on } [0,1] \text{ for all } i,\ j.$$

The curves $\Gamma_j^{(i)}$ are oriented such that the domain Ω is "on the left"; that is, the normal vector $(-\frac{d}{\theta}y_j^{(i)}(\theta), \frac{d}{\theta}x_j^{(i)}(\theta))$ points into Ω (cf. Fig. 1.2.1). The endpoints of the arc $\Gamma_j^{(i)}$ are the vertices $A_{j-1}^{(i)} = (x_j^{(i)}(0), y_j^{(i)}(0))$, $A_j^{(i)} = (x_j^{(i)}(1), y_j^{(i)}(1))$, and we set $A_0^{(i)} := A_{J_i}^{(i)}$. The internal angle at vertex $A_j^{(i)}$ is denoted $\omega_j^{(i)}$, and we exclude cusps by stipulating $0 < \omega_j^{(i)} < 2\pi$. In order to simplify the notation in this work, we assume without loss of generality that $N = 1$ and drop the superscript (i); i.e., we write $J = J_1$, $\Gamma_j = \Gamma_j^{(1)}$, $A_j = A_j^{(1)}$, etc. It is also convenient to write $\Gamma_0 = \Gamma_J$.
The remaining data appearing in (1.2.1) are assumed to be analytic: We suppose that $c \in \mathcal{A}(\overline{\Omega})$, $b \in \mathcal{A}(\overline{\Omega}, \mathbb{R}^2)$, and $A \in \mathcal{A}(\overline{\Omega}, \mathbb{S}^2_>)$; i.e., we stipulate the existence of C_A, C_b, C_c, and γ_A, γ_b, $\gamma_c > 0$ such that

$$\|\nabla^p A\|_{L^\infty(\Omega)} \le C_A \gamma_A^p p! \qquad \forall p \in \mathbb{N}_0, \tag{1.2.2a}$$
$$\|\nabla^p b\|_{L^\infty(\Omega)} \le C_b \gamma_b^p p! \qquad \forall p \in \mathbb{N}_0, \tag{1.2.2b}$$
$$\|\nabla^p c\|_{L^\infty(\Omega)} \le C_c \gamma_c^p p! \qquad \forall p \in \mathbb{N}_0. \tag{1.2.2c}$$

Furthermore, the matrix $A(x)$ is symmetric positive definite for each $x \in \Omega$ and there exists $\lambda_{min} > 0$ such that

$$A(x) \ge \lambda_{min} \qquad \forall x \in \Omega. \tag{1.2.2d}$$

We require the existence of $\mu > 0$ such that

$$-\frac{1}{2}(\nabla \cdot b)(x) + c(x) \ge \mu > 0 \qquad \forall x \in \Omega. \tag{1.2.2e}$$

The right-hand side f in (1.2.1) satisfies $f \in \mathcal{A}(\overline{\Omega})$, i.e., there are C_f, $\gamma_f > 0$ such that

$$\|\nabla^p f\|_{L^\infty(\Omega)} \le C_f \gamma_f^p p! \qquad \forall p \in \mathbb{N}_0. \tag{1.2.3}$$

Finally, the boundary data $g \in C(\partial\Omega)$ are assumed to be analytic on the arcs Γ_j: For each j, the function $g(x_j, y_j)$ is analytic on $[0,1]$, i.e., there are C_g, $\gamma_g > 0$ such that

$$\|D^p g(x_j(\cdot), y_j(\cdot))\|_{L^\infty((0,1))} \le C_g \gamma_g^p p! \qquad \forall p \in \mathbb{N}_0. \tag{1.2.4}$$

For most of our analysis, the singular perturbation parameter $\varepsilon \in (0,1]$ is assumed to be small, i.e., $\varepsilon \ll 1$.

Remark 1.2.1 The assumption of analyticity of the data on Ω can be relaxed. In fact, in most of the subsequent analysis, only piecewise analyticity of the data A, b, c, f, and g needs to be assumed. ▪

Solutions of (1.2.1) are understood in the weak sense; i.e., u_ε is the solution of the following problem:

$$\text{Find } u_\varepsilon \in H^1(\Omega) \text{ s.t. } u_\varepsilon|_{\partial\Omega} = g \text{ and } B_\varepsilon(u_\varepsilon, v) = F(v) \qquad \forall v \in H_0^1(\Omega). \tag{1.2.5}$$

Here, the bilinear form B_ε and the linear form F are defined by

$$B_\varepsilon(u,v) := \varepsilon^2 \int_\Omega (A(x)\nabla u)\cdot\nabla v + b(x)\cdot\nabla u v + c(x) u v\,dx, \tag{1.2.6a}$$

$$F(v) := \int_\Omega f(x) v\,dx. \tag{1.2.6b}$$

This bilinear form B_ε is closely connected with the *energy norm*

$$\|u\|_\varepsilon^2 := \varepsilon^2 \|\nabla u\|_{L^2(\Omega)}^2 + \|u\|_{L^2(\Omega)}^2 \tag{1.2.7}$$

as will become apparent in the subsequent Lemma 1.2.2. The bilinear form B_ε is coercive on the space $H_0^1(\Omega)$, and the variational formulation (1.2.5) has a unique solution even under the weaker assumptions $f \in L^2(\Omega)$ and $g \in H^{1/2}(\partial\Omega)$:

Lemma 1.2.2. *Let the coefficients A, b, c satisfy (1.2.2), $\varepsilon \in (0,1]$, and let $f \in L^2(\Omega)$, $g \in H^{1/2}(\partial\Omega)$. Then there exists a unique solution u_ε of (1.2.5) and a constant $C > 0$ independent of ε, f, and g such that*

$$\varepsilon\|\nabla u_\varepsilon\|_{L^2(\Omega)} + \|u_\varepsilon\|_{L^2(\Omega)} \le C\left[\|f\|_{L^2(\Omega)} + \|g\|_{H^{1/2}(\Omega)}\right].$$

Moreover, the bilinear form B_ε is coercive on the space $H_0^1(\Omega)$, and there holds

$$B_\varepsilon(u,u) \ge \varepsilon^2 \lambda_{min} \|\nabla u\|_{L^2(\Omega)}^2 + \mu \|u\|_{L^2(\Omega)}^2 \qquad \forall u \in H_0^1(\Omega). \tag{1.2.8}$$

The bilinear form B_ε is also continuous on the space $H^1(\Omega)$: There is $C > 0$ independent of A, b, c, and ε such that for all u, $v \in H^1(\Omega)$ we have:

$$|B_\varepsilon(u,v)| \le C\left[\|A\|_{L^\infty(\Omega)} + \|c\|_{L^\infty(\Omega)} + \|b\|_{L^\infty(\Omega)}\varepsilon^{-1}\right] \|u\|_\varepsilon \|v\|_\varepsilon. \tag{1.2.9}$$

Proof: (1.2.9) follows immediately from the Cauchy-Schwarz inequality.
As a first step, we show (1.2.8). We start by noting that for $u \in H_0^1(\Omega)$, an integration by parts gives

$$\int_\Omega (b\cdot\nabla u) u\,dx = -\int_\Omega (\nabla\cdot b) u^2 + u(b\cdot\nabla u)\,dx.$$

Therefore,

$$\int_\Omega (b\cdot\nabla u) u\,dx = -\frac{1}{2}\int_\Omega (\nabla\cdot b) u^2\,dx.$$

Combining this with assumption (1.2.2e) implies the coercivity of the bilinear form B_ε on the space $H_0^1(\Omega)$:

$$\begin{aligned} B_\varepsilon(u,u) &= \varepsilon^2 \int_\Omega (A(x)\nabla u)\cdot\nabla u + (b(x)\cdot\nabla u)u + c(x)u^2\,dx \\ &= \varepsilon^2 \int_\Omega (A(x)\nabla u)\cdot\nabla u + \left(c(x) - \frac{1}{2}\nabla\cdot b(x)\right)u^2\,dx \\ &\geq \varepsilon^2\lambda_{min}\|\nabla u\|^2_{L^2(\Omega)} + \mu\|u\|^2_{L^2(\Omega)}. \end{aligned}$$

This coercivity gives uniqueness of the solution of (1.2.5). In order to see existence of a solution, let $G \in H^1(\Omega)$ be an extension of g into Ω satisfying

$$G|_{\partial\Omega} = g, \qquad \|G\|_{H^1(\Omega)} \leq C\|g\|_{H^{1/2}(\partial\Omega)}$$

for some $C > 0$ depending only on Ω. The difference $\tilde{u} := u_\varepsilon - G$ must be the solution of the problem:

$$\text{Find } \tilde{u} \in H_0^1(\Omega) \text{ s.t. } B_\varepsilon(\tilde{u},v) = F(v) - B_\varepsilon(G,v) \qquad \forall v \in H_0^1(\Omega). \tag{1.2.10}$$

We see that for all $v \in H^1(\Omega)$

$$\begin{aligned} |F(v) - B_\varepsilon(G,v)| &\leq \|f\|_{L^2(\Omega)}\|v\|_{L^2(\Omega)} \\ &\quad + C\{\|G\|_{H^1(\Omega)}\varepsilon^2\|\nabla v\|_{L^2(\Omega)} + \|G\|_{H^1(\Omega)}\|v\|_{L^2(\Omega)} + \|G\|_{L^2(\Omega)}\|v\|_{L^2(\Omega)}\} \\ &\leq C\left\{\|f\|_{L^2(\Omega)} + \|G\|_{H^1(\Omega)}\right\}\|v\|_\varepsilon, \end{aligned}$$

where we assumed $\varepsilon \leq 1$. Therefore, by the classical Lax-Milgram Lemma, [36, 82], (1.2.10) indeed has a unique solution $\tilde{u}$ satisfying

$$\|\tilde{u}\|_\varepsilon \leq C\left[\|f\|_{L^2(\Omega)} + \|G\|_{H^1(\Omega)}\right].$$

Using $\varepsilon \leq 1$, we see that $u_\varepsilon := G + \tilde{u}$ satisfies the desired bounds. □

The greater part of our analysis will be done for the special case $b \equiv 0$; i.e., we consider the following singularly perturbed problem of elliptic-elliptic type:

$$-\varepsilon^2\nabla\cdot(A(x)\nabla u_\varepsilon) + c(x)u_\varepsilon = f \qquad \text{on } \Omega, \tag{1.2.11a}$$

$$u_\varepsilon = g \qquad \text{on } \partial\Omega, \tag{1.2.11b}$$

where assumption (1.2.2e) implies that $c \geq \mu > 0$ on Ω.

1.3 Principal results

The main result of the present work is the *robust exponential convergence* result Theorem 2.4.8 for high order finite element methods applied to (1.2.11). It is shown that with the proper choice of conforming subspaces V_N of dimension

$N \in \mathbb{N}$, the finite element method, i.e., Galerkin projection, yields approximants u_ε^N to the exact solution u_ε that satisfy

$$\|u_\varepsilon - u_\varepsilon^N\|_\varepsilon \leq Ce^{-bN^{1/3}}. \tag{1.3.1}$$

Here, the constants C, $b > 0$ are independent of ε; in fact, in our numerical experiments in Section 2.5 $b \approx 1$ and likewise $C = O(1)$. The finite element spaces V_N are given explicitly in Section 2.4. They consist of the usual piecewise polynomial spaces of degree p defined on meshes that are adapted to the length scale ε of the problem. Specifically, for the approximation with polynomial of degree p, these meshes are designed according to three principles:

1. near the edges of the domain, long, thin *needle* elements of width $O(p\varepsilon)$ are employed in order to capture boundary layer phenomena;
2. in an $O(p\varepsilon)$ neighborhood of the vertices a *geometric* mesh refinement is used in order to resolve corner singularities;
3. in the interior of the domain a standard coarse mesh is utilized for the resolution of smooth solution components.

It is worth stressing that the only information required for an application of these mesh design principles is the length scale ε of the problem, which is typically known in practice.

Let us compare our robust exponential convergence result with previous convergence analyses. Thus far only algebraic robust convergence results have been available, typical of low order methods. Robust algebraically convergent methods deliver approximate solutions u_ε^N from spaces V_N of dimension $N \in \mathbb{N}$ that satisfy error bounds of the form

$$\|u_\varepsilon - u_\varepsilon^N\|_\varepsilon \leq CN^{-\alpha}. \tag{1.3.2}$$

Here, C, $\alpha > 0$ are independent of ε. Even for optimally chosen meshes, $\alpha \leq 2$ is typical for two-dimensional problems. A good measure for comparing approximation results (1.3.1) and (1.3.2) is the alphanumerical work W required to compute the approximate solution u_ε^N. In the case of low order methods, an efficient iterative solver such as multigrid, [65], is essential for acceptable solution times. Such an optimal solution algorithm would solve the resulting linear system with linear complexity, i.e., $W = O(N)$. The best rate of convergence of these low order methods in terms of work W is therefore

$$\|u_\varepsilon - u_\varepsilon^N\|_\varepsilon \leq CW^{-\alpha}.$$

This work estimate, however, is based upon two strong assumptions. First, in order for α to be reasonable, e.g., $\alpha \approx 1$, the mesh has to be carefully designed so as to capture the relevant features of the solution. In particular, it has to contain highly anisotropic elements in the boundary layer. However, most state-of-the-art adaptive strategies do not allow for such elements: Their use of shape-regular elements precludes robustness, and the convergence rates visible in practice are

low, i.e., α is small. The second strong assumption made is the existence of multigrid methods (or, more generally, preconditioned iterative solvers) with linear complexity for meshes that contain anisotropic elements. Their construction is non-trivial, and, in fact, few results in this direction are available to date.
Let us turn to work estimates for high order approximations. Standard Gaussian elimination allows us to solve a linear system with work $W = \frac{1}{3}N^3 + O(N^2)$. Hence, if the linear system obtained in our high order method is so solved, our *hp*-approximation result takes the following form in the "error versus work" perspective:

$$\|u_\varepsilon - u_\varepsilon^N\|_\varepsilon \leq Ce^{-bW^{1/9}}. \tag{1.3.3}$$

Thus, even in terms of work, our exponential convergence result will (asymptotically) outperform methods with algebraic convergence rates. It should be pointed out that (1.3.3) is a rather crude estimate in that the estimate $W = \frac{1}{3}N^3+O(N^2)$ for the solution of the linear system does not make any sparsity assumptions on the matrix. However, even in high order methods, the resulting system matrices have structure and are sparse, which can be exploited by sophisticated modern direct solvers, [44, 52]. In practice, the work estimate (1.3.3) is therefore pessimistic.
Our robust exponential convergence result depends strongly on detailed regularity assertions for the solution of (1.2.1). A large portion of this work (Parts II, III) is therefore devoted to the derivation of new regularity results for (1.2.1). While our regularity assertions are interesting in their own right, they are derived in order to enable us to obtain a *priori* error bounds for piecewise polynomial approximation. This intended main application determines the type of the results and shapes their form. For our application it is essential to have bounds on higher order derivatives of the solution (or solution components) at any given point of the domain that are explicit in critical parameters such as the singular perturbation parameter, the distance to the vertices, and the distance to the boundary of the domain. For exponential convergence results it is furthermore necessary to have bounds on the derivatives of the solution that are explicit in the differentiation order. Finally, it is also important that these bounds depend on the given input data only, i.e., the coefficients of the differential equation, the geometry of the domain, the right-hand side, and the boundary data. This last requirement is closely connected with our goal to place assumptions on the input data that are typically met in practice and that can be checked explicitly. In contrast, existing convergence analyses are often done under the unrealistic assumption that certain – typically uncheckable – compatibility conditions are satisfied by the input data; we will elaborate this point in Sections 1.4.2 and 1.4.4.
Regularity assertions that meet all the above requirements are proved in the present work. The main regularity assertions proved here are the following:

1. A shift theorem in weighted spaces (Theorem 5.3.8), where the solution u_ε of (1.2.1) is shown to be in an appropriately weighted H^2 space.

2. A shift theorem in countably normed spaces (Theorem 5.3.10) where the growth of the derivatives of the solution at a point is controlled in terms of ε and the distance to the nearest vertex.
3. Complete asymptotics with error bounds for the solutions of (1.2.11) on curvilinear polygons. Analytic regularity results for all terms arising in the asymptotics—in particular the so-called corner layers—are provided.

Since these results are spread over Parts II, III, we collect the main results in Section 2.3 in Theorems 2.3.1, 2.3.4 in a form that enables us to prove robust exponential convergence of the *hp*-FEM.

1.4 Review of existing results

Numerical analysis and regularity theory are closely connected since the ability to characterize solution behavior precisely is essential for the design and analysis of efficient numerical methods. For this reason the present work is placed at the intersection of several fields, namely, regularity theory on non-smooth domains, asymptotic expansions methods, numerical methods for singular perturbation problems, and high order finite element methods. In the present section, we briefly review existing results in these areas. Sections 1.4.1 and 1.4.2 discuss two kinds of regularity results. Section 1.4.1 is concerned with regularity theory of elliptic problems on non-smooth domains. The classical theory discussed in that area does not address the issue of singular perturbation problems. Section 1.4.2 presents the approach of describing solution behavior through asymptotic expansions. There, the classical work is not concerned with problems posed on non-smooth domains. Finally, Sections 1.4.3 and 1.4.4 discuss high order methods for problems on non-smooth domains and numerical methods for singularly perturbed problems.

1.4.1 Elliptic problems in non-smooth domains

The main points of regularity theory for elliptic problems in non-smooth domains can already be understood for the following model problem:

$$-\Delta u = f \quad \text{on } \Omega \subset \mathbb{R}^2, \qquad u = 0 \quad \text{on } \partial\Omega. \tag{1.4.1}$$

If $\partial\Omega$ is smooth, then the classical elliptic shift theorem, [2–4], holds, i.e., $f \in H^{k-1}(\Omega)$, $k \geq 0$, implies $u \in H^{k+1}(\Omega)$. This shift theorem breaks down if the boundary $\partial\Omega$ fails to have sufficient regularity. For piecewise smooth boundaries, e.g., if Ω is a polygon, the shift theorem holds only for $k \in [0, k_0)$ where k_0 depends on the domain. Specifically, for (1.4.1) and polygonal domains Ω with interior angles $\omega_j \in (0, 2\pi)$, $k_0 = \pi/\max_j \omega_j$. It turns out that a "modified" shift theorem holds: For k beyond k_0, the classical shift theorem holds provided that a certain number of singular functions is subtracted. Let us associate with each vertex A_j of the polygon Ω polar coordinates (r_j, φ_j) (such that the lines

$\varphi_j = 0$ and $\varphi_j = \omega_j$ coincide with the edges meeting at A_j) and define singular functions S_{lj} by

$$S_{lj}(r_j, \varphi_j) = \begin{cases} r_j^{l\pi/\omega_j} \sin\left(\frac{l\pi}{\omega_j}\varphi_j\right) & \text{if } l\pi/\omega_j \notin \mathbb{N}, \\ r_j^{l\pi/\omega_j} \left\{\ln r_j \sin\left(\frac{l\pi}{\omega_j}\varphi_j\right) + \varphi_j \cos\left(\frac{l\pi}{\omega_j}\varphi_j\right)\right\} & \text{if } l\pi/\omega_j \in \mathbb{N}. \end{cases}$$

With the aid of these singular functions, we can formulate the modified shift theorem:

Proposition 1.4.1. *Let Ω be a polygon with vertices A_j, $j = 1, \ldots, J$, and interior angles $\omega_j \in (0, 2\pi)$. If $f \in H^k(\Omega)$, then the solution u of (1.4.1) can be decomposed as*

$$u = \sum_{j=1}^{J} \sum_{\substack{l \in \mathbb{N} \\ l\pi/\omega_j < k}} a_{lj}(f) S_{lj} + u_0 \tag{1.4.2}$$

for some $a_{lj}(f) \in \mathbb{R}$ and $u_0 \in H^{k+1}(\Omega)$. The coefficients $a_{lj}(\cdot)$ are in fact linear functionals, and for each k there exists $C_k > 0$ independent of f such that

$$\sum_{j=1}^{J} \sum_{\substack{l \in \mathbb{N} \\ l\pi/\omega_j < k}} |a_{lj}(f)| + \|u_0\|_{H^{k+1}(\Omega)} \leq C_k \|f\|_{H^{k-1}(\Omega)}. \tag{1.4.3}$$

Hence, up to finitely many singular components, the solution u of (1.4.1) satisfies a shift theorem. Forerunners of regularity assertions similar to Proposition 1.4.1 are [83,128]. The seminal paper, however, is [76,77]. This kind of regularity theory is by now very well developed, in that corresponding results for equations with variable coefficients, different kinds of boundary conditions, and elliptic systems are available. The monographs [39, 62, 63, 79] are classical in this area. The first decomposition of the type presented in Proposition 1.4.1 for the system of Lamé equations in polyhedral domains was presented in [126]. General results for ADN-elliptic systems can be found in [37]. We also mention the newer monographs [78, 91, 100] and the important papers [89, 90].

Proposition 1.4.1 is essentially concerned with *finite* regularity in the sense that for right-hand side f from a Sobolev space $H^{k-1}(\Omega)$, the solution is ascertained to be, up to finitely many singular components, in the Sobolev space $H^{k+1}(\Omega)$. Many problems of practical interest, however, have input data that are piecewise C^∞ or even piecewise analytic. Then, the solution u of (1.4.1) is piecewise C^∞ or piecewise analytic on Ω, [70, 98]. In particular, therefore, all derivatives of u exist. Decompositions of the form (1.4.2) are not an adequate means for control of *all* the derivatives as the constant C_k in (1.4.3) depends on k in an unspecified way. Especially in the context of high order approximation, it is important to be able to control the growth of the derivatives of the solutions to elliptic problems (cf. Section 1.4.3 below). These considerations motivated [15] to introduce so-called *countably normed spaces* $\mathcal{B}^l_\beta$. On a polygon Ω, the spaces $\mathcal{B}^l_\beta$ are defined as follows. With each vertex A_j, $j = 1, \ldots, J$, we associate a number $\beta_j \in [0, 1)$. We then set $\beta = (\beta_1, \ldots, \beta_J)$ and introduce for each $p \in \mathbb{N}_0$ the weight function

$$\Phi_{p,\beta}(x) := \prod_{j=1}^{J} (\text{dist}(x, A_j))^{p+\beta_j}.$$

A function u that is analytic on Ω is in $\mathcal{B}^l_\beta(\Omega, C_u, \gamma_u)$, if for C_u, $\gamma_u > 0$ and $l \in \mathbb{N}_0$ there holds

$$\|u\|_{H^{l-1}(\Omega)} \le C_u,$$
$$\|\Phi_{p,\beta} \nabla^{p+l} u\|_{L^2(\Omega)} \le C_u \gamma_u^p p! \qquad \forall p \in \mathbb{N}_0.$$

We note that such growth conditions on the derivatives impose conditions on the analytic functions. In particular, a function from a space $\mathcal{B}^l_\beta$ is analytic up to the boundary of the polygon Ω with the exception of the vertices A_j; i.e., for each $x \in \partial\Omega \setminus \cup_{j=1}^J A_j$ there is a neighborhood $B_r(x)$ to which u has an analytic extension. It can be checked that the singular functions S_{lj} introduced above are in the space $\mathcal{B}^2_\beta$ if $\beta_j \ge 0$ is chosen such that additionally $\beta_j \in (1 - \pi/\omega_j, 1)$. In the framework of these countably normed spaces, [15, 16] proved the following shift theorem for analytic functions:

Proposition 1.4.2. *Let $\beta_j > 0$ satisfy $\beta_j \in (1 - \pi/\omega_j, 1)$. Then for a right-hand side $f \in \mathcal{B}^0_\beta(\Omega, C_f, \gamma_f)$ with C_f, $\gamma_f > 0$, the solution u of (1.4.1) is in the countably normed space $\mathcal{B}^2_\beta(\Omega, C_u, \gamma_u)$ for some C_u, $\gamma_u > 0$.*

Proposition 1.4.2 thus characterizes the solution of (1.4.1) for analytic right-hand sides. The solutions are analytic on Ω, and the derivatives become singular at the vertices of the domain in a controlled way. This characterization is of essential importance in the proof of exponential convergence of *hp*-FEM on geometric meshes as will be discussed in greater detail below. As stated, Proposition 1.4.2 is only a typical representative of regularity results in countably normed spaces. Analogous results hold for strongly elliptic systems, [15,16] and the Lamé equations, [17].

[15] considered analytic input data (here: the right-hand side f) as they typically arise in computational practice. However, other kinds of regularity theory for C^∞ data have been considered in the literature. One such class of functions are Gevrey classes, in which less stringent conditions are placed on the growth of the derivatives. Shift theorems akin to Proposition 1.4.2 can also be formulated in such classes, [30].

The regularity theory in countably-normed spaces of [15] exemplified in Proposition 1.4.2 is not directly suited for an application to singular perturbation problems in polygonal domains because the spaces $\mathcal{B}^l_\beta$ have no explicit means of controlling the dependence of the solution on the perturbation parameter. The need for such explicit control motivates in the present work us to introduce parameter-dependent countably normed spaces $\mathcal{B}^l_{\beta,\varepsilon}$.

1.4.2 Regularity in terms of asymptotic expansions

In the preceding Section 1.4.1, we presented various approaches to the description of the behavior of solutions of elliptic problems in polygonal domains. For

singularly perturbed problems, a different kind of regularity theory is prevalent, namely, the use of *asymptotic expansions*. This approach has a long history, dating back at least to the middle of the 19th century.
We mention here [46,47,85] and [71,81] in which the method of *matched asymptotics* has been applied to a variety of singularly perturbed problems. Note that [71] is mostly concerned with these techniques for problems posed over *smooth* domains or rectangular domains, due to the fact that asymptotic expansions typically require smoothness of the input data in order to be defined. This point is best illustrated by the ensuing example, taken from [95].

An example of asymptotic expansions. We consider the problem

$$-\varepsilon^2 \Delta u_\varepsilon + u_\varepsilon = f \quad \text{on } \Omega \subset \mathbb{R}^2, \qquad u_\varepsilon = g \text{ on } \partial\Omega. \tag{1.4.4}$$

Here, $\varepsilon \in (0,1]$ is a small parameter and the data f, g, and $\partial\Omega$ are assumed to be C^∞. The solution u_ε exhibits boundary layers that are classically described with the aid of asymptotic expansions. These can be created by the classical method of matched asymptotic expansions, [71].
For the definition of the asymptotic expansions, we introduce *boundary-fitted coordinates*: Let $L > 0$ be the length of $\partial\Omega$ and let $(X(\theta), Y(\theta))$, $\theta \in [0, L)$, be the smooth, L-periodic parametrization of $\partial\Omega$ by arc length such that the normal vector $(-Y'(\theta), X'(\theta))$ always points into the domain Ω. Introduce the notation $\kappa(\theta)$ for the curvature of the boundary curve and denote by $\mathbb{T}_L$ the one-dimensional torus of length L, i.e., $\mathbb{R}/L\mathbb{Z}$, endowed with the usual topology. The functions X, Y and hence also κ are smooth on $\mathbb{T}_L$ by the smoothness of $\partial\Omega$. For $\rho_0 > 0$ sufficiently small, the mapping

$$\begin{aligned} \psi : [0, \rho_0] \times \mathbb{T}_L &\to \overline{\Omega} \\ (\rho, \theta) &\mapsto (X(\theta) - \rho Y'(\theta), Y(\theta) + \rho X'(\theta)) \end{aligned} \tag{1.4.5}$$

is smooth on (a neighborhood of) $[0, \rho_0] \times \mathbb{T}_L$. The function ψ maps the rectangle $(0, \rho_0) \times \mathbb{T}_L$ onto a half-tubular neighborhood Ω_0 of $\partial\Omega$. Furthermore the inverse ψ^{-1} exists and is also smooth on (a neighborhood of) the closed set $\overline{\Omega}_0$.
The first step in the method of *matched asymptotics* is to define the *outer expansion* w, which can be viewed as an approximation to a particular solution of (1.4.4). Here, it is obtained by making the formal ansatz

$$w(x, y) \sim \sum_{i=0}^{\infty} \varepsilon^i w_i(x, y)$$

and then inserting this ansatz into (1.4.4) to get a recurrence relation for the unknown functions w_i. For the present problem, we obtain

$$w(x, y) \sim \sum_{i=0}^{\infty} \varepsilon^{2i} \Delta^{(i)} f = f + \varepsilon^2 \Delta f + \varepsilon^4 \Delta\Delta f + \cdots . \tag{1.4.6}$$

For every $M \in \mathbb{N}_0$ the *outer expansion* of order $2M + 1$ is defined by

$$w_{2M+1} := \sum_{i=0}^{M} \varepsilon^{2i} \Delta^{(i)} f. \tag{1.4.7}$$

The function $u_\varepsilon - w_M$ then satisfies

$$L_\varepsilon(u_\varepsilon - w_{2M+1}) = f - L_\varepsilon w_M = \varepsilon^{2M+2} \Delta^{(M+1)} f. \tag{1.4.8}$$

Hence, asymptotically as ε tends to zero, the functions w_M satisfy the differential equation in Ω. However, the functions w_M do not satisfy the given boundary conditions g. We therefore introduce a boundary layer correction u^{BL} of w_M, which will lead to the inner expansion. For each M the correction u^{BL} is defined as the solution of

$$\begin{aligned} L_\varepsilon u^{BL} &= 0 \qquad \text{in } \Omega, \\ u^{BL} &= g - \sum_{i=0}^{M} \varepsilon^{2i} \left[\Delta^{(i)} f \right]_{\partial\Omega} \qquad \text{on } \partial\Omega. \end{aligned}$$

The *inner expansion* is an asymptotic expansion for this correction function u^{BL}. In order to define this expansion, we need to rewrite the differential operator L_ε in boundary-fitted coordinates (ρ, θ). With the curvature $\kappa(\theta)$ of $\partial\Omega$ and the function

$$\sigma(\rho, \theta) = \frac{1}{1 - \kappa(\theta)\rho} \tag{1.4.9}$$

we have (see, for example, [10])

$$\Delta u(\rho, \theta) = \partial_\rho^2 u - \kappa(\theta)\sigma(\rho, \theta)\partial_\rho u + \sigma^2(\rho, \theta)\partial_\theta^2 u + \rho\kappa'(\theta)\sigma^3(\rho, \theta)\partial_\theta u.$$

Introducing now the *stretched variable* notation $\widehat{\rho} = \rho/\varepsilon$, we have

$$L_\varepsilon = -\partial_{\widehat{\rho}}^2 + \mathrm{Id} + \varepsilon\kappa(\theta)\sigma(\varepsilon\widehat{\rho}, \theta)\partial_{\widehat{\rho}} - \varepsilon^2\sigma^2(\varepsilon\widehat{\rho}, \theta)\partial_\theta^2 - \varepsilon^3\widehat{\rho}\kappa'(\theta)\sigma^3(\varepsilon\widehat{\rho}, \theta)\partial_\theta .$$

Expanding in power series in ε, we can write the operator L_ε formally as

$$L_\varepsilon = \sum_{i=0}^{\infty} \varepsilon^i L_i, \tag{1.4.10}$$

where the operators L_i have the form

$$L_0 = -\partial_{\widehat{\rho}}^2 + \mathrm{Id}, \quad L_i = -\widehat{\rho}^{i-1} a_1^{i-1} \partial_{\widehat{\rho}} - \widehat{\rho}^{i-2} a_2^{i-2} \partial_\theta^2 - \widehat{\rho}^{i-2} a_3^{i-3} \partial_\theta , \quad i \geq 1,$$

and the coefficients a_j^i are given by

$$\begin{aligned} &a_1^i = -[\kappa(\theta)]^{i+1}, \ a_2^i = (i+1)[\kappa(\theta)]^i, \ a_3^i = \frac{(i+1)(i+2)}{2}[\kappa(\theta)]^i \kappa'(\theta), \ i \in \mathbb{N}_0, \\ &a_1^i = a_2^i = a_3^i = 0 \qquad \text{for } i < 0. \end{aligned}$$

Now, in order to define the inner expansion, we make the formal ansatz

$$u^{BL} = \sum_{i=0}^{\infty} \varepsilon^i \widehat{U}_i(\widehat{\rho}, \theta), \tag{1.4.11}$$

where the functions $\widehat{U}_i$ are to be determined. Setting $L_\varepsilon u^{BL} = 0$ in (1.4.10) yields

$$\sum_{i=0}^{\infty} \varepsilon^i \sum_{j=0}^{i} L_j \widehat{U}_{i-j} = 0.$$

Hence, upon setting the coefficients of this formal power series in ε to zero, we obtain a recurrence relation for the desired functions $\widehat{U}_i$:

$$-\partial_{\widehat{\rho}}^2 \widehat{U}_i + \widehat{U}_i = \widehat{F}_i = \widehat{F}_i^1 + \widehat{F}_i^2 + \widehat{F}_i^3, \qquad i = 0, 1, \ldots,$$

$$\widehat{F}_i^1 = \sum_{j=0}^{i-1} \widehat{\rho}^j a_1^j \partial_{\widehat{\rho}} \widehat{U}_{i-1-j}, \; \widehat{F}_i^2 = \sum_{j=0}^{i-2} \widehat{\rho}^j a_2^j \partial_\theta^2 \widehat{U}_{i-2-j}, \; \widehat{F}_i^3 = \sum_{j=0}^{i-3} \widehat{\rho}^{j+1} a_3^j \partial_\theta \widehat{U}_{i-3-j},$$

where we use the tacit convention that empty sums take the value zero. As we expect the boundary layer function u^{BL} to decay away from the boundary $\partial\Omega$ and we want to satisfy the boundary conditions, we supplement these ordinary differential equations for the $\widehat{U}_i$ with the boundary conditions

$$\widehat{U}_i \to 0 \qquad \text{as } \widehat{\rho} \to \infty, \tag{1.4.12}$$

$$[\widehat{U}_i]_{\partial\Omega} = G_i := \begin{cases} g - [f]_{\partial\Omega} & \text{if } i = 0 \\ -[\Delta^{(i/2)} f]_{\partial\Omega} & \text{if } i \in \mathbb{N} \text{ is even} \\ 0 & \text{if } i \in \mathbb{N} \text{ is odd.} \end{cases} \tag{1.4.13}$$

The *inner expansion* of order $2M+1$ is obtained by truncating the (formal) sum $\sum_{i=0}^{\infty} \varepsilon^i \widehat{U}_i(\widehat{\rho}, \theta)$ after $2M+1$ terms and transforming back to (x, y)-coordinates with the map ψ:

$$u_{2M+1}^{BL}(\rho, \theta) := \left(\sum_{i=0}^{2M+1} \varepsilon^i \widehat{U}_i(\widehat{\rho}, \theta) \right) \circ \psi^{-1} = \left(\sum_{i=0}^{2M+1} \varepsilon^i \widehat{U}_i(\rho/\varepsilon, \theta) \right) \circ \psi^{-1}. \tag{1.4.14}$$

It is not too difficult to see that the functions $\widehat{U}_i$ do decay exponentially away from $\partial\Omega$. They have the form

$$\widehat{U}_i(\widehat{\rho}, \theta) = \left\{ \sum_{j=0}^{i} \Theta_j(\theta) \widehat{\rho}^j \right\} e^{-\widehat{\rho}}, \tag{1.4.15}$$

where the functions $\Theta_j(\theta)$ are smooth functions of θ. We note that

$$u_{2M+1}^{BL} + w_{2M+1} = g \qquad \text{on } \partial\Omega \text{ for all } M.$$

It can be shown that for each M there exists C_M (depending on the data f and $\partial\Omega$) such that

$$|L_\varepsilon u_{2M+1}^{BL}| \le C_M \varepsilon^{2M+2} \qquad \text{in a neighborhood of } \partial\Omega.$$

This allows us to obtain error estimates for the difference between the exact solution u_ε and the expansion $u_{2M+1}^{BL} + w_{2M+1}$. Letting χ be an appropriate cut-off function supported by a neighborhood of $\partial\Omega$ that is identically one in a (smaller) neighborhood of $\partial\Omega$, we can define the *remainder* r_{2M+1} via the following equation

$$u_\varepsilon = w_M + \chi u_{2M+1}^{BL} + r_{2M+1}.$$

By construction, r_{2M+1} satisfies

$$L_\varepsilon r_{2M+1} = R_{2M+1} \quad \text{on } \Omega, \qquad r_{2M+1} = 0 \quad \text{on } \partial\Omega,$$

where the residual R_{2M+1} satisfies (for some $C_M > 0$ depending on the data f, $\partial\Omega$)

$$||R_{2M+1}||_{L^\infty(\Omega)} \le C_M \varepsilon^{2M+2}.$$

Thus, by standard energy estimates, we get

$$||r_{2M+1}||_\varepsilon = \varepsilon||\nabla r_{2M+1}||_{L^2(\Omega)} + ||r_{2M+1}||_{L^2(\Omega)} \le C_M \varepsilon^{2M+2}. \qquad (1.4.16)$$

Discussion of the example. The above example illustrates several points that are typically encountered when trying to describe solution behavior with the aid of asymptotic expansions.

The first point to note is the formal nature of the expansions: Asymptotic expansions such as (1.4.6), (1.4.11) are *formal* sums only and do in general not converge for a fixed ε. To give meaning to such approximations, the error (in our example: r_{2M+1}) has therefore to be estimated. The typical procedure is to derive an equation for the remainder and then to use *a priori* bounds for the solution operator of this equation. This was our procedure in the example and we arrived at (1.4.16), which justifies the approximation by asymptotic expansions since for fixed M and $\varepsilon \to 0$, $r_{2M+1} \to 0$. In practice, however, ε is given; then the question of the size of the bound $C_M \varepsilon^{2M+2}$ in (1.4.16) arises. In general, this bound diverges to ∞ as $M \to \infty$ because asymptotic expansions are typically divergent sums. The implications of this fact are twofold:

1. In general, one has to expect that, for fixed ε, the remainder $||r_{2M+1}||_\varepsilon$ can be bounded from below for all M. Thus, asymptotic expansions are only useful up to a certain error level, because the remainder r_{2M+1} cannot be made arbitrarily small. An implication is that expansion-based regularity assertions are not adequate for the description of the *asymptotic* behavior of a numerical method (such as the FEM), that is, they cannot describe its convergence behavior for fixed ε and a large number of unknowns (i.e., small mesh size h or large polynomial degree p). In this asymptotic regime, other forms of regularity assertions are required.
2. Given that the remainder r_{2M+1} cannot be made arbitrary small for fixed ε, the question arises of determining the "optimal" expansion order $2M+1$ for which the remainder is minimal. As the constant C_M depends in an almost

intractable way on higher order derivatives of the data f, $\partial\Omega$, this optimal choice is a formidable task in general. For certain classes of data, however, nearly optimal choices of M can be computed. It was shown in [95] that for *analytic* data f and $\partial\Omega$, the dependence of C_M on M can be made explicit: $C_M = C(2M+2)!\gamma^{2M+2}$ for some $C, \gamma > 0$ independent of ε and M. For analytic data f, $\partial\Omega$, the choice $M \sim \varepsilon^{-1}$ then minimizes the error bound $C_M\varepsilon^{2M+2}$, which takes the form $e^{-\gamma/\varepsilon}$ for some $\gamma > 0$ independent of ε.

The next point to note about the use of asymptotic expansions is that they require differentiability of the data. For example, from the representation of the outer expansions w_{2M+1}, we see that differentiability of order $2M$ is required of the right-hand side f. Likewise, the ability to define the inner expansion depends on differentiability properties of the parametrization of the boundary. This approach therefore fails for domains with non-smooth boundaries (such as polygons). The boundary layer expansions can of course be defined wherever the boundary is smooth, that is, for each edge of the polygon separately. Near a vertex, however, two boundary layer expansions meet in an incompatible way. To remove these incompatibilities, new functions, called corner layers, have to be introduced. These corner layers are solutions of appropriate auxiliary problems on cones. Being solutions of elliptic problems over cones, the corner layers are not smooth, as we just saw in Section 1.4.1, and thus they are not easily handled in the framework of asymptotic expansions, which heavily depends on differentiability.

These difficulties highlight the reason why asymptotic expansions for singular perturbation problems have typically been restricted to domains with smooth boundaries or particularly simple geometries such as the square. Regularity results for problems of the type (1.4.4) on the square can be found in [32,67,116]. The restriction to the square is largely due to the very special nature of the Laplace operator on that domain: The linear functionals a_{lj} of Proposition 1.4.1 are local in the sense that the functional a_{lj} depends on the behavior of the datum f at the vertex A_j only, i.e., on $f(A_j)$ and possibly higher derivatives $D^\alpha f(A_j)$. This coincidence allows for formulating explicit *compatibility* conditions on the data that suppress certain singularities. In the context of regularity theory for singular perturbations, this is a common occurrence.

Another difficulty encountered when trying to apply regularity results that include corner layers arises from the fact that these corner layers are typically defined implicitly as the solutions of auxiliary problems. While such a definition is mathematically convenient and removes some of the technical difficulties in defining expansions, it often has the disadvantage of not giving precise enough information about these corner layers for use in numerical schemes.

Results for problems of the type (1.4.4) on general polygons are scarce. The most notable one for the application to finite element methods is in [74,75]. However, even these results are not quite sharp enough for an application in low-order FEM as is discussed in [7, Sec. 25.1]. Furthermore, the results of [74,75] are cast in the classical framework of finite Sobolev regularity. The results are therefore not directly applicable to the *hp*-FEM, which requires control of all derivatives of the solution.

1.4.3 *hp* finite element methods

Most methods for the numerical treatment of partial differential equations are based on approximating the sought solution by piecewise polynomials. In Finite Element Methods (FEM) and Finite Volume Methods, the piecewise polynomial approximation is done explicitly, while in finite difference schemes it is done implicitly. Many methods employed today are so-called low order methods, in which the underlying polynomial degree p is fixed (typically $p \leq 2$) and convergence is achieved by decreasing the mesh size h. These methods are also called h-methods and have *algebraic* rates of convergence at best; i.e., error bounds of the form $O(N^{-\alpha})$ for some $\alpha > 0$ are typical (here, N stands for the number of unknowns of the resulting linear system and is some measure for the work required to compute the numerical solution). On the other hand, high order methods such as the *p*- and *hp*-version finite element methods in solid mechanics (see, e.g., [25, 112] and the references there) and spectral methods/spectral element methods in fluid mechanics (see, e.g., [33, 73] with their bibliographies) emerged in the early 1980s. In these methods, the polynomial degree p is increased while still keeping the option to perform mesh refinement. These high order methods are most appreciated for their high accuracy, for their rates of convergence are often very high and can in certain cases even be *exponential.*
Let us discuss the claim of high accuracy for high order methods and in particular high order FEM in more detail. The ability of (piecewise) polynomials to approximate a function that is analytic on an *open neighborhood* of the computational domain at an exponential rate has been known for a long time (see [40] for a nice exposition of this fact). However, most elliptic problems do not in practice have solutions that are analytic on such open neighborhoods. A large class of practical problems is posed on *piecewise* analytic domains in which case the solution becomes singular at certain boundary points only, as we saw in Section 1.4.1. In the context of the h-version on quasi-uniform meshes, the presence of these singularities downgrades the convergence rate. However, on so-called *radical meshes*, which are suitably refined toward the singularities, it is possible to recover the optimal algebraic rate of convergence achievable by fixed-order polynomial approximation: See the classical [21, 106] for two-dimensional results and the recent [7] for the corresponding results in three dimensions. We point out that regularity results of the type presented in Proposition 1.4.1 contain sufficient information to design these radical meshes and are essential for the proof of the optimal rate of convergence. Corresponding algebraic convergence results hold for the *p*-version where the mesh is fixed and the approximation order, i.e., the polynomial degree p, is increased. While the convergence rate of the *p*-version is still algebraic, it is twice that of the h-version (on quasi-uniform meshes) for problems with piecewise smooth input data, [22].
While both the h- and *p*-version achieve only algebraic rates of convergence in the presence of typical singularities, appropriate combinations of the two

leads to *exponential* rates of convergence. In the *hp*-version of the finite element method, this combination occurs via mesh refinement toward the singularities in conjunction with an increase of the polynomial degree in regions where the solution is smooth. The first such exponential convergence result was obtained in [19, 20], where it was shown that functions from the countably normed space $\mathcal{B}^2_\beta$ can be approximated at an exponential rate on so-called *geometrically refined* meshes. The regularity assertion Proposition 1.4.2 shows that this situation is met in practice.

Although the high accuracy of *hp*-finite element methods is their most striking feature, there are additional reasons for their increasing popularity, especially in solid mechanics. We mention here the issue of numerical *locking* in parameter-dependent problems, where high order methods tend to be more robust than their low-order counterparts, [24]. The hardest problems in linear finite element analysis where locking problems are rampant are shell problems. Numerical evidence of [54, 66, 104] for cylindrical shells shows that high order methods are indeed much better suited for dealing with these parameter-dependent problems than low order methods. Similar observations about the efficiency of high order methods were made for plate problems in [69, 132].

1.4.4 Numerical methods for singular perturbations

As mentioned previously, singular perturbation problems arise in many practical applications. However, the problems vary greatly in structure and a correspondingly large number of different numerical schemes has been devised—by far too large to be treated here in a comprehensive way, and the reader is therefore referred to the monographs [42,97,99,108] for an up-to-date overview. The presence of layers is a trademark of singularly perturbed problems. Numerical methods for such problems have therefore to address the issue of approximating such layers.

Layer approximation in one dimension. One of the most notable features of singularly perturbed problems is the presence of *layers*. In the example (1.4.4) they appear as *boundary layers*. These boundary layers are given by (1.4.15) and can be written in the form $S(\rho,\theta)e^{-\rho/\varepsilon}$, where S is a smooth function. This structure has motivated many authors to approximate such functions on meshes that have tensor product structure and to obtain mesh design principles from the analysis of the simpler one-dimensional case, that is, the approximation of the function $e^{-x/\varepsilon}$ by piecewise polynomials on the interval $(0, 1)$. It is clear that the mesh must depend on ε and that mesh points have to be condensed in a neighborhood of $x = 0$ in order to resolve the boundary layer function. The first such adapted grid was proposed in [26], where essentially half the grid points are concentrated in an $O(\varepsilon|\ln\varepsilon|)$ neighborhood of the point $x = 0$ and the remaining half form a uniform grid on rest of $(0, 1)$. We refer to [108, Sec. 2.4.2] for a nice exposition of these meshes and their many variations such as the Bakhvalov-type meshes of [127] and meshes of Gartland type, [51]. Closer to the type of meshes that we use for high order methods are so-called *Shishkin meshes*, [118], which are piecewise uniform meshes and again cluster half the grid points in a small neighborhood of the point $x = 0$. Specifically, for $N \in \mathbb{N}$ and a parameter $\tau > 0$, these meshes are piecewise uniform meshes defined by the set of nodes x_i as follows:

$$\sigma := \min\{1/2, \tau\varepsilon \ln N\},$$
$$x_i := \frac{i}{N}\sigma, \qquad i = 0, \dots, N,$$
$$x_{N+1+i} := \sigma + \frac{i}{N}(1-\sigma), \qquad i = 0, \dots, N.$$

The transition point σ, at which the mesh switches abruptly from a very fine one to a coarse one is chosen such that the boundary layer function $e^{-x/\varepsilon}$ is small on the coarse grid. In fact, it is not difficult to see that the difference between $u = e^{-x/\varepsilon}$ and the piecewise linear interpolant Iu on the Shishkin grid satisfies bounds of form $\|u - Iu\|_{L^\infty(0,1)} \leq C_\tau N^{-1} \ln N$ for each fixed $\tau \geq 1$. The constant C_τ depends on τ but is independent of ε and N. Similar approximation results hold, of course, for other norms as well.
The analog of this Shishkin mesh in the p-version FEM is the so-called "two-element" mesh of [113, 114]. For polynomial degree p, it is defined by the three nodes

$$x_0 = 0, \qquad x_1 = \min\{1/2, \tau p \varepsilon\}, \qquad x_2 = 1,$$

where $\tau > 0$ is a fixed parameter of size $O(1)$. Approximation of the boundary layer function $u = e^{-x/\varepsilon}$ by piecewise polynomials of degree p on this mesh can be achieved at a robust exponential rate; in other words, the rate is exponential (in p) and all constants are independent of ε, [113]. To illustrate this robust exponential approximability result, we point to the fact that it is not difficult to construct, for each $0 < \tau < e^{-1}$, a continuous, piecewise polynomial approximation Iu satisfying $\|u - Iu\|_{L^\infty((0,1))} \leq C_\tau \left[(\tau e)^p + e^{-\tau p}\right]$, where $C_\tau > 0$ depends on τ but is independent of ε and p. As in the case of Shishkin meshes, similar results can be obtained for the approximation in other norms as well.
The design of meshes and grids for the approximation of boundary layer functions in higher dimensions is guided by these ideas obtained from analyzing one-dimensional examples. Typically, the meshes are obtained through tensor product constructions using highly refined meshes of Bakhvalov or Shishkin type in the direction normal to the boundary and quasi-uniform meshes in the tangential direction. We refer to [7, 107] for good surveys on recent results concerning the use of such layer-adapted anisotropic grids.

Numerical methods for (1.2.11). The key properties of a numerical method are *stability* and *approximability*. The design of stable methods for non-symmetric problems (arising, for example, in convection-dominated fluid flow problems) is non-trivial and currently an active and extensive research area. In the present subsection, we restrict our discussion to numerical methods for symmetric problems of type (1.2.11). Typical FEM or finite difference discretizations are stable due to the symmetric positive-definite structure of the problem, and numerical analysis of this problem class is thus reduced to an approximability problem.

In the literature, equations (1.2.11) are typically simplified further, and the following model problem is analyzed:

$$-\varepsilon^2 \Delta u_\varepsilon + u_\varepsilon = f \quad \text{on } \Omega, \qquad u_\varepsilon = 0 \quad \text{on } \partial\Omega. \tag{1.4.17}$$

We now review existing work for the numerical solution of problems of type (1.4.17). An early low order finite element analysis was performed by [109] on quasi-uniform meshes. In the practical regime of mesh sizes $h \gg \varepsilon$, the boundary layer effects near $\partial\Omega$ cannot be captured, and the authors therefore focussed on local error estimates away from the boundary.
Methods devised to capture the boundary layer effects were presented in [117, 119] in a finite difference context on Shishkin-type meshes and in [8, 29, 84] in a first order finite-element context on Bakhvalov-type and Shishkin type meshes. Typically, these results were obtained on a special domain, the square, because their convergence proofs hinge on decompositions of the solution into smooth parts, boundary layer parts, and possibly corner layer parts. However, due to a lack of sufficiently precise knowledge of the regularity of the corner layers, their approximation is often avoided by stipulating compatibility conditions on the datum f so as to suppress corner layers or to show that they are sufficiently small. The ability to formulate such compatibility conditions rests on the fact that for the very special case of a square, the linear functionals describing the corner singularities are local. [7] gives results that are not restricted to a square by basing its convergence proof on some regularity assumptions. As pointed out in [7, Sec. 25.1], these assumptions are not rigorously proved. The present work provides a complete regularity analysis of the solutions of (1.4.17) that is suited for polynomial approximation results. Although we apply our regularity results to high order approximations, they could also be employed to rigorously establish these low order approximation results. We present such an approach in Sec. 2.6.
While a fair body of literature is available for lower order methods on meshes that can capture the boundary layer behavior of the solution, the situation is not so well developed for spectral and *hp*-methods. The first result in one dimension appears to be [113,114]. There, a two-element mesh was designed that allowed for the robust exponential approximation of boundary layer functions. This result, however, applies to the case of a constant-coefficients differential equation only. The first robust exponential approximation result in one dimension for general analytic input data (coefficients of the differential equation and right-hand side) is [92]. The first extension to two dimensions was done in [131, 133, 135]. On smooth domains, boundary-fitted tensor product meshes were employed and robust convergence of arbitrary order was proved for smooth input data. [131, 134] also considered the case of a square with simple right-hand side f for which, again due to the special nature of the domain and the data, robust convergence of arbitrary order was obtained. The first robust exponential convergence result for analytic input data f and $\partial\Omega$ was obtained in [94,95]. A further improvement of [94] over existing high order approximation results is that the meshes do not have to be boundary-fitted tensor product meshes. This is important in practice since in the presence of curved elements, boundary-fitted tensor product meshes

are not easily generated. The present work extends these results to the case of general curvilinear polygons and arbitrary analytic right-hand side.

1.5 Outline of the book

The book comprises three parts, each contributing to the main result of the work, the proof of robust exponential convergence of a properly designed *hp*-FEM for (1.2.11). While Part I focuses on *hp*-approximation on anisotropic meshes, Parts II and III provide two different types of regularity assertions for the solution that are required for proving robust exponential convergence.
Chapter 2 serves as an overview chapter to make the main results of this work more accessible. The main regularity assertions for solutions u_ε of (1.2.11), rigorously established in Parts II, III, are collected in Section 2.3: Theorem 2.3.1 extracts from Part II the assertion that u_ε is an element of a countably normed space $\mathcal{B}^2_{\beta,\varepsilon}$; Theorem 2.3.4 consolidates the analysis of Part III concerning the regularity of the description of u_ε by means of asymptotic expansions. In this overview chapter, we include the *hp*-FEM approximation result Theorem 2.4.8 together with numerical examples in Section 2.5 illustrating the robust exponential convergence result.
In order to motivate the regularity assertions Theorems 2.3.1, 2.3.4, we begin Chapter 2 with a discussion of the one-dimensional case. Also key features of *hp*-approximation on meshes that are suitable for the resolution of layer phenomena are examined in this one-dimensional setting and may serve as a motivation for the two-dimensional boundary layer meshes of Definition 2.4.4.

Part I: Finite element approximation. The main result of Part I is the robust exponential convergence result Theorem 2.4.8 for the boundary value problem (1.2.11). This result follows from the approximation result Theorem 3.4.8, where an approximant is explicitly constructed. Chapter 3 embeds Theorem 3.4.8 in a general discussion of *hp*-approximation on non-affine, anisotropic meshes. An important point to note is that the meshes for which robust exponential convergence is proved, the admissible boundary layer meshes of Definition 2.4.4, are essentially the minimal meshes with that property.
The application of the regularity assertions Theorems 2.3.1, 2.3.4 is not limited to the *hp*-FEM; we show in Section 2.6 that, based on these regularity assertions, robust algebraic convergence can be established for piecewise linear ansatz functions on Shishkin meshes.

Part II: Regularity in countably normed spaces. Part II is concerned with regularity assertions for solutions of (1.2.1) in countably normed spaces. The key feature is that the dependence on the parameter ε is made explicit.
In Chapter 4 the weighted spaces $H^{l,m}_{\beta,\varepsilon}$ and the countably normed spaces $\mathcal{B}^l_{\beta,\varepsilon}$ of analytic functions are introduced and their main properties are proved. The weights are functions of the perturbation parameter ε, the distance to the nearest vertex, and the differentiation order. The main result of Chapter 4 is Theorem 4.2.20, which states that membership in the countably normed spaces $\mathcal{B}^l_{\beta,\varepsilon}$

is invariant under analytic changes of variables. This property proves very useful in inferring regularity results for domains with curved boundaries. The spaces $\mathcal{B}^l_{\beta,\varepsilon}$ are defined through an L^2-based control of all derivatives. Pointwise control of all derivatives for functions from the spaces $\mathcal{B}^l_{\beta,\varepsilon}$ is given in Theorem 4.2.23. In Chapter 5 two kinds of regularity assertions for the solutions of (1.2.1) are proved. The first regularity statement is Theorem 5.3.8, which is a shift theorem in weighted spaces for (1.2.1). When applied to (1.2.11), it yields that for right-hand sides from the spaces $H^{0,0}_{\beta,\varepsilon}$, the solution is in $H^{2,2}_{\beta,\varepsilon}$. Theorem 5.3.10 represents a second kind of regularity result, where for right-hand sides from countably normed spaces, the solution of (1.2.1) is asserted to be in a countably normed space as well. For example, for (1.2.11), membership of the right-hand side in $\mathcal{B}^0_{\beta,\varepsilon}$ implies that the solution is an element of $\mathcal{B}^2_{\beta,\varepsilon}$.
Section 5.3 discusses the case of Dirichlet boundary conditions. The techniques employed are emenable to other kinds of boundary conditions: We analyze Neumann and Robin boundary conditions in Section 5.4 and additionally discuss a transmission problem.

Part III: Regularity in terms of asymptotic expansions. In Part III, we describe the regularity of the solutions u of (1.2.11) by means of asymptotic expansions. The definition of the so-called corner layer is done with the aid of a transmission problem with exponentially decaying data. Analytic regularity results for such transmission problems are therefore provided in Chapter 6. These regularity assertions are utilized in Chapter 7 to obtain analytic regularity results for the corner layers in the asymptotic expansions. Theorem 6.4.13 and Corollary 6.4.14 synthesize the results of Chapter 6: Theorem 6.4.13 shows that the solutions of the transmission problems considered are in exponentially weighted countably normed spaces, and Corollary 6.4.14 gives pointwise estimates for the growth of the derivatives of these solutions.

Part I

Finite Element Approximation

2. *hp*-FEM for Reaction Diffusion Problems: Principal Results

This part of the book is devoted to the finite element approximation to solutions of (1.2.11). The principal aim of the present Chapter 2 is the robust exponential convergence result Theorem 2.4.8, which is illustrated by numerical examples in Section 2.5. Essential for this robust exponential convergence result are detailed regularity assertions for the solution. For the convenience of the reader, the present chapter collects from Parts II, III the regularity results that are required for the proof of Theorem 2.4.8. The proofs of both the approximation result and the regularity assertions are very technical and therefore not included in this chapter.
In order to motivate the two-dimensional results of this chapter, we present the analogous results in the one-dimensional setting in Section 2.2. Technically, this setting is considerably simpler than the two-dimensional case, yet it exhibits many features that are relevant for the two-dimensional case.
We conclude this chapter with a discussion of a low-order method in Section 2.6, since the regularity assertions of Section 2.3 can also be employed to prove robust convergence of the *h*-FEM on Shishkin meshes.

2.1 Setting and Introduction

We consider the singularly perturbed problem (1.2.11) with $b \equiv 0$, i.e.,

$$-\nabla \cdot (A(x)\nabla u) + c(x)u = f \quad \text{in } \Omega, \tag{2.1.1a}$$
$$u = g \quad \text{on } \partial\Omega. \tag{2.1.1b}$$

The coefficients $A \in \mathcal{A}(\Omega, \mathbb{S}^2_>)$, $c \in \mathcal{A}(\Omega)$, and the right-hand side f are assumed to satisfy (1.2.2), (1.2.3), i.e.,

$$\|\nabla^p A\|_{L^\infty(\Omega)} \le C_A \gamma_A^p p! \qquad \forall p \in \mathbb{N}_0, \tag{2.1.2a}$$
$$\|\nabla^p c\|_{L^\infty(\Omega)} \le C_c \gamma_c^p p! \qquad \forall p \in \mathbb{N}_0, \tag{2.1.2b}$$
$$A(x) \ge \lambda_{min} > 0 \qquad \forall x \in \Omega, \tag{2.1.2c}$$
$$c(x) \ge \mu > 0 \qquad \forall x \in \Omega, \tag{2.1.2d}$$
$$\|\nabla^p f\|_{L^\infty(\Omega)} \le C_f \gamma_f^p p! \qquad \forall p \in \mathbb{N}_0. \tag{2.1.2e}$$

The boundary $\partial\Omega$ is assumed piecewise analytic as described in Section 1.2 and the Dirichlet data g are also piecewise analytic in the sense of (1.2.4). Solutions u_ε are understood in the weak sense, i.e., $u_\varepsilon \in H^1(\Omega)$ solves (2.1.1) if

$$u_\varepsilon|_{\partial\Omega} = g \qquad \text{and} \qquad B_\varepsilon(u_\varepsilon, v) = F(v) \quad \forall v \in H_0^1(\Omega), \tag{2.1.3}$$

where

$$B_\varepsilon(u,v) = \varepsilon^2 \int_\Omega (A(x)\nabla u)\cdot\nabla v + c(x)uv\,dx, \qquad F(v) = \int_\Omega f(x)v\,dx. \tag{2.1.4}$$

The energy norm $\|\cdot\|_\varepsilon$ is defined as in (1.2.7), i.e.,

$$\|u\|_\varepsilon^2 = \varepsilon^2\|\nabla u\|^2_{L^2(\Omega)} + \|u\|^2_{L^2(\Omega)}. \tag{2.1.5}$$

Due to the assumptions on the coefficients A, c, we have

$$\min\{\lambda_{min},\mu\}\|u\|_\varepsilon^2 \le B_\varepsilon(u,u) \le \max\{\|A\|_{L^\infty(\Omega)}, \|c\|_{L^\infty(\Omega)}\}\|u\|_\varepsilon^2 \quad \forall u \in H^1(\Omega).$$

Lemma 1.2.2 gives the existence and uniqueness of the solution u_ε to (2.1.1). As described in Section 1.1, the solution u_ε is analytic on Ω but has boundary layers near the boundary $\partial\Omega$ and corner singularities/corner layers near the vertices of the curvilinear polygon Ω. Theorems 2.3.1, 2.3.4 below collect the main regularity properties of the solution u_ε, which will enable us to prove robust exponential convergence of the *hp*-version of the finite element method applied to problems of type (2.1.1).

2.2 Prelude: the one-dimensional case

Many features of both regularity theory and high order approximation can be seen more clearly in one dimension. The present section is therefore devoted to an overview of the results in this considerably simpler setting.

2.2.1 Regularity in one dimension

To illustrate the main points of the regularity results of Theorems 2.3.1, 2.3.4, we recall a one-dimensional result.

Proposition 2.2.1. *Let $\Omega = (-1,1)$ and let c, f be analytic* [1] *on $\overline{\Omega}$ with $c \ge \underline{c} > 0$ on Ω. Then there exist C, γ, $\alpha > 0$ depending only on c, f such that for every $\varepsilon \in (0,1]$ the solution u_ε to*

$$-\varepsilon^2 u_\varepsilon''(x) + c(x)u_\varepsilon(x) = f(x) \quad \text{in } \Omega, \qquad u_\varepsilon(\pm 1) = 0 \tag{2.2.1}$$

satisfies:

1. *u_ε is analytic on $\overline{\Omega}$ and*

$$\|u_\varepsilon^{(n)}\|_{L^2(\Omega)} \le C\max\{n+1, \varepsilon^{-1}\}^n \qquad \forall n \in \mathbb{N}_0. \tag{2.2.2}$$

---

[1] A function $g : \overline{\Omega} \to \mathbb{C}$ is said to be analytic on the closed set $\overline{\Omega}$ if there exists an open neighborhood $\widetilde{\Omega}$ of $\overline{\Omega}$ and an analytic function $G : \widetilde{\Omega} \to \mathbb{C}$ with $G|_{\overline{\Omega}} = g$.

2. *For each* $\varepsilon \in (0,1]$, *the solution* u_ε *can be decomposed as*

$$u_\varepsilon = w_\varepsilon + u_\varepsilon^{BL} + r_\varepsilon, \tag{2.2.3}$$

where, upon setting $\rho(x) := \operatorname{dist}(x, \pm 1)$, *there holds for all* $n \in \mathbb{N}_0$

$$\begin{aligned} \|w_\varepsilon^{(n)}\|_{L^\infty(\Omega)} &\leq C\gamma^n n!, \\ \left|\left(u_\varepsilon^{BL}\right)^{(n)}(x)\right| &\leq C\gamma^n \max\{n+1, \varepsilon^{-1}\}^n \exp(-\alpha\rho(x)) \quad \forall x \in \Omega, \\ \|r_\varepsilon\|_{H^1(\Omega)} &\leq C\exp(-\alpha/\varepsilon). \end{aligned}$$

Furthermore, $r_\varepsilon(\pm 1) = 0$, r_ε *is smooth, and for each* $k \in \mathbb{N}_0$, *there exist* C_k, $\alpha_k > 0$ *such that*

$$\|r_\varepsilon^{(k)}\|_{L^\infty(\Omega)} \leq C_k \exp(-\alpha_k/\varepsilon), \qquad k = 0, 1, \ldots.$$

Proof: See [92] and the proof of Lemma 7.1.1. □

The decomposition (2.2.3) is obtained with the aid of the classical asymptotic expansions. It captures the main solution components: a smooth part w_ε, which is even analytic, and a boundary layer part u_ε^{BL} that decays very fast away from the boundary points ± 1. The remainder r_ε is ascertained to be small. For the application of the FEM, we will design the mesh such that the smooth part w_ε and the boundary layer part u_ε^{BL} can be approximated well. The small remainder r_ε is simply approximated by zero. However, for fixed ε, the remainder r_ε is small but finite. Thus, for FEM approximations with a required accuracy below $O(\exp(-\alpha/\varepsilon))$, the decomposition (2.2.3) cannot be employed; instead, the assertion (2.2.2) concerning the growth of the derivatives is employed. In this regime of accuracy, the bounds (2.2.2), although strongly ε-dependent, are sufficient. We will illustrate the interplay of these two types of regularity assertions in Section 2.2.2 below for the one-dimensional case; an analogous interplay takes place in the two-dimensional situation.

Remark 2.2.2 For the case of convection-diffusion equations, i.e., solutions to the equation $-\varepsilon^2 u''(x) + b(x)u'(x) + c(x)u(x) = f(x)$, the analog of Proposition 2.2.1 is available in [96]. ▪

2.2.2 *hp*-FEM in one dimension

We illustrate the *hp*-version of the FEM in the one-dimensional setting first, because some important features of the *hp*-FEM, in particular for the approximation of boundary layers, can already be seen in one dimension.

Abstract FEM in one dimension. We consider the approximation of the solution to (2.2.1). Upon writing $\Omega = (-1,1)$, the weak formulation is:

$$\text{Find } u_\varepsilon \in H_0^1(\Omega) \text{ s.t. } \quad B_\varepsilon(u_\varepsilon, v) = F(v) \quad \forall v \in H_0^1(\Omega), \tag{2.2.4}$$

where the bilinear form B_ε and the right-hand side functional F are given by

$$B_\varepsilon(u,v) = \int_\Omega \varepsilon^2 u' v' + c(x) uv \, dx, \qquad F(v) = \int_\Omega f(x) v \, dx.$$

For arbitrary subspace $V_N \subset H_0^1(\Omega)$, dim $V_N = N < \infty$, the FEM reads:

$$\text{Find } u_N \in V_N \text{ s.t. } \quad B_\varepsilon(u_N, v) = F(v) \quad \forall v \in V_N. \tag{2.2.5}$$

By Céa's Lemma, u_N exists and is the best-approximant in the energy norm $\|\cdot\|_\varepsilon$, i.e.,

$$\|u_\varepsilon - u_N\|_\varepsilon = \inf_{v \in V_N} \|u_\varepsilon - v\|_\varepsilon.$$

Here, the energy norm is given by

$$\|u\|_\varepsilon^2 = B_\varepsilon(u,u) \sim \varepsilon \|u'\|^2_{L^2(\Omega)} + \|u\|^2_{L^2(\Omega)}.$$

***hp*-FEM in one dimension.** In the *hp*-version of the FEM in one dimension, the spaces V_N are spaces of piecewise polynomials.
Let $-1 = x_0 < x_1 < \cdots < x_L = 1$ be mesh points and define the intervals $I_i := (x_{i-1}, x_i)$, $i = 1, \ldots, L$. The *mesh* $\mathcal{T}$ is then defined as $\mathcal{T} = \{I_i \,|\, i = 1, \ldots, L\}$. For a mesh and a polynomial degree $p \in \mathbb{N}$, we define the *hp*-FEM spaces $S^p(\mathcal{T})$, $S_0^p(\mathcal{T})$ as

$$\begin{aligned} S^p(\mathcal{T}) &:= \{u \in H^1(\Omega) \,|\, u|_{I_i} \text{ is a polynomial of degree } p \text{ for each } i = 1, \ldots, L\}, \\ S_0^p(\mathcal{T}) &:= S^p(\mathcal{T}) \cap H_0^1(\Omega). \end{aligned}$$

It is easy to check that dim $S_0^p(\mathcal{T}) = pL - 1$.

Remark 2.2.3 Analogously to the procedure in two dimensions, one could define the space $S^p(\mathcal{T})$ as

$$S^p(\mathcal{T}) = \{u \in H^1(\Omega) \,|\, u \circ M_i \in \mathcal{P}_p, \quad i = 1, \ldots, L\},$$

where $\mathcal{P}_p$ is the space of polynomials of degree p and the *element maps* M_i are given by

$$\begin{aligned} M_i : I &\to i_i \\ \xi &\mapsto x_{i-1} + h_i \xi, \qquad h_i = x_i - x_{i-1} \end{aligned}$$

for the *reference interval* $I = (0,1)$. ▪

We now focus on the question of how to choose the mesh $\mathcal{T}$ for the approximation of the solution u_ε of (2.2.1). Our aim is to find meshes $\mathcal{T}$ with as few elements as possible. For problems with boundary layers such as (2.2.1), [92, 113, 131] have proposed and analyzed the so-called *boundary layer mesh*, which yields robust exponential convergence with meshes consisting of three, judiciously chosen elements. We define boundary layer meshes as follows.

Definition 2.2.4 (boundary layer mesh). *For* $\kappa > 0$ *the* boundary layer mesh $\mathcal{T}_\kappa$ *is defined by the points*

$$x_0 = -1, \quad x_1 = -1 + \min\{0.5, \kappa\}, \quad x_2 = 1 - \min\{0.5, \kappa\}, \quad x_3 = 1.$$

Boundary layer meshes are suitable for the approximation of problems with boundary layers as is shown in the following proposition.

Proposition 2.2.5. *Let c, f satisfy the assumptions of Proposition 2.2.1. Denote by u_ε the solution to (2.2.1). Then there exist C, b, $\lambda_0 > 0$ independent of ε such that for every $\lambda \in (0, \lambda_0)$ and $p \in \mathbb{N}_0$*

$$\inf_{v \in S_0^p(\mathcal{T}_{\lambda p \varepsilon})} \|u_\varepsilon - v\|_{L^\infty(\Omega)} + \lambda p \varepsilon \|(u_\varepsilon - v)'\|_{L^\infty(\Omega)} \le C e^{-b\lambda p}.$$

In particular, therefore, if u_N denotes the hp-FEM solution of (2.2.5) based on $V_N = S_0^p(\mathcal{T}_{\lambda p\varepsilon})$, then there exist for each $\lambda \in (0, \lambda_0)$ constants C, $b' > 0$ independent of ε, p such that

$$\|u_\varepsilon - u_N\|_\varepsilon \le C e^{-b'N}, \qquad N = \dim S_0^p(\mathcal{T}_{\lambda p \varepsilon}).$$

Proof: This result was first obtained in [92]. We will, however, prove it here because the ideas will be encountered again in the proof of the technically more involved two-dimensional case, Theorem 2.4.8.
The required approximants to u_ε are constructed with the aid of the piecewise Gauss-Lobatto interpolation operator, $i_{\mathbf{p},\mathcal{T}}$. Specifically, for a mesh $\mathcal{T} = \{I_i \,|\, i = 1, \dots, L\}$ we associate a polynomial degree vector $\mathbf{p} := (p_i)_{i=1}^L \subset \{1, \dots, p\}$ and set

$$\begin{aligned} &i_{p,\mathcal{T}} : C(\overline{\Omega}) \to S^p(\mathcal{T}) \\ &\qquad u \mapsto i_{\mathbf{p},\mathcal{T}} u \quad \text{where} \quad i_{\mathbf{p},\mathcal{T}} u|_{I_i} = (i_{p_i}(u \circ M_i)) \circ M_i^{-1} \quad \forall I_i \in \mathcal{T}; \end{aligned}$$

here, $M_i : I \to I_i$ is the element map of element I_i and i_{p_i} denotes the Gauss-Lobatto interpolation operator $i_{p_i} : C(\overline{I}) \to \mathcal{P}_{p_i}$ on the reference element. The essential properties of the Gauss-Lobatto operator i_{p_i} can be found in Lemma 3.2.1 (stability) and Lemma 3.2.6 (approximation of analytic functions).
The proof employs both regularity assertions of Proposition 2.2.1.

Step 1: We start with the case $\lambda p \varepsilon \ge 1/2$. In this case, the boundary layer mesh $\mathcal{T}_{\lambda p \varepsilon}$ consists of the three elements $I_1 = (-1, -0.5)$, $I_2 = (-0.5, 0.5)$, $I_3 = (0.5, 1)$. From the regularity assertion (2.2.2) the pull-backs $\hat{u}_i := u_\varepsilon|_{I_i} \circ M_i$ satisfy for some C, $\gamma > 0$ independent of ε

$$\|\hat{u}_i^{(n)}\|_{L^2(I)} \le C\gamma^n \max\{n+1, \varepsilon^{-1}\}^n \qquad \forall n \in \mathbb{N}_0.$$

Hence, employing the Sobolev embedding $L^\infty(I) \subset H^1(I)$

$$\|\hat{u}_i^{(n)}\|_{L^\infty(I)} \le C\gamma^n \max\{n+1, \varepsilon^{-1}\}^{n+1} \qquad \forall n \in \mathbb{N}_0, \tag{2.2.6}$$

for some suitably chosen C, $\gamma > 0$. The term $\max\{n+1, \varepsilon^{-1}\}^{n+1}$ is brought to a more familiar form by noting

$$\begin{aligned} \max\{n+1, \varepsilon^{-1}\}^{n+1} &= \max\{(n+1)^{n+1}, \varepsilon^{-(n+1)}\} \\ &= (n+1)! \max\left\{\frac{(n+1)^{n+1}}{(n+1)!}, \frac{\varepsilon^{-(n+1)}}{(n+1)!}\right\} \\ &= (n+1)! \max\left\{e^{n+1}, e^{1/\varepsilon}\right\} \le C\,(n+1)!\, e^{1/\varepsilon}, \end{aligned} \tag{2.2.7}$$

where we used $\varepsilon \in (0,1]$ and Stirling's formula in the form $n^n \leq n!e^n$. Inserting this result in (2.2.6) we obtain after appropriately adjusting C, $\gamma > 0$

$$\|\hat{u}_i^{(n)}\|_{L^\infty(I)} \leq Ce^{1/\varepsilon}\gamma^n n! \qquad \forall n \in \mathbb{N}_0 .$$

From Lemma 3.2.6 (setting $C_u = Ce^{1/\varepsilon}$), we then obtain

$$\|\hat{u}_i - i_p\hat{u}_i\|_{L^\infty(I)} + \|\left(\hat{u}_i - i_p\hat{u}_i\right)'\|_{L^\infty(I)} \leq Ce^{1/\varepsilon}e^{-bp}$$

for some C, $b > 0$ independent of C, p. Mapping back to the elements I_i, we obtain using $\lambda p\varepsilon \geq 0.5$ and $\mathbf{p} = (p,p,p)$

$$\|u_\varepsilon - i_{\mathbf{p},\mathcal{T}}u_\varepsilon\|_{L^\infty(\Omega)} + \|\left(u_\varepsilon - i_{\mathbf{p},\mathcal{T}}u_\varepsilon\right)'\|_{L^\infty(\Omega)} \leq Ce^{1/\varepsilon}e^{-bp} \leq Ce^{(2\lambda - b)p} \leq Ce^{-(b/2)p}$$

provided that $\lambda < \lambda_0 \leq b/4$.

Step 2: For the case $\lambda p\varepsilon < 0.5$, we employ the decomposition (2.2.3). Denoting the pull-back of the smooth part w_ε to the reference element I by $\hat{w}_i := w_\varepsilon|_{I_i} \circ M_i$, we have

$$\|\hat{w}_i^{(n)}\|_{L^\infty(I)} \leq C(\gamma h_i)^n n! \qquad \forall n \in \mathbb{N}_0 ,$$

where $h_i = |I_i|$ is the length of I_i. Employing Lemma 3.2.6 and mapping back to the physical elements I_i, we get

$$\|w_\varepsilon - i_{\mathbf{p},\mathcal{T}}w_\varepsilon\|_{L^\infty(\Omega)} + \|\left(w_\varepsilon - i_{\mathbf{p},\mathcal{T}}w_\varepsilon\right)'\|_{L^\infty(\Omega)} \leq Ce^{-bp}$$

for some C, $b > 0$ independent of ε and p, where we set $\mathbf{p} = (p,p,p)$. Thus, the smooth part w_ε can be approximated in the desired fashion. We also note that, since the endpoints are sampling points for the Gauss-Lobatto interpolation operator, $w_\varepsilon(\pm 1) = i_{\mathbf{p},\mathcal{T}}w_\varepsilon(\pm 1)$.
We now turn to the approximation of the boundary layer part u_ε^{BL} by $i_{\mathbf{p},\mathcal{T}}u_\varepsilon^{BL}$, where we choose $\mathbf{p} = (p,1,p)$. Again, we introduce the pull-backs $\hat{u}_i^{BL} := u_\varepsilon^{BL} \circ M_i$. Using the results of Proposition 2.2.1, we have on $I_1 = (-1, -1 + \lambda p\varepsilon)$

$$\|\left(\hat{u}_1^{BL}\right)^{(n)}\|_{L^\infty(I)} \leq C\gamma^n \max\{n+1, \varepsilon^{-1}\}^n (\lambda p\varepsilon)^n \qquad \forall n \in \mathbb{N}_0 .$$

In the same way as in (2.2.7), we can estimate

$$\max\{n+1, \varepsilon^{-1}\}^n (\lambda p\varepsilon)^n \leq e\, n! \left(\frac{e}{2}\right)^n e^{\lambda p}.$$

Thus, Lemma 3.2.6 (with $C_u = Ce^{\lambda p}$) implies the existence of C, $b > 0$ independent of ε, p such that

$$\|\hat{u}_1^{BL} - i_p\hat{u}_1^{BL}\|_{L^\infty(I)} + \|\left(\hat{u}_1^{BL} - i_p\hat{u}_1^{BL}\right)'\|_{L^\infty(I)} \leq Ce^{\lambda p}e^{-bp}.$$

Again, assuming $\lambda < \lambda_0 \leq b/2$, we obtain after mapping back to I_1:

$$\|u_\varepsilon^{BL} - i_{p,\mathcal{T}}u_\varepsilon^{BL}\|_{L^\infty(I_1)} + \lambda p\varepsilon\|\left(u_\varepsilon^{BL} - i_{p,\mathcal{T}}u_\varepsilon^{BL}\right)'\|_{L^\infty(I_1)} \leq Ce^{-(b/2)p}.$$

By symmetry, the same estimate holds on $I_3 = (1-\lambda p\varepsilon, 1)$. It remains to consider $I_2 = (-1+\lambda p\varepsilon, 1-\lambda p\varepsilon)$. On I_2, we merely exploit the fact that u_ε^{BL} is small. Specifically, since $i_{1,\mathcal{T}} u_\varepsilon^{BL}$ reduces to the linear interpolant of u_ε^{BL} on the interval I_2 and the length of I_2 is bounded by $1 \leq |I_2| < 2$, we have

$$\begin{aligned}
&\|i_{1,\mathcal{T}} u_\varepsilon^{BL}\|_{L^\infty(I_2)} + \|\left(i_{1,\mathcal{T}} u_\varepsilon^{BL}\right)'\|_{L^\infty(I_2)} \\
&\quad \leq C\left[\left|u_\varepsilon^{BL}(-1+\lambda p\varepsilon)\right| + \left|u_\varepsilon^{BL}(1-\lambda p\varepsilon)\right|\right] \leq Ce^{-\alpha\lambda p}.
\end{aligned}$$

Hence we can bound

$$\begin{aligned}
\|u_\varepsilon^{BL} - i_{1,\mathcal{T}} u_\varepsilon^{BL}\|_{L^\infty(I_2)} &\leq \|u_\varepsilon^{BL}\|_{L^\infty(I_2)} + \|i_{1,\mathcal{T}} u_\varepsilon^{BL}\|_{L^\infty(I_2)} \\
&\leq Ce^{-\alpha\lambda p}, \\
\lambda p\varepsilon\|\left(u_\varepsilon^{BL} - i_{1,\mathcal{T}} u_\varepsilon^{BL}\right)'\|_{L^\infty(I_2)} &\leq \lambda p\varepsilon\|\left(u_\varepsilon^{BL}\right)'\|_{L^\infty(I_2)} + \lambda p\varepsilon\|\left(i_{1,\mathcal{T}} u_\varepsilon^{BL}\right)'\|_{L^\infty(I_2)} \\
&\leq C\lambda p e^{-\alpha\lambda p}.
\end{aligned}$$

By construction, we have $u_\varepsilon^{BL}(\pm 1) = i_{\mathbf{p},\mathcal{T}} u_\varepsilon^{BL}(\pm 1)$.
It remains to approximate r_ε. We simply approximate r_ε by zero, since

$$\|r_\varepsilon\|_{L^\infty(\Omega)} + \|r_\varepsilon'\|_{L^\infty(\Omega)} \leq Ce^{-\alpha/\varepsilon}$$

by Proposition 2.2.1. Our approximant coincides with u_ε in the endpoints ± 1 and has the desired approximation properties. □

Remark 2.2.6 From an application point of view, the draw-back of Proposition 2.2.5 is that λ_0, while independent of ε, is essentially left unspecified. In actual calculations, when good choices of λ are not available, one might use the following two ideas:

1. Since the dependence on λ is made explicit in the approximation result Proposition 2.2.5, it is possible to choose λ as a function of p; choosing for example $\lambda = 1/\ln p$ in the definition of the boundary layer mesh gives for the *hp*-FEM approximation

$$\|u_\varepsilon - u_N\|_\varepsilon \leq Ce^{-bp/\ln p},$$

with C, b independent of ε, p.
2. In view of Céa Lemma, one could consider the approximation from spaces $V_N \supset S_0^p(\mathcal{T}_{\lambda p\varepsilon})$. Good choices are meshes that are geometrically refined towards the endpoints ± 1. Specifically, for a *grading factor* $q \in (0,1)$ and a number of *layers* $L \in \mathbb{N}$, we define a geometric mesh $\mathcal{T}_L^{geo}$ by the points

$$\{-1,1\} \cup \{-1+q^i \,|\, i = 1,\ldots,L\} \cup \{1-q^i \,|\, i = 1,\ldots,L\}.$$

If L is chosen such that $q^L \approx \varepsilon$, then it is not difficult to see that

$$\inf_{v \in S_0^p(\mathcal{T}_L^{geo})} \|u - v\|_\varepsilon \leq Ce^{-bp}$$

for some C, $b > 0$ independent of ε and p. The condition $q^L \approx \varepsilon$ implies $L = O(|\ln \varepsilon|)$ and, since $\dim V_N = (2L+1)p - 1$, we obtain therefore in terms of the number of degrees of freedom $N = \dim V_N$

$$\inf_{v \in S_0^p(\mathcal{T}_L^{geo})} \|u - v\|_\varepsilon \leq C e^{-bp/|\ln \varepsilon|}.$$

While the hp-FEM on such a geometrically refined mesh with $O(|\ln \varepsilon|)$ elements is not robust in a strict sense, the ε-dependence is very weak. Note that a geometric mesh is shape-regular in the usual sense that neighboring elements are comparable in size, in contrast to the boundary layer mesh.

■

2.2.3 Numerical examples

Proposition 2.2.5 shows that the robust exponential convergence in the energy norm can be achieved on boundary layer meshes. The following numerical example shows that a) this robust exponential convergence is achieved in computational practice and that b) the small elements of size $O(p\varepsilon)$ in the layer are necessary to resolve the layers.

Example 2.2.7 We consider the approximation to the solution of

$$-\varepsilon^2 u_\varepsilon''(x) + u_\varepsilon(x) = \frac{1}{2 - x^2} \qquad \text{on } \Omega = (-1, 1), \qquad u_\varepsilon(\pm 1) = 0. \tag{2.2.8}$$

We apply the hp-FEM (2.2.5) based on the boundary layer mesh of Proposition 2.2.5. In the definition of the boundary layer mesh $\mathcal{T}_{\lambda p\varepsilon}$ we take $\lambda = 0.71$, remarking in passing that a more careful analysis in [92, 113] shows that in the present case, $\lambda_0 = 4/e$. We have $\dim S_0^p(\mathcal{T}_{\lambda p\varepsilon}) = 3p - 1$. For $\varepsilon = 10^{-2}$, $\varepsilon = 10^{-4}$, $\varepsilon = 10^{-6}$, $\varepsilon = 10^{-8}$, the left graph in Fig. 2.2.1 shows the error in energy (i.e., the square of the energy norm (2.1.5)) vs. $\dim S_0^p(\mathcal{T}_{\lambda p\varepsilon}) = 3p - 1$. In this linear-logarithmic plot, we observe robust exponential convergence, as the curves are almost straight lines and very close to each other, in spite of the very wide range of parameter values ε.
To show that the small elements of size $O(p\varepsilon)$ are necessary to achieve robust exponential convergence, we consider the performance of the p-version FEM on a single element in the right graph in Fig. 2.2.1. The p-version FEM is defined by (2.2.5) with

$$V_N^{\text{p-FEM}} = S_0^p(\mathcal{T}^{\text{p-FEM}}), \qquad \mathcal{T}^{\text{p-FEM}} = \{\Omega\}.$$

We note that $\dim V_N^{p-FEM} = p - 1$. Again, calculations are performed for $\varepsilon = 10^{-2}$, $\varepsilon = 10^{-4}$, $\varepsilon = 10^{-6}$, $\varepsilon = 10^{-8}$ and shown in the right part of Fig. 2.2.1. We observe very poor convergence for small ε (the error curves corresponding to $\varepsilon = 10^{-4}$, $\varepsilon = 10^{-6}$, $\varepsilon = 10^{-8}$ are practically on top of each other). In fact, in the regime of polynomial degrees shown ($p \leq 40$), robust exponential convergence is not visible but only robust convergence $O(p^{-1}\sqrt{1+\ln p})$, [113]. We therefore conclude that the small elements of size $O(p\varepsilon)$ in the layer are necessary for good performance of the *hp*-FEM. ∎

Example 2.2.8 In the preceding example, we saw that the small elements of size $O(p\varepsilon)$ in the layer are necessary for robust exponential convergence. In this example, we show that they may not be chosen too small. In fact, Proposition 2.2.5 suggests a deterioration of the approximation if $\lambda \to 0$. We now show numerically that such a deterioration arises indeed in computations. We consider the equation

$$-\varepsilon^2 u_\varepsilon'' + u_\varepsilon = 1 \quad \text{on } (0,1), \qquad u_\varepsilon(0) = 0, \quad u_\varepsilon'(1) = 0 \tag{2.2.9}$$

with solution $u_\varepsilon(y)$ is given by

$$u_\varepsilon(x) = 1 - \frac{\cosh\left((1-x)/\varepsilon\right)}{\cosh(1/\varepsilon)}.$$

We consider the p-version FEM based on a two-element mesh determined by the points $0 = y_0 < y_1 = a\varepsilon < y_2 = 1$. The performance for $\varepsilon = 10^{-3}$ and various choices of the parameter a is reported in Fig. 2.2.2. For fixed a, we note an initial exponential convergence which deteriorates if p becomes large. In fact, the exponential rate of convergence is visible until $p \approx a$. For $p > a$, the large element (which is $a\varepsilon$ away from the boundary point $y = 0$) dominates the overall possible error reduction. This can be seen as follows. As the boundary layer function in this particular case is essentially $e^{-y/\varepsilon}$, the function to be approximated on the large element is $e^{-a}e^{-(y-a\varepsilon)/\varepsilon}$. For small ε polynomial approximation of $e^{-(y-a\varepsilon)/\varepsilon}$ on the element $(a\varepsilon, 1)$ is quite poor as we have seen in Example 2.2.7, and the factor e^{-a} is comparatively large if a is small (relative to p). However, if a is large ($a \geq p$, say), then the boundary layer function $e^{-y/\varepsilon}$ on the large element is exponentially small (in p), and thus the contribution of the large element to the total error as well. We conclude therefore that for fixed a, the error on the large element is negligible for $p < a$, and the global error reduction is controlled by the error on the small element. In the regime $p > a$, the error on the large element dominates the global error. The choice of a variable mesh, i.e., taking $a = p$ balances the two errors; we see in Fig. 2.2.2 that this choice allows us to obtain exponential convergence. ∎

2.3 Regularity: the two-dimensional case

Our regularity assertions in the two-dimensional setting take a form similar to that of Proposition 2.2.1; that is, the regularity is described both in terms of bounds on the growth of the derivatives and in terms of decompositions that capture the main features of the solution. In the one-dimensional case of Proposition 2.2.1, the data c and f are smooth. In the two-dimensional case of (2.1.1),

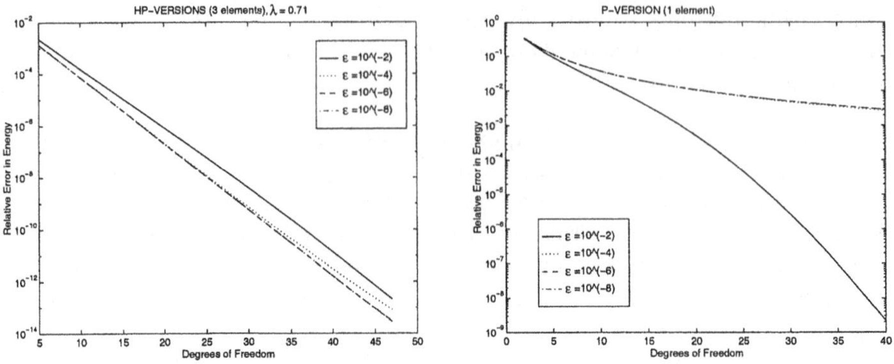

Fig. 2.2.1. Example 2.2.7: *hp*-FEM (left) and *p*-FEM (right) for a problem with boundary layers.

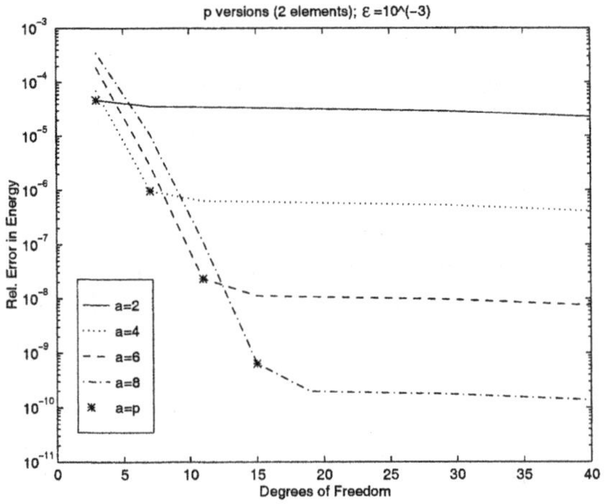

Fig. 2.2.2. Example 2.2.8 *p*-version for 1D example and various values of a; $\varepsilon = 10^{-3}$.

part of the data, namely, the boundary $\partial\Omega$ is only piecewise analytic. This introduces singularities into the solution. Thus, boundary layers as well as corner singularities have to be captured. The analog of the assertion on the growth of the derivatives of Proposition 2.2.1 is therefore replaced with estimates in weighted L^2-spaces. The analog of the decomposition of Proposition 2.2.1 includes an additional term that captures the corner singularities.
In order to be able to formulate the regularity assertions of Theorem 2.3.1, we need to introduce some notation. First we introduce the weight functions $\Phi_{p,\beta,\varepsilon}$ as follows. With each vertex A_j, $j = 1, \ldots, J$, we associate a number $\beta_j \in [0, 1)$, set $\beta = (\beta_1, \ldots, \beta_J)$, and write for $p \in \mathbb{N}_0$

$$\hat{\Phi}_{p,\beta_j,\varepsilon}(x) := \left(\min\left\{1, \frac{\operatorname{dist}(x, A_j)}{\min\{1, \varepsilon(p+1)\}}\right\}\right)^{p+\beta_j}.$$

The weight function $\Phi_{p,\beta,\varepsilon}$ is then defined as

$$\Phi_{p,\beta,\varepsilon}(x) := \prod_{j=1}^{J} \hat{\Phi}_{p,\beta_j,\varepsilon}(x).$$

Using the weight functions $\Phi_{p,\beta,\varepsilon}$ we can formulate the two-dimensional analog of (2.2.2):

Theorem 2.3.1 (Regularity in countably normed spaces). *Let Ω be a curvilinear polygon and A, c, f, g satisfy (2.1.2), (1.2.4). Then there exist C, $K > 0$, and a vector $\beta \in [0,1)^J$ independent of ε such that the solution u_ε of (2.1.1) satisfies $u_\varepsilon \in \mathcal{B}^2_{\beta,\varepsilon}(\Omega, C, K)$, i.e.,*

$$\|u_\varepsilon\|_\varepsilon \leq C$$
$$\|\Phi_{p,\beta,\varepsilon}\nabla^{p+2}u_\varepsilon\|_{L^2(\Omega)} \leq CK^p \max\{p+1, \varepsilon^{-1}\}^{p+2} \qquad \forall p \in \mathbb{N}_0.$$

In particular, for fixed neighborhoods $\mathcal{U}_j$ of the vertices A_j there holds for all $p \in \mathbb{N}_0$ and all $x \in \mathcal{U}_j \cap \Omega$

$$|\nabla^p (u(x) - u(A_j))| \leq CK^p\varepsilon^{-1} \min\left\{1, \frac{r_j}{\varepsilon}\right\}^{1-\beta_j} r_j^{-p} \max\left\{p+1, \frac{r_j}{\varepsilon}\right\}^{p+1},$$

where $r_j := \operatorname{dist}(x, A_j)$. In the interior $\Omega \setminus \cup_{j=1}^J \mathcal{U}_j$ there holds

$$|\nabla^p u(x)| \leq CK^p \max\{(p+1), \varepsilon^{-1}\}^{p+2} \qquad \forall p \in \mathbb{N}_0 \quad \forall x \in \Omega \setminus \cup_{j=1}^J \mathcal{U}_j.$$

Proof: The energy estimate $\|u_\varepsilon\|_\varepsilon$ follows from Lemma 1.2.2. The L^2-based bound for higher derivatives of u_ε is proved as Theorem 5.3.14, which asserts in particular that $u_\varepsilon \in \mathcal{B}^2_{\beta,\varepsilon}$. The pointwise bounds are then obtained from Theorem 4.2.23. An outline for the key ingredients of the proof can be found in Section 5.1. □

Remark 2.3.2

1. Theorem 2.3.1 merely asserts the existence of $\beta \in (0,1)^J$. The values of β_j, however, are available from the proof. For $A = \mathrm{Id}$, for example, one may choose any $\beta_j \in (0,1) \cap (1 - \pi/\omega_j, 1)$, where ω_j is the interior angle of the curvilinear polygon Ω at vertex A_j.
2. The assumption that f is analytic up to the boundary $\partial\Omega$ can be relaxed. Theorem 2.3.1 still holds true if $f \in \mathcal{B}^0_{\beta,\varepsilon}$, i.e., $\|\Phi_{p,\beta,\varepsilon}\nabla^p f\|_{L^2(\Omega)} \leq C_f\gamma_f^p p!$ for all $p \in \mathbb{N}_0$ and suitable constants C_f, $\gamma_f > 0$.
3. The assumption that c, f are real-valued is not essential; see Remark 5.3.5.
4. Theorem 2.3.1 can also be formulated for solutions to the differential equation (1.2.1) with $b \neq 0$. We refer to Theorem 5.3.14, where this case of covered.

5. The coefficients A, c, and the right-hand side f are assumed analytic on $\overline{\Omega}$. This is done for simplicity of notation. The results of this work can be extended to the case of piecewise analytic functions A, c, f.
6. In the case of analytic data c, f, and $\partial\Omega$, the assertion of Theorem 5.3.14 holds with the weight function $\Phi_{p,\beta,\varepsilon}$ replaced with $\Phi_{p,\beta,\varepsilon} \equiv 1$, [95].
7. The boundary value problem (1.2.11) is a Dirichlet problem. Regularity assertions analogous to that of Theorem 2.3.1 hold for other boundary conditions also as we show in Section 5.4 for Neumann, Robin boundary conditions, and a transmission problem.

■

The regularity assertion of Theorem 2.3.1 does not capture the boundary layer character of the solution. This is done classically with the aid of asymptotic expansions. Since additionally corner singularities are present in the solution u_ε, we present next a decomposition of the solution u_ε into a smooth part w_ε, a boundary layer part u_ε^{BL}, a corner layer part u_ε^{CL}, and a small remainder r_ε. In order to be able to formulate this decomposition, we need to introduce some notation.

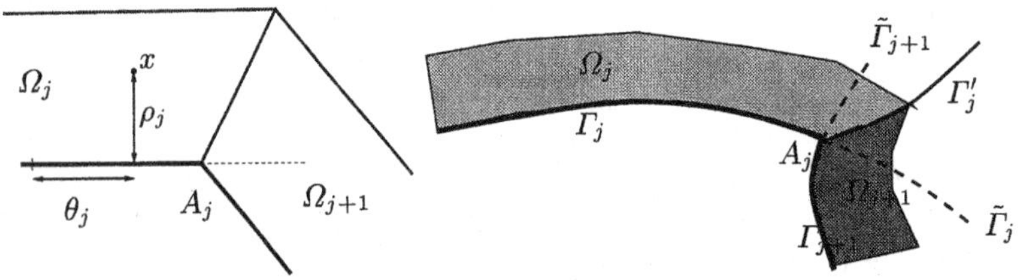

Fig. 2.3.1. Boundary fitted coordinates $\psi_j : (\rho_j, \theta_j) \mapsto x$.

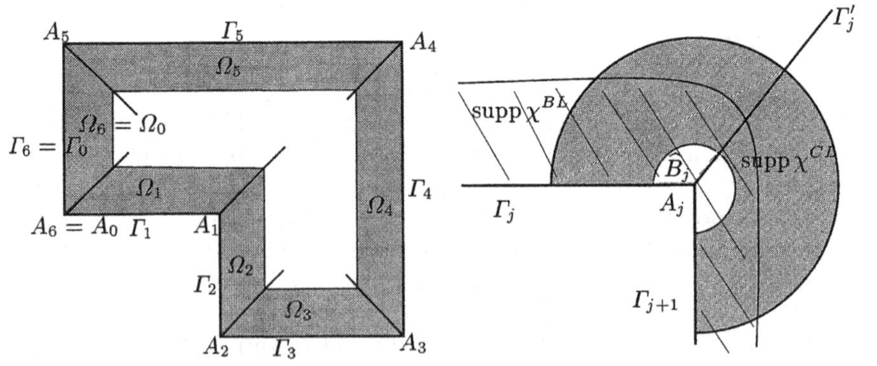

Fig. 2.3.2. Scheme of the supports of χ^{BL} and χ^{CL}.

Notation 2.3.3 *The notation introduced here is illustrated in Figs. 2.3.1, 2.3.2.*

1. (boundary fitted coordinates/ψ_j) *For each boundary arc Γ_j we introduce boundary-fitted coordinates (ρ_j, θ_j) as follows: Recalling that Γ_j*

$$\Gamma_j = \{x_j(\theta), y_j(\theta) \mid \theta \in (0,1)\}$$

is parametrized such that the normal vector $(-y'(\theta), x'(\theta))$ points into Ω, we define

$$\begin{aligned} \psi_j : \mathbb{R} \times (0,1) &\to \overline{\Omega} \\ (\rho, \theta) &\mapsto \begin{pmatrix} x_j(\theta) \\ y_j(\theta) \end{pmatrix} + \frac{\rho}{\sqrt{|x_j'(\theta)|^2 + |y_j'(\theta)|^2}} \begin{pmatrix} -y_j'(\theta) \\ x_j'(\theta) \end{pmatrix}. \end{aligned}$$

The maps ψ_j are real analytic and in fact invertible in a neighborhood of $\{0\} \times (0,1)$. Without loss of generality, we may assume that ρ_0 and $\Theta > 0$ are chosen so small that the analytic continuation of ψ_j (again denoted ψ_j) is real analytic and invertible on $(-2\rho_0, 2\rho_0) \times (-2\Theta, 1 + 2\Theta)$ for every $j \in \{1, \ldots, J\}$. The analytic continuation of the arc Γ_j is the arc $\tilde{\Gamma}_j := \{(x_j(\theta), y_j(\theta)) \mid \theta \in (-\Theta, 1 + \Theta)\}$. The inverse functions ψ_j^{-1} define boundary-fitted coordinates (ρ_j, θ_j) in a neighborhood of $\tilde{\Gamma}_j$ via $(\rho_j, \theta_j) = (\rho_j(x), \theta_j(x)) = \psi_j^{-1}(x)$. A geometric interpretation of ρ_j is $\rho_j(x) = \operatorname{dist}(x, \tilde{\Gamma}_j)$ (cf. Fig. 2.3.1).
2. (arcs Γ_j', subdomains Ω_j) *For each $j \in \{0, \ldots, J-1\}$ choose an analytic arc Γ_j' passing through vertex A_j such that the angles $\angle(\Gamma_j, \Gamma_j')$ and $\angle(\Gamma_{j+1}, \Gamma_j')$ at A_j are strictly less than $\pi/2$. Set $\Gamma_J' := \Gamma_0'$ and define Ω_j as the subset of $\psi_j((0, \rho_0) \times (-\Theta, 1 + \Theta))$ that is bordered by Γ_{j-1}', Γ_j, Γ_j' (see Fig. 2.3.2).*
3. (cut-off functions χ^{BL}, χ^{CL}) *Choose smooth cut-off functions χ^{CL}, χ^{BL} and numbers $\rho_1 < \rho_0$, $r_2 < r_1$ with the following properties:*

$$\begin{aligned} \operatorname{supp} \chi^{BL} &\subset \mathcal{U}_{\rho_0}(\partial\Omega), \\ \chi^{BL} &\equiv 1 \text{ on } \mathcal{U}_{\rho_1}(\partial\Omega), \\ \chi^{CL} &\equiv 1 \text{ on } \operatorname{supp} \chi^{BL} \cap \Gamma_j', \qquad j = 1, \ldots, J, \\ \operatorname{supp} \chi^{CL} &\subset \cup_{j=1}^J B_{r_1}(A_j), \\ \chi^{CL} &\equiv 1 \text{ on } \cup_{j=1}^J B_{r_2}(A_j). \end{aligned}$$

Finally, choose $r_3 < r_2$ so small that

$$\widehat{B}_j := B_{r_3}(A_j) \subset \{x \in \mathbb{R}^2 \mid \chi^{BL}(x) = 1\}.$$

These notations enable us to formulate the following regularity results, in which the solution u_ε is decomposed into a smooth part w_ε, a boundary layer part u_ε^{BL}, a corner layer part u_ε^{CL}, and a small remainder r_ε.

Theorem 2.3.4 (Regularity through asymptotic expansions). *Fix sets Ω_j, arcs Γ_j', and cut-off functions χ^{BL}, χ^{CL} as in Notation 2.3.3. Then there* *exist C, K, $\alpha > 0$, and $\beta \in [0,1)^J$ such that for each $\varepsilon \in (0,1]$ the solution u_ε of (2.1.1) can be decomposed as*

$$u_\varepsilon = w_\varepsilon + \chi^{BL} u_\varepsilon^{BL} + \chi^{CL} u_\varepsilon^{CL} + r_\varepsilon \tag{2.3.1}$$

with the following properties:

(i) *The* smooth part w_ε *is analytic on Ω and*

$$\|\nabla^p w_\varepsilon\|_{L^\infty(\Omega)} \le C\gamma^p p! \qquad \forall p \in \mathbb{N}_0 .$$

(ii) *The* remainder $r_\varepsilon \in H_0^1(\Omega) \cap H_{loc}^2(\Omega)$ *satisfies*

$$\|r_\varepsilon\|_{L^\infty(\Omega)} + \|r_\varepsilon\|_{H^1(\Omega)} + \|\Phi_{0,\beta,\varepsilon} \nabla^2 r_\varepsilon\|_{L^2(\Omega)} \le C e^{-\alpha/\varepsilon}.$$

(iii) *The* boundary layer u_ε^{BL} *satisfies on each Ω_j*

$$\sup_{\theta_j \in (-\Theta, 1+\Theta)} |(\partial_{\rho_j}^n \partial_{\theta_j}^m (u_\varepsilon^{BL} \circ \psi_j))(\rho_j, \theta_j)| \le C\varepsilon^{-n} \gamma^{n+m} m!\, e^{-\alpha\rho_j/\varepsilon}, \ \rho_j \ge 0.$$

(iv) *The* corner layer u_ε^{CL} *is analytic on $\widehat{\mathcal{U}}_j := (\Omega_j \cap \widehat{B}_j) \cup (\Omega_{j+1} \cap \widehat{B}_j)$ for each $j = 0, \ldots, J-1$ and satisfies for all $p \in \mathbb{N}_0$*

$$\|e^{\alpha r_j/\varepsilon} u_\varepsilon^{CL}\|_{L^2(\widehat{\mathcal{U}}_j)} + \varepsilon \|e^{\alpha r_j/\varepsilon} \nabla u_\varepsilon^{CL}\|_{L^2(\widehat{\mathcal{U}}_j)} \le C\varepsilon,$$
$$\|e^{\alpha r_j/\varepsilon} \Phi_{p,\beta,\varepsilon} \nabla^{p+2} u_\varepsilon^{CL}\|_{L^2(\widehat{\mathcal{U}}_j)} \le C\varepsilon\gamma^p \max\{p+1, \varepsilon^{-1}\}^{p+2}$$

together with the pointwise bounds (we abbreviate $r_j = \operatorname{dist}(x, A_j)$)

$$|\nabla^p u_\varepsilon^{CL}(x)| \le C\gamma^p p!\, \varepsilon^{\beta_j - 1} r_j^{1-p-\beta_j} e^{-\alpha r_j/\varepsilon} \quad \forall x \in \widehat{\mathcal{U}}_j \quad \forall p \in \mathbb{N}_0 .$$

Additionally, for $\mathcal{U}_j := (\Omega_j \cap B_{r_1}(A_j)) \cup (\Omega_{j+1} \cap B_{r_1}(A_j))$ we have

$$\|e^{\alpha r_j/\varepsilon} u_\varepsilon^{CL}\|_{L^2(\mathcal{U}_j)} + \varepsilon \|e^{\alpha r_j/\varepsilon} \nabla u_\varepsilon^{CL}\|_{L^2(\mathcal{U}_j)} \le C\varepsilon.$$

Proof: This result can be inferred from Corollary 7.4.6. An outline of the key steps of the construction of the decomposition an be found in Section 7.1. □

Remark 2.3.5 1. In the language of Part III, Theorem 2.3.4 asserts for the corner layer u^{CL} membership in an exponentially weighted countably normed space, namely, $u^{CL} \in \mathcal{B}^2_{\beta,\varepsilon,\alpha}(\widehat{\mathcal{U}}_j, C\varepsilon, \gamma)$ for each $j \in \{1, \ldots, J\}$.

2. Theorem 2.3.4 asserts differentiabilty properties for the corner layer on the sets $\widehat{\mathcal{U}}_j$. However, u^{CL} is also defined on $\mathcal{U}_j$, it is smooth on $\mathcal{U}_j$, and inspection of the proof in Section 7.4 reveals that for each $k \in \mathbb{N}_0$ there exists $C_k > 0$ (independent of ε) such that

$$|\nabla^k u^{CL}(x)| \le C_k \varepsilon^{\beta_j - 1} r_j^{1-k-\beta_j} e^{-\alpha r_j/\varepsilon} \quad \forall x \in \mathcal{U}_j .$$

We emphasize that $\cup_{j=1}^J \mathcal{U}_j \supset \operatorname{supp} \chi^{CL} \setminus \cup_{j=1}^J \Gamma_j'$.

3. Theorem 2.3.4 merely asserts $r_\varepsilon \in H^2_{loc}(\Omega)$. However, inspection of the procedure in Section 7.4 reveals that r_ε is obtained as the solution of an elliptic equations with piecewise smooth data; in can be shown that for each $k \in \mathbb{N}_0$ there exists $C_k > 0$ such that

$$\sum_{j=1}^{J} \|\Phi_{k,\beta,\varepsilon}\nabla^{k+2} r_\varepsilon\|_{L^2(\Omega_j)} + \|\Phi_{k,\beta,\varepsilon}\nabla^{k+2} r_\varepsilon\|_{L^2(\Omega\setminus\cup_{j=1}^J \Omega_j)} \le C_k \varepsilon^{-k} e^{-\alpha/\varepsilon}.$$

■

The decomposition whose existence is ascertained in Theorem 2.3.4 is constructed using the classical method of asymptotic expansions: First, the outer expansion is constructed that gives a particular solution to the differential equation. In the second step, the boundary conditions are corrected using the boundary layer functions. These are constructed on each of the subdomains Ω_j separately. In particular, the piecewise defined function $\chi^{BL} u_\varepsilon^{BL}$ may have jumps across the arcs Γ_j'. These jumps are corrected with the corner layer function u_ε^{CL}, which are solutions to transmission problems where the jump and the jump of the co-normal derivative across Γ_j' is prescribed such that $\chi^{BL} u_\varepsilon^{BL} + \chi^{CL} u_\varepsilon^{CL}$ is C^1-across the arcs Γ_j'. We formulated these observations in the following theorem for future reference:

Theorem 2.3.6 (Regularity through asymptotic expansions). *Assume the hypotheses of Theorem 2.3.4. Denote by L_ε the differential operator $L_\varepsilon v := -\nabla \cdot (A(x)\nabla v) + c(x)v$. With the constants C, $\alpha > 0$ of Theorem 2.3.4, the terms w_ε, u_ε^{BL}, u_ε^{CL} of the decomposition in Theorem 2.3.4 have the following additional properties:*

(i) $\|L_\varepsilon w_\varepsilon - f\|_{L^\infty(\Omega)} \le Ce^{-\alpha/\varepsilon}$.

(ii) $\chi^{BL} u_\varepsilon^{BL} + w_\varepsilon = g$ *on* $\partial\Omega$ *and for each* $j \in \{1, \ldots, J\}$ *there holds* $\|L_\varepsilon(\chi^{BL} u_\varepsilon^{BL})\|_{L^\infty(\Omega_j)} \le Ce^{-\alpha/\varepsilon}$.

(iii) $\chi^{CL} u_\varepsilon^{CL} = 0$ *on* $\partial\Omega$.

(iv) The function $u^C := \chi^{BL} u_\varepsilon^{BL} + \chi^{CL} u_\varepsilon^{CL}$ *is in* $C^1(\overline{\Omega} \setminus \cup_{j=1}^J \{A_j\})$ *and*

$$\|L_\varepsilon u^C\|_{L^\infty(\Omega\setminus\cup_{j=1}^J \Gamma_j')} \le Ce^{-\alpha/\varepsilon}.$$

Proof: The asserted properties follow from the construction of the decomposition in Section 7.4. Assertion (i) is shown in (7.4.12). The assertion (ii) follows from Lemma 7.4.7. The last assertion follows from the construction, viz., the definition of u_ε^{CL} in (7.4.33). □

Remark 2.3.7 The constants C, K in Theorems 2.3.4, 2.3.6, depend on the choice of the analytic arcs Γ_j'. In the proof of exponential convergence of the hp-FEM, the mesh has to be chosen such that at each vertex A_j, a mesh line can be chosen as an arc Γ_j' with the properties stipulated in Notation 2.3.3. ■

We conclude our discussion of the regularity properties of the solution u_ε with remarks on two special cases, namely, convex corners and smooth domains. In general, the boundary layer function u_ε^{BL} (and therefore also the corner layer function u_ε^{CL} and the remainder r_ε) is only piecewise smooth; near convex vertices, a special construction is possible that affords a smooth boundary layer u_ε^{BL}. This special case is formulated in the following theorem:

Theorem 2.3.8 (Regularity near convex vertices). *Assume the hypotheses of Theorem 2.3.4. Let A_j be convex vertex, i.e., $\angle(\Gamma_j, \Gamma_{j+1}) < \pi/2$. Then u_ε^{BL} and u_ε^{CL} can be constructed such that, in addition to the properties listed in Theorem 2.3.4, the following holds:*

(i) On $B_{\rho_0}(A_j)$ the function u_ε^{BL} is analytic and can be written as $u_\varepsilon^{BL} = u_{j,\varepsilon}^{BL} + u_{j+1,\varepsilon}^{BL}$, where the functions $\tilde{u}_j^{BL} := u_{j,\varepsilon}^{BL} \circ \psi_j$, $\tilde{u}_{j+1}^{BL} := u_{j+1,\varepsilon}^{BL} \circ \psi_{j+1}$ satisfy

$$\sup_{\theta \in (-\Theta, 1+\Theta)} |(\partial_\rho^n \partial_\theta^m \tilde{u}_j^{BL}(\rho, \theta)| \leq C\varepsilon^{-n}\gamma^{n+m} m! \exp(-\alpha\rho/\varepsilon), \quad \rho_j \geq 0,$$

$$\sup_{\theta \in (-\Theta, 1+\Theta)} |(\partial_\rho^n \partial_\theta^m \tilde{u}_{j+1}^{BL}(\rho, \theta)| \leq C\varepsilon^{-n}\gamma^{n+m} m! \exp(-\alpha\rho/\varepsilon), \quad \rho \geq 0.$$

(ii) The corner layer u_ε^{CL} is analytic on $\widehat{B}_j$ and satisfies for all $p \in \mathbb{N}_0$ and $x \in \widehat{B}_j$

$$|\nabla^p \left(u_\varepsilon^{CL}(x) - u_\varepsilon^{CL}(A_j)\right)| \leq C\gamma^p p! \, \varepsilon^{\beta_j - 1} \, r_j^{1-p-\beta_j} \exp(-\alpha r_j/\varepsilon),$$

where $r_j = \operatorname{dist}(x, A_j)$. Additionally, $\chi^{CL} u_\varepsilon^{CL} + \chi^{BL} u_\varepsilon = g$ on $\partial\Omega$.

Remark 2.3.9 In the construction of the corner layer u_ε^{CL} in the case of convex corners A_j, the parameter β_j can be characterized more precisely. For example, in the case $A = \text{Id}$, we may choose any $\beta_j \in (0, 1)$. ▪

Remark 2.3.10 In the case of analytic boundary curves $\partial\Omega$, a simplified expansion holds: The boundary layer function u_ε^{BL} is analytic on a neighborhood of $\partial\Omega$, where u_ε^{CL} may be chosen as $u_\varepsilon^{CL} = 0$ and the weight function $\Phi_{k,\beta,\varepsilon}$ in the estimates for the remainder r_ε may be replaced with $\Phi_{k,\beta,1} \equiv 1$. We refer to [95] for the details. ▪

2.4 *hp*-FEM approximation

2.4.1 *hp*-meshes and spaces

The reference square S and the reference triangle T are defined as

$$S = (0, 1) \times (0, 1), \qquad T = \{(x, y) \,|\, 0 < x < 1 \,|\, 0 < y < x\}.$$

We start by defining regular triangulations without "hanging nodes".

Definition 2.4.1. *A collection of triples* $\mathcal{T} = \{(K_i, M_i, \hat{K}_i) \,|\, i \in I(\mathcal{T})\}$ *is said to be a triangulation of a domain* Ω *if the subsets* $K_i \subset \Omega$*, the element maps* $M_i : \hat{K}_i \to \overline{K_i}$*, the reference elements* $\hat{K}_i \in \{S, T\}$*, and the index set* $I(\mathcal{T}) \subset \mathbb{N}$ *satisfy the following conditions (M1)–(M4):*

(M1) The elements K_i *partition the domain* Ω*, i.e.,* $\overline{\Omega} = \cup_i \overline{K_i}$*.*

(M2) For $i \neq j$*,* $\overline{K_i} \cap \overline{K_j}$ *is either empty, or a vertex or an entire edge (vertices and edges are the images of the vertices and edges of the reference elements under the maps* M_i*).*

(M3) The element maps $M_i : \hat{K}_i \to K_i$ *are analytic diffeomorphisms.*

(M4) The common edge of two neighboring elements has the same parametrization "from both sides": For two neighboring elements K_i*,* K_j*, let* $\gamma_{ij} = \overline{K_i} \cap \overline{K_j}$ *be the common edge with endpoints* P_1*,* P_2*. Then for any point* $P \in \gamma_{ij}$*, we have* $\operatorname{dist}(M_i^{-1}(P), M_i^{-1}(P_l))/L_i = \operatorname{dist}(M_j^{-1}(P), M_j^{-1}(P_l))/L_j$*,* $l = 1, 2$ *where* L_i *and* L_j *denote the lengths of the edges corresponding to* γ_{ij} *in the reference elements.*

If $\hat{K}_i = S$ *then* K_i *is said to be a (curvilinear) quadrilateral and if* $\hat{K}_i = T$ *then* K_i *is a (curvilinear) triangle.*

Once a triangulation $\mathcal{T}$ is chosen, one can define finite-element spaces $S^p(\mathcal{T})$ based on this triangulation.

Definition 2.4.2 (FE-spaces). *Given a triangulation* $\mathcal{T} = \{(K_i, M_i, \hat{K}_i) \,|\, i \in I(\mathcal{T})\}$*, the* H^1*-conforming finite element spaces* $S^p(\mathcal{T})$*,* $S_0^p(\mathcal{T})$ *of piecewise mapped polynomials are defined as*

$$S^p(\mathcal{T}) := \{u \in H^1(\Omega) \,|\, u|_{K_i} = \varphi_p \circ M_i^{-1} \quad \textit{for some } \varphi_p \in \Pi_p(\hat{K}_i)\}, \tag{2.4.1}$$

$$S_0^p(\mathcal{T}) := S^p(\mathcal{T}) \cap H_0^1(\Omega), \tag{2.4.2}$$

where the polynomial spaces $\Pi_p(\hat{K}_i)$ *are defined as*

$$\Pi_p(\hat{K}_i) = \begin{cases} \mathcal{Q}_p(S) & \textit{if } \hat{K}_i = S \\ \mathcal{P}_p(T) & \textit{if } \hat{K}_i = T, \end{cases}$$

$$\mathcal{Q}_p(S) := \operatorname{span}\{x^i y^j \,|\, 0 \le i, j \le p\},$$

$$\mathcal{P}_p(T) := \operatorname{span}\{x^i y^j \,|\, i + j \le p\}.$$

2.4.2 The minimal *hp*-mesh

The regularity assertion of Theorem 2.3.4 shows that we have to deal with two types of phenomena: boundary layers in the vicinity of the boundary curves Γ_j and corner singularities in neighborhoods of the vertices A_j. Meshes with special properties are required to resolve these two phenomena; namely, the meshes should contain *needle* elements near the boundary curves Γ_j to capture the boundary layers and they should include geometric refinement near the vertices A_j to catch the corner singularities. The (essentially) minimal mesh family that incorporates these two mesh design principles is characterized by our notion of *admissible boundary layer meshes* in Definition 2.4.4. An example of such an admissible mesh is presented in Fig. 2.4.1: The rectangles at the boundary are *boundary layer elements* of width $O(\kappa)$, the elements in the shaded regions are *corner layer elements*, and the remaining elements are *interior elements*. The parameter κ, which characterizes the width of the needle elements, will be chosen $\kappa = O(p\varepsilon)$, where p is the approximation order.

Our *hp*-FEM convergence results ahead rests on the decomposition of the exact solution into a smooth part, boundary layers, and corner layers given in Theorem 2.3.4. Our notion of admissible boundary layer meshes in Definition 2.4.4 reflects this by aligning mesh lines with the subdomains Ω_j introduced in Notation 2.3.3. Additionally, we employ the cut-off functions χ^{BL}, χ^{CL}, and the maps ψ_j providing the transfer from Cartesian coordinates to the boundary-fitted coordinates (ρ_j, θ_j) of Notation 2.3.3.

For the definition of admissible meshes, it is convenient to introduce the following two stretching maps:

Notation 2.4.3 *For $\kappa > 0$ and each $j \in \{1, \ldots, J\}$ introduce*

$$\begin{aligned} s_\kappa : \mathbb{R}_0^+ \times \mathbb{R} &\to \mathbb{R}_0^+ \times \mathbb{R} \\ (\rho, \theta) &\mapsto (\kappa\rho, \theta), \\ \tilde{s}_{j,\kappa} : \mathbb{R}^2 &\to \mathbb{R}^2 \\ (x, y) &\mapsto \kappa\left((x,y) - A_j\right) + A_j. \end{aligned}$$

The value $\kappa_0 > 0$ is such that on a κ_0-neighborhood of each arc Γ_j the cut-off function χ^{BL} is identically 1 and such that on a κ_0-neighborhood of the vertices A_j, the cut-off functions χ^{CL} are identically 1:

$$\mathcal{U}_{\kappa_0}(\Gamma_j) \cap \Omega_j \subset \{x \in \Omega \,|\, \chi^{BL}(x) = 1\}, \quad j = 1, \ldots, J, \tag{2.4.3a}$$
$$B_{\kappa_0}(A_j) \cap \Omega \subset \{x \in \Omega \,|\, \chi^{CL}(x) = 1\}, \quad j = 1, \ldots, J. \tag{2.4.3b}$$

where we set $\mathcal{U}_{\kappa_0}(\Gamma_j) := \{x \in \mathbb{R}^2 \mid \operatorname{dist}(x, \Gamma_j) < \kappa_0\}$.

The following definition introduces admissible boundary layer meshes, which includes needle elements of width $O(\kappa)$. The key requirement for the element maps M_i of such needle elements is that the "stretched" maps $\widetilde{M}_i := s_{\kappa^{-1}} \circ \psi_j \circ M_i$ satisfies the standard assumptions on element maps. We note that the anisotropic stretching $s_{\kappa^{-1}} \circ \psi_j$ corresponds to a stretching "normal" to the boundary, which correctly reflects the boundary layer behavior of the solution.

Definition 2.4.4 (admissible boundary layer mesh). *The two-parameter family $\mathcal{T}(\kappa, L) = \{(K_i, M_i, \hat{K}_i)\}_{\kappa,L}$, $(\kappa, L) \in (0, \kappa_0] \times \mathbb{N}_0$ of meshes satisfying (M1)–(M4) is said to be* admissible *if there are C, γ, $c_i > 0$, $i = 1, \ldots, 4$, $\sigma \in (0,1)$ and sets Ω_j, $j = 1, \ldots, J$, of the form described in Notation 2.3.3 such that each triple $(K_i, M_i, \hat{K}_i) \in \mathcal{T}(\kappa, L)$ falls into exactly one of the following three categories:*

(C1) K_i *is a* boundary layer element, *i.e., for some* $j \in \{1, \ldots, J\}$ *there holds* $K_i \subset \mathcal{U}_\kappa(\Gamma_j) \cap \Omega_j \setminus (B_{c_1\kappa}(A_{j-1}) \cup B_{c_1\kappa}(A_j))$, *and the element maps* M_i *satisfies*

$$\| (M_i')^{-1} \|_{L^\infty(\hat{K}_i)} \leq \frac{C}{\kappa},$$
$$\|D^\alpha \left(s_\kappa^{-1} \circ \psi_j^{-1} \circ M_i\right) \|_{L^\infty(\hat{K}_i)} \leq C\gamma^{|\alpha|}\alpha! \qquad \forall \alpha \in \mathbb{N}_0^2;$$

(C2) K_i *is a* corner layer element, *i.e., for some* $j \in \{0, \ldots, J-1\}$ *either* $K_i \subset B_\kappa(A_j) \cap \Omega_j$ *or* $K_i \subset B_\kappa(A_j) \cap \Omega_{j+1}$. *Additionally, denoting* $h_i = \operatorname{diam} K_i$, *the element map* M_i *satisfies*

$$\|D^\alpha \left(\tilde{s}_{j,h_i}^{-1} \circ M_i\right) \|_{L^\infty(\hat{K}_i)} \leq C\gamma^{|\alpha|}\alpha! \qquad \forall \alpha \in \mathbb{N}_0^2,$$
$$\| \left((\tilde{s}_{j,h_i}^{-1} \circ M_i)'\right)^{-1} \|_{L^\infty(\hat{K}_i)} \leq C;$$

furthermore, exactly one of the following situations is satisfied: either

$$A_j \in \overline{K_i} \text{ and } h_i \leq c_4 \kappa \sigma^L$$

or

$$A_j \notin \overline{K_i} \text{ and } c_3 h_i \leq \operatorname{dist}(A_j, K_i) \leq c_4 h_i.$$

(C3) K_i *is an* interior element, *i.e.,* $K_i \subset \Omega \setminus \mathcal{U}_{c_2\kappa}(\partial\Omega)$, *and the element map satisfies*

$$\|D^\alpha M_i\|_{L^\infty(\hat{K}_i)} \leq C\gamma^{|\alpha|}\alpha! \qquad \forall \alpha \in \mathbb{N}_0^2,$$
$$\| (M_i')^{-1} \|_{L^\infty(\hat{K}_i)} \leq \frac{C}{\kappa}.$$

The parameter κ in the definition of admissible meshes controls the width of the needle elements required to capture the boundary layer phenomena; in the *hp*-FEM approximation result Theorem 2.4.8, we will choose $\kappa = O(p\varepsilon)$, where p is the approximation order. The parameter L in admissible meshes represents the number of layers of geometric refinement towards the vertices in an $O(\kappa)$ neighborhood of the vertices. We refer to Fig. 2.4.1 for examples of admissible meshes.

Remark 2.4.5 Admissible boundary layer mesh families are essentially the minimal meshes that lead to robust exponential convergence for an *hp* FEM (see Theorem 2.4.8). In particular, highly distorted elements violating the maximal angle condition, [12], are admitted in our notion of admissible meshes: Minimal and maximal angles are allowed to be of sizes $O(\kappa)$ and $O(\pi - \kappa)$, cf. Fig. 2.4.2. ▪

Remark 2.4.6 The notion of admissible meshes of Definition 2.4.4 includes the standard *p*-FEM and *hp*-FEM meshes as special cases. For fixed κ and L, the

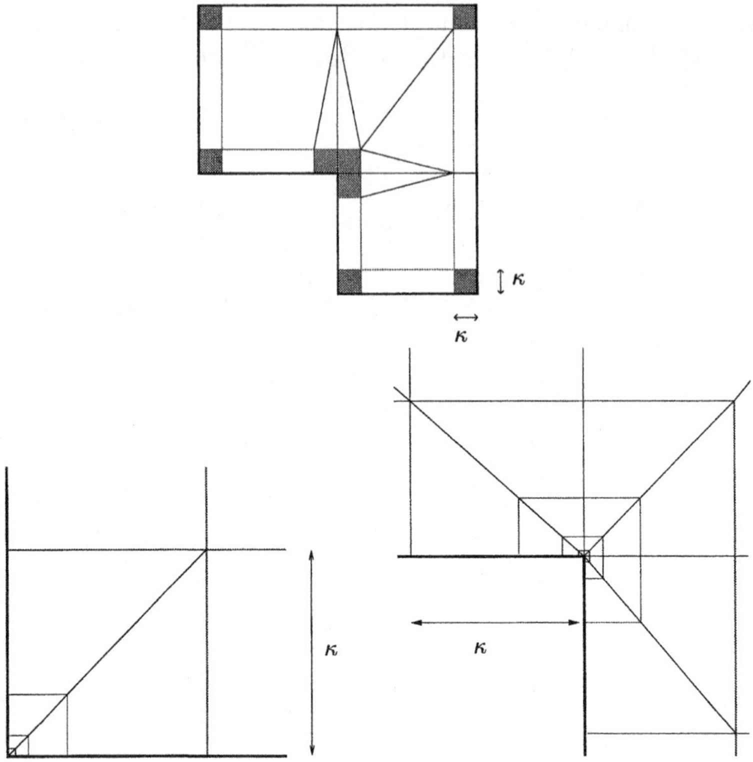

Fig. 2.4.1. Example of an admissible mesh $\mathcal{T}(\kappa, L)$ with $L = 3$: the domain (top) and zoom-ins at convex and concave vertices.

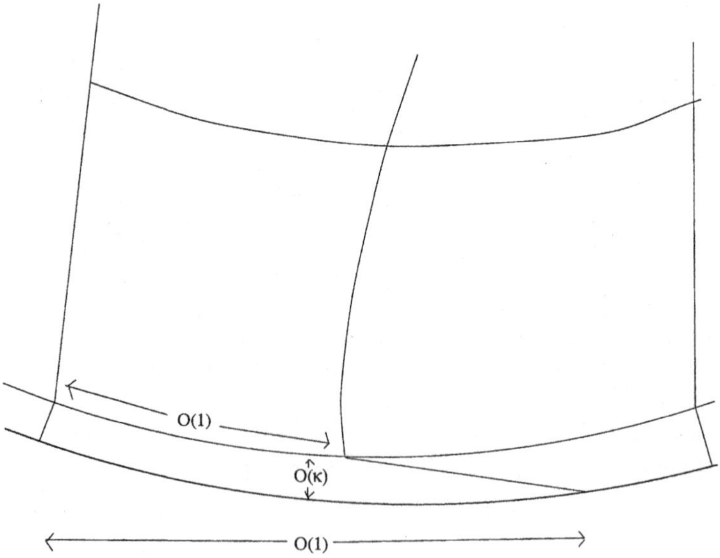

Fig. 2.4.2. Elements near the boundary may be distorted.

mesh $\mathcal{T}(\kappa, L)$ reduces to a standard p-version mesh with non-distorted elements. For fixed κ but variable L, the mesh $\mathcal{T}(\kappa, L)$ is a standard p-version outside a $O(\kappa)$ neighborhood of the vertices A_j. In the $O(\kappa)$ neighborhood of the vertices, a *geometric mesh* with $L+1$ layers is used. Finally, for fixed κ and fixed $\eta > 0$, the meshes $\mathcal{T}(\kappa, \lfloor \eta p \rfloor)$ are standard *hp*-FEM meshes with large elements in the interior of Ω and geometric mesh refinement with $O(p)$ layers toward the vertices as proposed in, e.g., [14, 112, 123]. ▪

Remark 2.4.7 As mentioned above, the elements in admissible meshes are allowed to be quite distorted in the sense that minimal and maximal angles may be close to 0 or π. For implementational reasons, it is advisable to be able to control the distortion of the elements, in particular the maximal angles. The introduction of *regular admissible meshes* in Definition 3.3.10 allows for this control, and we refer to Section 3.3 for a more detailed discussion of this issue. ▪

2.4.3 *hp*-FEM

With the *hp*-spaces $S^p(\mathcal{T})$ and the notion of admissible boundary layer meshes of Definition 2.4.4 in hand, we can proceed to the robust exponential approximation result for the approximation of solutions to (2.2.1).

The starting point of the *hp*-FEM for (2.1.1) is the weak formulation (2.1.3). In order to incorporate the Dirichlet boundary condition $u_\varepsilon|_{\partial\Omega} = g$, let $u_D \in H^1(\Omega)$ be an arbitrary function with $u_D|_{\partial\Omega} = g$. Then u_ε can be sought in the form $u_\varepsilon = u_D + u_0$ where u_0 is the solution to:

$$\text{Find } u_0 \in H_0^1(\Omega) \text{ s.t. } \quad B_\varepsilon(u_0, v) = F(v) - B_\varepsilon(u_D, v) \qquad \forall v \in H_0^1(\Omega).$$

Here, the bilinear form B_ε and the right-hand side F are defined in (2.1.4). In the *hp*-FEM, the infinite dimensional space $H_0^1(\Omega)$ is replaced with the finite dimensional space $S_0^p(\mathcal{T})$ and the function u_D is replaced with a suitable element of $S^p(\mathcal{T})$. For the sake of definiteness, we choose one particular form of enforcing the Dirichlet boundary conditions, namely, that of sampling the Dirichlet data g elementwise in the Gauss-Lobatto points: Let $u_{D,p} \in S^p(\mathcal{T})$ be any element of $S^p(\mathcal{T})$ satisfying

$$u_{D,p} = i_{p,\Gamma} u_\varepsilon \qquad \forall \text{ element edges } \Gamma \text{ with } \Gamma \subset \partial\Omega.$$

(See p. 88 for a precise definition of $i_{p,\Gamma}$.) Such an interpolant is easily constructed In fact, only the elements abutting on the boundary $\partial\Omega$ are affected. The finite element approximant u_N is then given by

$$u_N := u_{D,p} + u_{0,p}, \tag{2.4.4}$$

where the functions $u_{0,p} \in S_0^p(\mathcal{T})$ is the solution of the following problem:

$$\text{Find } u_{0,p} \in S_0^p(\mathcal{T}) \text{ s.t. } B_\varepsilon(u_{0,p}, v) = F(v) - B_\varepsilon(u_{D,p}, v) \quad \forall v \in S_0^p(\mathcal{T}). \tag{2.4.5}$$

The approximation error $u_\varepsilon - u_N$ can be controlled with Céa's Lemma, [35, 36] in the standard way: Noting that $u_\varepsilon - u_{D,p}$ satisfies the variational problem

$$B_\varepsilon(u_\varepsilon - u_{D,p}, v) = F(v) - B_\varepsilon(u_{D,p}, v) \qquad \forall v \in H_0^1(\Omega),$$

Céa's Lemma allows us to estimate

$$\|u_\varepsilon - u_N\|_\varepsilon = \|(u_\varepsilon - u_{D,p}) - u_{0,p}\|_\varepsilon = \inf_{\pi_p \in S_0^p(\mathcal{T})} \|(u_\varepsilon - u_{D,p}) - \pi_p\|_\varepsilon =$$
$$\inf\{\|u_\varepsilon - v\|_\varepsilon \,|\, v \in S^p(\mathcal{T}) \text{ s.t. } i_{p,\Gamma} u_\varepsilon = v \text{ for all edges } \Gamma \text{ of } \mathcal{T} \text{ with } \Gamma \subset \partial\Omega\}.$$

In Theorem 3.4.8, we construct a specific approximant to u_ε that leads to an upper bound in the last infimum. We then arrive at the following robust exponential convergence result:

Theorem 2.4.8. *Let Ω be a curvilinear polygon and $\mathcal{T}(\kappa, L)$ be a two-parameter family of admissible meshes in the sense of Definition 2.4.4. Let u_ε be the solution of (2.1.1) where the piecewise analytic Dirichlet data g satisfy (1.2.4) and the analytic function f satisfies (1.2.3). Then there are b, $\lambda_0 > 0$ independent of ε and p with the following property: For each $\lambda \in (0, \lambda_0)$ there exists $C > 0$ such that the finite element solution $u_N \in S^p(\mathcal{T}(\min\{\lambda p\varepsilon, 1\}, L))$ as defined by (2.4.4), (2.4.5) satisfies*

$$\|u - u_N\|_{L^2(\Omega)} + \varepsilon\|\nabla(u - u_N)\|_{L^2(\Omega)} \leq Cp^2(1 + \ln p)\left[e^{-b\lambda p} + \varepsilon p^3 e^{-b'L}\right].$$

Furthermore,

$$i_{p,\Gamma} u_\varepsilon = u_N|_\Gamma \qquad \forall \text{ edges } \Gamma \text{ of } \mathcal{T}(\min\{1, \lambda p\varepsilon\}, L) \text{ with } \Gamma \subset \partial\Omega.$$

In particular, if the number of elements $|\mathcal{T}(\min\{\lambda p\varepsilon, 1\}, L)| \sim L \sim p$, then $N = \dim S^p(\mathcal{T}(\min\{\lambda p\varepsilon, 1\}, L)) \sim p^3$ and therefore

$$\|u - u_N\|_\varepsilon \leq Ce^{-b' N^{1/3}},$$

where the constants C, b' are independent of ε (albeit dependent on λ).

Proof: The result follows by combining Céa's Lemma and the approximation result Theorem 3.4.8. A rigorous proof of Theorem 3.4.8 is involved. The main ideas, however, are similar to the one-dimensional case of Proposition 2.2.5. We outline in Subsection 3.1.3 the key steps of the proof. □

We have $|\mathcal{T}(\min\{\lambda p\varepsilon, 1\}, L)| \sim L$ in admissible boundary layer meshes that are typically chosen (cf. Fig. 2.4.1 and the meshes of the numerical examples in Section 2.5).

Remark 2.4.9 The p-dependence in the approximation result Theorem 2.4.8 is likely to be suboptimal. This is due to the our choice of the particular approximant of Theorem 3.4.8. ▪

2.5 Numerical Examples

The aim of the numerical examples of the present section is to illustrate and highlight several properties of the *hp*-FEM applied to (2.1.1). We discuss the following aspects:

1. Robust exponential convergence is indeed observed numerically in the practical range of values of polynomial degrees p.
2. Our numerical experiments confirm that the corner layer components of the solution of (1.2.11) are weak. Theorem 2.4.8 reflects this as, neglecting algebraic terms in p, we have

$$\|u_\varepsilon - u_N\|_\varepsilon \leq C\left\{e^{-b\lambda p} + \varepsilon e^{-b'L}\right\}.$$

   The first term, $e^{-b\lambda p}$, reflects the approximation of the smooth and the boundary layer parts whereas the second term, $\varepsilon e^{-b'L}$, is due to the approximation of corner layers. Balancing these two error contributions, we see that for small ε, the number L of layers of geometric refinement may be chosen small compared to p; that is, few layer of geometric refinement are sufficient for adequate resolution of the corner layers.
3. As noted in Remark 2.4.5 the finite element mesh may contain highly distorted elements with minimal angles of size $O(p\varepsilon)$ and maximal angles of size $\pi - O(p\varepsilon)$. In agreement with the predictions of Theorem 2.4.8, our numerical examples show robustness of the FEM with respect to mesh distortion as the presence of highly distorted elements has practically no effect on the energy convergence.

Several of these points have already been observed in previous numerical studies [94, 115, 131, 133, 134]. These studies considered the model problem

$$-\varepsilon^2 \Delta u_\varepsilon + u_\varepsilon = f \quad \text{on } \Omega, \qquad u_\varepsilon|_{\partial\Omega} = g. \tag{2.5.1}$$

For the case of smooth boundary curves, [131, 133] obtained robust algebraic rates of convergence of arbitrary order if the *hp*-FEM is based on boundary-fitted tensor product meshes with needle elements of width $O(p\varepsilon)$ in the layer. Robust exponential convergence was observed numerically and conjectured. In [94] this conjecture was rigorously established for the model problem (2.5.1) under the assumption of analyticity of the input data f, g, and $\partial\Omega$. [94] also removed the restriction to boundary-fitted tensor product meshes and allowed highly distorted elements as discussed in Remark 2.4.5.
Robust algebraic convergence of arbitrary order for (2.5.1) for the special case of Ω being *square* were obtained in [131, 134]. It was also observed numerically that corner layers are weak in the sense that few layer of geometric refinement are required for good resolution in the energy norm. As we pointed out above, Theorem 2.4.8 rigorously establishes this observation for arbitrary curvilinear polygons, and our numerical examples below corroborate this claim for the classical L-shaped domain.

2.5.1 The classical L-shaped domain

The numerical experiments that we show in this section illustrate that the *hp*-FEM achieves robust exponential convergence for (1.2.11) on polygonal domains as well. Additionally, we show that the corner layer is weak and that thus few layers of geometric refinement are sufficient for good convergence in the energy norm.

We consider the problem

$$-\varepsilon^2 \Delta u_\varepsilon + u_\varepsilon = f(x,y) \quad \text{on } \Omega, \qquad u_\varepsilon = 0 \quad \text{on } \partial\Omega. \tag{2.5.2}$$

Here, the domain Ω is the classical L-shaped domain, $\Omega = (-1,1)^2 \setminus (-1,0) \times (0,1)$. For our calculations, we chose $f(x,y) = \exp(x+y)$ so that the assumptions on the data of Theorem 2.4.8 are satisfied. The computations were done with the code CONCEPTS 1.4, [80]. The finite element mesh

$$\mathcal{T}(\kappa, L) \tag{2.5.3}$$

employed is depicted in Fig. 2.5.1 where the boundary layer elements are of width $\kappa = \min\{0.5, p\varepsilon\}$ and $L+1$ layers of geometric refinement with grading factor $\sigma = 0.5$ are employed in the vicinity of the vertices. The actual finite element spaces employed in our numerical examples were slightly different from those analyzed so far: We employed meshes with "hanging nodes" (cf. Fig. 2.5.1) and used the finite element space $\tilde{S}_0^p(\mathcal{T})$ based on the "trunk spaces" $\mathcal{Q}'_p(S)$ (cf., e.g., [123]). The precise definition of $\tilde{S}_0^p(\mathcal{T})$ is

$$\tilde{S}_0^p(\mathcal{T}) := \{u \in H_0^1(\Omega) \,|\, u \circ M_i \in \mathcal{Q}'_p(S)\},$$
$$\mathcal{Q}'_p(S) := \operatorname{span}\{x^p y, xy^p, x^i y^j \,|\, 0 \le i+j \le p\}.$$

The finite element spaces $\tilde{S}_0^p(\mathcal{T})$ are thus $H_0^1(\Omega)$-conforming finite element spaces, and this conformity constraint on meshes with hanging nodes is properly treated during the assembly procedure. It can be shown that an approximation result analogous to that of Theorem 2.4.8 could be formulated for the spaces $\tilde{S}_0^p(\mathcal{T})$ on meshes with hanging nodes as well. Our numerical results are presented in Figs. 2.5.2, 2.5.3, 2.5.4 for $\varepsilon = 1$, $\varepsilon = 10^{-3}$, $\varepsilon = 10^{-6}$, where the energy error $\|u_\varepsilon - u_N\|_\varepsilon^2$ is graphed versus the number of degrees of freedom $N = \dim \tilde{S}_0^p(\mathcal{T})$. In each figure, lines correspond to meshes with a fixed number of $L+1$ layers of geometric refinement and polynomial degree p increasing from 1 to 15. Comparing Figs. 2.5.2–2.5.4 we first observe robustness of the FEM based on admissible meshes as the energy errors do not depend on ε.

We now discuss the numerical results of Figs. 2.5.2–2.5.4 in more detail and start with the case $\varepsilon = 1$ in Fig. 2.5.2. Each line in Fig. 2.5.2 corresponds to the p-version on a fixed mesh with $L+1$ layers of geometric refinement toward the vertices. Fig. 2.5.2 exhibits the typical convergence pattern of the p-version FEM: On fixed meshes, there is an initial phase of exponential convergence (visible for larger values of L in Fig. 2.5.2) followed by the asymptotic algebraic convergence. The asymptotic convergence rate is $p^{-4/3}$ in the energy norm (cf. [22]), i.e., a

convergence of $N^{-4/3}$ in the energy. Indeed, the asymptotic slopes in Fig. 2.5.2 are very close to the predicted value 4/3. "True" exponential convergence can only be observed if the number of layers of geometric refinement is taken proportional to the polynomial degree p, i.e., if $L = \lfloor \kappa p \rfloor$ for some fixed $\kappa > 0$.
We now illustrate that the corner layers are indeed rather weak. This weakness manifests itself numerically in the fact that for fixed L, the pre-asymptotic exponential convergence is visible for a large range of values of p for small ε. Put differently, for fixed L, the onset of the asymptotic algebraic convergence occurs at larger values of p as ε decreases. We see this pattern by comparing lines corresponding to the same value of L in Figs. 2.5.2, 2.5.3, and 2.5.4. In particular, in Fig. 2.5.4, the asymptotic algebraic convergence occurs below machine accuracy (16 digits).
This extension of the pre-asymptotic exponential convergence behavior can be explained with the aid of Theorem 2.4.8. Theorem 2.4.8 states that the FEM-error behaves like (ignoring the algebraic factors involving powers of p)

$$e^{-b\lambda p} + \varepsilon e^{-b' L} \tag{2.5.4}$$

for some b, $b' > 0$ independent of ε, p, L. The proof of Theorem 3.4.8 shows that the factor $\varepsilon e^{-b' L}$ is due to the approximation of the corner layers. For small ε and fixed L, (2.5.4) suggests an exponential decrease (in p) for the finite element error. This exponential decay is visible until the two terms are equilibrated, i.e., until

$$e^{-b\lambda p} \approx \varepsilon e^{-b' L}.$$

This observation explains why the pre-asymptotic exponential convergence phase of the FEM increases as ε becomes small. (2.5.4) suggests another characteristic of the FEM applied to (2.1.1): For small ε the FEM is quite insensitive to an increase of the number of layers L of geometric refinement near the vertices. This is indeed visible in Fig. 2.5.3 for $p = 1, 2, 3$, and in Fig. 2.5.4 for p ranging from 1 to 9 since the energy error is effectively constant for all values of L.

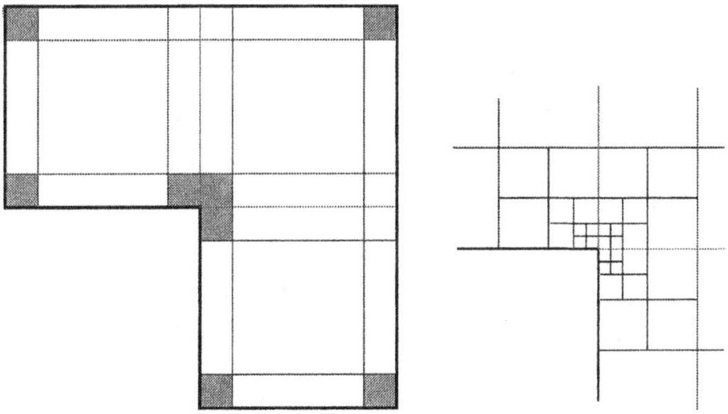

Fig. 2.5.1. The mesh and the local refinement near the reentrant corner ($L = 3$).

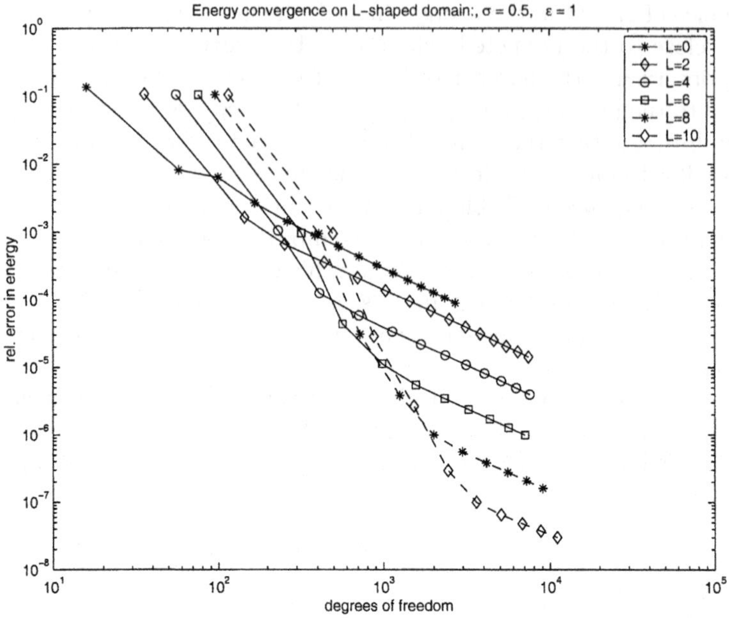

Fig. 2.5.2. Rel. energy error vs. DOF on mesh of Fig. 2.5.1; $\varepsilon = 1$.

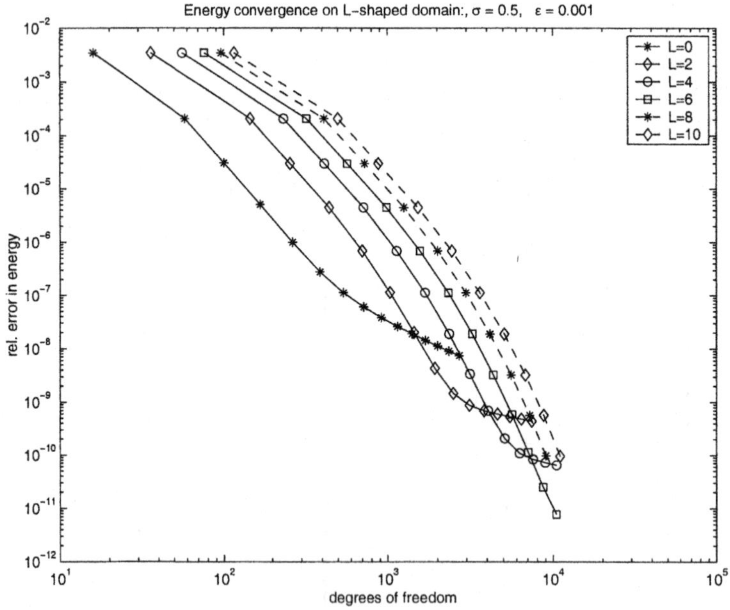

Fig. 2.5.3. Rel. energy error vs. DOF on mesh of Fig. 2.5.1; $\varepsilon = 10^{-3}$.

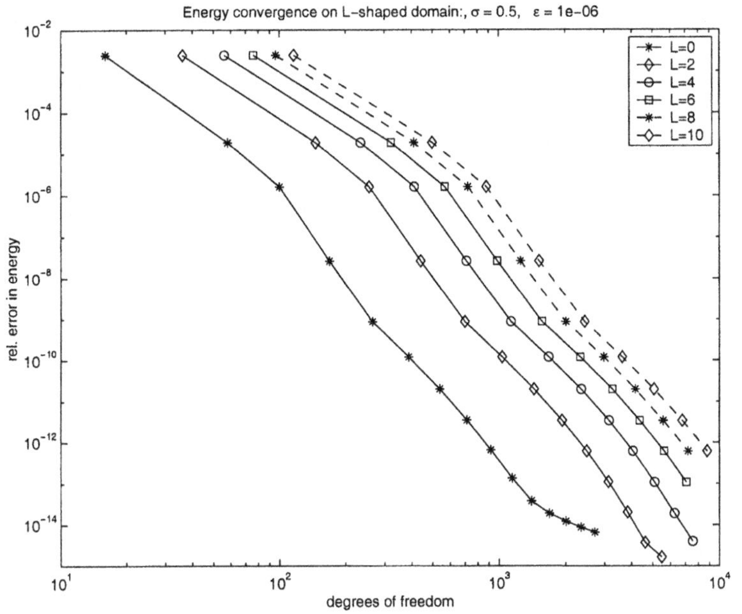

Fig. 2.5.4. Rel. energy error vs. DOF on mesh of Fig. 2.5.1; $\varepsilon = 10^{-6}$.

2.5.2 Robustness with respect to mesh distortion

Our next numerical example illustrates robustness of the FEM on admissible boundary layer meshes with respect to mesh distortion. Theorem 2.4.8 states that robust exponential convergence is achieved on admissible boundary layer meshes, which may be highly distorted as discussed in Remark 2.4.5. To show that the presence of highly distorted elements does not affect the energy convergence, we consider the following quasi one-dimensional model problem.

$$-\varepsilon^2 \Delta u_\varepsilon + u_\varepsilon = 1 \qquad \text{on } \Omega := (0,1)^2, \tag{2.5.5a}$$

$$u_\varepsilon = 0 \qquad \text{on } \Gamma_D := \{(x,y) \in \partial\Omega \,|\, y = 0\}, \tag{2.5.5b}$$

$$\partial_n u_\varepsilon = 0 \qquad \text{on } \Gamma_N := \partial\Omega \setminus \Gamma_D. \tag{2.5.5c}$$

The solution of this problem is given by

$$u_\varepsilon(x,y) = 1 - \frac{\cosh((1-y)/\varepsilon)}{\cosh(1/\varepsilon)}. \tag{2.5.6}$$

We note that u_ε has a boundary layer at Γ_D and no corner singularities although Ω is a square. We therefore do not need any geometric refinement near the vertices of Ω, and needle elements are required only near Γ_D.
For $\varepsilon = 10^{-3}$ our numerical calculations were performed with the commercial code STRESS CHECK, [48], a p-version code with highest polynomial degree $p_{max} = 8$. The meshes consisted of straight triangles and quadrilaterals

(see Figs. 2.5.5– 2.5.7), where the parameter b determines the distortion of the meshes.

On a fixed quadrilateral mesh as depicted in Fig. 2.5.5 the tensor product spaces $\mathcal{Q}_p(S)$ with p ranging from 1 to p_{max} were used. The relative error in energy $\|u_\varepsilon - u_N\|_\varepsilon^2$ versus the square root of the number of degrees of freedom (DOF) is reported in Fig. 2.5.5. In the case $b = 0.5$ all quadrilaterals satisfy a maximum and minimum angle condition (even as ε tends to zero). For the case $b = 0.25$ the maximum angle is $\pi - O(\varepsilon)$ and the minimum angle is $O(\varepsilon)$, i.e., the mesh is highly distorted. Nevertheless, the error curves in Fig. 2.5.5 are practically on top of each other showing the robustness with respect to mesh distortion of the approximation properties of admissible meshes. The situation is completely analogous for triangular meshes: the error graph in Fig. 2.5.6 shows the performance of the p version on the triangular mesh also shown in Fig. 2.5.6; again, the convergence is not visibly affected by the presence of highly distorted elements in the boundary layer.

According to Theorem 2.4.8, the needle elements in the boundary layer should have width $O(p\varepsilon)$, i.e., the mesh should depend on ε as well as on p. However, for practical purposes, it is more convenient to fix a mesh and to increase p. The question arises then what the appropriate width of the needle elements is. If only one layer of needle elements is used, we advocate the use of needle elements of width $O(p_{max}\varepsilon)$. The following numerical example supports this choice. In Fig. 2.5.7, we show the relative error in energy versus the number of degrees of freedom for the mesh also shown in Fig. 2.5.7. While robustness with respect to mesh distortion is again clearly visible as the choice of the parameter b has practically no effect, we note that the error curves in Fig. 2.5.7 level off at an error of about 10^{-7} corresponding to $p = 6$. Actually, already for $p = 5$, some deterioration of the rate of convergence is visible. This is due to the fact that the width of the needle elements is fixed at 4ε instead of $8\varepsilon = p_{max}\varepsilon$ (In the meshes of Figs. 2.5.5, 2.5.6, the width was at least 10ε, which exceeds $p_{max}\varepsilon$). As demonstrated in the one-dimensional setting in Example 2.2.8, the large elements are too close to Γ_D and dominate the global error reduction.

2.5.3 Examples with singular right-hand side

So far, the right-hand side f was chosen analytic in $\overline{\Omega}$. This implies that, as $\varepsilon \to 0$, the "limit solution" $u_0(x) := \lim_{\varepsilon \to 0} u_\varepsilon(x) = f(x)/c(x)$ is smooth as well and does not exhibit corner singularities. For many singularly perturbed problems such as the Reissner-Mindlin plate model on polygonal domains, the limit solution has corner singularities as well. From an approximation point of view the mesh should be designed to capture the boundary and corner layers as well as the behavior of the limit solution. A good mesh design strategy is therefore to combine two types of meshes, namely, a) the meshes presented above, i.e., meshes capable of resolving boundary layer and corner singularities with length scale $O(\varepsilon)$, and b) meshes with classical geometric refinement towards the corners, which allows for resolving corner singularities present in the limit solution.

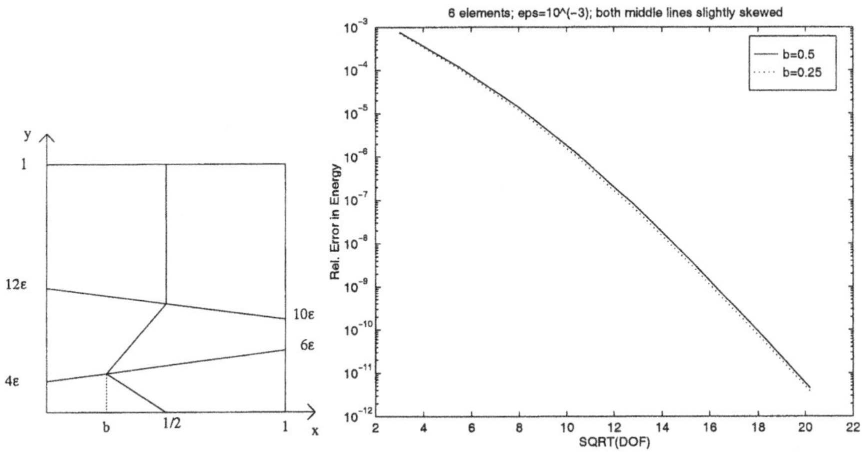

Fig. 2.5.5. Left: mesh with parameter b determining distortion (not drawn to scale); right: p-version on that mesh, $\varepsilon = 10^{-3}$.

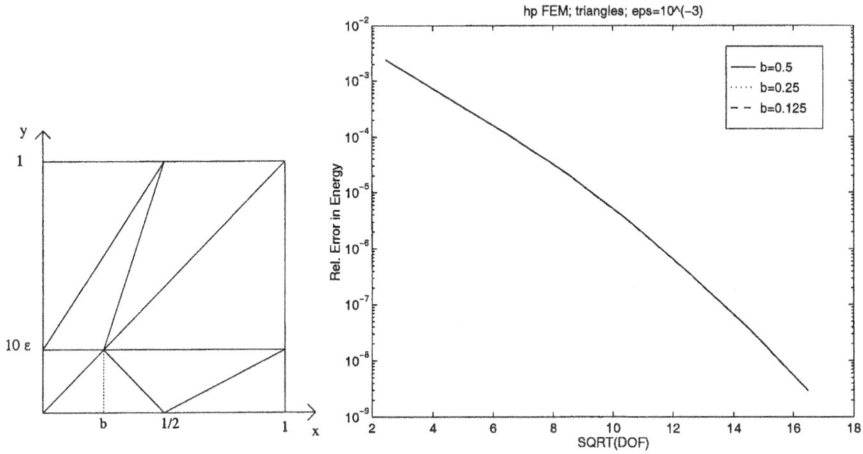

Fig. 2.5.6. Left: mesh with parameter b determining distortion (not drawn to scale); right: p-version on that mesh, $\varepsilon = 10^{-3}$.

Example 2.5.1 This example shows two points. First, combining geometric mesh refinement toward singularities with boundary layer meshes is very effective. Second, as proposed in Remark 2.2.6, geometric mesh refinement can also be used to replace boundary layer meshes, albeit at the expense of losing robustness in a strict sense. We consider

$$-\varepsilon^2 u_\varepsilon''(x) + u_\varepsilon(x) = f(x) := (1+x)^{-0.45} \quad \text{on } (-1,1), \qquad u_\varepsilon(\pm 1) = 0.$$

The mesh employed consists of the points

$$\{-1,1\} \cup \{x_i = -1 + q^{L+1-i} \,|\, i = 1,\ldots,L\} \cup \{1 - \lambda p \varepsilon\}, \qquad q = 0.15, \quad \lambda = 0.71,$$

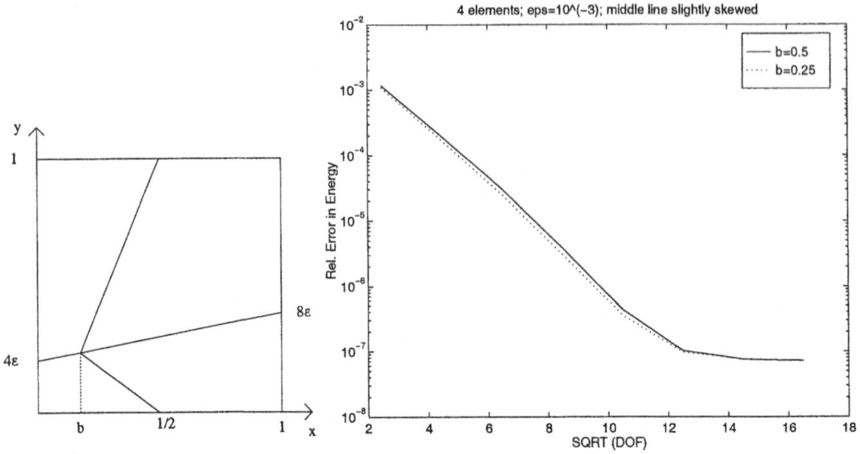

Fig. 2.5.7. left: mesh with parameter b determining distortion (not drawn to scale); right: p-version on that mesh, $\varepsilon = 10^{-3}$.

where the number of layers L of geometric refinement is a parameter of our numerical investigations. The boundary layer at the right endpoint $x = 1$ is resolved with the aid of the small element of size $\lambda p\varepsilon$ (cf. Example 2.2.7). Fig. 2.5.8 show the performance of this choice of mesh for $\varepsilon = 10^{-4}$, $\varepsilon = 10^{-6}$ and different values of L. While the mesh is designed to resolve the small-scale features at the right endpoint, we have to choose L such that $q^L \approx \varepsilon$ in order to resolve the features at the left endpoint $x = -1$. For our choices of ε, this happens for the moderate values $L = 2$ for the case $\varepsilon = 10^{-4}$ and $L = 4$ in the case $\varepsilon = 10^{-6}$. This shows that, as already discussed in Remark 2.2.6, geometric mesh refinement toward an endpoint is a very effective means of generating elements that reach size $O(\varepsilon)$ very quickly. Indeed, the numerical experiments in Figs. 2.5.8 confirm this observation.

In an $O(\varepsilon)$-neighborhood of the endpoint $x = -1$, the solution u_ε has a singularity; in a fact, a calculation reveals

$$u_\varepsilon(x) = -\frac{1}{(1-0.45)(2-0.45)}\varepsilon^{-2}(x+1)^{2-0.45} + o(1).$$

Singular functions of this type can be approximated well in the context of the hp-FEM by geometric meshes of exactly the type considered here. Outside a small neighborhood of $x = -1$, we expect the solution u_ε to be close to $u_0(x) := f(x)$. Again, the geometric mesh of the type considered here is very suitable for the approximation of the function u_0. On such geometric meshes, exponential convergence can be achieved if the number of layers L is chosen proportional to p, the approximation order. For fixed L, asymptotically, only algebraic convergence can be expected, which is visible in Fig. 2.5.8. Also visible is a preasymptotic exponential convergence, and we note that for fairly moderate values of L, the asymptotic algebraic convergence sets in at very small error levels. ▪

Example 2.5.2 On the L-shaped domain $\Omega t = (-1,1)^2 \setminus (-1,0)^2$, we consider the boundary value problem

$$-\varepsilon^2 \Delta u_\varepsilon + u_\varepsilon = f(x) = \frac{e^{x+y}}{\sqrt{x^2+y^2}} \quad \text{in } \Omega, \qquad u_\varepsilon = 0 \quad \text{on } \partial\Omega.$$

The meshes employed are depicted in Fig. 2.5.9. They consist of a combination of the meshes $\mathcal{T}(\min\{0.5, p\varepsilon\}, L_2)$ of (2.5.3) combined with a classical geometric mesh

$$\mathcal{T}^{geo}(L_1, \sigma_1)$$

with L_1 layers of refinement towards the origin and grading factor $\sigma_1 \in (0,1)$. We refer to Fig. 2.5.9 for an example with $L_1 = 2$; the inset figure shows the refinement with $\sigma_2 = 0.5$ and $L_2 = 2$ layers of refinement in the small $O(p\varepsilon)$-neighborhood of the origin.
The computations are performed with $\sigma_1 = 0.05$ and $\varepsilon = 10^{-4}$. We compare the case $L_1 = 0$, i.e., simply the mesh $\mathcal{T}(\min\{0.5, p\varepsilon\}, L_2)$ with the case $L_1 = 2$, which represents a mesh that is suitable for the approximation of u_0. The polynomial degree varies from $p = 1$ to $p = 15$. In Fig. 2.5.10 we note that for the case $L_1 = 2$, a polynomial degree $p \approx 4$ leads to the same accuracy as the case $L_1 = 0$ with $p = 15$. ▪

Remark 2.5.3 To combine geometric mesh refinement toward the corners starting at distance $O(1)$ with boundary layer meshes has also been proposed and successfully employed for Reissner-Mindlin plate calculations in [132]. ▪

Remark 2.5.4 For elliptic systems, the additional problem of stable discretizations arises, especially on meshes with highly anisotropic elements. We mention here [5, 110, 111, 125] for Stokes' equations. An additional issue is locking; we refer to [112, Sec. 6.3] and the references therein for a discussion of this issue in the context of p- and hp-FEM. ▪

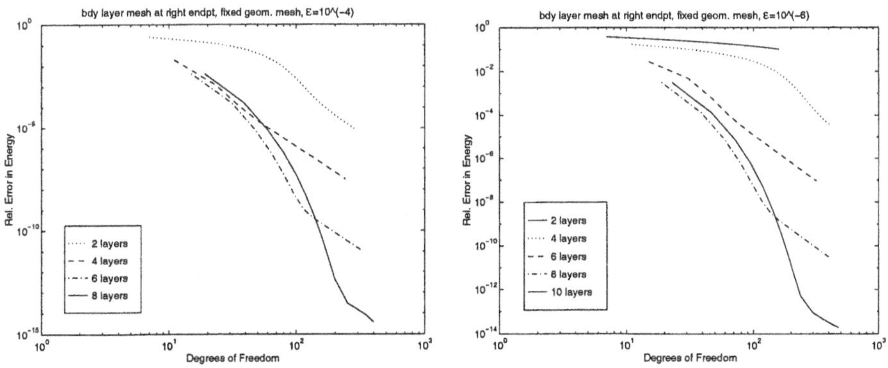

Fig. 2.5.8. Example 2.5.1, $\varepsilon = 10^{-4}$ (left) and $\varepsilon = 10^{-6}$ (right).

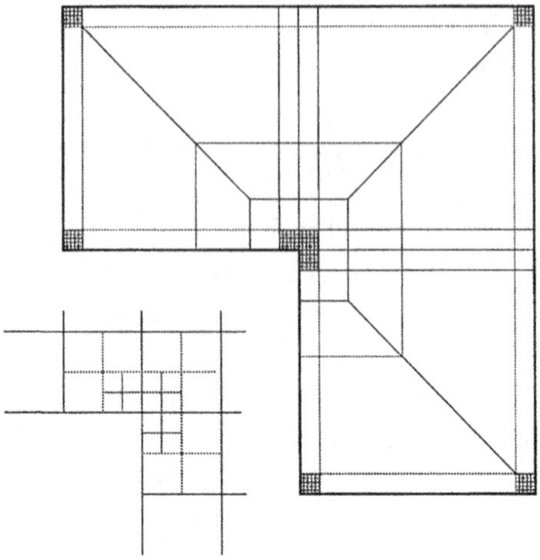

Fig. 2.5.9. Example 2.5.2: meshes.

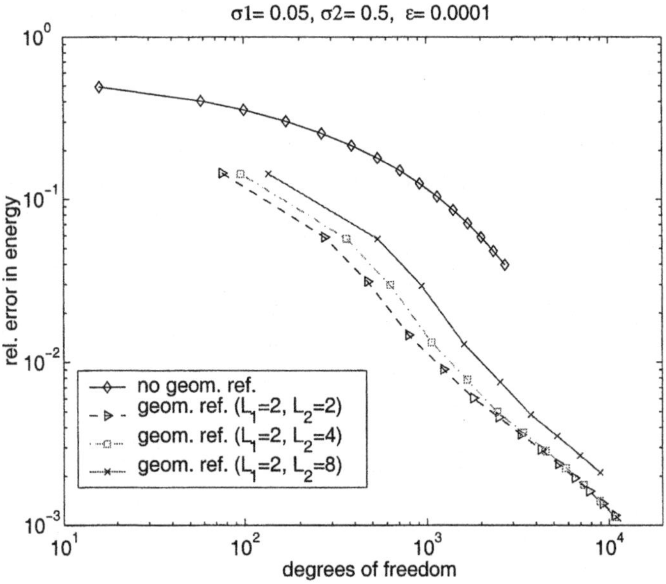

Fig. 2.5.10. Example 2.5.2: Convergence behavior.

2.6 h-FEM approximation

The regularity assertions of Section 2.3 can also be utilized to obtain *a priori* estimates in the context of low order methods. Robust algebraic convergence can be achieved on suitably designed meshes. We present these ideas in the context of Shishkin meshes. As in the case of the hp-FEM, we start with the one-dimensional case in Section 2.6.1. We then move to the two-dimensional case in Section 2.6.3, where we introduce a class of Shishkin-type meshes and analyze the performance of the h-FEM on these meshes.

2.6.1 Approximation on Shishkin meshes in one dimension

The boundary layer mesh of Definition 2.2.4 can be viewed as the *hp*-version analog of the Shishkin mesh, [118]. In the context of the approximation of the solution u_ε to (2.2.1), the Shishkin meshes may be defined as follows:

Definition 2.6.1 (Shishkin mesh in one dimension). *For $N \in \mathbb{N}\backslash\{1\}$ and a transition point $\kappa \in (0, 1/2)$ define*

$$\begin{aligned}\Omega_1 &:= (-1, -1 + \min\{\kappa, 0.5\}),\\ \Omega_2 &:= (-1 + \min\{\kappa, 0.5\}, 1 - \min\{\kappa, 0.5\}),\\ \Omega_3 &:= (1 - \min\{\kappa, 0.5\}, 1).\end{aligned}$$

The Shishkin mesh $\mathcal{T}^{Shishkin}_{\kappa,N}$ *is then given by the piecewise uniform mesh obtained by placing N elements of equal size* $\min\{\kappa, 0.5\}/N$ *in Ω_1, Ω_3 and N elements of equal size* $(2 - \min\{\kappa, 0.5\})/N$ *in Ω_3.*

The h-FEM on a Shishkin mesh is then given by (2.2.5) with ansatz space $V_N = S^1_0(\mathcal{T}^{Shishkin}_{\kappa,N})$. The analog of Proposition 2.2.1 reads

Proposition 2.6.2. *Let c, f satisfy the assumptions of Proposition 2.2.1, and let u_ε be the solution to (2.2.1) for $\varepsilon \in (0, 1]$. Then there exist C, $\lambda_0 > 0$ such that for every $\lambda > \lambda_0$, $N \in \mathbb{N}$, and transition point* $\kappa := \min\{1/2, \lambda\varepsilon \ln N\}$

$$\begin{aligned}\inf_{v \in S^1_0(\mathcal{T}^{Shishkin}_{\kappa,N})} &\|u_\varepsilon - v\|_{L^2(\Omega)} + \varepsilon\|(u_\varepsilon - v)'\|_{L^2(\Omega)}\\ &\leq C\left[\varepsilon^{1/2} N^{-1}(\lambda \ln N)^{3/2} + N^{-2}(\lambda \ln N)^2\right].\end{aligned}$$

Furthermore, $\dim S^1_0(\mathcal{T}^{Shishkin}_{\kappa,N}) = 3N - 1$.

Proof: Results of the type presented in Proposition 2.6.2 are by now classical. Proofs for the case $c \equiv 1$ can be found, e.g., in the monographs [97, 108]. In order to stress the fact that the same regularity results as in the proof of Proposition 2.2.5 are really key to the proof, we sketch the main procedure. The statement $\dim S^1_0(\mathcal{T}^{Shishkin}_{\kappa,N}) = 3N - 1$ follows immediately from the fact that the mesh $\mathcal{T}^{Shishkin}_{\kappa,N}$ has $3N - 1$ internal nodes.

1. Step: The case $\lambda\varepsilon \ln N \geq 0.5$. The mesh $\mathcal{T}^{Shishkin}_{\kappa,N}$ consists of a quasi-uniform mesh with mesh size $h \sim 1/N$. Proposition 2.2.1 asserts $\|u''_\varepsilon\|_{L^2(\Omega)} \leq C\varepsilon^{-2}$. Thus, standard estimates for the piecewise linear interpolant I give

$$\begin{aligned}\|u_\varepsilon - Iu_\varepsilon\|_{L^2(\Omega)} &\leq Ch^2\|u''_\varepsilon\|_{L^2(\Omega)} \leq CN^{-2}\varepsilon^{-2} \leq CN^{-2}(\lambda \ln N)^2,\\ \varepsilon\|(u_\varepsilon - Iu_\varepsilon)'\|_{L^2(\Omega)} &\leq C\varepsilon h\|u''_\varepsilon\|_{L^2(\Omega)} \leq C\varepsilon^{1/2}N^{-1}\varepsilon^{-3/2} \leq C\varepsilon^{1/2}N^{-1}(\lambda \ln N)^{3/2},\end{aligned}$$

which is the desired bound.

2. Step: The case $\lambda\varepsilon \ln N < 0.5$. We employ the decomposition $u_\varepsilon = w_\varepsilon + u_\varepsilon^{BL} + r_\varepsilon$ of (2.2.3). We choose

$$\lambda_0 := \frac{2}{\alpha} \tag{2.6.7}$$

where $\alpha > 0$, which measures how fast the boundary layer part decays away from the boundary, is given in Proposition 2.2.1. We approximate each of the terms w_ε, u_ε^{BL}, r_ε in turn. The standard estimates for the piecewise linear interpolant allows us to infer for the smooth part w_ε

$$\|w_\varepsilon - Iw_\varepsilon\|_{L^2(\Omega)} \leq h^2 \|w_\varepsilon''\|_{L^2(\Omega)} \leq CN^{-2},$$
$$\|(w_\varepsilon - Iw_\varepsilon)'\|_{L^2(\Omega)} \leq h \|w_\varepsilon''\|_{L^2(\Omega)} \leq CN^{-1},$$

where we exploited $h = \max_i h_i \leq CN^{-1}$. Thus, for a $C > 0$ (which depends on λ_0 but is independent of ε, N) we have for any $\lambda \geq \lambda_0$:

$$\|w_\varepsilon - Iw_\varepsilon\|_{L^2(\Omega)} + \varepsilon \|(w_\varepsilon - Iw_\varepsilon)'\|_{L^2(\Omega)} \leq C \left[\varepsilon \lambda^{3/2} N^{-1} + \lambda^2 N^{-2}\right].$$

For the remainder r_ε we note that the assumption $\lambda\varepsilon \ln N < 0.5$ implies

$$\alpha/\varepsilon \geq 2\lambda\alpha \ln N.$$

Since $\|r_\varepsilon\|_{H^1(\Omega)} \leq Ce^{-\alpha/\varepsilon}$, we obtain for $\lambda \geq \lambda_0$ in view of (2.6.7)

$$\|r_\varepsilon\|_{L^2(\Omega)} + \|r_\varepsilon'\|_{L^2(\Omega)} \leq Ce^{-2\lambda\alpha \ln N} \leq Ce^{-2\lambda_0\alpha \ln N} = CN^{-4}.$$

We finally approximate the boundary layer term u_ε^{BL}. To that end, we consider the approximation on the subdomains Ω_1, Ω_2, and Ω_3 separately. Noting that the mesh size on Ω_1, Ω_3 is $h = \lambda\varepsilon \ln N/N$, we get

$$\|u_\varepsilon^{BL} - Iu_\varepsilon^{BL}\|_{L^2(\Omega_1)} \leq Ch \|\left(u_\varepsilon^{BL}\right)'\|_{L^2(\Omega_1)} \leq Ch\varepsilon^{-1}\sqrt{|\Omega_1|},$$
$$\varepsilon \|\left(u_\varepsilon^{BL} - Iu_\varepsilon^{BL}\right)'\|_{L^2(\Omega_1)} \leq C\varepsilon h \|\left(u_\varepsilon^{BL}\right)''\|_{L^2(\Omega_1)} \leq Ch\varepsilon^{-1}\sqrt{|\Omega_1|}$$

and therefore

$$\begin{aligned}
&\|u_\varepsilon^{BL} - Iu_\varepsilon^{BL}\|_{L^2(\Omega_1)} + \varepsilon \|\left(u_\varepsilon^{BL} - Iu_\varepsilon^{BL}\right)'\|_{L^2(\Omega_1)} \\
&\quad \leq Ch\varepsilon^{-1}\sqrt{|\Omega_1|} = Ch\varepsilon^{-1}\sqrt{\lambda\varepsilon \ln N} = C\varepsilon^{1/2} N^{-1} (\lambda \ln N)^{3/2}.
\end{aligned}$$

By symmetry, a completely analogous estimate holds for Ω_3. On the large subdomain Ω_2, we note that $h \sim N^{-1}$ and that $\operatorname{dist}(\Omega_2, \partial\Omega) \geq \lambda\varepsilon \ln N$. Hence, using again (2.6.7) and a standard inverse estimate for piecewise linear functions:

$$\begin{aligned}
\|u_\varepsilon^{BL} - Iu_\varepsilon^{BL}\|_{L^2(\Omega_2)} &\leq C\left[\|u_\varepsilon^{BL}\|_{L^\infty(\Omega_2)} + \|Iu_\varepsilon^{BL}\|_{L^\infty(\Omega_2)}\right] \\
&\leq Ce^{-\alpha\lambda \ln N} \leq CN^{-\alpha\lambda_0} = CN^{-2}, \\
\varepsilon \|\left(u_\varepsilon^{BL} - Iu_\varepsilon^{BL}\right)'\|_{L^2(\Omega_2)} &\leq \varepsilon \|\left(u_\varepsilon^{BL}\right)'\|_{L^\infty(\Omega_2)} + \frac{\varepsilon}{h}\|Iu_\varepsilon^{BL}\|_{L^2(\Omega_2)} \\
&\leq Ce^{-\alpha\lambda \ln N}[1 + \varepsilon N] \leq CN^{-2}[1 + \varepsilon N].
\end{aligned}$$

Combining the estimates for $w_\varepsilon - Iw_\varepsilon$, $u_\varepsilon^{BL} - Iu_\varepsilon^{BL}$, and r_ε gives the desired result. □

2.6.2 h-FEM meshes

In any h-FEM, approximation is achieved by increasing the number of elements, i.e., by reducing the size of the elements. To that end, the notion of "element size" needs to be properly introduced, which is the purpose of the following Definition 2.6.3.

Definition 2.6.3 (normalizable). *An invertible C^2-map $M : \overline{\hat{K}} \to M(\overline{\hat{K}}) \subset \mathbb{R}^2$ is said to be (C_M, κ, h)-*normalizable *if it can be factored as*

$$M = G \circ S,$$

where the affine stretching map S satisfies

$$S' = \begin{pmatrix} h & 0 \\ 0 & h \end{pmatrix}, \tag{2.6.8}$$

and the C^2-map G satisfies

$$\|\nabla^p G_i\|_{L^\infty(R)} \le C_M, \qquad p \in \{0,1,2\}, \tag{2.6.9a}$$

$$\|(G')^{-1}\|_{L^\infty(R)} \le \frac{C_M}{\kappa}. \tag{2.6.9b}$$

Here, the set R is the image of the reference element $\hat{K}$ under the affine map S, i.e., $R = S(\hat{K})$.

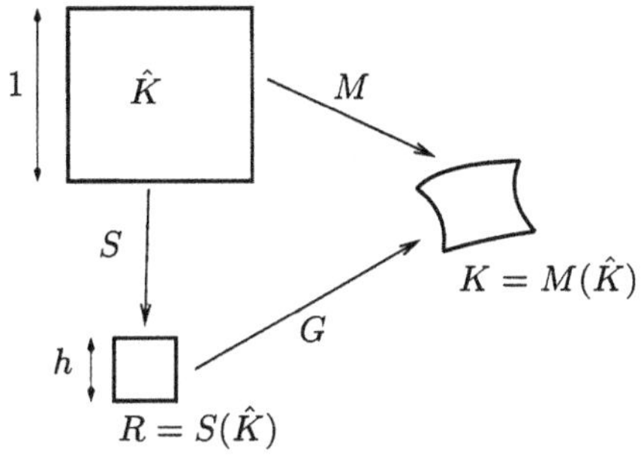

Fig. 2.6.11. Factorization of a normalizable element map M as $M = G \circ S$.

Several comments concerning Definition 2.6.3 are in order.

Remark 2.6.4 1. Since in this section we are interested in the approximation from $S^1(\mathcal{T})$, analyticity of the element maps is not required; Definition 2.6.3 therefore restricts the regularity requirement to C^2.

2. Definition 2.6.3 allows us to introduce the notion of "element size" for non-affine elements: For triangulations $\mathcal{T} = (K_i, M_i, \hat{K}_i)$ where, for some fixed $C_\mathcal{T}$, each element map M_i is $(C_\mathcal{T}, 1, h_i)$-normalizable for some suitably chosen $h_i > 0$, it is meaningful to speak of h_i as the element size, since $h_i \sim \operatorname{diam} K_i$.

3. The chain rule allows us to infer the existence of C' depending only on C_M such that the element map M of a (C_M, κ, h)-normalizable map satisfies

$$\|M'\|_{L^\infty(\hat{K})} \leq C', \qquad \|(M')^{-1}\|_{L^\infty(\hat{K})} \leq C' \frac{1}{\kappa h}. \tag{2.6.10}$$

■

Singularities of the type arising in solutions to elliptic boundary value problems posed on domains with piecewise smooth boundary can be effectively treated numerically with meshes that are refined towards the singularities. The appropriate refinement strategy is as follows: A mesh $\mathcal{T}$ is radically refined towards a point A with refinement exponent $\mu \in [0,1)$, if the elements K_i, which are of size h_i, satisfy the following dichotomy:

Either $A \in \overline{K_i}$ (i.e., K_i abuts on the point A) and $h_i \sim h^{1/(1-\mu)}$
or $A \notin \overline{K_i}$, in which case the element satisfies $h_i \sim h\,(\text{dist}(K_i, A))^\mu$.

A more formal definition of such radical meshes is given in Definition 2.6.5 below. It differs slightly from the standard notion of radical meshes because it is formulated in view of a later application in the context of Shishkin meshes. There, mesh refinement is required in small neighborhoods of the vertices only; therefore, our notion of radical meshes contains a parameter $\kappa > 0$, which controls the size of the region in which mesh refinement takes place.
The Shishkin meshes that are the final goal of this section are meshes whose elements have very different character depending on the location of the element. It is therefore of interest to introduce the notion of a collection of elements: A collection of triples $(K_i, M_i, \hat{K}_i)_{i\in I}$, where I is some index set, is said to be a *collection of elements*, if conditions (M2)–(M4) of Definition 2.4.1 are satisfied. In the following definition, the reader will recognize the "standard" notion of radical meshes for the special case $\kappa = 1$; the parameter μ controls the strength of the refinement near A:

Definition 2.6.5 (radical mesh). *Let $\mathcal{T}(h) = \{(K_i, M_i, \hat{K}_i)_{i\in I(h)}\}$ be a one-parameter family (parametrized by $h \in (0,1)$) of collections of elements, i.e., for each fixed h the triples $\{(K_i, M_i, \hat{K}_i)_{i\in I(h)}\}$ satisfy (M2)–(M4) of Definition 2.4.1. Let $A \in \mathbb{R}^2$, $C_\mathcal{T}$, $c_{rad} > 0$, $\kappa \in (0,1]$, $\mu \in [0,1)$ be given. Then $\mathcal{T}(h)$ is said to be a $(C_\mathcal{T}, c_{rad}, \kappa, \mu)$-*radically refined, *if for each h the elements satisfy the following conditions:*

1. M_i is $(C_\mathcal{T}, 1, h_i)$-normalizable for some suitably chosen $h_i > 0$; h_i is called the size *of element K_i.*
2. $K_i \subset B_{c_{rad}\kappa}(A)$ for all $i \in I(h)$, i.e., the elements K_i are in a neighborhood of A.
3. The following dichotomy holds:
Either

$$A \in \overline{K_i} \quad \textit{and} \quad c_{rad}^{-1} h^{1/(1-\mu)} \leq \frac{h_i}{\kappa} \leq c_{rad}\, h^{1/(1-\mu)} \tag{2.6.11a}$$

or

$$A \notin \overline{K_i} \quad and \quad c_{rad}^{-1} h \sup_{x \in K_i} \widetilde{\Phi}_{\mu,\kappa}(x) \leq \frac{h_i}{\kappa} \leq c_{rad} h \inf_{x \in K_i} \widetilde{\Phi}_{\mu,\kappa}(x), \qquad (2.6.11b)$$

where

$$\widetilde{\Phi}_{\mu,\kappa}(x) = \min\left\{1, \frac{\operatorname{dist}(x, A)}{\kappa}\right\}^{\mu}.$$

The parameter μ controls the refinement near the point A. Choosing $\mu = 0$ corresponds to no refinement near A. One way to construct radical meshes is illustrated in the following example:

Example 2.6.6 Radical meshes are meshes that are refined toward a boundary point can be constructed by mapping uniform meshes. This is illustrated in Fig. 2.6.12 where a radical mesh on $(0,1)^2$ is obtained as the image of a uniform mesh under the map $x \mapsto x\|x\|_\infty^{-1+1/(1-\mu)}$; the exponent μ is chosen as $\mu = 2/3$ in Fig. 2.6.12. ▪

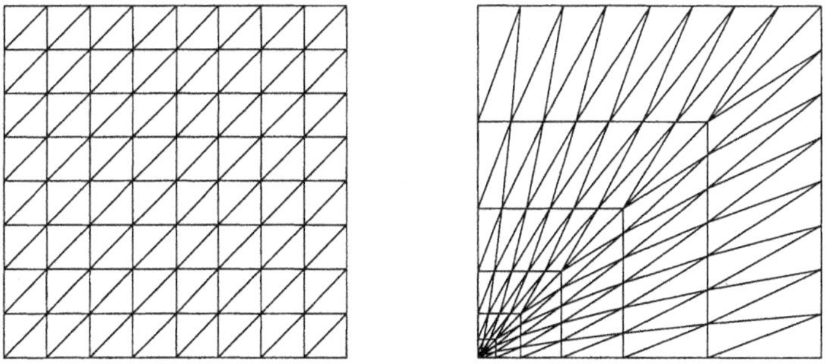

Fig. 2.6.12. (see Example 2.6.6) Radical meshes obtained by mapping uniform meshes: points of uniform mesh (left) are mapped under $x \mapsto x\|x\|_\infty^2$ (right).

Essential properties of radical meshes are collected in the following lemma:

Lemma 2.6.7. *Let the collection $\mathcal{T}(h)$ of elements be $(C_\mathcal{T}, c_{rad}, \mu, h)$-radically refined in the sense of Definition 2.6.5. Then there exists a constant $C > 0$ depending only on $C_\mathcal{T}$, c_{rad}, μ, such that for each $h \in (0,1]$:*

$$\max_{i \in I(h)} \operatorname{diam} K_i \leq C\kappa h,$$

$$|\mathcal{T}(h)| \leq Ch^{-2},$$

$$\operatorname{dist}(K_i, A) \geq C^{-1}\kappa h^{1/(1-\mu)} \qquad \text{if } A \notin \overline{K_i}.$$

Here, $|\mathcal{T}(h)|$ denote the number of elements in $\mathcal{T}(h)$.

Proof: The third estimate follows immediately from Definition 2.6.5. The second estimate follows from idea of [21]. □

An important tool for the approximation from the space $S^1(\mathcal{T})$ is the piecewise linear interpolant, which we define here for completeness' sake:

Definition 2.6.8 (linear interpolation operator I). *Let $\mathcal{T}$ be a mesh on a domain Ω. The linear operator*

$$I : C(\overline{\Omega}) \to S^1(\mathcal{T}), \qquad u \mapsto Iu,$$

is uniquely defined by the condition that $u(V) = (Iu)(V)$ for all vertices V of the mesh (vertices of the mesh are the images of the vertices of the reference elements under the element maps).

The approximation properties of radical meshes is well understood (see, e.g., [21]). Nevertheless, we formulate an approximation result here in a form that will be convenient for our convergence analysis on Shishkin meshes in Section 2.6.3.

Proposition 2.6.9. *Let the collection $\mathcal{T}(h)$ of elements be $(C_{\mathcal{T}}, c_{rad}, \mu, h)$-radically refined in the sense of Definition 2.6.5. Let $\beta \in [0,1)$ and set $\Omega_h := \cup_{i\in I(h)} K_i$. Then the piecewise linear interpolant Iu of a function u satisfies*

$$\|u - Iu\|_{L^2(\Omega_h)} \leq C\kappa^2 h^{1+\delta}\|\widetilde{\Phi}_{\beta,\kappa}\nabla^2 u\|_{L^2(\Omega_h)} + C(\kappa h)^2\|u\|_{H^1(\Omega_h)},$$
$$\|\nabla(u - Iu)\|_{L^2(\Omega_h)} \leq C\kappa h^{\delta}\|\widetilde{\Phi}_{\beta,\kappa}\nabla^2 u\|_{L^2(\Omega_h)} + C(\kappa h)\|u\|_{H^1(\Omega_h)},$$

provided the function u is such that the right-hand sides are finite. Here,

$$\delta = \min\left\{1, \frac{1-\beta}{1-\mu}\right\}.$$

Proof: For fixed h, we decompose the set $I(h)$ as $I(h) = I^{int} \cup I_A$, where

$$I^{int} = \{i \in I(h) \,|\, A \notin \overline{K_i}\}, \qquad I_A = \{i \in I(h) \,|\, A \in \overline{K_i}\} = I(h) \setminus I^{int}.$$

Next, we observe that Definition 2.6.5 implies that all elements K_i satisfy $K_i \subset B_{c_{rad}\kappa}(A)$ and

$$\widetilde{\Phi}_{\beta,\kappa}(x) \sim \left(\frac{\text{dist}(x,A)}{\kappa}\right)^{\beta} \qquad \forall x \in B_{c_{rad}\kappa}(A), \tag{2.6.12}$$

with implied constants independent of x and κ.
The assumption that the element maps M_i are $(C_{\mathcal{T}}, 1, h_i)$ normalizable implies that each M_i can be factored as $M_i = G_i \circ S_i$, where the affine map $S_i : \hat{K}_i \to S_i(\hat{K}_i)$ satisfies $S_i' = h_i$ and the map G_i is a C^2-diffeomorphism between $R_i := S_i(\hat{K}_i)$ and K_i with implied constants depending only on $C_{\mathcal{T}}$. We set $u_i := u \circ G_i$. For simplicity of notation, we assume that the affine maps S_i are of the form $S_i(x) = h_i x$; note that this implies that the origin 0 is a vertex of R_i. Additionally, we require that $A \in \overline{K_i}$ implies $G_i(0) = A$.
The fact that the maps G_i are C^2-diffeomorphisms gives the existence of $C > 0$ (depending only on $C_{\mathcal{T}}$ and μ) such that:

$$\|\nabla^2 u_i\|_{L^2(R_i)} \leq C\left[\|\nabla^2 u\|_{L^2(K_i)} + \|u\|_{H^1(K_i)}\right] \qquad \forall i \in I^{int}, \tag{2.6.13a}$$
$$\|r^{\beta}\nabla^2 u_i\|_{L^2(R_i)} \leq C\left[\kappa^{\beta}\|\widetilde{\Phi}_{\beta,\kappa}\nabla^2 u\|_{L^2(K_i)} + h_i^{\beta}\|u\|_{H^1(K_i)}\right] \; \forall i \in I_A; \tag{2.6.13b}$$

here, we employed the shorthand $r = r(x) = |x|$.
Since $R_i = S_i(\hat{K}_i)$ is square or a triangle of diameter $O(h_i)$, standard interpolation estimates (see, e.g., Proposition 3.2.21 for the case $p = 1$ together with a scaling argument) yield for the error of the linear interpolant

$$\|u_i - Iu_i\|_{L^2(R_i)} + h_i \|\nabla(u_i - Iu_i)\|_{L^2(R_i)} \leq C \begin{cases} h_i^2 \|\nabla^2 u_i\|_{L^2(R_i)} & , i \in I^{int} \\ h_i^{2-\beta} \|r^\beta \nabla^2 u_i\|_{L^2(R_i)} & , i \in I_A. \end{cases}$$

Exploiting the fact that the maps G_i are C^1-diffeomorphisms gives us:

$$\|\nabla(u - Iu)\|^2_{L^2(\Omega_h)} \leq C \sum_{i \in I^{int}} \|\nabla(u_i - Iu_i)\|^2_{L^2(R_i)} + C \sum_{i \in I_A} \|\nabla(u_i - Iu_i)\|^2_{L^2(R_i)}.$$

We treat the two sums separately. For the second sum, we get using (2.6.12)

$$\begin{aligned} \sum_{i \in I_A} \|\nabla(u_i - Iu_i)\|^2_{L^2(R_i)} &\leq C \sum_{i \in I_A} h_i^{2(1-\beta)} \|r^\beta \nabla^2 u_i\|^2_{L^2(R_i)} \\ &\leq C \sum_{i \in I_A} \kappa^{2\beta} h_i^{2(1-\beta)} \|\widetilde{\Phi}_{\beta,\kappa} \nabla^2 u\|^2_{L^2(K_i)} + C \sum_{i \in I_A} h_i^2 \|u\|^2_{H^1(K_i)}. \end{aligned}$$

Since the assumptions on the mesh imply $h_i \leq C\kappa h^{1/(1-\mu)} \leq C\kappa h$ for $i \in I_A$, we get

$$\begin{aligned} &\sum_{i \in I_A} \|\nabla(u_i - Iu_i)\|^2_{L^2(R_i)} \\ &\leq C \left[\kappa^2 h^{2(1-\beta)/(1-\mu)} \|\widetilde{\Phi}_{\beta,\kappa} \nabla^2 u\|^2_{L^2(\Omega_h)} + \kappa^2 h^2 \|u\|^2_{H^1(\Omega_h)} \right] \\ &\leq C\kappa^2 h^{2\delta} \|\widetilde{\Phi}_{\beta,\kappa} \nabla^2 u\|_{L^2(\Omega_h)} + C\kappa^2 h^2 \|u\|^2_{H^1(\Omega_h)}. \end{aligned}$$

For $\sum_{i \in I^{int}} \|\nabla(u_i - Iu_i)\|^2_{L^2(R_i)}$ we bound

$$\begin{aligned} \sum_{i \in I^{int}} \|\nabla(u_i - Iu_i)\|^2_{L^2(R_i)} &\leq C \sum_{i \in I^{int}} h_i^2 \|\nabla^2 u_i\|^2_{L^2(R_i)} \\ &\leq C \sum_{i \in I^{int}} h_i^2 \left[\|\nabla^2 u\|^2_{L^2(K_i)} + \|u\|^2_{H^1(K_i)} \right] \\ &\leq C \sum_{i \in I^{int}} h_i^2 \|\nabla^2 u\|^2_{L^2(K_i)} + C\kappa^2 h^2 \|u\|^2_{H^1(\Omega_h)} \\ &\leq C \sum_{i \in I^{int}} \frac{h_i^2}{\inf_{x \in K_i} |\widetilde{\Phi}_{\beta,\kappa}(x)|^2} \|\widetilde{\Phi}_{\beta,\kappa} \nabla^2 u\|^2_{L^2(K_i)} + C\kappa^2 h^2 \|u\|^2_{H^1(\Omega)}. \end{aligned}$$

In view of (2.6.12) and Lemma 2.6.7, we can estimate

$$\sup_{x \in K_i} \frac{\widetilde{\Phi}_{\mu,\kappa}(x)}{\widetilde{\Phi}_{\beta,\kappa}(x)} \leq Ch^{-1+\delta}.$$

Thus, we get

$$\frac{h_i^2}{\inf_{x\in K_i} |\widetilde{\Phi}_{\beta,\kappa}(x)|^2} \leq \left(\frac{h_i^2}{\inf_{x\in K_i} |\widetilde{\Phi}_{\mu,\kappa}(x)|^2}\right)\left(\sup_{x\in K_i} \frac{\widetilde{\Phi}_{\mu,\kappa}(x)}{\widetilde{\Phi}_{\beta,\kappa}(x)}\right)^2 \leq c_{rad} h^2 h^{-2+2\delta} = c_{rad} h^{2\delta}.$$

Combining the above estimate yields the bound for $\|\nabla(u - Iu)\|_{L^2(\Omega_h)}$. Analogous reasoning yields

$$\|u - Iu\|_{L^2(\Omega_h)} \leq C\left[\kappa^2 h^{1+\delta} \|\widetilde{\Phi}_{\beta,\kappa}\nabla^2 u\|_{L^2(\Omega_h)} + (\kappa h)^2 \|u\|_{H^1(\Omega_h)}\right].$$

□

2.6.3 h-FEM boundary layer meshes

Our aim is the introduction of meshes $\mathcal{T}$ such that robust approximation of solution of (2.1.1) from the h-FEM spaces $S^1(\mathcal{T})$ is possible. Such meshes $\mathcal{T}$ need to provide the capability to resolve boundary layers and corner singularities. As in the case of admissible meshes in Definition 2.4.4, the approximation of boundary layers is made possible with anisotropic boundary layer elements, whose aspect ratio is controlled by a parameter κ; the resolution of corner singularities is achieved with radical meshes, where the refinement is near a vertex A_j is controlled by a parameter $\mu_j \in [0,1)$.

Definition 2.6.10 (h-FEM boundary layer mesh). *Consider a three-parameter family $\mathcal{T}(\kappa, h, \mu) = \{(K_i, M_i, \hat{K}_i)\}$, where the three parameters satisfy $\kappa > 0$, $h > 0$, and $\mu \in [0,1)^J$. This family $\mathcal{T}$ is said to be* of boundary layer type *if there are c_i, $i = 1, \ldots, 4$, $\sigma \in (0,1)$, $C_M > 0$, and sets Ω_j, $j = 1, \ldots, J$, of the form given in Notation 2.3.3 such that the elements K_i of the meshes fall into exactly one of the following three categories:*

(C1) K_i is a boundary layer element, *i.e., for some $j \in \{1, \ldots, J\}$ we have the inclusion $K_i \subset \mathcal{U}_\kappa(\Gamma_j) \cap \Omega_j \setminus (B_{c_1\kappa}(A_{j-1}) \cup B_{c_1\kappa}(A_j))$; additionally the map*

$$\tilde{G}_i := s_\kappa^{-1} \circ \psi_j^{-1} \circ M_i$$

is $(C_M, 1, h)$-normalizable in the sense of Definition 2.6.3.

(C2) K_i is a corner layer element, *i.e., for some $j \in \{1, \ldots, J\}$, the element K_i satisfies $K_i \subset B_\kappa(A_j) \cap \Omega_j$ or $K_i \subset B_\kappa(A_j) \cap \Omega_{j+1}$, and M_i is $(C_M, 1, h_i)$-normalizable in the sense of Definition 2.6.3, where the parameter $h_i =$* diam $K_i > 0$ *is the diameter of the element K_i. Additionally, the following dichotomy holds:*
Either

$$A_j \in \overline{K_i} \quad \text{and} \quad \frac{h_i}{\kappa} \leq c_4 h^{1/(1-\mu_j)}$$

or

$$A_j \notin \overline{K_i} \quad \text{and} \quad c_3 h \sup_{x\in K_i} \Phi_{0,\mu,\kappa}(x) \leq \frac{h_i}{\kappa} \leq c_4 h \sup_{x\in K_i} \Phi_{0,\mu,\kappa}(x).$$

(C3) K_i is an interior element, *i.e., $K_i \subset \Omega \setminus \mathcal{U}_{c_2\kappa}(\partial\Omega)$ and M_i is (C_M, κ, h)-normalizable in the sense of Definition 2.6.3.*

The conditions placed on the different types of elements in Definition 2.6.10 are best understood by considering the following examples of meshes.

Example 2.6.11 For fixed $\kappa_0 > 0$ and $\mu_0 \in [0,1)^J$, a one-parameter family $\mathcal{T}(\kappa_0, h, \mu_0)$ of meshes coincides with a "standard" radical mesh for the approximation of solutions of elliptic boundary value problems posed on domains with piecewise smooth boundary in the following sense: The mesh family $\mathcal{T}(\kappa_0, h, \mu_0)$ can be viewed as a generalization to the context of non-affine element maps of standard radical meshes of [21,106], which consist of shape-regular elements and the refinement toward the vertices of the domain is such that the element size h_i satisfies (2.6.11). ▪

Example 2.6.12 The mesh shown in the left part of Fig. 2.6.13 is the tensor-product of the one-dimensional Shishkin mesh of Definition 2.6.1. It is an h-FEM boundary layer mesh in the sense of Definition 2.6.10 for the special case $\mu_j = 0$, $j = 1, \ldots, 4$. This special choice of refinement parameters μ_j corresponds to not performing mesh grading in the neighborhood of the vertices. For the present case of convex corners and an intended approximation by piecewise linear/bilinear functions, this is completely adequate as will be shown in Theorem 2.6.15 and the discussion following Theorem 2.6.15. ▪

Example 2.6.13 The right part of Fig. 2.6.13 shows an example of a boundary layer mesh for an L-shaped domain. In the shaded regions near the re-entrant corner, refined meshes are inserted that can be constructed as illustrated in Example 2.6.6 by mapping a uniform mesh with mesh size h on $(0,1)^2$ under the map $x \mapsto x\|x\|_\infty^{-1+1/(1-\mu)}$ and afterwards inserting scaled (by κ) and rotated versions into the three shaded regions. ▪

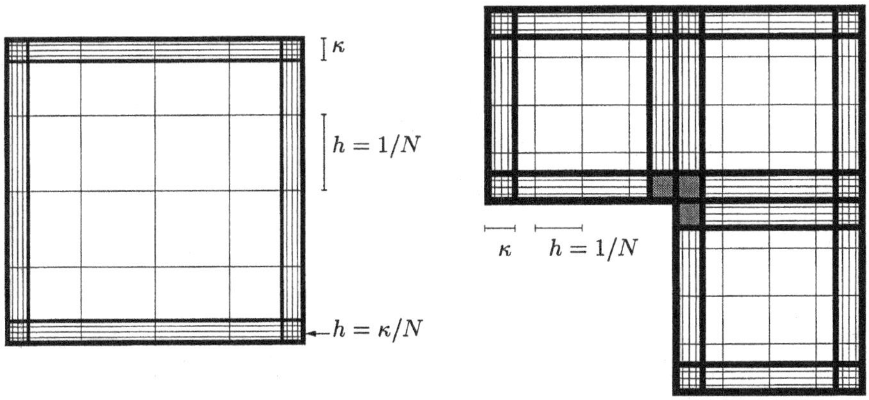

Fig. 2.6.13. Left: Shishkin-type mesh on a square with $\mu_j = 0$, $j = 1, \ldots, 4$. Right: Shishkin-type mesh on L-shaped domain.

From the regularity result Theorem 2.3.4, we can extract the following simplified version that is suitable for approximating with piecewise linear functions:

Proposition 2.6.14. *Let the subdomains Ω_j, $j = 1, \ldots, J$, and the boundary fitted coordinates (ρ_j, θ_j) be as in Theorem 2.3.4. Write*

$$r(x) = \min_{j=1,\ldots,J} \operatorname{dist}(x, A_j).$$

Let $\beta \in [0,1)^J$ be given by the statement of Theorem 2.3.4 and write $1-\beta$ for the vector $(1-\beta_1, \ldots, 1-\beta_J)$. Then there exist constants C, $\alpha > 0$ such that for each $\varepsilon \in (0,1]$ the solution u_ε of (2.1.1) can be decomposed as

$$u_\varepsilon = w_\varepsilon + \tilde{u}_\varepsilon^{BL} + \tilde{u}_\varepsilon^{CL} + r_\varepsilon,$$

with the following properties:

(i) $w_\varepsilon \in C^2(\overline{\Omega})$ *and*

$$\|w_\varepsilon\|_{L^\infty(\Omega)} + \|\nabla w_\varepsilon\|_{L^\infty(\Omega)} + \|\nabla^2 w_\varepsilon\|_{L^\infty(\Omega)} \leq C.$$

(ii) On Ω_j the function $\tilde{u}_\varepsilon^{BL}$ satisfies

$$|(\partial_{\rho_j}^k \partial_{\theta_j}^m (\tilde{u}_\varepsilon^{BL} \circ \psi_j))(\rho_j, \theta_j)| \leq C\varepsilon^{-k} e^{-\alpha\rho_j/\varepsilon}, \qquad k, m \in \{0,1,2\}.$$

(iii) On Ω_j, $j = 1, \ldots, J$, the function $\tilde{u}_\varepsilon^{CL}$ satisfies

$$|\tilde{u}_\varepsilon^{CL}(x)| \leq C r^{-k} \Phi_{0,1-\beta,\varepsilon}(x)\, e^{-\alpha r/\varepsilon}, \quad k \in \{0,1,2\}.$$

(iv) $u^C := \tilde{u}_\varepsilon^{BL} + \tilde{u}_\varepsilon^{CL}$ is in $C^1(\overline{\Omega} \setminus \cup_{j=1}^J \{A_j\})$ and for all $x \in \Omega$:

$$|\nabla^k u^C(x)| \leq C \left[\varepsilon^{-k} e^{-\alpha \operatorname{dist}(x,\partial\Omega)/\varepsilon} + r^{-k} \Phi_{0,1-\beta,\varepsilon}(x)\, e^{-\alpha r/\varepsilon}\right], \quad k \in \{0,1\}.$$

Additionally, $u^C \in H^2_{loc}(\Omega)$ and in neighborhoods of the vertices A_j, we have for $\kappa > 0$ and $j \in \{1, \ldots, J\}$:

$$\|u^C\|_{L^2(\Omega \cap B_\kappa(A_j))} + \varepsilon \|\nabla u^C\|_{L^2(\Omega \cap B_\kappa(A_j))} \leq C\left[\sqrt{\kappa\varepsilon} + \varepsilon\right],$$
$$\|\Phi_{0,\beta,\varepsilon} \nabla^2 u^C\|_{L^2(\Omega \cap B_\kappa(A_j))} \leq C\left[\sqrt{\kappa}\varepsilon^{-3/2} + \varepsilon^{-1}\right].$$

(v) $r_\varepsilon \in H^2_{loc}(\Omega) \cap H^1_0(\Omega)$ and

$$\|r_\varepsilon\|_\varepsilon + \|\Phi_{0,\beta,\varepsilon} \nabla^2 r_\varepsilon\|_{L^2(\Omega)} \leq C e^{-\alpha/\varepsilon}.$$

Proof: The proposition is a corollary to Theorem 2.3.4. Let w_ε, r_ε be the functions given in Theorem 2.3.4. Assertions (i), (v) follow immediately from Theorem 2.3.4. Next, we define

$$\tilde{u}_\varepsilon^{BL} := \chi^{BL} u_\varepsilon^{BL}, \qquad \tilde{u}_\varepsilon^{CL} := \chi^{CL} u_\varepsilon^{CL},$$

where the cut-off functions χ^{BL}, χ^{CL} and the functions u_ε^{BL}, u_ε^{CL} are defined in Theorem 2.3.4. The result then follows from the properties of u_ε^{BL}, u_ε^{CL} ascertained in Theorem 2.3.4. For example, for the bound on $\|u^C\|_{L^2(B_\kappa(A_j))}$ we compute

$$\|u^C\|_{L^2(B_\kappa(A_j))} \leq \|\tilde{u}_\varepsilon^{BL}\|_{L^2(B_\kappa(A_j))} + \|\tilde{u}_\varepsilon^{CL}\|_{L^2(B_\kappa(A_j))} \leq \|\tilde{u}_\varepsilon^{BL}\|_{L^2(B_\kappa(A_j))} + C\varepsilon,$$

where we employed Theorem 2.3.4 to bound $\|\tilde{u}_\varepsilon^{CL}\|_{L^2(B_\kappa(A_j))} \leq C\varepsilon$. Next, using again Theorem 2.3.4, we estimate

$$\|\tilde{u}_\varepsilon^{BL}\|^2_{L^2(B_\kappa(A_j))} \leq C \int_0^\kappa \int_0^\kappa e^{-\alpha\rho/\varepsilon}\, d\rho \leq C\kappa\varepsilon,$$

which leads to the desired bound. □

On boundary layer meshes we can formulate the following approximation result, which is the h-FEM analog of Theorem 2.4.8:

Theorem 2.6.15. *Let $\mathcal{T}(\kappa, h, \mu)$ be a family of boundary layer meshes in the sense of Definition 2.6.10. Let u_ε be the solution to (2.1.1) and let $\beta \in [0,1)^J$ be given by Proposition 2.6.14. Then there exist $\lambda_0 > 0$ and $C > 0$ independent of ε and $h \in (0, 1/2)$ with the following properties: Setting, for each $\lambda \geq \lambda_0$ and $\mu \in [0,1)^J$,*

$$\kappa := \min\{1, \lambda\varepsilon|\ln h|\},$$
$$\delta := \min_{j=1,\dots,J} \left\{1, \frac{1-\beta_j}{1-\mu_j}\right\}$$

there exists $v \in S^1(\mathcal{T}(\kappa, h, \mu))$ with $v = Iu_\varepsilon$ on $\partial\Omega$ and

$$\|u_\varepsilon - v\|_{L^2(\Omega)} + \kappa h \|\nabla(u_\varepsilon - v)\|_{L^2(\Omega)}$$
$$\leq C\left[\varepsilon h^{1+\delta}|\lambda \ln h|^3 + \varepsilon^{1/2} h^2 |\lambda \ln h|^{5/2} + h^2|\lambda \ln h|\right].$$

Proof: As in the one-dimensional case, Proposition 2.6.2, we distinguish the asymptotic case $\lambda\varepsilon|\ln h| > 1$ and the preasymptotic case $\lambda\varepsilon|\ln h| \leq 1$.

The case $\lambda\varepsilon|\ln h| > 1$: For fixed $\mu \in [0,1)^J$, we consider boundary layer meshes $\mathcal{T}(1, h, \mu)$, which is a one-parameter family of meshes as described in Example 2.6.11. Proposition 2.6.9 then yields for the piecewise linear interpolant Iu_ε of u_ε:

$$\|u_\varepsilon - Iu_\varepsilon\|_{L^2(\Omega)} \leq C\left[h^{1+\delta}\|\Phi_{0,\beta,1}\nabla^2 u_\varepsilon\|_{L^2(\Omega)} + h^2\|u_\varepsilon\|_{H^1(\Omega)}\right],$$
$$\|\nabla(u_\varepsilon - Iu_\varepsilon)\|_{L^2(\Omega)} \leq C\left[h^{\delta}\|\Phi_{0,\beta,1}\nabla^2 u_\varepsilon\|_{L^2(\Omega)} + h\|u_\varepsilon\|_{H^1(\Omega)}\right].$$

Noting that $\varepsilon\lambda|\ln h| > 1$ implies $\kappa = 1$, we get using $\Phi_{0,\beta,1}(x) \leq \Phi_{0,\beta,\varepsilon}(x)$ and the estimate of Theorem 2.3.1

$$\begin{aligned}
\|u_\varepsilon - Iu_\varepsilon\|_{L^2(\Omega)} + \kappa h\|\nabla(u_\varepsilon - Iu_\varepsilon)\|_{L^2(\Omega)} \\
&\leq C\left[h^{1+\delta}\|\Phi_{0,\beta,\varepsilon}\nabla^2 u_\varepsilon\|_{L^2(\Omega)} + h^2\|u_\varepsilon\|_{H^1(\Omega)}\right] \\
&\leq C\left[h^{1+\delta}\varepsilon^{-2} + h^2\varepsilon^{-1}\right] \leq C\left[\varepsilon h^{1+\delta}\varepsilon^{-3} + h^2\varepsilon^{-1}\right] \\
&\leq C\left[\varepsilon h^{1+\delta}|\lambda \ln h|^3 + h^2|\lambda \ln h|\right],
\end{aligned}$$

where, in the last step we used the assumption $\lambda|\ln h| \geq \varepsilon^{-1}$.

The case $\lambda\varepsilon|\ln h| \leq 1$*:* In this case, we employ the decomposition given by Proposition 2.6.14, i.e.,

$$u_\varepsilon = w_\varepsilon + \tilde{u}_\varepsilon^{BL} + \tilde{u}_\varepsilon^{CL} + r_\varepsilon. \tag{2.6.14}$$

We approximate the function r_ε by zero; in view of Proposition 2.6.14 we obtain with $\kappa = \lambda\varepsilon|\ln h|$

$$\begin{aligned}
\|r_\varepsilon\|_{L^2(\Omega)} + \kappa h\|\nabla r_\varepsilon\|_{L^2(\Omega)} &\leq (1 + h\lambda|\ln h|)\|r_\varepsilon\|_\varepsilon \leq C(1 + h\lambda|\ln h|)e^{-\alpha/\varepsilon} \\
&\leq C(1 + h\lambda|\ln h|)e^{-\alpha\lambda|\ln h|} \leq C(1 + h\lambda|\ln h|)h^{\alpha\lambda} \\
&\leq Ch^2
\end{aligned}$$

for $\lambda \geq \lambda_0$ sufficiently large.
It remains to approximate the function $w_\varepsilon + \tilde{u}_\varepsilon^{BL} + \tilde{u}_\varepsilon^{CL}$. To that end, we denote

$$u^C := \tilde{u}_\varepsilon^{BL} + \tilde{u}_\varepsilon^{CL}$$

and consider the error of the piecewise linear interpolant,

$$e := (w_\varepsilon + u^C) - I(w_\varepsilon + u^C) = (w_\varepsilon - Iw_\varepsilon) + (u^C - Iu^C) \tag{2.6.15}$$

$$= (w_\varepsilon - Iw_\varepsilon) + (\tilde{u}_\varepsilon^{BL} - I\tilde{u}_\varepsilon^{BL}) + (\tilde{u}_\varepsilon^{CL} - I\tilde{u}_\varepsilon^{CL}). \tag{2.6.16}$$

We will use both representations of the error e. To that end, we consider the approximation on each the three types of elements, namely, boundary layer elements, corner layer elements, and interior elements, separately. For notational convenience, we write the set $I_\mathcal{T}$ of element indices i as the pairwise disjoint union

$$I_\mathcal{T} = I^{int} \cup I^{BL} \cup I^{CL}$$

of indices corresponding to interior elements, boundary layer elements, and corner layer elements, respectively.
1. step: boundary layer elements. Let K_i be a boundary layer element with $K_i \subset \Omega_j$. Our aim is to show

$$\|w_\varepsilon - Iw_\varepsilon\|_{L^\infty(K_i)} + \kappa h\|\nabla(w_\varepsilon - Iw_\varepsilon)\|_{L^\infty(K_i)} \leq Ch^2, \tag{2.6.17a}$$

$$\|\tilde{u}_\varepsilon^{BL} - I\tilde{u}_\varepsilon^{BL}\|_{L^\infty(K_i)} + \kappa h\|\nabla(\tilde{u}_\varepsilon^{BL} - I\tilde{u}_\varepsilon^{BL})\|_{L^\infty(K_i)} \leq Ch^2|\lambda \ln h|^2, \tag{2.6.17b}$$

$$\|\tilde{u}_\varepsilon^{CL} - I\tilde{u}_\varepsilon^{CL}\|_{L^\infty(K_i)} + \kappa h\|\nabla(\tilde{u}_\varepsilon^{CL} - I\tilde{u}_\varepsilon^{CL})\|_{L^\infty(K_i)} \leq Ch^2. \tag{2.6.17c}$$

By assumption the element map M_i is such that the maps $s_\kappa^{-1} \circ \psi_j^{-1} \circ M_i$ is $(C_\mathcal{T}, 1, h)$-normalizable, i.e., $s_\kappa^{-1} \circ \psi_j^{-1} \circ M_i = G_i \circ S_i$, where G_i is a C^2-diffeomorphism and $S_i' = h_i$. The formula $M_i = (\psi_j \circ s_\kappa \circ G_i) \circ S_i$ shows that M_i is (C', κ, h)-normalizable with a constant $C' > 0$ that is independent of κ and i.

We start with proving (2.6.17a). Using the fact that M_i is (C', κ, h) -normalizable and that $w_\varepsilon \in C^2(\overline{\Omega})$ (cf. Proposition 2.6.14), we get

$$\|w_\varepsilon \circ M_i - I(w_\varepsilon \circ M_i)\|_{W^{1,\infty}(\hat{K}_i)} \leq Ch^2. \tag{2.6.18}$$

Returning to K_i, we get in view of the fact that M_i is (C', κ, h)-normalizable,

$$\|w_\varepsilon - Iw_\varepsilon\|_{L^\infty(K_i)} + \kappa h \|\nabla(w_\varepsilon - Iw_\varepsilon)\|_{L^\infty(K_i)} \leq Ch^2, \tag{2.6.19}$$

which is (2.6.17a).
We now turn to the proof of (2.6.17b). On the reference element $\hat{K}_i$, we have

$$\tilde{u}_\varepsilon^{BL} \circ M_i = (\tilde{u}_\varepsilon^{BL} \circ \psi_j \circ s_\kappa) \circ (s_\kappa^{-1} \circ \psi_j^{-1} \circ M_i) =: \tilde{u}_i \circ (G_i \circ S_i).$$

Proposition 2.6.14 implies

$$\|\nabla^k \tilde{u}_i\|_{L^\infty(\tilde{R}_i)} \leq C\{1 + (\kappa/\varepsilon)^k\}, \qquad k \in \{0, 1, 2\}, \tag{2.6.20}$$

where $\tilde{R}_i = (G_i \circ S_i)(\hat{K}_i)$. Since G_i is a C^2-diffeomorphism, we get

$$\|\nabla^2(\tilde{u}_\varepsilon^{BL} \circ M_i)\|_{L^\infty(\hat{K})} \leq Ch^2 \sum_{k=0}^{2} \{1 + (\kappa/\varepsilon)^k\}.$$

Inserting $\kappa = \lambda\varepsilon |\ln h|$ gives for $\lambda \geq \lambda_0$

$$\|\nabla^2(\tilde{u}_\varepsilon^{BL} \circ M_i)\|_{L^\infty(\hat{K})} \leq Ch^2(\lambda \ln h)^2.$$

Therefore for the interpolation error on the reference element $\hat{K}_i$

$$\|\tilde{u}_\varepsilon^{BL} \circ M_i - I(\tilde{u}_\varepsilon^{BL} \circ M_i)\|_{W^{1,\infty}(\hat{K}_i)} \leq Ch^2(\lambda \ln h)^2; \tag{2.6.21}$$

thus, on K_i we get, again due to the fact that M_i is (C', κ, h)-normalizable,

$$\|\tilde{u}_\varepsilon^{BL} - I\tilde{u}_\varepsilon^{BL}\|_{L^\infty(K_i)} + \kappa h \|\nabla(\tilde{u}_\varepsilon^{BL} - I\tilde{u}_\varepsilon^{BL})\|_{L^\infty(K_i)} \leq Ch^2(\lambda \ln h)^2,$$

which is (2.6.17b).
We finally turn to (2.6.17c), where we simply exploit the fact that $\tilde{u}_\varepsilon^{CL}$ is small away from the vertices. We start with the observation that the fact that M_i is (C', κ, h)-normalizable and that a standard inverse estimate for linear functions on the reference element gives

$$\|\nabla I\tilde{u}_\varepsilon^{CL}\|_{L^\infty(K_i)} \leq C\frac{1}{\kappa h}\|\tilde{u}_\varepsilon^{CL}\|_{L^\infty(K_i)}.$$

This bound allows us to estimate

$$\|\tilde{u}_\varepsilon^{CL} - I\tilde{u}_\varepsilon^{CL}\|_{L^\infty(K_i)} \leq 2\|\tilde{u}_\varepsilon^{CL}\|_{L^\infty(K_i)},$$
$$\|\nabla(\tilde{u}_\varepsilon^{CL} - I\tilde{u}_\varepsilon^{CL})\|_{L^\infty(K_i)} \leq \|\nabla\tilde{u}_\varepsilon^{CL}\|_{L^\infty(K_i)} + C\frac{1}{\kappa h}\|\tilde{u}_\varepsilon^{CL}\|_{L^\infty(K_i)}.$$

Proposition 2.6.14 together with $\operatorname{dist}(K_i, A_j) \geq c\kappa$ for all $j \in \{1, \dots, J\}$ imply

$$\|\tilde{u}_\varepsilon^{CL} - I\tilde{u}_\varepsilon^{CL}\|_{L^\infty(K_i)} + \kappa h \|\nabla(\tilde{u}_\varepsilon^{CL} - I\tilde{u}_\varepsilon^{CL})\|_{L^\infty(K_i)} \leq C e^{-\alpha'\kappa/\varepsilon}$$

for some C, $\alpha' > 0$ independent of ε, h. Taking $\lambda_0 > 0$ sufficiently large, we get

$$\|\tilde{u}_\varepsilon^{CL} - I\tilde{u}_\varepsilon^{CL}\|_{L^\infty(K_i)} + \kappa h \|\nabla(\tilde{u}_\varepsilon^{CL} - I\tilde{u}_\varepsilon^{CL})\|_{L^\infty(K_i)} \leq C h^2,$$

which is (2.6.17c).
Since all boundary layer elements are contained in an $O(\kappa)$-neighborhood of $\partial\Omega$, we get from the L^∞-estimates of (2.6.17) for the error e of (2.6.16) by summing over all elements:

$$\left(\sum_{i \in I^{BL}} \|e\|_{L^2(K_i)}^2 + (\kappa h)^2 \|\nabla e\|_{L^2(K_i)}^2\right)^{1/2} \leq C\sqrt{\kappa} h^2 (\lambda \ln h)^2 \leq C\sqrt{\varepsilon} h^2 |\lambda \ln h|^{5/2}.$$

This is the desired bound on boundary layer elements.
2. step: corner layer elements. On corner layer elements, we use Proposition 2.6.9. For $w_\varepsilon + (\tilde{u}_\varepsilon^{BL} + \tilde{u}_\varepsilon^{CL}) =: w_\varepsilon + u^C$, we note that Proposition 2.6.14 yields

$$\|w_\varepsilon + u^C\|_{H^1(B_\kappa(A_j)\cap\Omega)} \leq C\left[\sqrt{\kappa/\varepsilon} + 1\right],$$

$$\|\Phi_{0,\beta,\varepsilon} \nabla^2 (w_\varepsilon + u^C)\|_{L^2(B_\kappa(A_j)\cap\Omega)} \leq C\left[\sqrt{\kappa}\varepsilon^{-3/2} + \varepsilon^{-1} + \kappa\right] \leq C\left[\sqrt{\kappa}\varepsilon^{-3/2} + \varepsilon^{-1}\right]$$

where we used the hypotheses $\kappa \leq 1$ and $\varepsilon \leq 1$. Proposition 2.6.9 then allows us to estimate for the error e

$$\begin{aligned}\left(\sum_{i \in I^{CL}} \|e\|_{L^2(K_i)}^2 + (\kappa h)^2 \|\nabla e\|_{L^2(K_i)}^2\right)^{1/2} &\leq \kappa^2 h^{1+\delta}\left[\sqrt{\kappa}\varepsilon^{-3/2} + \varepsilon^{-1}\right] \\ &\leq C\varepsilon h^{1+\delta} |\lambda \ln h|^{5/2},\end{aligned}$$

where we inserted $\kappa = \lambda\varepsilon|\ln h|$ and used the assumptions $\lambda \geq \lambda_0$, $h \leq 1/2$.
3. step: interior elements. Let K_i be an interior element. Reasoning exactly as in the proof of (2.6.17a), we get

$$\|w_\varepsilon - Iw_\varepsilon\|_{L^\infty(K_i)} + \kappa h \|\nabla(w_\varepsilon - Iw_\varepsilon)\|_{L^\infty(K_i)} \leq C h^2.$$

Likewise, reasoning as in the proof of (2.6.17c), we can estimate for u^C in view of the fact that $\operatorname{dist}(K_i, \partial\Omega) \geq c\kappa$:

$$\|u^C - Iu^C\|_{L^\infty(K_i)} + \kappa h \|\nabla(u^C - Iu^C)\|_{L^\infty(K_i)} \leq C h^2 \tag{2.6.22}$$

for suitably chosen $\lambda_0 > 0$. Squaring and summing these two estimates gives

$$\left(\sum_{i \in I^{int}} \|e\|_{L^2(K_i)}^2 + (\kappa h)^2 \|\nabla e\|_{L^2(K_i)}^2\right)^{1/2} \leq C h^2.$$

Combining the estimates for boundary layer elements, corner layer elements, and interior element proves the theorem. □

A few comments concerning Theorem 2.6.15 are in order.

Remark 2.6.16 1. The corner singularities are weak: The regularity assertion of Proposition 2.6.14 shows that the corner singularities are restricted to an $O(\varepsilon)$ neighborhood of the vertices. The factor ε in front of the term $h^{1+\delta}$ reflects this. Note that if no refinement in the $O(\kappa)$-neighborhoods is performed, i.e., a quasi-uniform mesh with mesh size $h\kappa$ is used, this corresponds to $\mu = 0$, i.e., $\delta < 1$. Nevertheless, the factor ε in the term $\varepsilon h^{1+\delta}$ mitigates this neglect of refinement.
2. Results analogous to Theorem 2.6.15 have been obtained in [8] and [7, Chap. 5], where also numerical examples can be found.

■

We now show that this estimate can be improved if further assumptions on the mesh are made. We note that the interior elements of h-FEM boundary layer meshes of Definition 2.6.10 may be quite distorted. An inspection of the proof of Theorem 2.6.15 reveals that additional assumptions on the element maps for interior elements allow us to improve the approximation of the smooth part w_ε. This observation is formulated in the Corollary 2.6.19 below. The key observation is that the notion of normalizable element maps of Definition 2.6.3 can be generalized to anisotropic elements in the following form:

Definition 2.6.17. *An invertible C^2-map $M : \overline{\hat{K}} \to M(\overline{\hat{K}}) \subset \mathbb{R}^2$ is said to be* anisotropically (C_M, h_x, h_y)-normalizable *if it can be factored as*

$$M = G \circ S,$$

where the affine stretching map S satisfies

$$S' = \begin{pmatrix} h_x & 0 \\ 0 & h_y \end{pmatrix},$$

and the C^2-map G satisfies

$$\begin{aligned} \|\nabla^p G_i\|_{L^\infty(R)} &\le C_M, \qquad p \in \{0,1,2\}, \\ \|(G')^{-1}\|_{L^\infty(R)} &\le C_M \end{aligned}$$

Here, the set R is the image of the reference element $\hat{K}$ under the affine map S, i.e., $R = S(\hat{K})$.

For the approximation on anisotropic elements in the interior, we will need the following lemma concerning the interpolation error on anisotropic elements:

Lemma 2.6.18. *Let h_x, $h_y > 0$ and let R be either the rectangle $(0, h_x) \times (0, h_y)$ or the triangle with vertices $(0,0)$, $(h_x, 0)$, $(0, h_y)$. Then there exists $C > 0$ independent of h_x, h_y such that for every $u \in C^2(\overline{R})$ the linear/bilinear interpolation Iu in the vertices of R leads to the following errors:*

$$\|u - Iu\|_{L^\infty(R)} \le Ch^2 \|\nabla^2 u\|_{L^\infty(R)},$$
$$\|\partial_x(u - Iu)\|_{L^\infty(R)} + \|\partial_y(u - Iu)\|_{L^\infty(R)} \le Ch\|\nabla^2 u\|_{L^\infty(R)},$$

where $h = \max\{h_x, h_y\}$.

Proof: Approximation results of this type can be found in [7, Chap. 2]. For the sake of completeness, however, we include a proof of the present simple case. We define the function $\hat{u}$ by $\hat{u} = u \circ S$, where S is the anisotropic stretching given by $S : (x, y) \mapsto (xh_x, yh_y)$. Set $\hat{K} := S^{-1}(R)$, which is the reference square $(0,1)^2$ if R is a rectangle and the reference triangle $\{(x,y) \,|\, 0 < x < 1, 0 < y < 1 - x\}$ if R is a triangle. We claim the following interpolation error for $\hat{u}$:

$$\|\hat{u} - I\hat{u}\|_{L^\infty(\hat{K})} \le C\|\nabla^2 \hat{u}\|_{L^\infty(\hat{K})}, \tag{2.6.24}$$
$$\|\partial_x(\hat{u} - I^y \circ I^x \hat{u})\|_{L^\infty(\hat{K})} \le C\left[\|\partial_x^2 \hat{u}\|_{L^\infty(\hat{K})} + \|\partial_x\partial_y \hat{u}\|_{L^\infty(\hat{K})}\right], \tag{2.6.25}$$

and an analogous estimate for $\|\partial_y(\hat{u} - I^y \circ I^x \hat{u})\|_{L^\infty(\hat{K})}$. The desired bounds follows from (2.6.24), (2.6.25) from the fact $\hat{u} = u \circ S$ and $Iu = (I\hat{u}) \circ S^{-1}$.
It remains to show (2.6.24), (2.6.25). The bound (2.6.24) is standard. For the bound (2.6.24), we consider the cases of $\hat{K}$ being the reference square and $\hat{K}$ being the reference triangle separately.
$\hat{K}$ is the reference square: We note the following bounds for univariate functions:

$$\|w - Iw\|_{L^\infty((0,1))} \le C\|w'\|_{L^\infty((0,1))} \qquad \forall w \in C^1([0,1]), \tag{2.6.26a}$$
$$\|w - Iw\|_{L^\infty((0,1))} \le C\|w''\|_{L^\infty((0,1))} \qquad \forall w \in C^2([0,1]), \tag{2.6.26b}$$
$$\|(w - Iw)'\|_{L^\infty((0,1))} \le C\|w''\|_{L^\infty((0,1))} \qquad \forall w \in C^2([0,1]). \tag{2.6.26c}$$

The interpolation operator I has the form $I = I^y \circ I^x$, where I^x, I^y denote the one-dimensional linear interpolation operator acting on the x- and the y-variable, respectively. We then get with the triangle inequality and the fact that the operators I^y, ∂_x commute:

$$\|\partial_x(\hat{u} - I^y \circ I^x \hat{u})\|_{L^\infty(\hat{K})} \le \|\partial_x \hat{u} - I^y \partial_x \hat{u}\|_{L^\infty(\hat{K})} + \|I^y(\partial_x(\hat{u} - I^x \hat{u}))\|_{L^\infty(\hat{K})}.$$

Next, exploiting the one-dimension estimates (2.6.26) allows us to bound

$$\begin{aligned}
\|\partial_x \hat{u} - I^y \partial_x \hat{u}\|_{L^\infty(\hat{K})} &= \sup_{x\in(0,1)} \sup_{y\in(0,1)} |\partial_x \hat{u}(x,y) - I^y \partial_x \hat{u}(x,y)| \\
&\le C \sup_{x\in(0,1)} \|\partial_y \partial_x \hat{u}(x,\cdot)\|_{L^\infty((0,1))}, \\
\|I^y(\partial_x(\hat{u} - I^x \hat{u}))\|_{L^\infty(\hat{K})} &\le \sup_{y\in(0,1)} \|\partial_x(\hat{u}(\cdot,y) - I^x \hat{u}(\cdot,y))\|_{L^\infty((0,1))} \\
&\le C \sup_{y\in(0,1)} \|\partial_x \partial_x \hat{u}(\cdot,y)\|_{L^\infty((0,1))},
\end{aligned}$$

which is (2.6.25).

$\hat{K}$ is the reference triangle: The derivative $\partial_x I\hat{u}$ is a constant, which is computed in (3.2.49) as

$$\partial_x I\hat{u} = \int_0^1 \partial_x \hat{u}(x,0)\,dx.$$

From the mean value theorem, we conclude that there exists $\xi \in (0,1)$ such that $\partial_x I\hat{u} = \partial_x\hat{u}(\xi,0)$. For arbitrary $(x,y) \in \hat{K}$, the mean value theorem applied to the function $t \mapsto \partial_x\hat{u}(tx+(1-t)\xi, ty)$ then gives in view of the convexity of $\hat{K}$ the existence of $t^* \in (0,1)$ such that

$$|\partial_x\hat{u}(x,y) - \partial_x I\hat{u}| = |\partial_x\hat{u}(x,y) - \partial_x\hat{u}(\xi,0)| \leq \|\partial_x^2\hat{u}\|_{L^\infty(\hat{K})} + \|\partial_y\partial_x\hat{u}\|_{L^\infty(\hat{K})}.$$

Taking the supremum over all $(x,y) \in \hat{K}$ gives (2.6.25). □

We now come to the improvement of Theorem 2.6.15 when the element maps of interior elements are anisotropically $(C_M, h, h\kappa)$-normalizable:

Corollary 2.6.19. *Assume the hypotheses and notations of Theorem 2.6.15. Assume additionally that the mesh family $\mathcal{T}(\kappa,h,\mu)$ satisfies the following condition: each* interior element *is anisotropically $(C_{\mathcal{M}}, h, h\kappa)$-normalizable in the sense of Definition 2.6.17. Then there exist C, $\lambda_0 > 0$ independent of ε and $h \in (0,1/2)$ such that for every $\lambda \geq \lambda_0$ and $\kappa = \min\{1, \lambda\varepsilon|\ln h|\}$ there exists $v \in \mathcal{T}(\kappa,h,\mu)$ with $v = Iu_\varepsilon$ on $\partial\Omega$ and*

$$\begin{aligned}&\|u_\varepsilon - v\|_{L^2(\Omega)} + \kappa\|\nabla(u_\varepsilon - v)\|_{L^2(\Omega)}\\ &\quad\leq C\left[\varepsilon h^\delta|\lambda\ln h|^3 + \varepsilon^{1/2}h|\lambda\ln h|^{5/2} + h^2\right].\end{aligned}$$

Proof: We first note that the assumption that interior elements be anisotropically $(C_M, h, h\kappa)$-normalizable implies that they are (C_M, κ, h)-normalizable in the sense of Definition 2.6.3 since $\kappa \leq 1$. Thus, most of the arguments in the proof of Theorem 2.6.15 can be employed in the present situation as well; we will therefore merely highlight the main differences.
The case $\varepsilon\lambda|\ln h| > 1$: Inspection of the arguments in the proof of Theorem 2.6.15 leads to

$$\begin{aligned}\|u_\varepsilon - Iu_\varepsilon\|_{L^2(\Omega)} + \kappa\|\nabla(u_\varepsilon - Iu_\varepsilon)\|_{L^2(\Omega)} &\leq C\left[h^\delta\varepsilon^{-2} + h\varepsilon^{-1} + h^{1+\delta}\varepsilon^{-2} + h^2\varepsilon^{-2}\right]\\ &\leq C\varepsilon h^\delta\varepsilon^{-3} \leq C\varepsilon h^\delta|\lambda\ln h|^3.\end{aligned}$$

The case $\varepsilon\lambda|\ln h| \leq 1$: Arguing as in the proof of Theorem 2.6.15 allows us to bound for suitably chosen λ_0:

$$\|r_\varepsilon\|_{L^2(\Omega)} + \kappa\|\nabla r_\varepsilon\|_{L^2(\Omega)} \leq Ch^2.$$

Concerning the boundary layer elements, we see that (2.6.17) holds; therefore arguing as in the proof of Theorem 2.6.15, we obtain for the error e given by (2.6.16)

$$\left(\sum_{i\in I^{BL}} \|e\|^2_{L^2(K_i)} + \kappa^2\|\nabla e\|^2_{L^2(K_i)}\right)^{1/2} \leq C\sqrt{\kappa}h(\lambda \ln h)^2 \leq C\sqrt{\varepsilon}h|\lambda \ln h|^{5/2}.$$

For the corner layer elements, we check the proof of Theorem 2.6.15 to see that

$$\left(\sum_{i\in I^{CL}} \|e\|^2_{L^2(K_i)} + \kappa^2\|\nabla e\|^2_{L^2(K_i)}\right)^{1/2} \leq C\varepsilon h^\delta|\lambda \ln h|^{5/2}.$$

We finally turn to the interior elements K_i. Since the corresponding element map M_i is anisotropically $(C_M, h, h\kappa)$-normalizable, it can be factored as $M_i = G_i \circ S_i$, where G_i is a C^2-diffeomorphism and S_i is affine with

$$S_i' = \begin{pmatrix} h & 0 \\ 0 & h\kappa \end{pmatrix}.$$

We define $R_i := S_i(\hat{K}_i)$. If $\hat{K}_i$ is the unit square, then R_i is congruent to the rectangle with vertices $(0,0)$, $(h,0)$, $(0,h\kappa)$, $(h,h\kappa)$; if $\hat{K}_i$ is the reference triangle, then R_i is congruent to the triangle with vertices $(0,0)$, $(h,0)$, $(0,h\kappa)$. Since the function $w_i := w_\varepsilon \circ G_i$ satisfies $\|\nabla^2 w_i\|_{L^\infty(R_i)} \leq C$ for a constant $C > 0$ independent of ε and the element K_i, we conclude with Lemma 2.6.18 for the interpolation error

$$\|w_i - Iw_i\|_{L^\infty(R_i)} + h\|\nabla(w_i - Iw_i)\|_{L^\infty(R_i)} \leq Ch^2;$$

transforming this result to the element K_i gives, since G_i is a C^2-diffeomorphism

$$\|w_\varepsilon - Iw_\varepsilon\|_{L^\infty(K_i)} + h\|\nabla(w_\varepsilon - Iw_\varepsilon)\|_{L^\infty(K_i)} \leq Ch^2. \qquad (2.6.27)$$

Inspection of the arguments leading to (2.6.22) reveals that λ_0 can be chosen such that the factor h^2 can be replaced with h^3 (in fact, arbitrary powers of h can be obtained), i.e.,

$$\|u^C - Iu^C\|_{L^\infty(K_i)} + \kappa h\|\nabla(u^C - Iu^C)\|_{L^\infty(K_i)} \leq Ch^3.$$

We therefore obtain for the error e

$$\left(\sum_{i\in I^{int}} \|e\|^2_{L^2(K_i)} + \kappa^2\|\nabla e\|^2_{L^2(K_i)}\right)^{1/2} \leq C\left[h^2 + \kappa h\right] \leq C\left[h^2 + \varepsilon h|\lambda \ln h|\right],$$

which can be bounded in the desired fashion. □

3. *hp* Approximation

3.1 Motivation and outline

3.1.1 General overview of Chapter 3

The aim of the present chapter is Theorem 3.4.8, where an *hp*-approximant to solutions of (1.2.11) on the minimal meshes of Definition 2.4.4 is constructed. The *hp*-FEM approximation result Theorem 2.4.8 then follows from Theorem 3.4.8. The actual proof of the approximation result Theorem 3.4.8 is very technical; in Section 3.1.3, we therefore sketch the key steps of the proof.

The scope of the present chapter, however, goes beyond proving the approximation result Theorem 3.4.8 on admissible boundary layer meshes. This chapter addresses several issues pertinent to the design of *hp*-FEMs in general and to *hp*-FEMs for singularly perturbed problems in particular. Two lines of ideas are developed in this chapter in parallel:

1. The main line of ideas is concerned with the design of *minimal meshes* and the analysis of the *hp*-FEM for our model problem (1.2.11) on these meshes. It can be traced through the introduction of the polynomial projection operator Π_p^∞ in Theorem 3.2.20, the projector $\Pi_{p,\mathcal{T}}^\infty$ in (3.3.3), the notion of admissible boundary layer meshes in Definition 2.4.4, the analysis of the approximation properties of the operator $\Pi_{p,\mathcal{T}}^\infty$ on admissible boundary layer meshes in Section 3.4, and finally its application to the FEM in Section 2.4.3.
2. In the second line of ideas, the construction and analysis of anisotropic meshes that have more structure than the minimal meshes is explored. The development of these ideas can be seen in the definition of the projection operator $\Pi_p^{1,\infty}$ and the notion of regular admissible boundary layer meshes in Definition 3.3.10. This line of thought culminates in the introduction of *mesh patches* in Sections 3.3.2–3.3.4.

Let us discuss the first line of ideas in more detail. Central is the notion of admissible boundary layer meshes already introduced in Definition 2.4.4. These meshes are motivated by the description of the solution behavior in Section 2.3 (cf. Theorems 2.3.1, Theorem 2.3.4). In particular, in Theorem 2.3.4 we characterized the solution behavior in terms of asymptotic expansions. This decomposition suggests the main features of the *hp*-FEM to be used: thin needle elements near the boundary to capture the boundary layer behavior and geometric refinement near the vertices to resolve the corner layers. The admissible meshes of Definition 2.4.4 reflect these requirements by containing needle elements of width $O(\kappa)$ near the boundary and geometric mesh refinement with $L+1$ layers near the vertices; in the robust exponential approximation result Theorem 3.4.8 (and hence, also in the robust exponential convergence result Theorem 2.4.8), this parameter κ is then chosen as $\kappa = O(p\varepsilon)$, where p is the polynomial degree. Admissible meshes can be quite distorted, i.e., minimal angles can be very small (they may be of size $O(\kappa)$) and maximal angles may be very large (they may be $\pi - O(\kappa)$). Due to the distortion of the elements, the standard polynomial approximation results, which are essentially based on H^1-projectors on the reference elements, do not lead to robust bounds. We therefore base our approximation theory on an operator that is essentially an L^∞ projector. The main advantage of this approach is the invariance of the L^∞ norm under changes of variables, which allows us to avoid some of the difficulties associated with distorted meshes. Our technical tool is the operator Π_p^∞ defined on the reference element (see Theorem 3.2.20) and the corresponding operator $\Pi_{p,\mathcal{T}}^\infty$ defined on the triangulation $\mathcal{T}$ as the element-by-element application of Π_p^∞ (see (3.3.3)). The operators Π_p^∞ and $\Pi_{p,\mathcal{T}}^\infty$ are constructed so as to interpolate the given function in the Gauss-Lobatto points of the edges of the triangulations. This is done mostly for convenience's sake as this permits constructing H^1-conforming approximations in a truly element-by-element fashion (see Remark 3.3.9). Section 3.4 is then devoted to the analysis of the error $u_\varepsilon - \Pi_{p,\mathcal{T}}^\infty u_\varepsilon$ on admissible meshes. A direct consequence of this approximation result is the robust exponential convergence result Theorem 2.4.8 for the *hp*-FEM.

Let us now turn to a discussion of the second line of ideas in this chapter. The minimal meshes of Definition 2.4.4 may be very distorted and have little structure. From implementational considerations, minimal meshes have the following disadvantages:

1. The presence of distorted elements may increase the sensitivity of the FEM to quadrature error.
2. The presence of thin needle elements may affect the conditioning of the resulting stiffness matrix. If the mesh has some structure, one may selectively condense out degrees of freedom locally to improve the conditioning of the matrix and/or devise preconditioners that can handle these needle elements.

The ability to control mesh distortion is captured with the notion of *regular admissible meshes* in Definition 3.3.10. In essence, Definition 3.3.10 stipulates that needle elements be the images of reference needle rectangles (or triangles). Thus, maximal angles of elements cannot degenerate to π; we note that we have used a similar idea in the context of the *h*-FEM in Definition 2.6.3. A step further in the direction of structured meshes is taken with the notion of *mesh patches*. This idea is closely related to domain decomposition and substructuring. Meshes are created in two steps: In a first step, the computational domain is covered by a fixed coarse mesh, the "patches". In a second step, the final mesh is constructed by mapping reference configurations (see Section 3.3.3 for some reference configurations relevant for the resolution of boundary layer and corner layer phenomena) on the reference elements to physical space with the patch maps. The reference configurations can be chosen to reflect the solution behavior, for example, boundary layers and corner singularities. The main advantage of this approach is that practically only few typical configurations can occur and that the resulting mesh has considerable structure. These mesh patches represent natural divisions for parallel implementations and domain decomposition techniques.

The distinction between admissible and regular admissible meshes can be embedded in a larger context, namely, defining element size for curved anisotropic elements. We present two approaches to this issue with the notions of (C_M, γ_M)-*regular* triangulations in Definition 3.3.1 and (C_M, γ_M)-*normalizable* triangulations in Definition 3.3.3. In both approaches, the element size (isotropic or anisotropic) is encoded in an affine stretching map A_i. In an (C_M, γ_M)-regular triangulation, an affine maps A_i is associated with each element K_i such that the concatenations $A_i^{-1} \circ M_i$ can be controlled uniformly in i. (C_M, γ_M)-normalizable triangulations are more restrictive than (C_M, γ_M)-regular triangulations because uniform control of $M_i \circ A_i$ for appropriate affine stretching maps A_i is required. Conceptually, admissible meshes are regular meshes while regular admissible meshes and meshes generated with mesh patches are normalizable meshes.

Regular admissible meshes (and hence also meshes generated with mesh patches) are also admissible meshes (Proposition 3.3.11). Hence, the approximation theory developed for admissible meshes applies to these meshes as well. However, regular admissible meshes have more structure and therefore sharper polynomial approximation results could be obtained. In the context of the h-FEM in Section 2.6.3, such an additional structure allowed us to improve Theorem 2.6.15 to Corollary 2.6.19. While we do not develop a complete theory for polynomial approximation on normalizable meshes, we do provide two essential tools for doing so, namely, a) results in Section 3.3.5 that show how regularity results on the physical domain can be transferred to the reference configuration; and b) the operator $\Pi_p^{1,\infty}$ of Theorem 3.2.24 together with the approximation result Proposition 3.2.25 which show how the regularity results on the reference patch obtained with the tools of Section 3.3.5 can be used for obtaining sharp polynomial approximation on the reference patch.

3.1.2 Outline of Chapter 3

We begin this chapter with a sketch of the key steps of the proof of Theorem 3.4.8 in Section 3.1.3.

As mentioned at the beginning of this chapter, this chapter develops in parallel two lines of thought on polynomial approximation. We will outline their development separately.

We start with the line of ideas connected with minimal meshes. In Section 3.2, we provide polynomial approximation results on the reference element, i.e., on the reference square S and the reference triangle T. Our approximation results on minimal meshes are based on the projector Π_p^∞ defined in Section 3.2.4. The

essential result about the projector Π_p^∞ is formulated in Proposition 3.2.21. After these approximation results on the reference square and reference triangle, Section 3.4 is then devoted to the proof of polynomial approximation results on minimal meshes, culminating in Theorem 3.4.8, where robust exponential approximability of solutions to (1.2.11) is shown on minimal meshes. For a rigorous proof of Theorem 3.4.8, both forms of regularity results for the solution u_ε of (1.2.11) are required: In the "pre-asymptotic" range, in which the polynomial degree p is small compared with ε^{-1}, we employ the asymptotic expansions of Theorem 2.3.4. However, as can be seen in Theorem 2.3.4, such an asymptotic expansions can describe the solution u_ε only up to a certain error level, since the remainder r_ε is not arbitrarily small. Hence, in the "asymptotic" range in which the polynomial degree p is large compared with ε^{-1}, we resort to the $\mathcal{B}^2_{\beta,\varepsilon}$ regularity results of Theorem 2.3.1.
The second line of ideas can be traced through the following sections. We introduce in Section 3.2.5 the projector $\Pi_p^{1,\infty}$ that is suitable for approximation on anisotropic meshes and normalizable triangulations in particular. We recall that regular admissible meshes and meshes generated via mesh patches fall in this category. The essential approximation properties of $\Pi_p^{1,\infty}$ are collected in Proposition 3.2.25. The approximation of analytic functions based on the projector $\Pi_p^{1,\infty}$ relies on polynomial approximation results obtained in Section 3.2.2. The main ideas concerning mesh patches are developed in Section 3.3.2. Patches that are required for the resolution of boundary and corner layers are collected in Section 3.3.3, and a formal definition of meshes generated by mesh patches is given in Section 3.3.4. Of more general nature is Section 3.3.5, in which results are obtained concerning the regularity of functions pulled back to the reference configuration with the patch map.

3.1.3 Robust exponential convergence: key ingredients of proof

In this subsection, we briefly present the key ingredients of the proof of the robust exponential convergence result, Theorems 2.4.8 and 3.4.8. Structurally, the proof is similar to that of Proposition 2.2.5 in the one-dimensional situation in that the distinction $\lambda p\varepsilon \geq 1$ and $\lambda p\varepsilon < 1$ is made. For the "asymptotic case" $\lambda p\varepsilon \geq 1$ the regularity assertions of Theorem 2.3.1 are employed; for the "pre-asymptotic case" $\lambda p\varepsilon < 1$ the decomposition result of Theorem 2.3.4 is used. We start with the preasymptotic case $\lambda p\varepsilon < 1$.

hp-approximation in the pre-asymptotic regime $\lambda p\varepsilon < 1$.
For piecewise polynomial approximation by polynomials of degree p, the meshes $\mathcal{T}(\min\{\lambda p\varepsilon, 1\}, L)$ have two characteristic features:

1. thin needle elements of width $O(p\varepsilon)$ are employed in the boundary layer to capture the solution's boundary layer components;
2. a geometric mesh refinement in an $O(p\varepsilon)$ neighborhood of the vertices is used for corner singularity resolution.

The aim of the present section is to illustrate the main mechanisms at work in the piecewise polynomial approximation of solutions to (1.2.11) using meshes that contain thin needle elements of width $O(p\varepsilon)$ in the boundary layers and geometric refinement in an $O(p\varepsilon)$ neighborhood of the vertices as stated above. In Theorem 2.3.4 we decomposed u_ε into a smooth part w_ε, a boundary layer part u_ε^{BL}, a corner layer part u_ε^{CL}, and a (small) remainder r_ε. The finite element mesh on which piecewise polynomial approximation of u_ε is performed has to be designed such that each of these components can be approximated well. The smooth part w_ε is easily approximated by (piecewise) polynomials. The remainder r_ε, being exponentially small in ε, may be neglected. Approximability of the boundary and corner layer contributions therefore dictate the finite element mesh design. In the following, we illustrate for simple model situations the mesh design principles that allow for robust exponential approximability of these two solution components.

Boundary layer approximation. We start with the boundary layer approximation. We consider the domain $\Omega := (0,1)^2 = \{(\rho,\theta)\,|\,0<\rho<1, \quad 0<\theta<1\}$. We say that the function $u = u(\rho,\theta)$ is of boundary layer type with length scale $\varepsilon \in (0,1]$ if there are constants α, C, $\gamma > 0$ such that

$$|\partial_\rho^s \partial_\theta^t u(\rho,\theta)| \le C_u \gamma_u^{s+t} t! \max\{s+1,\varepsilon^{-1}\}^s e^{-\alpha\rho/\varepsilon} \qquad \forall (s,t) \in \mathbb{N}_0^2. \qquad (3.1.1)$$

We easily recognize this to be a generalization of the regularity assertions about u_ε^{BL} in Theorem 2.3.4. We now wish to approximate such functions of boundary layer type on Ω by piecewise polynomials. This can be done very efficiently on a two-element mesh, i.e., a mesh containing a long thin needle element in the layer and one large element away from the layer. Specifically, for $\kappa > 0$ we define the "two-element" mesh $\mathcal{T}_\kappa$ as (cf. Fig. 3.1.1)

$$\begin{aligned} &\mathcal{T}_\kappa := \{(K_1, M_1, \hat{K}), (K_2, M_2, \hat{K})\}, && \text{where} \\ &K_1 = (0, \min\{\kappa, 0.5\}) \times (0,1), && K_2 = (\min\{\kappa, 0.5\}, 1) \times (0,1)\}, \end{aligned}$$

and the element maps M_i are simply affine maps mapping $\hat{K} = (0,1)^2$ onto K_i. We now show that the choice $\kappa = O(p\varepsilon)$ allows for robust exponential

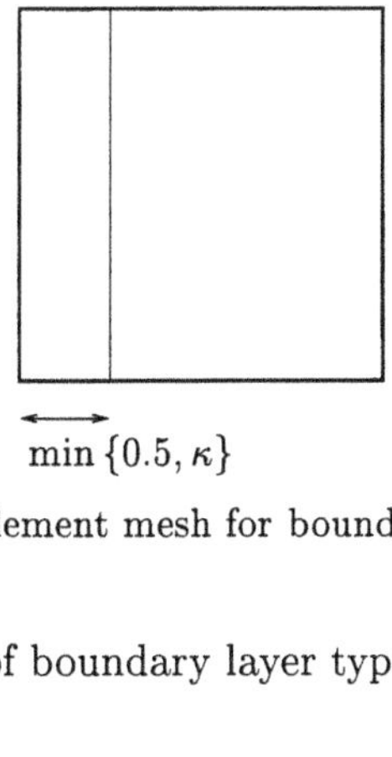

Fig. 3.1.1. Two-element mesh for boundary layer resolution.

approximability of function of boundary layer type from the space $S^p(\mathcal{T}_\kappa)$:

Lemma 3.1.1. *Let u satisfy (3.1.1) on $\Omega = (0,1)^2$. Then there are C, $b > 0$ and $\lambda_0 > 0$ depending only on γ_u and α such that for all $\lambda \in (0, \lambda_0)$ and all $p \in \mathbb{N}$*

$$\inf_{\pi_p \in S^p(\mathcal{T}_{\lambda p \varepsilon})} \lambda p \varepsilon \|\nabla(u - \pi_p)\|_{L^\infty(\Omega)} + \|u - \pi_p\|_{L^\infty(\Omega)} \leq C(1 + \ln p)^2(1 + \varepsilon p^2)e^{-b\lambda p}.$$

Proof: A slightly sharper version of this lemma is proved in [55]; a similar result is used implicitly in [94]. Lemma 3.1.1 is not formulated in its sharpest form as its aim is to expose the basic mechanics of *hp*-approximation of functions of boundary layer type.

We construct the approximant π_p explicitly for each element K_1, K_2.

1. Step: Approximation on K_1. Let us consider the needle element K_1. For $\kappa = \lambda p \varepsilon$, the element map M_1 may be assumed to be of the form $M_1(x, y) = (\min\{\lambda p \varepsilon, 0.5\}x, y)$. Thus, the pull-back $\hat{u}$ of $u|_{K_1}$ to the reference element $\hat{K}$, i.e., $\hat{u} := u \circ M_1$, satisfies on $\hat{K}$

$$\|\partial_x^s \partial_y^t \hat{u}(x,y)\|_{L^\infty(\hat{K})} \leq C_u \min\{0.5, \lambda p \varepsilon\}^s \gamma_u^{s+t} \max\{s+1, \varepsilon^{-1}\}^s t! \quad \forall (s,t) \in \mathbb{N}_0^2.$$

We now claim that there are C', $\gamma > 0$ depending only on γ_u such that

$$\|\partial_x^s \partial_y^t \hat{u}(x,y)\|_{L^\infty(\hat{K})} \leq C' C_u e^{2\lambda p} \gamma^{s+t} s! t! \qquad \forall (s,t) \in \mathbb{N}_0^2. \tag{3.1.2}$$

In order to see this, we consider the case $\min\{0.5, \lambda p \varepsilon\} = \lambda p \varepsilon$ and the converse case $\min\{0.5, \lambda p \varepsilon\} = 0.5$ separately. Consider first the case $\min\{0.5, \lambda p \varepsilon\} = \lambda p \varepsilon$.

$$\begin{aligned} \min\{0.5, \lambda p \varepsilon\}^s \max\{s+1, \varepsilon^{-1}\}^s &= (\lambda p \varepsilon)^s \max\{s+1, \varepsilon^{-1}\}^s \\ &= \max\{(\lambda p \varepsilon)^s (s+1)^s, (\lambda p)^s\} \leq \max\{0.5^s (s+1)^s, \frac{(\lambda p)^s}{s!} s!\} \\ &\leq \max\{0.5^s (s+1)^s, e^{\lambda p} s!\} \leq C \gamma^s e^{\lambda p} s! \end{aligned} \tag{3.1.3}$$

for some appropriate C, $\gamma > 0$ independent of λ, p, and ε. Let us now consider the second case, $\min\{0.5, \lambda p \varepsilon\} = 0.5$. Then we bound by similar reasoning

$$\begin{aligned} &\min\{0.5, \lambda p \varepsilon\}^s \max\{s+1, \varepsilon^{-1}\}^s = 0.5^s \max\{s+1, \varepsilon^{-1}\}^s \\ &\leq 0.5^s \max\{(s+1)^s, (2\lambda p)^s\} \leq 0.5^s \max\{(s+1)^s, \frac{(2\lambda p)^s}{s!} s!\} \\ &\leq C \gamma^s e^{2\lambda p} s!. \end{aligned} \tag{3.1.4}$$

Combining (3.1.3), (3.1.4) then gives (3.1.2). We now apply polynomial approximation results on $\hat{K}$, which are proved in detail in Section 3.2.2. For example, the two-dimensional Gauss-Lobatto interpolation operator j_p (cf. Section 3.2.1) applied to the function $\hat{u}$ yields the existence of C, $b > 0$ depending only on γ and C' of (3.1.2) such that

$$\|\hat{u} - j_p \hat{u}\|_{L^\infty(\hat{K})} + \|\nabla(\hat{u} - j_p \hat{u})\|_{L^\infty(\hat{K})} \leq C C_u e^{2\lambda p} e^{-bp}.$$

(A rigorous proof of this approximation result is obtained by combining Theorem 3.2.19 with Theorem 3.2.20 and noting that the operator Π_p^∞ of Theorem 3.2.20 coincides with j_p in the present case of approximation on the reference square.) Observe the presence of the factor $e^{2\lambda p}$. This factor is already present in the bounds on $\hat{u}$ and reappears in this bound due to the linearity of the Gauss-Lobatto interpolation operator j_p. As the constant b is independent of λ, p, we choose $\lambda_0 := b/4$ to get

$$\|\hat{u} - j_p\hat{u}\|_{L^\infty(\hat{K})} + \|\nabla(\hat{u} - j_p\hat{u})\|_{L^\infty(\hat{K})} \le CC_u e^{-(b/2)p}.$$

Mapping back to K_1 gives the desired bound on K_1.
2. Step: Approximation on K_2. We note that $\operatorname{dist}(K_2, \{\rho = 0\}) \ge \kappa = \min\{0.5, \lambda p\varepsilon\}$. The assumptions on u therefore imply the existence of C, $b > 0$ (depending only on γ and α) such that

$$\|u\|_{L^\infty(K_2)} \le CC_u e^{-b\lambda p}, \qquad \|\nabla u\|_{L^\infty(K_2)} \le CC_u\varepsilon^{-1}e^{-b\lambda p}. \tag{3.1.5}$$

Thus, u is already exponentially small (in p) and fairly crude polynomial approximations suffice. For definiteness' sake, we will use again the Gauss-Lobatto interpolant $j_p u$. We start by noting that the element map M_2 satisfies

$$\|M_i'\|_{L^\infty(\hat{K})} \le 1, \qquad \|\left(M_i'\right)^{-1}\|_{L^\infty(\hat{K})} \le 2.$$

From this, it is easy to see that the pull-back $\hat{u} := u|_{K_2} \circ M_2$ satisfies bounds analogous to (3.1.5):

$$\|\hat{u}\|_{L^\infty(\hat{K})} \le CC_u e^{-b\lambda p}, \qquad \|\nabla\hat{u}\|_{L^\infty(\hat{K})} \le CC_u\varepsilon^{-1}e^{-b\lambda p},$$

where, in fact, only the constant C may have changed by a factor 2. From basic properties of the Gauss-Lobatto interpolation operator (Lemma 3.2.1) and inverse estimates for polynomials (Lemma 3.2.2), we have

$$\begin{aligned}
\|\hat{u} - j_p\hat{u}\|_{L^\infty(\hat{K})} &\le \|\hat{u}\|_{L^\infty(\hat{K})} + \|j_p\hat{u}\|_{L^\infty(\hat{K})} \le C(1+\ln p)^2\|\hat{u}\|_{L^\infty(\hat{K})},\\
\|\nabla(\hat{u} - j_p\hat{u})\|_{L^\infty(\hat{K})} &\le \|\nabla\hat{u}\|_{L^\infty(\hat{K})} + \|\nabla j_p\hat{u}\|_{L^\infty(\hat{K})},\\
&\le \|\nabla\hat{u}\|_{L^\infty(\hat{K})} + Cp^2\|j_p\hat{u}\|_{L^\infty(\hat{K})}\\
&\le \|\nabla\hat{u}\|_{L^\infty(\hat{K})} + Cp^2(1+\ln p)^2\|\hat{u}\|_{L^\infty(\hat{K})}.
\end{aligned}$$

These estimates imply after an adjustment of the constant b:

$$\begin{aligned}
\|\hat{u} - j_p\hat{u}\|_{L^\infty(\hat{K})} &\le C(1+\ln p)^2 e^{-b\lambda p},\\
\lambda p\varepsilon\|\nabla(\hat{u} - j_p\hat{u})\|_{L^\infty(\hat{K})} &\le C\left[1 + \varepsilon p^2(1+\ln p)^2\right]e^{-b\lambda p}.
\end{aligned}$$

Mapping this last estimate back to K_2 gives the desired bound. □

A few comments concerning the proof of Lemma 3.1.1 are in order. First, the algebraic powers of p in front of the term $e^{-b\lambda p}$ are suboptimal. They were accepted in the proof of Lemma 3.1.1 in order to be able to concentrate on the essential mechanisms in the proof:

1. On the element K_2 that is $O(p\varepsilon)$ away from the line $\rho = 0$, the function u to be approximated is exponentially small (in p) and thus crude approximations (e.g., by the zero-function) are sufficient.
2. The main mechanism for the approximation on the element K_1 is better seen by writing the proof in a slightly different way as follows. Introducing the anisotropic stretching map

$$s_\kappa : (\rho, \theta) \to (\kappa\rho, \theta),$$

we can write the pull-back $\hat{u} := u \circ M_1$ to the reference element as $\hat{u} = (u \circ s_\kappa) \circ (s_\kappa^{-1} \circ M_1)$ where $\kappa = \min\{0.5, \lambda p\varepsilon\}$. Next, we define the auxiliary set $\widetilde{K} := s_\kappa^{-1}(K_1)$. Paralleling the arguments in the proof of Lemma 3.1.1, we obtain for $u \circ s_\kappa$ on $\widetilde{K}$:

$$\|\nabla^n (u \circ s_\kappa)\|_{L^\infty(\tilde{K})} \leq C e^{\lambda p} \gamma^n n! \qquad \forall n \in \mathbb{N}_0, \tag{3.1.6}$$

where C, γ are independent of ε and p. In order to get bounds on the derivatives of $\hat{u}$, we require control of the map $s_\kappa^{-1} \circ M_1$. We observe that there exists $C > 0$ independent of ε such that

$$\|\nabla^n \left(s_\kappa^{-1} \circ M_1\right)\|_{L^\infty(\hat{K})} \leq C, \qquad n \in \{0, 1\}, \tag{3.1.7a}$$

$$\|\left((s_\kappa^{-1} \circ M_1)'\right)^{-1}\|_{L^\infty(\hat{K})} \leq C. \tag{3.1.7b}$$

As $s_\kappa^{-1} \circ M_1$ is affine, (3.1.7a) in fact holds trivially for all $n \in \mathbb{N}_0$. As we will see later on, it suffices to stipulate that the map $s_\kappa^{-1} \circ M_1$ be an analytic diffeomorphism whose constants of analyticity C, γ can be bounded independently of ε:

$$\|\nabla^n \left(s_\kappa^{-1} \circ M_1\right)\|_{L^\infty(\hat{K})} \leq C \gamma^n n! \qquad \forall n \in \mathbb{N}_0, \tag{3.1.8a}$$

$$\|\left((s_\kappa^{-1} \circ M_1)'\right)^{-1}\|_{L^\infty(\hat{K})} \leq C. \tag{3.1.8b}$$

Combining (3.1.6) and (3.1.7) gives (3.1.2). Key to the present approach is the existence of anisotropic stretching maps s_κ such that both $u \circ s_\kappa$ and $s_\kappa^{-1} \circ M_1$ can can be controlled uniformly in ε as ascertained in (3.1.6), (3.1.8).

Corner layer approximation. We now turn to the presentation of the main points for the approximation of corner layers. For the purpose of this discussion, we will say that a function $u(x, y)$ defined on $\Omega = (0,1)^2$ is of corner layer type, if there are C_u, γ_u, $\alpha > 0$, $\beta \in [0, 1)$ such that

$$\|e^{\alpha|x|/\varepsilon} u\|_{L^2(\Omega)} + \varepsilon \|e^{\alpha|x|/\varepsilon} \nabla u\|_{L^2(\Omega)} \leq \varepsilon C_u \tag{3.1.9a}$$

$$\varepsilon^2 \left\| e^{\alpha|x|/\varepsilon} \min\left\{1, \left(\frac{|x|}{\varepsilon}\right)^\beta\right\} \nabla^2 u \right\|_{L^2(\Omega)} \leq \varepsilon C_u, \tag{3.1.9b}$$

$$|\nabla^n u(x)| \leq C_u \gamma_u^n |x|^{-n} \left(\frac{|x|}{\varepsilon}\right)^{1-\beta} e^{-\alpha|x|/\varepsilon} \qquad \forall n \in \mathbb{N}_0. \tag{3.1.9c}$$

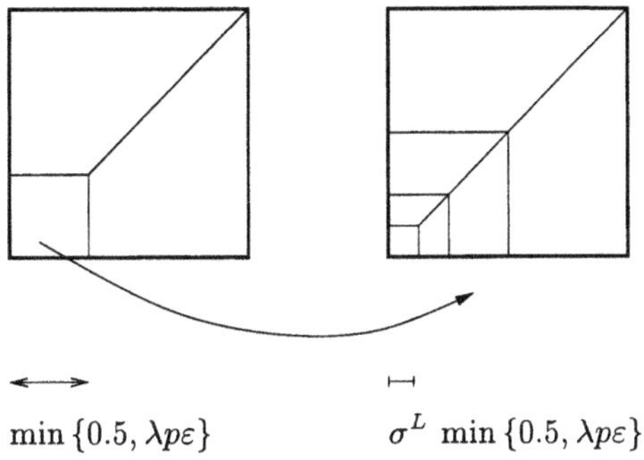

Fig. 3.1.2. Geometrically refined mesh with $L+1$ (here: $L=3$) layers in $O(p\varepsilon)$ neighborhood of origin for corner layer resolution.

We recognize these bounds to be typical of the corner layer functions u_ε^{CL} in Theorem 2.3.4. We now discuss meshes $\mathcal{T}$ on Ω that are appropriate for the approximation of such functions of corner layer type. We note that (3.1.9c) shows that u is exponentially small (in p) outside an $O(p\varepsilon)$ neighborhood of the origin $x=0$. Thus, just as in the proof of Lemma 3.1.1, fairly crude polynomial approximations suffice outside this neighborhood. It remains to consider the approximation in this $O(p\varepsilon)$ neighborhood of the origin. There, we propose the use of a geometrically refined mesh to resolve the singularity as is standard in *hp*-FEM. For illustration purposes, we consider the approximation on meshes $\mathcal{T}=\{(K_i, M_i, \hat{K})\}$ as depicted in Fig. 3.1.2. Note that the corresponding element maps M_i are *bilinear* functions, i.e.,

$$\nabla^3 M_i = 0 \qquad \forall i. \tag{3.1.10}$$

It is also convenient to introduce the element size h_i as

$$h_i = \operatorname{diam} K_i.$$

The essential features of the meshes $\mathcal{T}$ as depicted in Fig. 3.1.2 are:

1. Outside an $O(p\varepsilon)$ neighborhood of the origin, few elements are employed (here: 2).
2. In an $O(p\varepsilon)$ neighborhood of the origin, a *geometrically* refined mesh is employed with *grading factor* $\sigma \in (0,1)$ and $L+1 \in \mathbb{N}$ layers of geometric refinement. This means:
   a) For all elements K_i with $\operatorname{dist}(K_i, 0) > 0$ there holds the two-sided bound $C^{-1}h_i \le \operatorname{dist}(K_i, 0) \le Ch_i$.
   b) If $\operatorname{dist}(K_i, 0) = 0$, then $h_i \le C\lambda p\varepsilon\sigma^L$.

This list decomposes the elements into three categories; correspondingly, we can decompose the set of indices of the elements of such an mesh $\mathcal{T}$ into three sets:

$$I_{org} := \{i \,|\, 0 \in \overline{K_i}\}, \tag{3.1.11a}$$
$$I_{geom} := \{i \,|\, C^{-1}h_i \le \operatorname{dist}(K_i, 0) \le Ch_i\}, \tag{3.1.11b}$$
$$I_{out} := \{i \,|\, i \notin I_{org} \text{ and } i \notin I_{geom}\}. \tag{3.1.11c}$$

We note self-similarity properties of the elements in the geometric refinement in Fig. 3.1.2 (i.e., the elements K_i with $i \in I_{geom}$), which entail certain uniformity properties of the element maps. In order to describe those, it is convenient to introduce for $\kappa > 0$ the following affine stretching maps:

$$s_\kappa(x) := \kappa x \qquad \forall x \in \mathbb{R}^2.$$

By self-similarity of most elements, it is not difficult to see that the maps $s_{h_i}^{-1} \circ M_i$ given by

$$\begin{aligned} \hat{K} &\to \widetilde{K}_i := s_{h_i}^{-1}(K_i) \\ x &\mapsto \left(s_{h_i}^{-1} \circ M_i\right)(x) \end{aligned}$$

are analytic diffeomorphisms between $\hat{K}$ and the stretched elements $\widetilde{K}_i$ and satisfy in fact for some $C, \gamma > 0$

$$\|\nabla^n (s_{h_i}^{-1} \circ M_i)\|_{L^\infty(\hat{K})} \leq C\gamma^n n! \qquad \forall n \in \mathbb{N}_0, \tag{3.1.12a}$$

$$\| \left((s_{h_i}^{-1} \circ M_i)'\right)^{-1} \|_{L^\infty(\hat{K})} \leq C. \tag{3.1.12b}$$

Note that (3.1.10) in fact implies that $\nabla^n(s_{h_i}^{-1} \circ M_i) = 0$ for $n \geq 3$ so that (3.1.12) is not difficult to check.

An additional important property of the elements of the geometric refinement is that for each $\delta > 0$, there exists $C_{\delta,\sigma}$ (depending on δ and the grading factor σ) such that

$$\sum_{i \in I_{geom}} h_i^{2\delta} \leq C_{\delta,\sigma}(\lambda p \varepsilon)^{2\delta}. \tag{3.1.13}$$

This estimate follows easily from the fact that the elements of the geometric mesh are self-similar and that hence the element size decays in a geometric progression as the elements approach the origin; a formal and more general proof can be found in Lemma 3.4.6.

After these preparatory considerations, we formulate the following approximation result for functions of corner layer type:

Lemma 3.1.2. *Let u satisfy (3.1.9) and let $\mathcal{T}$ be the mesh with $L+1$ layers of geometric refinement and grading factor $\sigma \in (0,1)$ in an $O(p\varepsilon)$ neighborhood as in Fig. 3.1.2. Then there exist C, $b > 0$ independent of ε and p such that*

$$\begin{aligned} &\inf_{\pi_p \in S^p(\mathcal{T})} \|u - \pi_p\|_{L^2(\Omega)} + \varepsilon \|\nabla(u - \pi_p)\|_{L^2(\Omega)} \\ &\qquad \leq C(1 + \ln p)^2 \left[(1 + \varepsilon p^2)e^{-b\lambda p} + \varepsilon p^5 \sigma^{(1-\beta)L}\right]. \end{aligned}$$

Proof: As in the proof of Lemma 3.1.1, the approximant π_p is constructed as the piecewise Gauss-Lobatto interpolant. The main advantage of this choice is that it automatically takes care of the correct interelement continuity requirement and yields elements of $S^p(\mathcal{T})$ (cf. Lemma 3.3.7 for the details).

We distinguish the three kinds of elements corresponding to the three sets I_{out}, I_{geom}, I_{org} of (3.1.11): The two elements K_i outside the $O(p\varepsilon)$ neighborhood of the origin where u is exponentially small, the elements K_i in the $O(p\varepsilon)$ neighborhood with $\operatorname{dist}(K_i,0) > 0$, and finally the element with $\operatorname{dist}(K_i,0) = 0$. The proof is therefore divided into three steps.
1. Step: Elements outside $O(p\varepsilon)$ neighborhood of origin. From Fig. 3.1.2, these are the elements K_i with $\operatorname{dist}(K_i,0) \geq \min\{0.5, \lambda p\varepsilon\} \geq \lambda p\varepsilon$. From (3.1.9c) and the fact that we can bound for suitable $C_{\alpha,\beta} > 0$

$$x^{1-\beta}e^{-\alpha x} \leq C_{\alpha,\beta}e^{-(\alpha/2)x} \qquad \forall x > 0,$$

we obtain for appropriate C, $b > 0$ independent of ε, p, and λ:

$$\|\nabla^n u\|_{L^\infty(K_i)} \leq C(\lambda p\varepsilon)^{-n}e^{-b\lambda p}, \qquad n \in \{0,1\}.$$

We observe that these bounds are the same as those on boundary layer functions on the element K_2 in the proof of Lemma 3.1.1. We may therefore conclude in the same manner that the approximation $\pi_p|_{K_i}$ defined by $\pi_p|_{K_i} = (j_p(u \circ M_i)) \circ M_i^{-1}$ satisfies

$$\|u - \pi_p\|_{L^\infty(K_i)} + \lambda p\varepsilon\|\nabla(u-\pi_p)\|_{L^\infty(K_i)} \leq C(1+\ln p)^2(1+\varepsilon p^2)e^{-b\lambda p}$$

for some appropriate C, $b > 0$ independent of ε, p, and λ.
2. Step: Elements in $O(p\varepsilon)$ neighborhood of origin with $\operatorname{dist}(K_i,0) > 0$. We consider polynomial approximation of the pull-back $\hat{u} := u|_{K_i} \circ M_i$ on the reference element $\hat{K}$. It is not difficult to see from (3.1.9c) and the property $C^{-1}h_i \leq \operatorname{dist}(K_i,0) \leq Ch_i$ that on the stretched element $\widetilde{K}_i := s_{h_i}^{-1}(K_i)$ the function $u \circ s_{h_i}$ satisfies for some appropriate C, $\gamma > 0$:

$$\|\nabla^n(u \circ s_{h_i})\|_{L^\infty(\widetilde{K}_i)} \leq Ch_i^n\|\nabla^n u\|_{L^\infty(K_i)} \leq C\left(\frac{h_i}{\varepsilon}\right)^{1-\beta}\gamma^n n! \qquad \forall n \in \mathbb{N}_0.$$

Writing $\hat{u} = (u \circ s_{h_i}) \circ (s_{h_i}^{-1} \circ M_i)$ and using (3.1.12), we infer from Lemma 4.3.4 that there are C, $\gamma > 0$ independent of ε and i such that

$$\|\nabla^n\hat{u}\|_{L^\infty(\hat{K})} \leq C\left(\frac{h_i}{\varepsilon}\right)^{1-\beta}\gamma^n n! \qquad \forall n \in \mathbb{N}_0.$$

From the approximation results of Section 3.2.2, we can then infer that the Gauss-Lobatto interpolant $j_p\hat{u}$ satisfies for some C, $b > 0$:

$$\|\hat{u} - j_p\hat{u}\|_{L^\infty(\hat{K})} + \|\nabla(\hat{u} - j_p\hat{u})\|_{L^\infty(\hat{K})} \leq C\left(\frac{h_i}{\varepsilon}\right)^{1-\beta}e^{-bp}, \qquad p = 1,2,\ldots.$$

Mapping this approximation result back to the physical element K_i, we obtain by setting $\pi_p|_{K_i} = (j_p\hat{u}) \circ M_i^{-1}$:

$$\|u - \pi_p\|_{L^\infty(K_i)} + h_i\|\nabla(u-\pi_p)\|_{L^\infty(K_i)} \leq C\left(\frac{h_i}{\varepsilon}\right)^{1-\beta}e^{-bp}, \qquad p = 1,2,\ldots.$$

Integrating over K_i and summing on $i \in I_{geom}$ gives

$$\sum_{i \in I_{geom}} \|u - \pi_p\|^2_{L^2(K_i)} \le C e^{-2bp} \varepsilon^{-2(1-\beta)} \sum_{i \in I_{geom}} h_i^{2+2(1-\beta)},$$

$$\sum_{i \in I_{geom}} \|\nabla(u - \pi_p)\|^2_{L^2(K_i)} \le C e^{-2bp} \varepsilon^{-2(1-\beta)} \sum_{i \in I_{geom}} h_i^{2(1-\beta)}.$$

The bound (3.1.13) gives

$$\sum_{i \in I_{geom}} \|u - \pi_p\|^2_{L^2(K_i)} \le C e^{-2bp} \varepsilon^{-2(1-\beta)} (\lambda p \varepsilon)^{2+2(1-\beta)},$$

$$\sum_{i \in I_{geom}} \|\nabla(u - \pi_p)\|^2_{L^2(K_i)} \le C e^{-2bp} \varepsilon^{-2(1-\beta)} (\lambda p \varepsilon)^{2(1-\beta)},$$

which in turn implies

$$\sum_{i \in I_{geom}} \|u - \pi_p\|^2_{L^2(K_i)} + (\lambda p \varepsilon)^2 \|\nabla(u - \pi_p)\|^2_{L^2(K_i)} \le C e^{-2bp} \varepsilon^2 (\lambda p)^{2+2(1-\beta)}.$$

Absorbing the λ-dependence into the constant C gives an expression that can be bounded in the desired fashion.

3. Step: The element K_i abutting on the origin. Polynomial approximation on the element abutting on the origin just exploits that the element size is exponentially small in the number L of layers of geometric refinement, i.e., $h_i \le C \lambda p \varepsilon \sigma^L$. Fairly crude polynomial approximation therefore suffices. Again, polynomial approximation is done on the reference element $\hat{K}$ and we consider $\hat{u} = u|_{K_i} \circ M_i$. We exploit the following fact about the linear interpolant $j_1 \hat{u}$ (see Lemma 3.2.4)

$$\|\hat{u} - j_1 \hat{u}\|_{L^\infty(\hat{K})} + \|\nabla(\hat{u} - j_1 \hat{u})\|_{L^2(\hat{K})} \le C \| \, |x|^\beta \nabla^2 \hat{u}\|_{L^2(\hat{K})}. \tag{3.1.14}$$

In order to estimate $\| \, |x|^\beta \nabla^2 u\|_{L^2(\hat{K})}$, let us assume for notational convenience that $M_i(0) = 0$. Observing that for our mesh of Fig. 3.1.2 the element map M_i is affine and has the form $M_i(x) = h_i x$, scaling yields

$$\begin{aligned} \| \, |x|^\beta \nabla^2 \hat{u}\|_{L^2(\hat{K})} &= h_i^{1-\beta} \| \, |x|^\beta \nabla^2 u\|_{L^2(K_i)} \\ &= h_i^{1-\beta} \varepsilon^\beta \| \, (|x|/\varepsilon)^\beta \nabla^2 u\|_{L^2(K_i)}. \end{aligned} \tag{3.1.15}$$

To continue the estimates, we use $|x| \le h_i$ to get

$$\frac{|x|}{\varepsilon} = \min\left\{1, \frac{|x|}{\varepsilon}\right\} \max\left\{1, \frac{|x|}{\varepsilon}\right\} \le \min\left\{1, \frac{|x|}{\varepsilon}\right\} \left(1 + \frac{h_i}{\varepsilon}\right) \tag{3.1.16}$$

and then employ (3.1.9b) to arrive at

$$\begin{aligned} h_i^{1-\beta} \varepsilon^\beta \| \, (|x|/\varepsilon)^\beta \nabla^2 u\|_{L^2(K_i)} &\le h_i^{1-\beta} \varepsilon^\beta \left(1 + \frac{h_i}{\varepsilon}\right) \left\| \min\left\{1, (|x|/\varepsilon)^\beta\right\} \nabla^2 u\right\|_{L^2(K_i)} \\ &\le C h_i^{1-\beta} \varepsilon^{\beta-1} \left(1 + \frac{h_i}{\varepsilon}\right) \le C \left(\frac{h_i}{\varepsilon}\right)^{1-\beta} \left(1 + \frac{h_i}{\varepsilon}\right). \end{aligned}$$

As we want to approximate $\hat{u}$ by $j_p\hat{u}$ rather than $j_1\hat{u}$, we use stability properties of the Gauss-Lobatto interpolation operator j_p (Lemma 3.2.1), inverse estimates (Lemma 3.2.2), and the fact that $j_p(j_1\hat{u}) = j_1\hat{u}$ to arrive at

$$\begin{aligned}
\|\hat{u} - j_p\hat{u}\|_{H^1(\hat{K})} &= \|\hat{u} - j_1\hat{u} - j_p\left(\hat{u} - j_1\hat{u}\right)\|_{H^1(\hat{K})} \\
&\le \|\hat{u} - j_1\hat{u}\|_{H^1(\hat{K})} + Cp^2\|j_p(\hat{u} - j_1\hat{u})\|_{L^2(\hat{K})} \\
&\le \|\hat{u} - j_1\hat{u}\|_{H^1(\hat{K})} + Cp^2(1+\ln p)^2\|\hat{u} - j_1\hat{u}\|_{L^\infty(\hat{K})} \qquad (3.1.17)\\
&\le Cp^2(1+\ln p)^2\left(\frac{h_i}{\varepsilon}\right)^{1-\beta}\left(1+\frac{h_i}{\varepsilon}\right).
\end{aligned}$$

Exploiting now the assumption $h_i \le C\lambda p\varepsilon\sigma^L$, we finally get

$$\|\hat{u} - j_p\hat{u}\|_{H^1(\hat{K})} \le Cp^4(1+\ln p)^2\sigma^{(1-\beta)L}.$$

Upon setting $\pi_p|_{K_i} = (j_p\hat{u}) \circ M_i^{-1}$ we obtain by mapping back to the physical element K_i

$$\begin{aligned}
\|u - \pi_p\|_{L^2(K_i)} &\le Ch_ip^4(1+\ln p)^2\sigma^{(1-\beta)L} \le C\varepsilon p^5(1+\ln p)^2\sigma^{(1-\beta)L},\\
\lambda p\varepsilon\|\nabla(u - \pi_p)\|_{L^2(K_i)} &\le C\varepsilon p^5(1+\ln p)^2\sigma^{(1-\beta)L}.
\end{aligned}$$

□

***hp*-approximation in the asymptotic regime $\lambda p\varepsilon \ge 1$.**
The key key point is that the assumption $\varepsilon^{-1} \le \lambda p$ allows us to replace negative powers of ε with powers of the polynomial degree p. We illustrate the main mechanisms with the following lemma, where we consider the approximation of a function u with the regularity properties given by Theorem 2.3.1 on geometrically refined meshes:

Lemma 3.1.3. *Let $\Omega = (0,1)^2$ and let $\mathcal{T}$ be a mesh on Ω with L layers of refinement and grading factor $\sigma \in (0,1)$ as depicted in Fig. 3.1.2. Assume that u is analytic on Ω and satisfies for some C, $K > 0$, $\beta \in [0,1)$ and all $n \in \mathbb{N}_0$*

$$\begin{aligned}
|\nabla^n u(x)| &\le CK^n\varepsilon^{-1}|x|^{-n}\left(|x|/\varepsilon\right)^{1-\beta}\max\left\{n+1,\varepsilon^{-1}\right\}^{n+1},\\
\|\min\left\{1,|x|/\varepsilon\right\}^{\beta}\nabla^2 u\|_{L^2(\Omega)} &\le C\varepsilon^{-2}.
\end{aligned}$$

Then there exist C, b, $\lambda_0 > 0$ independent of $\varepsilon \in (0,1]$ such that for every $\lambda \in (0,\lambda_0)$ and every p satisfying

$$\lambda p\varepsilon \ge 1 \qquad (3.1.18)$$

there holds

$$\inf_{\pi_p \in S^p(\mathcal{T})}\|u - \pi_p\|_{H^1(\Omega)} \le Cp^2(1+\ln p)^2(\lambda p)^4\left[e^{-bp} + \varepsilon\sigma^{L(1-\beta)}\right].$$

Before we prove the lemma, we point out that the regularity assumptions on u are slightly weaker than those ascertained in Theorem 2.3.1 for the solutions u_ε of (1.2.1). We remark that the powers of p in the statement of Lemma 3.1.3 are not optimal and chosen so as to keep the exposition simple.
Proof: The approximant is taken as the piecewise Gauss-Lobatto interpolant. We distinguish the elements K_i that are away from the origin and the (single) element K_i that touches the origin. The proof is very similar to that of Lemma 3.1.2, and we will merely sketch the main differences. We introduce the two sets of indices

$$I_{geom} := \{i \,|\, 0 \notin \overline{K_i}\}, \qquad I_{org} := \{i \,|\, 0 \in \overline{K_i}\},$$

where in view of Fig. 3.1.2 the set I_{org} consists of a single element.
1. Step: Elements K_i with $\operatorname{dist}(K_i, 0) > 0$*:* We parallel the second step of the proof of Lemma 3.1.2. On the stretched element $\widetilde{K}_i := s_{h_i}^{-1}(K_i)$, the function $u \circ s_{h_i}$ satisfies

$$\|\nabla^n(u \circ s_{h_i})\|_{L^\infty(\widetilde{K}_i)} \le C h_i^n \|\nabla^n u\|_{L^\infty(K_i)} \le C\gamma^n \varepsilon^{-2+\beta} h_i^{1-\beta} \max\{n+1, h_i/\varepsilon\}^{n+1}.$$

Simplifying further

$$\begin{aligned} \max\{n+1, h_i/\varepsilon\}^{n+1} &= \max\{(n+1)^{n+1}, (h_i/\varepsilon)^{n+1}\} \\ &\le \max\{(n+1)^{n+1}, (n+1)! e^{h_i/\varepsilon}\} \le C\gamma^n n! e^{h_i/\varepsilon} \end{aligned}$$

for some appropriate C, $\gamma > 0$ independent of ε, we get

$$\|\nabla^n(u \circ s_{h_i})\|_{L^\infty(\widetilde{K}_i)} \le C\varepsilon^{-2} h_i^{1-\beta} \gamma^n n! e^{1/\varepsilon}$$

for some suitably chosen constants $C, \gamma > 0$ independent of ε. Using Lemma 4.3.4, we can then conclude for the function $\tilde{u} = (u \circ s_{h_i}) \circ (s_{h_i}^{-1} \circ M_i)$

$$\|\nabla^n \tilde{u}\|_{L^\infty(\hat{K})} \le C h_i^{1-\beta} \varepsilon^{-2} \gamma^n n! e^{1/\varepsilon} \qquad \forall n \in \mathbb{N}_0$$

for some suitably chosen C, $\gamma > 0$. Proceeding as in the in second step of the proof of Lemma 3.1.2, we conclude that the function $\pi_p = (j_p \tilde{u}) \circ M_i^{-1}$ satisfies

$$\|u - \pi_p\|_{L^\infty(K_i)} + h_i \|\nabla(u - \pi_p)\|_{L^\infty(K_i)} \le C h_i^{1-\beta} \varepsilon^{-2} e^{1/\varepsilon} e^{-bp} \tag{3.1.19}$$

for some $C, b > 0$ independent of ε. Paralleling the arguments leading to (3.1.13), we get

$$\sum_{i \in I_{geom}} h_i^{2(1-\beta)} \le C\,(\operatorname{diam} \Omega)^{2(1-\beta)} \le C$$

for some suitable constant $C > 0$. Hence, integrating the estimates (3.1.19) over K_i and then summing over $i \in I_{geom}$ gives

$$\Big(\sum_{i \in I_{geom}} \|u - \pi_p\|^2_{H^1(K_i)} \Big)^{1/2} \le C\varepsilon^{-2} e^{1/\varepsilon} e^{-bp}. \tag{3.1.20}$$

2. Step: The element K_i abutting on 0: The element K_i abutting on 0 is treated as in Lemma 3.1.2. Combining the estimates (3.1.14), (3.1.17), (3.1.16), (3.1.15), we conclude that $\pi_p|_{K_i} = (j_p(u \circ M_i)) \circ M_i^{-1}$ satisfies

$$\begin{aligned}
&\|u - \pi_p\|_{L^\infty(K_i)} + \|\nabla(u - \pi_p)\|_{L^2(K_i)} \\
&\quad \leq Cp^2(1+\ln p)^2 h_i^{1-\beta}\varepsilon^\beta(1 + h_i/\varepsilon)\|\min\{1, |x|/\varepsilon\}^\beta \nabla^2 u\|_{L^2(K_i)} \\
&\quad \leq Cp^2(1+\ln p)^2 \sigma^{(1-\beta)L}\varepsilon^{-3}. \qquad (3.1.21)
\end{aligned}$$

Combining (3.1.20) and (3.1.21) allows us to conclude that

$$\begin{aligned}
\|u - \pi_p\|_{H^1(\Omega)} &\leq Cp^2(1+\ln p)^2\varepsilon^{-4}\left[e^{1/\varepsilon}e^{-bp} + \varepsilon\sigma^{L(1-\beta)}\right] \\
&\leq Cp^2(1+\ln p)^2(\lambda p)^4\left[e^{(\lambda-b)p} + \varepsilon\sigma^{L(1-\beta)}\right],
\end{aligned}$$

where we employed in the last step the assumption $\lambda p\varepsilon \geq 1$. Choosing now λ_0 so small that $\lambda - b \leq \lambda_0 - b \leq -b/2$ allows us to conclude the proof. □

3.2 Polynomial approximation results

3.2.1 Notation and properties of polynomials

We start with some notation: $I = (0,1)$, $S = I \times I = (0,1)^2$, $T = \{(x,y)\,|\,x \in I, 0 < y < 1-x\}$. On the interval I, the reference square S, and the reference triangle T, we introduce the space of polynomials $\mathcal{P}_p(I)$, the *tensor product space* $\mathcal{Q}_p(S)$, and the space of polynomials $\mathcal{P}_p(T)$ by

$$\mathcal{P}_p(I) := \operatorname{span}\{x^i\,|\,i = 0,\dots,p\}, \qquad (3.2.1)$$
$$\mathcal{Q}_p(S) := \mathcal{P}_p(I) \otimes \mathcal{P}_p(I) = \operatorname{span}\{x^i y^j\,|\,0 \leq i,j \leq p\}, \qquad (3.2.2)$$
$$\mathcal{P}_p(T) := \mathcal{P}_p(S) := \operatorname{span}\{x^i y^j\,|\,i,j \in \mathbb{N}_0, i+j \leq p\}. \qquad (3.2.3)$$

We introduce the following shorthand for the spaces $\mathcal{P}_p(T)$ and $\mathcal{Q}_p(S)$:

$$\Pi_p(K) := \begin{cases} \mathcal{Q}_p(S) & \text{if } K = S, \\ \mathcal{P}_p(T) & \text{if } K = T. \end{cases} \qquad (3.2.4)$$

We will be interested in two types of polynomial approximation operators: The Gauss-Lobatto interpolant and the L^2 projection. As usual, L_p denotes the Legendre polynomial of degree p normalized to satisfy $L_p(1) = 1$ (cf., e.g., [61]). As we work on $I = (0,1)$ rather than $(-1,1)$, we introduce the scaled polynomials $\tilde{L}_p(x) := L_p(2x-1)$. Next, for given p, let

$$\mathcal{GL}_p := \{x_i\,|\,i = 0,\dots,p\} \qquad (3.2.5)$$

be the zeros of the polynomial $x \mapsto x(1-x)\tilde{L}_p'(x)$. It is a well-known fact (see, e.g., [27]) that this polynomial has $p+1$ distinct zeros lying in $[0,1]$. Clearly,

$\{0,1\} \subset \mathcal{GL}_p$ and by symmetry properties of the Legendre polynomials the set $\mathcal{GL}_p$ is symmetric with respect to the midpoint $1/2$ of the interval I. Using the Gauss-Lobatto points $\mathcal{GL}_p = \{x_i \,|\, i = 0,\dots,p\}$ we can then define the Gauss-Lobatto interpolation operator $i_p : C(I) \to \mathcal{P}_p(I)$ by interpolation in the $(p+1)$ Gauss-Lobatto nodes $\mathcal{GL}_p$, i.e.,

$$i_p : C(\overline{I}) \to \mathcal{P}_p(I)$$
$$f \mapsto (i_p f)(x) := \sum_{i=0}^{p} f(x_i) l_i^{(p)}(x),$$

where the Lagrange polynomials $l_i^{(p)} \in \mathcal{P}_p(I)$ are defined as

$$l_i^{(p)}(x) = \prod_{\substack{j=0\\ j\neq i}}^{p} \frac{x - x_j}{x_i - x_j}.$$

Similarly, we define the two dimensional Gauss-Lobatto interpolation operator $j_p : C(\overline{S}) \to \mathcal{Q}_p(S)$ by interpolation in the $(p+1)^2$ nodes obtained by taking the tensor product of the one-dimensional nodes, i.e., $j_p = i_p^x \circ i_p^y = i_p^y \circ i_p^x$, where we denoted by i_p^x, i_p^y the one-dimensional Gauss-Lobatto interpolation operators with respect to the x and y variable. Finally, for edges Γ of S or T we introduce the one-dimensional Gauss-Lobatto interpolant operators $i_{p,\Gamma}$ by identifying the edge with the interval I and using i_p. More specifically, if P_1, P_2 are the two endpoints of Γ, then $\gamma : \overline{I} \to \overline{\Gamma}$ given by $\gamma(t) := tP_1 + (1-t)P_2$ parametrizes Γ, and for a function u defined on $\overline{\Gamma}$ we can set $i_{p,\Gamma} u := i_p(u \circ \gamma) \circ \gamma^{-1}$. We note that for a square S there holds for all four edges Γ $i_{p,\Gamma} u|_\Gamma = (j_p u)|_\Gamma$.
For future reference we state the following stability result.

Lemma 3.2.1. *There is $C > 0$ independent of p such that*

$$\|i_p f\|_{L^\infty(I)} \leq C(1+\ln p)\|f\|_{L^\infty(I)} \qquad \forall f \in C(\overline{I}), \tag{3.2.6}$$
$$\|j_p f\|_{L^\infty(S)} \leq C(1+\ln p)^2\|f\|_{L^\infty(S)} \qquad \forall f \in C(\overline{S}), \tag{3.2.7}$$
$$\|(i_p f)'\|_{L^\infty(I)} \leq Cp\|f'\|_{L^\infty(I)} \qquad \forall f \in W^{1,\infty}(I). \tag{3.2.8}$$

Proof: The first two estimates are due to [121, 122]. For the last estimate, we first note an inverse estimate, [105] (cf. also [112, Thm. 3.92]),

$$\|f\|_{L^\infty(I)} \leq 4p\|f\|_{L^2(I)} \qquad \forall f \in \mathcal{P}_p.$$

This bound is employed in the following way:

$$\|(i_p f)'\|_{L^\infty(I)} \leq 4p\|(i_p f)'\|_{L^2(I)} \leq Cp\|i_p f\|_{H^1(I)} \leq Cp\|f\|_{H^1(I)}, \tag{3.2.9}$$

where we employed the H^1-stability of the Gauss-Lobatto interpolation operator that is asserted in Theorem A.3.1. Applying this last estimate to the function $f - c$ with $c \in \mathbb{R}$ arbitrary yields

$$\|(i_p f)'\|_{L^\infty(I)} \le Cp \inf_{c\in\mathbb{R}} \|f - c\|_{H^1(I)} \le Cp\|f'\|_{L^2(I)} \le Cp\|f'\|_{L^\infty(I)}.$$

□

Next, we recall the following inverse estimates of Markov type (see, e.g., [112]).

Lemma 3.2.2. *There is $C > 0$ such that for all $p \in \mathbb{N}$*

$$\|\pi_p'\|_{L^\infty(I)} \le 2p^2 \|\pi_p\|_{L^\infty(I)} \qquad \forall \pi_p \in \mathcal{P}_p(I), \tag{3.2.10}$$

$$\|\nabla \pi_p\|_{L^\infty(S)} \le Cp^2 \|\pi_p\|_{L^\infty(S)} \qquad \forall \pi_p \in \mathcal{Q}_p(S), \tag{3.2.11}$$

$$\|\nabla \pi_p\|_{L^\infty(T)} \le Cp^2 \|\pi_p\|_{L^\infty(T)} \qquad \forall \pi_p \in \mathcal{P}_p(T). \tag{3.2.12}$$

Lemma 3.2.3. *Let $K = T$ or $K = S$. Then there is $C > 0$ such that the following holds. For each $f \in C(\partial K)$ that is a polynomial of degree p on each edge of K, there exists $F \in \Pi_p(K)$ such that*

$$F|_{\partial K} = f,$$

$$\|F\|_{L^\infty(K)} + p^{-2}\|\nabla F\|_{L^\infty(K)} \le C\|f\|_{L^\infty(\partial K)}.$$

Moreover, the extension map $E : f \mapsto F$ is in fact a bounded linear operator.

Proof: Since the inverse estimate of Lemma 3.2.2 implies $p^{-2}\|\nabla F\|_{L^\infty(K)} \le C\|F\|_{L^\infty(K)}$, it suffices to prove the bound $\|F\|_{L^\infty(K)} \le C\|f\|_{L^\infty(\partial K)}$. For this pointwise estimates of F, we proceed as follows. After subtracting a linear (if $K = T$) or a bilinear (if $K = S$) interpolant, we may assume that f vanishes at the vertices. Next, we may assume without loss of generality that f vanishes on all sides but one. This side may be taken to be $\Gamma := \{(x, 0) \mid x \in I\}$. For $K = S$, we set $F(x, y) = (1 - y)f(x)$. The pointwise estimate follows immediately. For $K = T$, we observe that, since f is assumed to vanish at the vertices of K, it can be written in the form

$$f(x) = x(1 - x)\widetilde{f}(x)$$

for some suitable $\widetilde{f} \in \mathcal{P}_{p-2}$. We then set

$$F(x, y) := x(1 - x - y)\widetilde{f}(x) \in \mathcal{P}_p$$

and bound

$$\begin{aligned}\sup_{(x,y)\in T} |F(x, y)| &= \sup_{x\in(0,1)} \sup_{y\in(0,1-x)} x(1 - x - y)|\widetilde{f}(x)| \\ &= \sup_{x\in(0,1)} \sup_{y\in(0,1-x)} x(1 - x)|\widetilde{f}(x)| = \sup_{x\in(0,1)} |f(x)|.\end{aligned}$$

□

We finally conclude this section with a result concerning the approximation properties of the linear/bilinear interpolant. For reasons that will become clear in the following subsection, the linear/bilinear interpolation operator is denoted by Π_1^∞.

Lemma 3.2.4. *Let $K = T$ or $K = S$ and let A be one of the vertices of K. Let $\beta \in [0,1)$. Then there is $C > 0$ depending only on β such that the difference $u - \Pi_1^\infty u$ between u and its linear/bilinear interpolant $\Pi_1^\infty u$ satisfies*

$$\begin{aligned}\|u - \Pi_1^\infty u\|_{L^\infty(K)} &\leq C\left[\|u - \Pi_1^\infty u\|_{H^1(K)} + \|\,|x-A|^\beta \nabla^2(u - \Pi_1^\infty u)\|_{L^2(K)}\right] \\ &\leq C\|\,|x-A|^\beta \nabla^2 u\|_{L^2(K)}.\end{aligned}$$

Proof: The set K is a sector with apex A. From Lemma 4.2.9, we get $\|u - \Pi_1^\infty u\|_{L^\infty(K)} \leq C\|u - \Pi_1^\infty u\|_{H^{2,2}_{\beta,1}(K)}$, which is the first inequality. The second bound is taken directly from [112, Lemmata 4.16, 4.25]. □

3.2.2 Approximation of analytic functions: intervals and squares

Approximation on the interval *I*. For $\rho > 1$ we denote by $\mathcal{E}_\rho$ the ellipse (in the complex plane) with foci ± 1 and sum of semi-axes ρ, i.e.,

$$\mathcal{E}_\rho = \{z \in \mathbb{C} \,|\, |z-1| + |z+1| < \rho + \rho^{-1}\}. \tag{3.2.13}$$

Remark 3.2.5 An important property of the ellipses $\mathcal{E}_\rho$ is the fact that $\partial\mathcal{E}_\rho$ is the image of $\partial B_\rho(0) \subset \mathbb{C}$ under the conformal map $z \mapsto w = \frac{1}{2}(z + z^{-1})$. ▪

We start with the following one-dimensional approximation result.

Lemma 3.2.6. *Let u be analytic on I and satisfy for some C_u, $\gamma > 0$ and $h \in (0,1]$*

$$\|D^n u\|_{L^\infty(I)} \leq C_u (\gamma h)^n n! \qquad \forall n \in \mathbb{N}. \tag{3.2.14}$$

Then there are C, $\sigma > 0$ depending only on γ such that the Gauss-Lobatto interpolant $i_p u$ satisfies

$$\|u - i_p u\|_{L^\infty(I)} + \|(u - i_p u)'\|_{L^\infty(I)} \leq C C_u \left(\frac{h}{h+\sigma}\right)^{p+1} \qquad \forall p \in \mathbb{N}.$$

Proof: The proof proceeds in three steps.
1. step: Let $\overline{u} \in \mathbb{R}$ be the average of u, i.e., $\overline{u} = \int_I u\,dx$. By the mean value theorem, there is $\xi \in I$ with $\overline{u} = u(\xi)$. Thus, $\widetilde{u}(x) := u(x) - \overline{u}$ satisfies

$$\|D^n \widetilde{u}\|_{L^\infty(I)} \leq \max\{1, 2\gamma\} C_u (\gamma h)^n n! \qquad \forall n \in \mathbb{N}_0.$$

Next, we observe that these bounds on the derivatives of $\widetilde{u}$ imply the existence of σ, $C > 0$ depending only on γ such that $\widetilde{u}$ is holomorphic on $\mathcal{E}_\rho$ with $\rho \geq 1 + \sigma/h$; additionally, it satisfies on $\mathcal{E}_\rho$

$$\|\widetilde{u}\|_{L^\infty(\mathcal{E}_\rho)} \leq C C_u.$$

2. step: From [40, Thm. 12.4.7], we get the existence of C such that (after appropriately adjusting σ)

$$\widetilde{u}(x) = \sum_{i=0}^{\infty} u_i \tilde{L}_i(x) \qquad \text{uniformly on } I,$$
$$|u_i| \leq CC_u(1+\sigma/h)^{-i} \qquad \forall i \in \mathbb{N}_0,$$

where for the standard Legendre polynomials L_i we set $\tilde{L}_i(x) := L_i(2x-1)$. We now define $u_p(x) := \sum_{i=0}^{p} u_i \tilde{L}_i(x) - \overline{u} \in \mathcal{P}_p(I)$ and bound with (3.2.10)

$$\begin{aligned}\|(u-u_p)'\|_{L^\infty(I)} &\leq \|(\widetilde{u} - \sum_{i=0}^{p} u_i \tilde{L}_i)'\|_{L^\infty(I)} \leq \sum_{i=p+1}^{\infty} |u_i|\, \|\tilde{L}_i'\|_{L^\infty(I)} \\ &\leq CC_u \sum_{i=p+1}^{\infty} i^2 (1+\sigma/h)^{-i} \leq CC_u \left(\frac{h}{h+\sigma'}\right)^{p+1}\end{aligned}$$

for some $\sigma' < \sigma$ and $C > 0$. An analogous result holds for $\|u - u_p\|_{L^\infty(I)}$. Thus, we have proved

$$\|u - u_p\|_{L^\infty(I)} + \|(u-u_p)'\|_{L^\infty(I)} \leq CC_u \left(\frac{h}{h+\sigma}\right)^{p+1}. \tag{3.2.15}$$

3. step: We employ the stability result Lemma 3.2.1 in order to obtain bounds for $u - i_p u$:

$$\begin{aligned}\|u - i_p u\|_{L^\infty(I)} &\leq \|u - u_p\|_{L^\infty(I)} + \|u_p - i_p u\|_{L^\infty(I)} \\ &\leq \|u - u_p\|_{L^\infty(I)} + \|i_p(u_p - u)\|_{L^\infty(I)} \\ &\leq C(1+\ln p)\|u - u_p\|_{L^\infty(I)}, \\ \|(u - i_p)'\|_{L^\infty(I)} &\leq \|(u-u_p)'\|_{L^\infty(I)} + \|(u_p - i_p u)'\|_{L^\infty(I)} \\ &\leq \|(u-u_p)'\|_{L^\infty(I)} + p^2\|u_p - i_p u\|_{L^\infty(I)} \\ &\leq \|(u-u_p)'\|_{L^\infty(I)} + Cp^2(1+\ln p)\|u - u_p\|_{L^\infty(I)}.\end{aligned}$$

Inserting (3.2.15) gives the desired bounds on $u - i_p u$ after appropriately adjusting the constant σ. □

Approximation on the square S. We now turn to the two-dimensional case. For the case of a square, a natural approach is to proceed by tensorizing the 1D arguments. This leads to approximants in $\mathcal{Q}_p(S)$. However, since we want to utilize the approximation results of this section for the approximation on the triangle below, it is more convenient to construct approximant that lie in $\mathcal{P}_p(T)$. The approximants are obtained as truncated Legendre series. To do so, we introduce the L^2-projector $\Pi_p^{L^2} : L^2(S) \to \mathcal{P}_p(T)$ by expanding u into a Legendre series $u(x,y) = \sum_{i,j\in\mathbb{N}_0} u_{ij} \tilde{L}_i(x)\tilde{L}_j(y)$ (convergence being understood in $L^2(S)$) and then setting

$$\Pi_p^{L^2} u(x,y) := \sum_{\substack{i,j\in\mathbb{N}_0 \\ i+j\leq p}} u_{ij} \tilde{L}_i(x)\tilde{L}_j(y) \in \mathcal{P}_p(T). \tag{3.2.16}$$

Lemma 3.2.7. *Let u be analytic on S and satisfy for some C_u, $\gamma > 0$, h_x, $h_y \in (0,1]$*

$$\|D^\alpha u\|_{L^\infty(S)} \leq C_u h_x^{\alpha_1} h_y^{\alpha_2} \gamma^{|\alpha|} \alpha! \qquad \forall \alpha = (\alpha_1, \alpha_2) \in \mathbb{N}_0^2 \setminus (0,0). \tag{3.2.17}$$

Then u can be expanded in a Legendre series on S, and there are C, $\sigma > 0$ depending only on γ such that

$$u(x,y) = \sum_{i,j=0}^{\infty} u_{ij} \tilde{L}_i(x) \tilde{L}_j(y) \qquad \textit{uniformly on } S,$$

$$|u_{ij}| \leq C_u C \left(1 + \sigma/h_x\right)^{-i} \left(1 + \sigma/h_y\right)^{-j}, \qquad (i,j) \neq (0,0).$$

Proof: The proof is very similar to the one-dimensional one in Lemma 3.2.6. Defining $\overline{u} = \int_S u(x,y)\, dxdy$ one considers $\widetilde{u}(x,y) := u(x,y) - \overline{u}$. From the mean value theorem one has the existence of $(x', y') \in S$ such that $\overline{u} = u(x', y')$ and hence one gets

$$\|D^\alpha \widetilde{u}\|_{L^\infty(S)} \leq \max\{1, 2\sqrt{2}\gamma\} C_u h_x^{\alpha_1} h_y^{\alpha_2} \gamma^{|\alpha|} \alpha! \qquad \forall \alpha \in \mathbb{N}_0^2.$$

We note that $\widetilde{u}$ can be extended holomorphically to $\mathcal{E}_{\rho_x} \times \mathcal{E}_{\rho_y} \subset \mathbb{C} \times \mathbb{C}$ for $\rho_x = 1 + \sigma/h_x$, $\rho_y = 1 + \sigma/h_y$, where $\sigma > 0$ depends only on γ. Furthermore there is $C > 0$ (again depending only on γ) such that the extended function (again denoted $\widetilde{u}$) satisfies

$$\|\widetilde{u}\|_{L^\infty(\mathcal{E}_{\rho_x} \times \mathcal{E}_{\rho_y})} \leq C C_u.$$

The remainder of the proof is now a straightforward extension to 2 dimensions of the argument given in [40, Thm. 12.4.7]. □

The decay properties of the Legendre coefficients u_{ij} of in Lemma 3.2.7 allow us to obtain exponential rates of convergence for the L^2-projection operator $\Pi_p^{L_2}$:

Proposition 3.2.8. *Under the hypotheses of Lemma 3.2.7 there exist constants C, $\sigma > 0$ depending only on $\gamma > 0$ such that*

$$\|u - \Pi_p^{L^2} u\|_{L^\infty(S)} \leq C C_u \left[\left(\frac{h_x}{h_x + \sigma}\right)^{p+1} + \left(\frac{h_y}{h_y + \sigma}\right)^{p+1}\right],$$

$$\|\partial_x (u - \Pi_p^{L^2} u)\|_{L^\infty(S)} \leq C C_u \frac{h_x}{h_x + \sigma} \left[\left(\frac{h_x}{h_x + \sigma}\right)^{p} + \left(\frac{h_y}{h_y + \sigma}\right)^{p}\right],$$

$$\|\partial_y (u - \Pi_p^{L^2} u)\|_{L^\infty(S)} \leq C C_u \frac{h_y}{h_y + \sigma} \left[\left(\frac{h_x}{h_x + \sigma}\right)^{p} + \left(\frac{h_y}{h_y + \sigma}\right)^{p}\right].$$

Proof: We will only prove the bound for $\|\partial_x (u - \Pi_p^{L^2} u)\|_{L^\infty(S)}$ as the other ones are proved similarly. We have to bound

$$\|\partial_x (u - \Pi_p^{L^2} u)\|_{L^\infty(S)} \leq \sum_{\substack{i \geq 1,\ j \geq 0 \\ i+j \geq p+1}} |u_{ij}|\, \|\tilde{L}_i'(x) \tilde{L}_j(y)\|_{L^\infty(S)}.$$

By Markov's inequality (3.2.10) we have $\|\tilde{L}_i'\|_{L^\infty((0,1))} \leq 2i^2 \|\tilde{L}_i\|_{L^\infty((0,1))}$. Thus, upon setting $q_x = h_x/(h_x+\sigma)$, $q_y = h_y/(h_y+\sigma)$ we get from Lemma 3.2.7 (after appropriately decreasing σ in order to absorb the factor i^2)

$$
\begin{aligned}
C \sum_{\substack{i\geq 1,\, j\geq 0\\ i+j\geq p+1}} q_x^i q_y^j &= C \sum_{i=1}^{p+1} \sum_{j=p+1-i}^{\infty} q_x^i q_y^j + C \sum_{i=p+2}^{\infty} \sum_{j=0}^{\infty} q_x^i q_y^j \\
&= C(1-q_y)^{-1} \sum_{i=1}^{p+1} q_x^i q_y^{p+1-i} + C(1-q_y)^{-1} q_x^{p+2} \\
&\leq C q_x \left[\sum_{i=0}^{p} q_x^i q_y^{p-i} + q_x^{p+1} \right] = C q_x \left[q_x^p + q_y^p + \sum_{i=1}^{p-1} q_x^i q_y^{p-i} + q_x^{p+1} \right].
\end{aligned}
$$

Next, Young's inequality gives

$$
q_x^i q_y^{p-i} \leq \frac{i}{p} q_x^p + \frac{p-i}{p} q_y^p, \qquad i = 1, \ldots, p-1,
$$

and we obtain therefore

$$
\|\partial_x (u - \Pi_p^{L^2} u)\|_{L^\infty(S)} \leq C q_x p \left[q_x^p + q_y^p \right].
$$

Decreasing again the value of σ in order to absorb the factor p, we get the desired result. □

3.2.3 Approximation of analytic functions on triangles

The main result of this section is Theorem 3.2.19 where we show that an approximation result analogous to Proposition 3.2.8 holds on triangles as well. The approximation result Proposition 3.2.8 relied on truncted Legendre expansions and estimates for the decay of the coefficients in this expansion. This could conveniently be done since tensor products of Legendre polynomials are orthogonal polynomials on the square S. The general approach in the case of approximation on triangles is similar: The polynomials $\psi_{p,q}$ of (3.2.23) are orthogonal polynomials for the triangle, and our main aim in this section is to estimate the decay of the coefficients $u_{p,q}$ in the expansion $u = \sum_{p,q} u_{p,q} \psi_{p,q}$.

For the approximation on triangles, it will be convenient to employ the Jacobi polynomials $P_p^{(\alpha,\beta)}$, which have the following orthogonality properties in weighted spaces (see, e.g., [124, eq. (4.3.3)], [61, eq. 7.391]):

$$
\int_{-1}^{1} (1-t)^\alpha (1+t)^\beta P_p^{(\alpha,\beta)}(t) P_q^{(\alpha,\beta)}(t)\, dt = \qquad (3.2.18)
$$

$$
\delta_{pq} \frac{2^{\alpha+\beta+1}}{\alpha+\beta+1+2p} \frac{\Gamma(\alpha+p+1)\Gamma(\beta+p+1)}{p!\,\Gamma(\alpha+\beta+p+1)}.
$$

Due to the fact that the Jacobi polynomials are defined on the reference interval $(-1,1)$ instead of $(0,1)$ we adopt in the present section the following convention concerning the reference square $\underline{S}$ and the reference triangle $\underline{T}$:

$$\underline{S} = (-1,1)^2, \qquad \underline{T} = \{(x,y) \mid -1 < x < 1, -1 < y < -x\}. \tag{3.2.19}$$

The following transformation (sometimes referred to as the Duffy transformation, [45]) maps $\underline{S}$ onto $\underline{T}$:

$$\begin{aligned} D &: \underline{S} \to \underline{T} \\ (\eta_1,\eta_2) &\mapsto (\xi_1,\xi_2) = \left(\tfrac{(1+\eta_1)(1-\eta_2)}{2} - 1, \eta_2\right) \end{aligned} \tag{3.2.20}$$

with inverse map

$$\begin{aligned} D^{-1} &: \underline{T} \to \underline{S} \\ (\xi_1,\xi_2) &\mapsto (\eta_1,\eta_2) = \left(2\tfrac{1+\xi_1}{1-\xi_2} - 1, \xi_2\right). \end{aligned} \tag{3.2.21}$$

In terms of the Jacobi polynomials, we define the following polynomials on $\underline{S}$:

$$\tilde{\psi}_{p,q}(\eta_1,\eta_2) := P_p^{(0,0)}(\eta_1)\left(\frac{1-\eta_2}{2}\right)^p P_q^{(2p+1,0)}(\eta_2), \qquad p,q \in \mathbb{N}_0. \tag{3.2.22}$$

The functions

$$\psi_{p,q} := \tilde{\psi}_{p,q} \circ D^{-1} \tag{3.2.23}$$

defined on $\underline{T}$, are orthogonal polynomials of degree $p+q$ as ascertained in the following lemma, which is due to Dubiner, [43]:

Lemma 3.2.9 (orthogonal polynomials on the triangle). *The functions $\psi_{p,q}$ of (3.2.23) satisfy $\psi_{p,q} \in \mathcal{P}_{p+q}(\underline{T})$, they are $L^2(\underline{T})$-orthogonal, and*

$$\int_{\underline{T}} \psi_{p,q}(\xi_1,\xi_2)\psi_{p',q'}(\xi_1,\xi_2)\,d\xi_1 d\xi_2 = \delta_{pp'}\delta_{qq'}\frac{2}{2p+1}\frac{2}{2p+2q+2}.$$

Proof: We start with the assertion that $\psi_{p,q}$ is a polynomial of degree $p+q$. With D^{-1} defined in (3.2.21), we get

$$\psi_{p,q}(\xi_1,\xi_2) = \tilde{\psi}_{p,q}(2\tfrac{1+\xi_1}{1-\xi_2} - 1, \xi_2) = P_p^{(0,0)}(2\tfrac{1+\xi_1}{1-\xi_2} - 1)\left(\frac{1-\xi_2}{2}\right)^p P_q^{(2p+1,0)}(\xi_2).$$

Expanding the Legendre polynomial $P^{(0,0)}$ as $P_p^{(0,0)}(x-1) = \sum_{k=0}^p c_k x^k$, we get

$$\psi_{p,q}(\xi_1,\xi_2) = \sum_{k=0}^{p} c_k 2^{k-p}(1+\xi_1)^k(1-\xi_2)^{p-k}P_q^{(2p+1,0)}(\xi_2);$$

since $P_q^{(2p+1,0)}$ is a polynomial of degree q, we get $\psi_{p,q} \in \mathcal{P}_{p+q}(\underline{T})$.
We next demonstrate the orthogonality. By transforming to $\underline{S}$ we get using (3.2.18) twice

$$\int_{\underline{T}} \psi_{p,q}(\xi_1,\xi_2)\psi_{p',q'}(\xi_1,\xi_2)\, d\xi_1 d\xi_2 = \int_{\underline{S}} \tilde{\psi}_{p,q}(\eta_1,\eta_2)\tilde{\psi}_{p',q'}(\eta_1,\eta_2)\frac{1-\eta_2}{2} d\eta_1 d\eta_2$$
$$= \int_{-1}^{1}\int_{-1}^{1} P_p^{(0,0)}(\eta_1)P_{p'}^{(0,0)}(\eta_1)\left(\frac{1-\eta_2}{2}\right)^{p+p'+1} P_q^{(2p+1,0)}(\eta_2)P_{q'}^{(2p'+1,0)}(\eta_2)\, d\eta_1 d\eta_2$$
$$= \frac{2}{2p+1}\delta_{pp'}2^{-(2p+1)}\int_{-1}^{1}(1-\eta_2)^{2p+1}P_q^{(2p+1,0)}(\eta_2)P_{q'}^{(2p+1,0)}(\eta_2)d\eta_2$$
$$= \frac{2}{2p+1}\delta_{pp'}\delta_{qq'}\frac{2}{2p+2q+2}.$$

□

We are interested in estimating the decay of the coefficients $u_{p,q}$ when expanding an $L^2(\underline{T})$-function in terms of the orthogonal basis $(\psi_{p,q})_{p,q\in\mathbb{N}_0}$ of $L^2(\underline{T})$. To do that, it will be important to expand the function $t \mapsto 1/(w-t)$ in Jacobi polynomials, which is done in the following lemma:

Lemma 3.2.10. *Let α, $\beta > -1$. Then for every $q \in \mathbb{N}_0$ the function*

$$w \mapsto \tilde{Q}_q^{(\alpha,\beta)}(w) = \int_{-1}^{1}(1-t)^{\alpha}(1-t)^{\beta}\frac{P_q^{(\alpha,\beta)}(t)}{w-t}\, dt \tag{3.2.24}$$

is holomorphic on $\mathbb{C}\setminus[-1,1]$. Writing $w = \frac{1}{2}(z+z^{-1})$ with $z \in \mathbb{C}$, $|z| > 1$, $\tilde{Q}_q^{(\alpha,\beta)}(w)$ can be written as

$$\tilde{Q}_q^{(\alpha,\beta)}(w) = \sum_{m=q}^{\infty}\sigma_{m,q,\alpha,\beta}\, z^{-(m+1)},$$

where the coefficients $\sigma_{m,q,\alpha,\beta} \in \mathbb{R}$ are independent of z and satisfy

$$|\sigma_{m,q,\alpha,\beta}| \le 2^{\alpha+\beta+2}\sqrt{m+1}$$
$$\times\sqrt{\frac{\Gamma(\alpha+1)\Gamma(\beta+1)}{(\alpha+\beta+1)\Gamma(\alpha+\beta+1)}}\sqrt{\frac{\Gamma(\alpha+q+1)\Gamma(\beta+q+1)}{(\alpha+\beta+1+2q)q!\Gamma(\alpha+\beta+q+1)}}.$$

For the case $\beta = 0$, we have the particular bounds

$$|\sigma_{m,q,\alpha,0}| \le 2^{\alpha+2}\frac{\sqrt{m+1}}{\sqrt{2q+\alpha+1}\sqrt{\alpha+1}},$$
$$|\sigma_{m,q,0,0}| \le 2\pi.$$

Proof: The second kind Chebyshev polynomials $U_m \in \mathcal{P}_m$ are defined on the interval $[-1,1]$ by the relation

$$U_m(\cos\theta) = \frac{\sin(m+1)\theta}{\sin\theta}, \qquad \theta \in [0,\pi].$$

With $w = \frac{1}{2}(z+z^{-1})$ we can then write in view of [61, eq. 8.945]

$$\frac{1}{w-t} = \frac{1}{(z+z^{-1})/2-t} = \frac{2}{z}\frac{1}{1-2tz^{-1}+z^{-2}} = \frac{2}{z}\sum_{m=0}^{\infty} U_m(t) z^{-m}.$$

Inserting this into the definition of $\tilde{Q}_q^{(\alpha,\beta)}$ and exploiting $U_m \in \mathcal{P}_m$ together with the orthogonality properties of the Jacobi polynomials $P_m^{(\alpha,\beta)}$, we get

$$\begin{aligned}
&\int_{-1}^{1} (1-t)^\alpha (1+t)^\beta \frac{P_q^{(\alpha,\beta)}(t)}{w-t}\, dt \\
&\quad = 2\sum_{m=0}^{\infty} z^{-(m+1)} \int_{-1}^{1} (1-t)^\alpha (1+t)^\beta P_q^{(\alpha,\beta)}(t) U_m(t)\, dt \\
&\quad = 2\sum_{m=q}^{\infty} z^{-(m+1)} \int_{-1}^{1} (1-t)^\alpha (1+t)^\beta P_q^{(\alpha,\beta)}(t) U_m(t)\, dt.
\end{aligned}$$

We then get the desired representation by setting

$$\sigma_{m,q,\alpha,\beta} := 2\int_{-1}^{1} (1-t)^\alpha (1+t)^\beta P_q^{(\alpha,\beta)}(t) U_m(t)\, dt.$$

For the bound on $\sigma_{m,q,\alpha,\beta}$, we employ the Cauchy-Schwarz inequality to get

$$\begin{aligned}
|\sigma_{m,q,\alpha,\beta}| &= 2\left|\int_{-1}^{1} (1-t)^\alpha (1+t)^\beta P_q^{(\alpha,\beta)}(t) U_m(t)\, dt\right| \le 2\|U_m\|_{L^\infty((-1,1))} \\
&\times \left\{\int_{-1}^{1} (1-t)^\alpha (1+t)^\beta\, dt\right\}^{1/2} \left\{\int_{-1}^{1} (1-t)^\alpha (1+t)^\beta \left(P_q^{(\alpha,\beta)}(t)\right)^2 dt\right\}^{1/2}.
\end{aligned}$$

The observation $P_0^{(\alpha,\beta)} = 1$ and the formula (3.2.18) then allows us to obtain the desired bound, since elementary considerations reveal $\|U_m(t)\|_{L^\infty((-1,1))} \le (m+1)$.
The first bound for the particular case $\beta = 0$ follows immediately from the general case by setting $\beta = 0$. For the special case $\alpha = \beta = 0$, we employ the fact that $|P^{(0,0)}(t)| \le 1$ for all $t \in [-1,1]$ to bound

$$\begin{aligned}
|\sigma_{m,q,0,0}| &= \left|2\int_{-1}^{1} P_q^{(0,0)}(t) U_m(t)\, dt\right| = 2\left|\int_0^\pi P_q^{(0,0)}(\cos\theta)\frac{\sin(m+1)\theta}{\sin\theta}\sin\theta\, d\theta\right| \\
&\le 2\pi.
\end{aligned}$$

□

In the following, we will merely require the special case $\beta = 0$ in Lemma 3.2.10:

Corollary 3.2.11. *For $\rho > 1$, the functions $\tilde{Q}_q^{(\alpha,0)}$ of (3.2.24) satisfy*

$$\left|\tilde{Q}_q^{(0,0)}(w)\right| \le \frac{2\pi}{1-1/\rho}\rho^{-(q+1)} \quad \forall w \in \partial\mathcal{E}_\rho,$$

$$\left|\tilde{Q}_q^{(\alpha,0)}(w)\right| \le \frac{2^{\alpha+2}}{\alpha+1}\frac{(q+2)}{(1-1/\rho)^2}\rho^{-(q+1)} \quad \forall w \in \partial\mathcal{E}_\rho.$$

Proof: We will only show the second estimate. By Remark 3.2.5 we can write $w \in \partial \mathcal{E}_\rho$ as $w = \frac{1}{2}(z + z^{-1})$ for a $z \in \mathbb{C}$ with $|z| = \rho$. Lemma 3.2.10 then implies

$$\left|\tilde{Q}_q^{(\alpha,0)}(w)\right| \leq \sum_{m=q}^{\infty} |\sigma_{\alpha,0,q,m}|\rho^{-(m+1)} \leq \frac{2^{\alpha+2}}{\alpha+1} \sum_{m=q}^{\infty} \sqrt{m+1}\rho^{-(m+1)}$$

$$\leq \frac{2^{\alpha+2}}{\alpha+1} \sum_{m=q}^{\infty} (m+2)\rho^{-(m+1)} \leq \frac{2^{\alpha+2}}{\alpha+1} \frac{(q+2)}{(1-1/\rho)^2}\rho^{-(q+1)},$$

where the last step follows from the fact that the power series $\sum_{m=q}^{\infty}(m+2)x^{m+1}$ can evaluated in closed form. □

We seek to approximate functions u that are analytic on the closure of $\underline{T}$. The following lemma analyzes the domain of analyticity of the function $\tilde{u} = u \circ D$. The key observation is that the transformation D is degenerate in the sense that the line $\{(\eta_1, 1) \,|\, \eta_1 \in \mathbb{R}\}$ is mapped to the single point $(-1, 1)$; this implies that the domain of holomorphy of the function $\eta_1 \mapsto \tilde{u}(\eta_1, \eta_2)$ is very large for η_2 close to 1.

Lemma 3.2.12. *Let D be defined in (3.2.20) and let u be analytic on the closure of $\underline{T}$. Then there exist C, $\delta > 0$, $\rho > 1$ depending only on u such that:*

1. *The function $\tilde{u} := u \circ D$ is analytic on the closure of $\underline{S}$ and can be extended holomorphically to $\mathcal{E}_\rho \times \mathcal{E}_\rho$. Denoting this extension again by $\tilde{u}$ there holds*

$$\|\tilde{u}\|_{L^\infty(\mathcal{E}_\rho \times \mathcal{E}_\rho)} \leq C. \tag{3.2.25}$$

2. *For each $\eta_2 \in (-1, 1)$ the function $\eta_1 \mapsto \tilde{u}(\eta_1, \eta_2)$ is holomorphic on $\mathcal{E}_{1+\delta/(1-\eta_2)}$ with*

$$\sup_{\eta_2 \in (-1,1)} \|\tilde{u}(\cdot, \eta_2)\|_{L^\infty(\mathcal{E}_{1+\delta/(1-\eta_2)})} \leq C. \tag{3.2.26}$$

For each $p \in \mathbb{N}_0$ the function $\eta_2 \mapsto U_p(\eta_2) := \int_{-1}^{1} \tilde{u}(\eta_1, \eta_2) P_p^{(0,0)}(\eta_1)\, d\eta_1$ has the following properties:

1. *U_p is holomorphic on $\mathcal{E}_\rho$ and has a zero of multiplicity p at $\eta_2 = 1$;*
2. *$|U_p(\zeta_2)| \leq C\rho^{-(p+1)} \quad \forall \zeta_2 \in \mathcal{E}_\rho$.*

Proof: Since u is analytic on the closure of $\underline{T}$, there exists a complex neighborhood $T' \subset \mathbb{C}^2$ of the closure $\operatorname{cl} \underline{T}$ of $\underline{T}$ such that u is holomorphic on T'. We may additionally assume that T' is chosen such that u is holomorphic on a neighborhood of the closure of T'. This assumption implies $\|u\|_{L^\infty(T')} < \infty$. Next, we can find a (complex) neighborhood S' of $\underline{S}$ such that $D(S') \subset T'$. By the continuity of D, we can find $\rho > 1$ such that $\mathcal{E}_\rho \times \mathcal{E}_\rho \subset S'$; hence the first claim concerning the domain of holomorphy of $\tilde{u}$ is proved.
Setting for $\delta > 0$

$$G_\delta := \{(\zeta_1, \zeta_2) \,|\, \zeta_2 \in (-1, 1), \zeta_1 \in \mathcal{E}_{1+\delta/(1-\zeta_2)}\},$$

a direct calculation shows that for $\delta > 0$ sufficiently small we have $D(G_\delta) \subset T'$. Thus, the claim about the domain of holomorphy of the function $\eta_1 \mapsto \tilde{u}(\eta_1, \eta_2)$ is proved. Since $D(G_\delta) \subset T'$, the bound (3.2.26) follows also.
We now turn to the statements concerning U_p. Since there exists $\rho > 1$ such that $\tilde{u}$ is holomorphic on $\mathcal{E}_\rho \times \mathcal{E}_\rho$, standard results (see, e.g., [103, Chap. 2, Thm. 1.1]) give that U_p is holomorphic on $\mathcal{E}_\rho$. In order to show that U_p has a zero of multiplicity p at $\eta_2 = 1$, it suffices to show $|U_p(\eta_2)| \leq C(1-\eta_2)^p$ for $\eta_2 \in (-1, 1)$. To that end,we employ Cauchy's integral representation formula, the fact that $\pi\rho_2 \leq \text{length}(\partial\mathcal{E}_{\rho_2}) \leq 4\rho_2$, and Corollary 3.2.11, to get with the abbreviation $\rho_2 := 1 + \delta/(1-\eta_2)$

$$\begin{aligned}
|U_p(\eta_2)| &= \left| \int_{-1}^{1} \frac{1}{2\pi \mathbf{i}} \int_{\zeta_1 \in \partial\mathcal{E}_{\rho_2}} \frac{\tilde{u}(\zeta_1, \eta_2)}{\zeta_1 - \eta_1} P_p^{(0,0)}(\eta_1)\, d\zeta_1\, d\eta_1 \right| \\
&= \left| \frac{1}{2\pi \mathbf{i}} \int_{\zeta_1 \in \partial\mathcal{E}_{\rho_2}} \tilde{u}(\zeta_1, \eta_2) \tilde{Q}_p^{(0,0)}(\zeta_1) d\zeta_1 \right| \\
&\leq \frac{\text{length}(\partial\mathcal{E}_{\rho_2})}{2\pi} \|\tilde{u}\|_{L^\infty(G_\delta)} \|\tilde{Q}_p^{(0,0)}\|_{L^\infty(\mathcal{E}_{\rho_2})} \\
&\leq \frac{4\rho_2}{1 - 1/\rho_2} \|\tilde{u}\|_{L^\infty(G_\delta)} \rho_2^{-(p+1)} \leq C \left(\frac{1-\eta_2}{\delta + (1-\eta_2)} \right)^p ,
\end{aligned}$$

For the last bound, we proceed similarly. The function $\tilde{u}$ is holomorphic on $\mathcal{E}_\rho \times \mathcal{E}_\rho$ and bounded there. We get with Cauchy's integral representation

$$\begin{aligned}
|U_p(\zeta_2)| &\leq \left| \frac{1}{2\pi \mathbf{i}} \int_{\zeta_1 \in \partial\mathcal{E}_\rho} \tilde{u}(\zeta_1, \zeta_2) \int_{-1}^{1} \frac{P_p^{(0,0)}(\eta_1)}{\zeta_1 - \eta_1}\, d\eta_1 d\zeta_1 \right| \\
&\leq \frac{\text{length}(\partial\mathcal{E}_\rho)}{2\pi} \|\tilde{u}\|_{L^\infty(\mathcal{E}_\rho \times \mathcal{E}_\rho)} \|\tilde{Q}_p^{(0,0)}\|_{L^\infty(\mathcal{E}_\rho)} \leq C\rho^{-p}.
\end{aligned}$$

□

Remark 3.2.13 The proof of Lemma 3.2.12 shows that ρ depends only on the domain of holomorphy of u.

Proposition 3.2.14. *Let $\underline{T}$ and the polynomials $\psi_{p,q}$ be defined in (3.2.19), (3.2.23). Let u be analytic on the closure of $\underline{T}$. Then there exist C, $b > 0$ such that u can be expanded as*

$$u = \sum_{p,q \in \mathbb{N}_0} u_{p,q} \psi_{p,q} \quad \textit{in } L^2(\underline{T}),$$

where the coefficients $u_{p,q} \in \mathbb{R}$ satisfy

$$|u_{p,q}| \leq C e^{-b(p+q)}.$$

Furthermore, $b > 0$ depends only on the domain of holomorphy of (the holomorphic extension) of u.

Proof: The polynomials $\psi_{p,q}$ are $L^2(\underline{T})$-orthogonal; hence, the coefficients $u_{p,q}$ are given by

$$\begin{aligned} u_{p,q} &= \frac{1}{\|\psi_{p,q}\|_{L^2(\underline{T})}} \int_{\underline{T}} u(\xi_1,\xi_2)\psi_{p,q}(\xi_1,\xi_2)\, d\xi_1 d\xi_2 \\ &= \frac{\sqrt{2p+1}\sqrt{2p+2q+2}}{2} \int_{\underline{T}} u(\xi_1,\xi_2)\psi_{p,q}(\xi_1,\xi_2)\, d\xi_1 d\xi_2 . \end{aligned}$$

Writing $\tilde{u} = u \circ D$, we have to estimate

$$\begin{aligned} \tilde{u}_{p,q} &:= \int_{\underline{T}} u(\xi_1,\xi_2)\psi_{p,q}(\xi_1,\xi_2)\, d\xi_1 d\xi_2 \\ &= \int_{-1}^{1}\int_{-1}^{1} P_p^{(0,0)}(\eta_1)\left(\frac{1-\eta_2}{2}\right)^{p+1} P_q^{(2p+1,0)}(\eta_2)\tilde{u}(\eta_1,\eta_2)\, d\eta_1 d\eta_2 . \end{aligned}$$

To that end, we proceed in several steps.
1. step: We define the function

$$U_p(\eta_2) := \int_{-1}^{1} \tilde{u}(\eta_1,\eta_2) P_p^{(0,0)}(\eta_1)\, d\eta_1 . \tag{3.2.27}$$

By Lemma 3.2.12 there exist C, $\rho > 1$ independent of p such that

$$\sup_{\eta_2 \in \mathcal{E}_\rho} |U_p(\eta_2)| \le C\rho^{-p}.$$

This bound together with (3.2.18) implies

$$\begin{aligned} |\tilde{u}_{p,q}| &= \left| \int_{-1}^{1} U_p(\eta_2)\left(\frac{1-\eta_2}{2}\right)^{p+1} P_q(\eta_2) d\eta_2 \right| \\ &\le \left\{ \int_{-1}^{1} |U_p(\eta_2)|^2 \left(\frac{1-\eta_2}{2}\right) d\eta_2 \right\}^{1/2} \left\{ \int_{-1}^{1} \left(\frac{1-\eta_2}{2}\right)^{2p+1} \left(P_q^{(2p+1,0)}(\eta_2)\right)^2 d\eta_2 \right\}^{1/2} \\ &\le C\rho^{-p} \frac{2}{2p+2q+2} \le C\rho^{-p}. \end{aligned} \tag{3.2.28}$$

We require a second estimate for $\tilde{u}_{p,q}$ that decays exponentially in q. To that end, we note that Lemma 3.2.12 states that U_p is holomorphic on $\mathcal{E}_\rho$ and has a zero of multiplicity p at $\eta_2 = 1$. Hence, we may apply Cauchy's integral theorem to the holomorphic function $\eta_2 \mapsto U_p(\eta_2)/(1-\eta_2)^p$ to arrive at

$$\begin{aligned} \tilde{u}_{p,q} &= \int_{-1}^{1} U_p(\eta_2)\left(\frac{1-\eta_2}{2}\right)^{p+1} P_q^{(2p+1,0)}(\eta_2)\, d\eta_2 \\ &= \frac{1}{2\pi \mathbf{i}} \int_{\zeta_2 \in \partial\mathcal{E}_\rho} \frac{U_p(\zeta_2)}{((1-\zeta_2)/2)^p} \int_{-1}^{1} \left(\frac{1-\eta_2}{2}\right)^{2p+1} \frac{P_q^{(2p+1,0)}(\eta_2)}{\zeta_2 - \eta_2} d\eta_2 d\zeta_2 . \end{aligned}$$

Corollary 3.2.11 and Lemma 3.2.12 then imply the existence of C, $\tilde{\rho} \in (1,\rho)$ independent of p such that

$$|\tilde{u}_{p,q}| \leq C\rho^{-p}\left(\operatorname{dist}(\partial\mathcal{E}_\rho, 1)\right)^{-p} 2^{-p} \|\tilde{Q}_q^{(2p+1,0)}\|_{L^\infty(\mathcal{E}_\rho)} \leq C\gamma^p\tilde{\rho}^{-q}. \tag{3.2.29}$$

2. step: We now combine the estimates (3.2.28), (3.2.29) in order to get the desired result. To that end, we will consider the cases $p \leq \lambda q$ and $p > \lambda q$ separately for a $\lambda \in (0,1)$ to be chosen shortly.
For $p \leq \lambda q$, we have, assuming without loss of generality $\gamma > 1$ in (3.2.29),

$$|\tilde{u}_{p,q}| \leq C\gamma^p\tilde{\rho}^{-q} \leq C\gamma^{\lambda q}\tilde{\rho}^{-q} \leq C\hat{\rho}^{-q}, \tag{3.2.30}$$

for some $\hat{\rho} \in (1,\tilde{\rho})$, if we choose λ sufficiently small in dependence on γ, $\tilde{\rho}$. Since $\lambda < 1$ and $\lambda p \leq q$, we can estimate $q = \frac{1}{2}q + \frac{1}{2}q \geq \frac{1}{2}q + \frac{\lambda}{2}p \geq \lambda(p+q)/2$. Inserting this in (3.2.30) gives

$$|\tilde{u}_{p,q}| \leq C\tilde{\rho}^{-\lambda(p+q)/2} \qquad \forall (p,q) \in \mathbb{N}_0 \quad \text{such that } p \leq \lambda q.$$

In the converse case $p > \lambda q$ we reason in the same way to get $p \geq \frac{\lambda}{2}(p+q)$ and

$$|\tilde{u}_{p,q}| \leq C\rho^{-p} \leq C\rho^{-(p+q)\lambda/2} \quad \forall (p,q) \in \mathbb{N}_0^2 \quad \text{such that } p > \lambda q. \tag{3.2.31}$$

Since $\hat{\rho} < \tilde{\rho} < \rho$, combining (3.2.29) with (3.2.31) gives the desired bound by setting $b = (\lambda/2)\ln\hat{\rho}$. The statement that b depends only on the domain of holomorphy of u stems from the observation that b is determined by ρ of Lemma 3.2.12, which in turn depends only on the domain of holomorphy of u by Remark 3.2.13. □

Proposition 3.2.14 shows that exponential convergence can be achieved for the $L^2(\underline{T})$-projection onto spaces of polynomials. In order to obtain exponential convergence of the $L^2(\underline{T})$-projection in stronger norms such as the $L^\infty(\underline{T})$-norm, we need a result that allows us to control the orthogonal polynomials $\psi_{p,q}$ on (complex) neighborhoods of the closure of $\underline{T}$:

Lemma 3.2.15. *Let $\psi_{p,q}$ be the orthogonal polynomials polynomials defined in (3.2.23). Then for every $\rho > 1$ there exist C and a complex neighborhood $T' \subset \mathbb{C}^2$ of the closure $\operatorname{cl}\underline{T}$ of $\underline{T}$ such that*

$$\|\psi_{p,q}\|_{L^\infty(T')} \leq C\rho^{p+q} \qquad \forall p,q \in \mathbb{N}_0.$$

Proof: We start by recalling the Bernstein estimate for univariate polynomials of degree p:

$$\forall \rho > 1 \quad \forall u \in \mathcal{P}_p \qquad \|u\|_{L^\infty(\mathcal{E}_\rho)} \leq \rho^p \|u\|_{L^\infty((-1,1))} \tag{3.2.32}$$

(see, e.g., [41, Chap. 4, Thm. 2.2]). By tensor product arguments, it is easy to see that for bivariate polynomials $u \in \mathcal{Q}_p(S)$ we have

$$\|u\|_{L^\infty(\mathcal{E}_\rho \times \mathcal{E}_\rho)} \leq \rho^{2p}\|u\|_{L^\infty(\underline{S})}.$$

Put differently, for any $\rho > 1$, there exists a complex neighborhood $S' \subset \mathbb{C}^2$ of the closure $\operatorname{cl}\underline{S}$ of the square $\underline{S}$ such that

$$\|u\|_{L^\infty(S')} \le \rho^p \|u\|_{L^\infty(\underline{S})} \quad \forall u \in \mathcal{Q}_p(S).$$

With the aid of affine changes of variables, it follows then that for every closed parallelogram P and every $\rho > 1$ there exists a complex neighborhood P' of P such that

$$\|u\|_{L^\infty(P')} \le \rho^p \|u\|_{L^\infty(P)} \quad \forall u \in \mathcal{P}_p(T).$$

Since the triangle $\underline{T}$ can be covered by finitely many parallelograms, we conclude that for every $\rho > 1$ there exists a complex neighborhood T' of the triangle $\underline{T}$ such that

$$\|u\|_{L^\infty(T')} \le \rho^p \|u\|_{L^\infty(\underline{T})} \quad \forall u \in \mathcal{P}_p(T). \tag{3.2.34}$$

In order to conclude the proof, we require $\|\psi_{p,q}\|_{L^\infty(\underline{T})}$. By a standard inverse estimate (see, e.g., [112, Thm. 4.76]) and Lemma 3.2.9, we can bound with some $C > 0$ independent of p, q

$$\begin{aligned} \|\psi_{p,q}\|_{L^\infty(\underline{T})} &\le C(1+p+q)^2 \|\psi_{p,q}\|_{L^2(\underline{T})} \qquad (3.2.35)\\ &= C(1+p+q)^2 \left(\frac{2}{2p+1}\frac{2}{2p+2q+2}\right)^{1/2} \le C(1+p+q)^{3/2}. \end{aligned}$$

The claim of the lemma now follows by combining (3.2.35) with (3.2.34). □

Proposition 3.2.16. *Let u be analytic on the closure $\operatorname{cl}\underline{T}$ of $\underline{T}$. Then there exist C, $b > 0$ and a complex neighborhood T' of $\operatorname{cl}\underline{T}$ with the following properties: u can extended holomorphically to T' (the extension being again denoted u) and*

$$\inf_{v \in \mathcal{P}_p(T)} \|u - v\|_{L^\infty(T')} \le C e^{-bp} \quad \forall p \in \mathbb{N}_0 .$$

Moreover, the constant $b > 0$ depends only on the domain of holomorphy of u.

Proof: By the analyticity of u on $\operatorname{cl}\underline{T}$ there exists a complex neighborhood $T'' \subset \mathbb{C}^2$ of $\operatorname{cl}\underline{T}$ such that u can be extended holomorphically to T''. Next, Proposition 3.2.14 gives

$$u = \sum_{p,q \in \mathbb{N}_0} u_{p,q} \psi_{p,q} \quad \text{in } L^2(\underline{T}), \tag{3.2.36}$$

where the coefficients $u_{p,q} \in \mathbb{R}$ satisfy for some C, $b > 0$

$$|u_{p,q}| \le C e^{-b(p+q)}. \tag{3.2.37}$$

Choosing $\rho > 1$ so small that $\ln \rho - b \le -b/2$, we get from Lemma 3.2.15 the existence of $C > 0$ and a closed complex neighborhood $T' \subset T''$ of $\operatorname{cl}\underline{T}$ such that

$$\|\psi_{p,q}\|_{L^\infty(T')} \le C \rho^{p+q} \le C e^{-(b/2)(p+q)}. \tag{3.2.38}$$

Hence, the bounds (3.2.37), (3.2.38) imply that the series in (3.2.36) convergences in $L^\infty(T')$ and by holomorphy to u. The desired bound is easily ascertained for the choice $v = \sum_{i,j:i+j\le p} u_{i,j}\psi_{i,j} \in \mathcal{P}_p(T)$. □

Corollary 3.2.17. *Let u be analytic on the closure of $\underline{T}$ and let $k \in \mathbb{N}_0$. Then there exist C, $b > 0$ such that*

$$\inf_{v\in\mathcal{P}_p(T)} \|u - v\|_{W^{k,\infty}(\underline{T})} \le Ce^{-bp} \qquad \forall p \in \mathbb{N}_0.$$

Proof: The result follows from the Cauchy integral representation of derivatives and Proposition 3.2.16. □

Remark 3.2.18 Exponential polynomial approximability on triangles is in principle known. Previous literature has usually made the assumption that the function to be approximated can be extended analytically to a square containing the triangle—Proposition 3.2.16 removes this restriction and makes minimal assumptions on the domain of analyticity of the function u to be approximated.

The following result is the analog of Proposition 3.2.8 for the case of triangles.

Theorem 3.2.19 (approximation on triangles and squares). *Let K be the reference square or the reference triangle. Let u be analytic on K and satisfy for some C_u, $\gamma_u > 0$ and h_x, $h_y \in (0, 1]$*

$$\|D^\alpha u\|_{L^\infty(K)} \le C_u\gamma_u^{|\alpha|}\alpha! h_x^{\alpha_1} h_y^{\alpha_2} \qquad \forall\alpha \in \mathbb{N}_0^2 \setminus (0,0).$$

Then there exist C, $b > 0$ depending only on γ_u, and a sequence $v_p \in \mathcal{P}_p(T)$ of polynomials such that

$$\|u - v_p\|_{L^\infty(K)} \le CC_u \left[\left(\frac{h_x}{h_x + b}\right)^{p+1} + \left(\frac{h_y}{h_y + b}\right)^{p+1}\right],$$

$$\|\partial_x(u - v_p)\|_{L^\infty(K)} \le CC_u\frac{h_x}{h_x + b}\left[\left(\frac{h_x}{h_x + b}\right)^{p} + \left(\frac{h_y}{h_y + b}\right)^{p}\right],$$

$$\|\partial_y(u - v_p)\|_{L^\infty(K)} \le CC_u\frac{h_y}{h_y + b}\left[\left(\frac{h_x}{h_x + b}\right)^{p} + \left(\frac{h_y}{h_y + b}\right)^{p}\right].$$

Proof: For $K = S$, the result is proved in Proposition 3.2.8. The case $K = T$ can be seen as follows. There exists $h_0 < 1$ depending only on γ_u such that if $h_x < h_0$ or $h_y < h_0$ then u can be extended analytically to S. Let us assume that $h_x < h_0$. The analytic extension of u to S (again denoted u) then satisfies

$$\|\partial_x^{n_1}\partial_y^{n_2} u\|_{L^\infty(S)} \le CC_u\widetilde{\gamma}_u^{n_1+n_2}(n_1 + n_2)! h_0^{-n_1} h_x^{n_1} h_y^{n_2} \qquad \forall(n_1, n_2) \in \mathbb{N}_0^2 \setminus (0, 0)$$

for some C, $\widetilde{\gamma}_u$ depending only on γ_u. The result now follows from the analysis for squares of Proposition 3.2.8. Thus, Theorem 3.2.19 is proved if $h_x < h_0$ or $h_y < h_0$. It remains to consider the case where simultaneously $h_x \ge h_0$ and $h_y \ge h_0$. Then, the result follows immediately from Proposition 3.2.16. □

3.2.4 The projector Π_p^∞

Our ultimate goal is the construction of polynomial approximants in an element-by-element fashion. To that end, it is convenient to construct approximants that coincide with the Gauss-Lobatto interpolant on the edges. On squares S, it is therefore natural to consider the Gauss-Lobatto interpolation operator j_p. On triangles T, the construction of an interpolation operator that coincides with the Gauss-Lobatto interpolants on the edges is not as straight forward. This is accomplished in the ensuing theorem.

Theorem 3.2.20 (Definition & Properties of Π_p^∞). *Let $K = S$ or $K = T$. There is $C > 0$ such that for every $p \geq 1$ there is a bounded linear operator $\Pi_p^\infty : C(\overline{K}) \to \Pi_p(K)$ with the following properties:*

$$i_{p,\Gamma} u = \Pi_p^\infty u|_\Gamma \qquad \forall \text{ edges } \Gamma \text{ of } K, \tag{3.2.39}$$

$$\|u - \Pi_p^\infty u\|_{L^\infty(K)} \leq C C_K(p) \inf_{v \in \Pi_p(K)} \|u - v\|_{L^\infty(K)}, \tag{3.2.40}$$

$$\Pi_p^\infty v = v \qquad \text{for all } v \in \Pi_p(K). \tag{3.2.41}$$

Here, the constant $C_K(p)$ is given by

$$C_K(p) := \begin{cases} (1 + \ln p)^2 & \text{if } K = S, \\ p(1 + \ln p) & \text{if } K = T. \end{cases} \tag{3.2.42}$$

In particular, for $p = 1$, the operator Π_1^∞ coincides with the linear/bilinear interpolation operator.
For functions $u \in H^1(K) \cap C(\overline{K})$ there holds

$$\|\nabla(u - \Pi_p^\infty u)\|_{L^2(K)} \leq \inf_{v \in \Pi_p(K)} \left\{ \|\nabla(u - v)\|_{L^2(K)} + C p^2 C_K(p) \|u - v\|_{L^\infty(K)} \right\}, \tag{3.2.43}$$

and for functions $u \in W^{1,\infty}(K)$ we have

$$\|\nabla(u - \Pi_p^\infty u)\|_{L^\infty(K)} \leq \inf_{v \in \Pi_p(K)} \left\{ \|\nabla(u - v)\|_{L^\infty(K)} + C p^2 C_K(p) \|u - v\|_{L^\infty(K)} \right\}. \tag{3.2.44}$$

Proof: We proceed in two steps. First, we construct the bounded linear operator $\Pi_p^\infty : C(\overline{K}) \to \Pi_p(K)$ satisfying (3.2.39), (3.2.41), and

$$\|\Pi_p^\infty\|_{C(\overline{K}) \to L^\infty} \leq C C_K(p) \tag{3.2.45}$$

for some $C > 0$ independent of p; $C_K(p)$ is given in the statement of the theorem. In a second step, the estimates (3.2.40) and (3.2.43), (3.2.44) are obtained. For the case $K = S$, we set $\Pi_p^\infty := j_p$ and note that Lemma 3.2.1 gives readily (3.2.45). It is easy to see that j_p also satisfies (3.2.39), (3.2.41). We now turn to the definition of the operator Π_p^∞ on triangles T. For $u \in C(\overline{T})$ let $i_{p,\partial T} u$ be the (edgewise) Gauss-Lobatto polynomial interpolant of u. With the extension operator E of Lemma 3.2.3 we then obtain with the stability properties of the Gauss-Lobatto interpolation operator (Lemma 3.2.1)

$$\|E(i_{p,\partial T}u)\|_{L^\infty(T)} \leq C\|i_{p,\partial T}u\|_{L^\infty(\partial T)} \leq C(1+\ln p)\|u\|_{L^\infty(\partial T)}.$$

Next, we introduce the closed subspace $V := \{u \in \mathcal{P}_p(T) \mid u = 0 \text{ on } \partial T\}$ and let Π^∞ be the bounded linear operator from $C(\overline{T})$ onto V given by Corollary A.4.2. The desired operator Π_p^∞ is then taken as

$$\Pi_p^\infty u := E(i_{p,\partial T}u) + \Pi^\infty(u - E(i_{p,\partial T}u)).$$

By the bound $\|\Pi^\infty\|_{L^\infty \to L^\infty} \leq (p+1)$, it is clear that $\|\Pi_p^\infty\|_{C(\overline{K})\to L^\infty} \leq Cp(1+\ln p) = CC_K(p)$ with $C_K(p)$ given in the statement of the theorem for $K = T$. By construction, (3.2.39) holds, and it is easy to see that Π_p^∞ satisfies (3.2.41). In the second step, we obtain (3.2.40). As Π_p^∞ reduces to the identity on $\Pi_p(K)$, we have for all $v \in \Pi_p(K)$

$$\|u - \Pi_p^\infty u\|_{L^\infty(K)} = \|(u-v) - \Pi_p^\infty(u-v)\|_{L^\infty(K)} \leq CC_K(p)\|u - v\|_{L^\infty(K)}.$$

Taking the infimum over all $v \in \Pi_p(K)$ gives (3.2.40). (3.2.43) and (3.2.44) are proved in a similar way, and we therefore only show (3.2.43). For every $v \in \Pi_p(K)$, we have

$$\begin{aligned}
&\|\nabla(u - \Pi_p^\infty u)\|_{L^2(K)} = \|\nabla((u-v) - \Pi_p^\infty(u-v))\|_{L^2(K)} \\
&\leq \|\nabla(u-v)\|_{L^2(K)} + \|\nabla(\Pi_p^\infty u - v)\|_{L^2(K)} \\
&\leq \|\nabla(u-v)\|_{L^2(K)} + Cp^2\|\Pi_p^\infty u - v\|_{L^2(K)} \\
&\leq \|\nabla(u-v)\|_{L^2(K)} + Cp^2\left[\|u - \Pi_p^\infty u\|_{L^\infty(K)} + \|u - v\|_{L^2(K)}\right] \\
&\leq \|\nabla(u-v)\|_{L^2(K)} + Cp^2 C_K(p)\|u - v\|_{L^\infty(K)},
\end{aligned}$$

where in the last step, we employed (3.2.40). Taking the infimum over all $v \in \Pi_p(K)$ gives the desired bound. □

For future reference, we collect some approximation results for the projector Π_p^∞ in the following proposition.

Proposition 3.2.21. *Let $K = S$ or $K = T$ and let A be one vertex of K. Let $\beta \in [0,1)$. Then there holds*

$$\begin{aligned}
&\|u - \Pi_p^\infty u\|_{L^\infty(K)} + \|u - \Pi_p^\infty u\|_{L^2(K)} + p^{-2}\|\nabla(u - \Pi_p^\infty u)\|_{L^2(K)} \\
&\leq Cp(1+\ln p)\|\,|x - A|^\beta \nabla^2 u\|_{L^2(K)}.
\end{aligned} \tag{3.2.46}$$

For functions u that are analytic on K and satisfy for some C_u, $\gamma_u > 0$

$$\|\nabla^p u\|_{L^\infty(K)} \leq C_u \gamma_u^p p! \qquad \forall p \in \mathbb{N},$$

there are C, $b > 0$ depending only on γ_u such that

$$\|u - \Pi_p^\infty u\|_{L^\infty(K)} + \|\nabla(u - \Pi_p^\infty u)\|_{L^\infty(K)} \leq CC_u e^{-bp}. \tag{3.2.47}$$

Proof: The first estimate follows from inserting the result of Lemma 3.2.4 into the bounds of Theorem 3.2.20. Inserting the results of Theorem 3.2.19 into the statement of Theorem 3.2.20 gives the second assertion. □

Remark 3.2.22 For analytic functions u, Proposition 3.2.21 shows that the projector Π_p^∞ yields exponential convergence. In the case of finite regularity, i.e., $u \in H^k$ for some $k > 1$, estimating (3.2.43) requires simultaneous approximation in different norms, which is provided in [6]. The resulting estimates, however, are not p-optimal due to our using inverse estimates. An alternative projector leading to p-optimal estimates is presented in Section 3.2.6. ▪

3.2.5 Anisotropic projection operators: $\Pi_p^{1,\infty}$

In order to motivate the introduction of the second interpolation operator $\Pi_p^{1,\infty}$, let us consider the approximation of a (smooth) function u on a rectangle $R = (0,h_x) \times (0,h_y)$ whose aspect ratio may be very large. On the reference square S, one has to approximate the function $\hat{u} := u(h_x x, h_y y)$. The operator Π_p^∞ is essentially an L^∞ projector on the reference square. Because the L^∞-norm is invariant under bijective mappings, it also yields good approximation results for u on R in the L^∞ norm. However, the operator Π_p^∞ does not allow for good anisotropic gradient estimates. The aim of the present section is the construction of an operator $\Pi_p^{1,\infty}$ that shares with Π_p^∞ the property that it reduces to a Gauss-Lobatto interpolation operator on the edges of the domain (square S or triangle T) and at the same time reflects anisotropy of the gradient of the function to be interpolated.
The anisotropic behavior of the gradient behavior will be captured by the anisotropic norm $|||\cdot|||$. For given h_x, $h_y > 0$ we introduce on the space $W^{1,\infty}(K)$ the following weighted norm

$$|||u||| := \|u\|_{L^\infty(K)} + h_x^{-1}\|\partial_x u\|_{L^\infty(K)} + h_y^{-1}\|\partial_y u\|_{L^\infty(K)}. \tag{3.2.48}$$

In order to construct the interpolation operator $\Pi_p^{1,\infty}$ that is stable uniformly in h_x, h_y we need an lemma:

Lemma 3.2.23. *Let $K = S$ or $K = T$. Then there is $C > 0$ such that the following holds. For every h_x, $h_y > 0$ and every $p \in \mathbb{N}$ there is a bounded linear operator $E^{1,\infty} : W^{1,\infty}(K) \to \Pi_p(K)$ with the following properties:*

$$E^{1,\infty}u|_\Gamma = i_{p,\Gamma}\, u|_\Gamma \qquad \forall \text{ edges } \Gamma \text{ of } K,$$
$$|||E^{1,\infty}u||| \le C C_K(p)|||u|||,$$

where $C_K(p)$ is given by

$$C_K(p) := \begin{cases} p(1+\ln p) & \text{if } K = S \\ p & \text{if } K = T. \end{cases}$$

Furthermore, in the case $K = S$, the operator $E^{1,\infty}$ may be taken as j_p.

Proof: We start with the case $K = S$. Then we set $E^{1,\infty} := j_p$, the tensor-product Gauss-Lobatto interpolation operator. We have from Lemma 3.2.1 $\|E^{1,\infty}u\|_{L^\infty(K)} \leq C(1+\ln p)^2\|u\|_{L^\infty(K)} \leq Cp(1+\ln p)\|u\|_{L^\infty(K)}$. For the derivatives, we recall $j_p = i_p^y \circ i_p^x$ and estimate, using Lemma 3.2.1 twice,

$$\|\partial_x(i_p^y \circ i_p^x u)\|_{L^\infty(S)} = \|i_p^y \partial_x i_p^x u\|_{L^\infty(S)} \leq C(1+\ln p)\|\partial_x i_p^x u\|_{L^\infty(S)}$$
$$\leq Cp(1+\ln p)\|\partial_x u\|_{L^\infty(S)}.$$

An analogous estimate holds for $\partial_y j_p u$, which proves the lemma if $K = S$.
We now turn to the case $K = T$. We first show that for the case $p = 1$, we may take the linear interpolant given by

$$Iu(x,y) = yu(0,1) + xu(1,0) + (1-x-y)u(0,0).$$

Obviously $\|Iu\|_{L^\infty(T)} \leq \|u\|_{L^\infty(T)}$. For the derivatives, we compute

$$\partial_x Iu = u(1,0) - u(0,0) = \int_0^1 \partial_x u(\xi,0)\,d\xi, \tag{3.2.49}$$

and therefore $\|\partial_x Iu\|_{L^\infty(T)} \leq \|\partial_x u\|_{L^\infty(T)}$. Since an analogous estimate holds for $\partial_y Iu$, we conclude

$$|||Iu||| \leq |||u|||.$$

For $p > 1$, we may therefore construct the extension operator $E^{1,\infty}$ for functions u that vanish at the vertices of T. Furthermore, without loss of generality we may assume

$$h_y \leq h_x. \tag{3.2.50}$$

Let $u \in W^{1,\infty}$ vanish at the vertices of T. We construct $E^{1,\infty}u$ by considering the edge $\Gamma_1 = \{(0,y)\,|\,y \in (0,1)\}$ first and then the two remaining edges. Set

$$\pi_p^1(x,y) := \frac{1-x-y}{1-y}h_1(y), \qquad h_1(y) := i_p^y u(0,y).$$

Since $u(0,1) = 0$, the function $h_1 \in \mathcal{P}_p(I)$ vanishes at $y = 1$ and thus $\pi_p^1 \in \mathcal{P}_p(T)$. We note furthermore that on Γ_1 the polynomial π^1 coincides with the Gauss-Lobatto interpolant of u, and π^1 vanishes on the two remaining edges of T.
For the auxiliary function h_1, we obtain with the aid of Lemma 3.2.1

$$|h_1(y)| \leq C(1+\ln p)\|u\|_{L^\infty(T)}, \tag{3.2.51}$$
$$\left|\frac{h_1(y)}{1-y}\right| = \left|\frac{h_1(y)-h_1(1)}{1-y}\right| \leq \|h_1'\|_{L^\infty((0,1))} \leq Cp\|\partial_y u\|_{L^\infty(T)} \tag{3.2.52}$$

for all $y \in (0,1)$. We compute the derivatives

$$\partial_x \pi_p^1(x,y) = -\frac{h_1(y)}{1-y},$$
$$\partial_y \pi_p^1(x,y) = -\frac{x}{1-y}\frac{h_1(y)}{1-y} + \frac{1-x-y}{1-y}h_1'(y).$$

The bounds

$$\left|\frac{x}{1-y}\right| \le 1, \quad \left|\frac{1-x-y}{1-y}\right| \le 1 \qquad \forall (x,y) \in T \tag{3.2.53}$$

together with (3.2.51), (3.2.52), and the assumption (3.2.50) allow us to estimate

$$\begin{aligned}
\|\pi_p^1\|_{L^\infty(T)} &\le \|h_1\|_{L^\infty((0,1))} \le C(1+\ln p)\|u\|_{L^\infty(T)} \le Cp|||u|||,\\
h_x^{-1}\|\partial_x \pi_p^1\|_{L^\infty(T)} &\le C h_x^{-1} p \|\partial_y u\|_{L^\infty(T)} \le C p h_y^{-1}\|\partial_y u\|_{L^\infty(T)} \le Cp|||u|||,\\
h_y^{-1}\|\partial_y \pi_p^1\|_{L^\infty(T)} &\le C p h_y^{-1}\|\partial_y u\|_{L^\infty(T)} \le Cp|||u|||.
\end{aligned}$$

It remains to correct the other two edges. We introduce the one dimensional interpolants

$$h_1(x) = i_p^x u(x,0), \qquad h_2(x) = i_p^x u(x, 1-x)$$

and define

$$\pi_p^2(x,y) = h_1(x) + \frac{y}{1-x}\left(h_2(x) - h_1(x)\right).$$

Because $h_1(1) = h_2(1) = 0$, we have $\pi_p \in \mathcal{P}_p(T)$. We note again that π_p vanishes on the edge $x = 0$ and equals $i_{p,\Gamma}^x u$ for the remaining two edges Γ of T. Analogously to the bounds (3.2.51), (3.2.52), we have for all $x \in (0,1)$:

$$\begin{aligned}
|h_1(x)| + |h_2(x)| &\le C(1+\ln p)\|u\|_{L^\infty(T)}, &(3.2.54)\\
\|h_1'\|_{L^\infty((0,1))} &\le Cp\|\partial_x u\|_{L^\infty(T)}, &(3.2.55)\\
\|h_2'\|_{L^\infty((0,1))} &\le Cp\left[\|\partial_x u\|_{L^\infty(T)} + \|\partial_y u\|_{L^\infty(T)}\right]. &(3.2.56)
\end{aligned}$$

In order to bound $(h_2(x) - h_1(x))/(1-x)$, we note that, since $h_2(1) = h_1(1)$, we have $(h_2(x) - h_1(x))/(1-x) \in \mathcal{P}_{p-1}$. Since this polynomial coincides in the Gauss-Lobatto points with the function $(u(x,0) - u(x,1-x))/(1-x)$, we conclude

$$\frac{h_2(x) - h_1(x)}{1-x} = i_p^x\left(\frac{u(x,0) - u(x,1-x)}{1-x}\right).$$

Estimating

$$|u(x,0) - u(x,1-x)| = \left|\int_0^{1-x} \partial_y u(x,t)\,dt\right| \le (1-x)\|\partial_y u\|_{L^\infty(T)},$$

we obtain with the stability of the Gauss-Lobatto interpolation operator

$$\begin{aligned}
\sup_{x\in(0,1)}\left|\frac{h_2(x) - h_1(x)}{1-x}\right| &\le C(1+\ln p) \sup_{x\in(0,1)}\left|\frac{u(x,0) - u(x,1-x)}{1-x}\right|\\
&\le C(1+\ln p)\|\partial_y u\|_{L^\infty(T)}. \qquad (3.2.57)
\end{aligned}$$

A calculation reveals

$$\begin{aligned}
\partial_x \pi_p^2(x,y) &= \frac{1-x-y}{1-x} h_1'(x) + \frac{y}{1-x} h_2'(x) + \frac{y}{1-x}(h_2(x) - h_1(x)),\\
\partial_y \pi_p^2(x,y) &= \frac{h_2(x) - h_1(x)}{1-x}.
\end{aligned}$$

Hence, we can bound using (3.2.54)–(3.2.57), (3.2.53), and (3.2.50)

$$\|\pi_p^2\|_{L^\infty(T)} \le C(1+\ln p)\|u\|_{L^\infty(T)} \le C(1+\ln p)|||u|||,$$
$$h_x^{-1}\|\partial_x \pi_p^2\|_{L^\infty(T)} \le Cph_x^{-1}\left[\|\partial_x u\|_{L^\infty(T)} + \|\partial_y u\|_{L^\infty(T)}\right] \le Cp|||u|||,$$
$$h_y^{-1}\|\partial_y \pi_p^2\|_{L^\infty(T)} \le C(1+\ln p)h_y^{-1}\|\partial_y u\|_{L^\infty(T)} \le C(1+\ln p)|||u|||.$$

The claim of the lemma now follows. □

We are now in position to define the projection operator $\Pi_p^{1,\infty}$.

Theorem 3.2.24 (Definition & Properties of $\Pi_p^{1,\infty}$). *Let $K = S$ or $K = T$ and h_x, $h_y > 0$. Let $|||\cdot|||$ be given by (3.2.48). Then there exists $C > 0$ independent of h_x, h_y, and p, and there exists a bounded linear operator $\Pi_p^{1,\infty} : W^{1,\infty}(K) \to \Pi_p(K)$ with the following properties:*

$$i_{p,\Gamma} u = \Pi_p^{1,\infty} u|_\Gamma \qquad \forall \text{ edges } \Gamma \text{ of } K, \tag{3.2.58}$$
$$|||u - \Pi_p^{1,\infty} u||| \le C C_K(p) \inf_{v \in \Pi_p(K)} |||u - v|||, \tag{3.2.59}$$
$$\Pi_p^{1,\infty} v = v \qquad \text{for all } v \in \Pi_p(K). \tag{3.2.60}$$

Here, the constant $C_K(p)$ is given by

$$C_K(p) := \begin{cases} p(1+\ln p) & \text{if } K = S, \\ p^2 & \text{if } K = T. \end{cases} \tag{3.2.61}$$

Proof: For $K = S$, we choose $\Pi_p^{1,\infty} = j_p$ and the statement follows from Lemma 3.2.23. It remains to define $\Pi_p^{1,\infty}$ for triangles. We proceed as in the proof of Theorem 3.2.20. We set $V := \{u \in \mathcal{P}_p(T) \,|\, u = 0 \text{ on } \partial T\}$. Let $E^{1,\infty}$ be the bounded linear operator of Lemma 3.2.23 and let $\Pi^{1,\infty}$ be the bounded linear operator of Corollary A.4.3 mapping $W^{1,\infty}(K)$ onto V. We then set

$$\Pi_p^{1,\infty} u := E^{1,\infty} u + \Pi^{1,\infty}\left(u - E^{1,\infty} u\right).$$

Clearly, $\Pi_p^{1,\infty}$ is linear, it satisfies (3.2.58), and it is bounded:

$$|||\Pi_p^{1,\infty} u||| \le Cp^2 |||u||| \qquad \forall u \in C^1(\overline{T}).$$

Furthermore, $\Pi_p^{1,\infty}$ reduces to the identity operator on the space $\mathcal{P}_p(T)$. The desired final bound (3.2.59) now follows as in the proof of Theorem 3.2.20. □

We can therefore get the following approximation result for analytic functions with anisotropic bounds on the derivatives.

Proposition 3.2.25. *Let $K = S$ or $K = T$. Assume that a function u is analytic on K and satisfies, for some C_u, $\gamma_u > 0$, h_x, $h_y \in (0,1]$*

$$\|\partial_x^n \partial_y^m u\|_{L^\infty(K)} \le C_u h_x^n h_y^m \gamma_u^{n+m} n! m! \qquad \forall (p,q) \in \mathbb{N}_0^2 \setminus (0,0).$$

Then there exist C, $\sigma > 0$ depending only on γ_u such that the difference $e := u - \Pi_p^{1,\infty} u$ satisfies with $h := \max\{h_x, h_y\}$

$$\|e\|_{L^\infty(K)} + \frac{h}{h_x}\|\partial_x e\|_{L^\infty(K)} + \frac{h}{h_y}\|\partial_y e\|_{L^\infty(K)} \le C C_u \left(\frac{h}{h+\sigma}\right)^{p+1}.$$

Proof: The result follows by choosing in the infimum in Theorem 3.2.24 the approximant of Theorem 3.2.19. □

Remark 3.2.26 The growth conditions on the derivatives of the analytic function u in the statement of Proposition 3.2.25 appear quite naturally if so-called *normalizable* triangulations are considered, cf. Definition 3.3.3 ahead. ▪

3.2.6 An optimal error estimate for an H^1-projector

For analytic functions u, the projectors Π_p^∞, $\Pi_p^{1,\infty}$ yield approximants that are exponentially close to u. The situation is different if u has finite regularity, e.g., $u \in H^k(K)$ for some $k > 1$. Then the projectors Π_p^∞, $\Pi_p^{1,\infty}$ do not yield the expected optimal rate $p^{-(k-1)}$ when measuring the error in the H^1-norm (cf., e.g., (3.2.43)). While the projectors Π_p^∞, $\Pi_p^{1,\infty}$ are sufficient for the purposes of this work, it is interesting to note that it is possible to construct a projector Π^{H^1} that realizes the optimal rate of convergence and at the same time is constrained to coincide with the Gauss-Lobatto interpolation operator on the boundary.

Proposition 3.2.27. *Let $K = S$ or $K = T$, $k > 3/2$. Then there exist a constant $C(k) > 0$ and a projector $\Pi^{H^1} : H^k(K) \to \Pi_p(K)$ such that:*

$$i_{p,\Gamma} u = \Pi^{H^1} u|_{\partial K} \qquad \forall \text{edges } \Gamma \text{ of } K, \tag{3.2.62}$$

$$\Pi^{H^1} u = u \qquad \forall u \in \Pi_p(K), \tag{3.2.63}$$

$$\|u - \Pi^{H^1} u\|_{H^1(K)} \le C(k) p^{-(k-1)} |u|_{H^k(K)}. \tag{3.2.64}$$

Proof: By [22] there exists a bounded linear operator $\Pi : H^k(K) \to \Pi_p(K)$ with the following properties:

$$\|u - \Pi u\|_{H^t(K)} \le C p^{-(k-t)} \|u\|_{H^k(K)}, \qquad t \in \{0,1\},$$
$$\Pi u(A) = u(A) \qquad \forall \text{ vertices } A \text{ of } K,$$
$$\|u - \Pi u\|_{H^t(\Gamma)} \le C p^{-(k-1/2-t)} \|u\|_{H^k(K)} \qquad \forall \text{ edges } \Gamma \text{ of } K, \quad t \in \{0,1\}.$$

We conclude by interpolation

$$\|u - \Pi u\|_{H^{1/2}_{00}(\Gamma)} \le C p^{-(k-1)} \|u\|_{H^k(K)} \qquad \forall \text{edges } \Gamma \text{ of } K.$$

Next, by Lemma 3.2.28 below and the trace theorem, we have for each edge Γ of K the bound

$$\|u - i_{p,\Gamma}u\|_{H^{1/2}_{00}(\Gamma)} \leq Cp^{-(k-1)}\|u\|_{H^k(K)}.$$

By construction $\Pi u|_{\partial K} - i_{p,\partial K}u$ vanishes at the vertices of K, it is a polynomial of degree p on each edge, and

$$\|i_{p,\partial K} - \Pi u\|_{H^{1/2}_{00}(\Gamma)} \leq Cp^{-(k-1)}\|u\|_{H^k(K)} \qquad \forall \text{ edges } \Gamma \text{ of } K. \tag{3.2.65}$$

By [13] there exists a linear map

$$E : \{f \in C(\partial K) \,|\, f|_\Gamma \in \mathcal{P}_p \text{ for each edge } \Gamma\} \to \Pi_p(K)$$

with the property that

$$\|Ef\|_{H^1(K)} \leq C\|f\|_{H^{1/2}(\partial K)}$$

for some $C > 0$ independent of p. We conclude with (3.2.65)

$$\|E(i_{p,\partial K}u - \Pi u)\|_{H^1(K)} \leq C \sum_{\Gamma \subset \partial K} \|i_{p,\Gamma}u - \Pi u\|_{H^{1/2}_{00}(\Gamma)} \leq Cp^{-(k-1)}\|u\|_{H^k(K)}.$$

Hence, the map

$$\Pi' : u \mapsto \Pi u + E(i_{p,\partial K}u - \Pi u)$$

satisfies (3.2.62) and

$$\|u - \Pi' u\|_{H^1(K)} \leq Cp^{-(k-1)}\|u\|_{H^k(K)}.$$

Next, we adjust Π' such that (3.2.63) is satisfied. To that end, we define the best approximation operator $\Pi'' : H^1(K) \to \Pi_p(K) \cap H^1_0(K)$ by

$$\|u - \Pi'' u\|_{H^1(K)} = \min\{\|u - q\|_{H^1(K)} \,|\, q \in \Pi_p(K) \cap H^1_0(K)\}.$$

Π'' is a bounded linear map with $\|\Pi'' u\|_{H^1(K)} \leq \|u\|_{H^1(K)}$ and $\Pi'' u = u$ for all $u \in \Pi_p(K) \cap H^1_0(K)$. We then define

$$\Pi^{H^1} := \Pi' + \Pi''(\mathrm{Id} - \Pi').$$

It is easy to see that this operator satisfies (3.2.62) (3.2.63) and

$$\|u - \Pi^{H^1} u\|_{H^1(K)} \leq Cp^{-(k-1)}\|u\|_{H^1(K)}.$$

Since $\Pi^{H^1} u = u$ for all $u \in \Pi_p(K)$, we may replace the full H^k-norm on the right-hand side by the H^k-semi norm in the standard way. □

Lemma 3.2.28. *Let $I = (0, 1)$, $k \geq 1$. Then there exists $C(k) > 0$ such that the Gauss-Lobatto interpolation operator i_p satisfies*

$$\|u - i_p u\|_{H^1(I)} \leq Cp^{-(k-1)}\|u\|_{H^k(I)}, \tag{3.2.66}$$

$$\|u - i_p u\|_{L^2(I)} \leq Cp^{-k}\|u\|_{H^k(I)}. \tag{3.2.67}$$

By interpolation therefore

$$\|u - i_p u\|_{H^{1/2}_{00}(I)} \leq Cp^{-(k-1/2)}\|u\|_{H^k(I)}. \tag{3.2.68}$$

Proof: The estimate (3.2.66) follows directly from the H^1-stability of the Gauss-Lobatto interpolation, Theorem A.3.1. Likewise, the L^2-bound (3.2.67) follow from a stability estimate of Theorem A.3.1:

$$\|u - i_p u\|_{L^2(I)} \leq C \inf_{q\in\mathcal{P}_p} \|u-q\|_{L^2(I)} + p^{-1}\|u-q\|_{H^1(I)} \leq Cp^{-k}\|u\|_{H^k(I)}.$$

Since u and $i_p u$ coincide at the endpoints of I, the operator $\mathrm{Id} - i_p$ maps in fact into $H^1_0(I)$; that is, we have $\mathrm{Id} - i_p : H^k(I) \to H^1_0(I)$ with norm bounded by $Cp^{-(k-1)}$ and $\mathrm{Id} - i_p : H^k(I) \to L^2(I)$ with norm bounded by Cp^{-k}. Interpolation then gives the desired bound (3.2.68). □

Remark 3.2.29 The projector Π^{H^1} of Proposition 3.2.27 is such that the approximant coincides with the function to be approximated in the Gauss-Lobatto points on the boundary. For *a priori* estimates in the p-version FEM, this allows for the construction of approximants in an element-by-element fashion. Different approaches, which are also able to handle the case $k \in (1, 3/2)$, have been taken in [18, 22, 23]. ▪

3.3 Admissible boundary layer meshes and finite element spaces

We recall that the reference square S and the reference triangle T are defined as $S = (0,1) \times (0,1)$ and $T = \{(x,y) \,|\, 0 < x < 1 \,|\, 0 < y < x\}$. We also refer the reader to Definition 2.4.1 for the precise notion of triangulations $\mathcal{T}$.
We now consider the problem of introducing the notion of element size for curved elements. We will not develop a general theory for characterizing the size of an element but in the next two definitions, we present two different approaches to this issue.

Definition 3.3.1 ((C_M, γ_M)-regular triangulation). *A triangulation $\mathcal{T} = \{(K_i, M_i, \hat{K}_i)\}$ of a domain Ω is called (C_M, γ_M)-regular if for each element K_i there is an affine map A_i with $A_i' = h_i$, $h_i := \mathrm{diam} K_i$, such that*

$$\begin{aligned} \| \left((A_i^{-1} \circ M_i)'\right)^{-1} \|_{L^\infty(\hat{K}_i)} &\leq C_M, \\ \|\nabla^p (A_i^{-1} \circ M_i)\|_{L^\infty(\hat{K}_i)} &\leq C_M \gamma_M^p p! \qquad \forall p \in \mathbb{N}_0 . \end{aligned}$$

Remark 3.3.2 As stated, Definition 3.3.1 allows only for isotropic elements of size h_i. The definition could be extended to introduce the notion of anisotropic elements—this would require that a Cartesian coordinate (possibly different for each element) be chosen with respect to which anisotropic stretching is done. The reader will recognize that this approach is taken implicitly in the definition of boundary layer elements in Definition 2.4.4. ▪

In the approach of Definition 3.3.1, the notion of element size is introduce by stipulating the ability to control $A_i^{-1} \circ M_i$ where A_i is some appropriate stretching map. Clearly, one could also think of stipulating the ability to control $M_i \circ A_i$ for some stretching map A_i. This approach is taken in the notion of *normalizable* triangulations that we introduce now.

Definition 3.3.3 (**(C_M, γ_M)-normalizable triangulation).** *A triangulation* $\mathcal{T} = (K_i, M_i, \hat{K}_i)$ *is called* (C_M, γ_M)-normalizable *if for each element* K_i *there are* $h_{x,i}$, $h_{y,i} \in (0,1]$ *such that the element map* M_i *can be written in the form*

$$M_i = G_i \circ A_i,$$

where A_i *is an* affine *map satisfying*

$$A_i' = \begin{pmatrix} h_{x,i} & 0 \\ 0 & h_{y,i} \end{pmatrix},$$

and the analytic map G_i *satisfies*

$$\begin{aligned} \|\nabla^p G_i\|_{L^\infty(R_i)} &\le C_M \gamma_M^p p! \qquad \forall p \in \mathbb{N}_0, \\ \|(G_i')^{-1}\|_{L^\infty(R_i)} &\le C_M. \end{aligned}$$

Here, the sets R_i *are the images of the reference element* $\hat{K}_i$ *under the affine maps* A_i, *i.e.,* $R_i = A_i(\hat{K}_i)$.

Remark 3.3.4 Some comments concerning Definition 3.3.3 are in order. Firstly, the notion of normalizable triangulations introduces in a very natural way anisotropic elements, and we employed a similar notion in the context of Shishkin meshes in Definition 2.6.17. Secondly, the requirement for a triangulation to be (C_M, γ_M)-normalizable is stronger than that to be (C_M, γ_M)-regular (see Proposition 3.3.5 and the closely related Proposition 3.3.11). We point out that the stronger notion of normalizable triangulations lends itself more easily to error bounds in approximation theory; in fact, Propositions 3.2.25, 3.2.8 are formulated so as to fit into the framework of normalizable triangulations. ▪

Proposition 3.3.5. *Let* $\mathcal{T} = \{(K_i, M_i, \hat{K}_i)\}$ *be a* (C_M, γ_M)*-normalizable triangulation and assume that for each element* K_i *there holds* $h_{x,i} = h_{y,i} = h_i = \operatorname{diam} K_i$. *Then there are* C, $\gamma > 0$ *depending only on* C_M, γ_M *such that* $\mathcal{T}$ *is a* (C, γ)*-regular triangulation in the sense of Definition 3.3.1*

Proof: Consider an element K_i. By assumption, the element map M_i has the form $M_i = G_i \circ A_i$. The affine map whose existence is stipulated in Definition 3.3.1 is now taken as $\tilde{A}_i x := A_i(x - M_i(0))$. The assumptions on G_i then easily imply the desired result. □

Once a triangulation $\mathcal{T}$ is chosen, one can defined finite-element spaces $S^p(\mathcal{T})$ based on this triangulation.

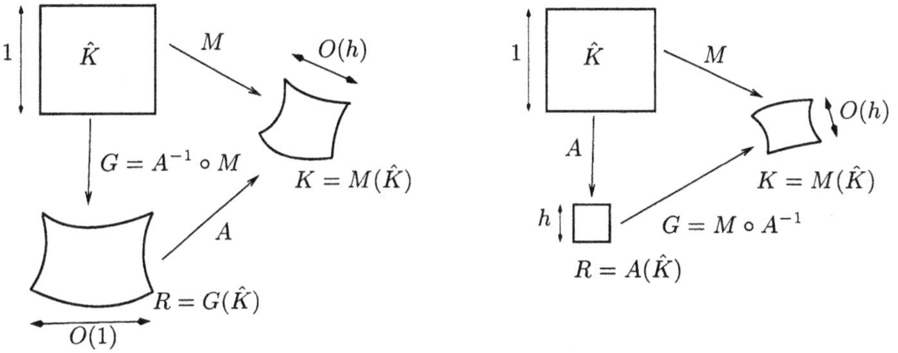

Fig. 3.3.1. Two different approaches to the concept of element size: via Def. 3.3.1 (left) and Def. 3.3.3 (right).

Definition 3.3.6 (FE-spaces). *Given a triangulation $\mathcal{T} = \{(K_i, M_i, \hat{K}_i) \,|\, i \in I(\mathcal{T})\}$, the H^1-conforming finite element spaces $S^p(\mathcal{T})$, $S^p_0(\mathcal{T})$ of piecewise mapped polynomials are defined as*

$$S^p(\mathcal{T}) := \{u \in H^1(\Omega) \,|\, u|_{K_i} = \varphi_p \circ M_i^{-1} \ \text{ for some } \varphi_p \in \Pi_p(\hat{K}_i)\}, \tag{3.3.1}$$

$$S^p_0(\mathcal{T}) := S^p(\mathcal{T}) \cap H^1_0(\Omega) \tag{3.3.2}$$

with spaces $\Pi_p(\hat{K}_i)$ defined in (3.2.4).

In Section 3.2.4, we introduced the interpolation operator Π^∞_p on the reference square S and the reference triangle T. For a mesh $\mathcal{T}$ we can then define the operator $\Pi^\infty_{p,\mathcal{T}}$ by an elementwise application of Π^∞_p:

$$\Pi^\infty_{p,\mathcal{T}} u|_{K_i} := \left(\Pi^\infty_p(u \circ M_i)\right) \circ M_i^{-1} \qquad \forall \text{ elements } K_i. \tag{3.3.3}$$

For a mesh $\mathcal{T}$ in the sense of Definition 2.4.1 we can naturally speak about vertices and edges as the images of the vertices and edges of the references elements. Hence, we can also introduce the edge-wise Gauss-Lobatto interpolation operator $i_{p,\Gamma}$ as follows: Let Γ be an edge of an element K_i, let $\hat{\Gamma} := M_i^{-1}(\Gamma)$, and set

$$i_{p,\Gamma} u := \left(i_{p,\hat{\Gamma}}(u \circ M_i)\right) \circ M_i^{-1}. \tag{3.3.4}$$

We note that the operator $i_{p,\Gamma}$ is well-defined: An edge Γ of a triangulation $\mathcal{T}$ is shared (in general) by two elements K_i, K_j. However, assumption (M4) of Definition 2.4.1 guarantees that the parametrization of Γ induced by both M_i and M_j coincides, thus making the definition of $i_{p,\Gamma}$ in (3.3.4) well-defined. We now show that the operator $\Pi^\infty_{p,\mathcal{T}}$ in fact is an operator from $C(\overline{\Omega})$ into $S^p(\mathcal{T})$:

Lemma 3.3.7. *Let $\mathcal{T} = \{(K_i, M_i, \hat{K}_i) \,|\, i \in I(\mathcal{T})\}$ be a triangulation of a domain Ω in the sense of Definition 2.4.1. Let $u \in C(\overline{\Omega})$. For each element K_i let $\pi_{p,i} \in \Pi_p(\hat{K}_i)$ satisfy*

$$i_{p,\hat{\Gamma}}(u \circ M_i) = \pi_{p,i}|_{\hat{\Gamma}} \qquad \forall \text{ edges } \hat{\Gamma} \text{ of } \hat{K}_i. \tag{3.3.5}$$

Then π defined on Ω by $\pi|_{K_i} := \pi_{p,i} \circ M_i^{-1}$ is an element of $S^p(\mathcal{T})$. Furthermore, if $u = 0$ on $\partial\Omega$, then $\pi \in S_0^p(\mathcal{T})$.

Proof: One has to check that π defined elementwise is in fact continuous across the edges. This follows from a) assumption (M4) of Definition 2.4.1 that shared edges of the triangulation have the same parametrization; b) that the Gauss-Lobatto interpolation points are distributed symmetrically; and c) the uniqueness of the Gauss-Lobatto interpolation process. □

Of particular use to us will the following result.

Corollary 3.3.8. *Let $\mathcal{T} = \{(K_i, M_i, \hat{K}_i) \,|\, i \in I(\mathcal{T})\}$ be a triangulation of a domain Ω in the sense of Definition 2.4.1. Then the operator $\Pi^{\infty}_{p,\mathcal{T}} : C(\overline{\Omega}) \to S^p(\mathcal{T})$ defined in (3.3.3) is a linear operator. Furthermore, if $u = 0$ on $\partial\Omega$, then $\Pi^{\infty}_{p,\mathcal{T}} u \in S_0^p(\mathcal{T})$. The same result holds if the operator Π^{∞}_p in (3.3.3) is replaced with $\Pi^{1,\infty}_p$.*

Remark 3.3.9 The operator $\Pi^{\infty}_{p,\mathcal{T}}$ constructs approximations from $S^p(\mathcal{T})$ in a strictly elementwise fashion. This procedure is different from the classical approach of Babuška & Suri, [22]. There, approximants are constructed in two steps: In the first step, a discontinuous approximant is constructed in an element-by-element fashion. In the second step, the interelement jumps across edges are removed with a polynomial trace-lifting. This procedure is not convenient for our purposes as we expressly want to consider meshes which may be distorted and in which thin needle elements may share an edge with large elements, i.e., where two adjacent elements may be structurally very different. Our elementwise approach with the operator $\Pi^{\infty}_{p,\mathcal{T}}$ avoids having to consider two adjacent elements at the same time. It should be mentioned, however, that the construction of interpolants using $\Pi^{\infty}_{p,\mathcal{T}}$ does not lead to optimal p-version approximation results in the case of finite Sobolev regularity, i.e., if the function u to be approximated is only known to be in a Sobolev space $H^k(\Omega)$ for some $k \geq 1$. The projector Π^{H^1} constructed in Section 3.2.6 overcomes this difficulty. ▪

3.3.1 *hp*-meshes for the approximation of boundary and corner layers

The regularity assertion of Theorem 2.3.4 shows that we have to deal with two types of phenomena: boundary layers in the vicinity of the boundary curves Γ_j and corner singularities in neighborhoods of the vertices A_j. In order to resolve these two phenomena, one needs meshes that contain *needle* elements near the boundary curves Γ_j and geometric meshes near the vertices A_j. The (essentially) minimal mesh family is one with needle elements near the boundary and geometric mesh refinement near the vertices. An example of such an admissible mesh is presented in Fig. 2.4.2: The rectangles at the boundary are *boundary layer elements* of width $O(\kappa)$, the elements in the shaded regions are *corner layer elements*, and the remaining elements are *interior elements*. The precise requirements for these three element types were presented in Definition 2.4.4.

Admissible boundary layer meshes as defined in Definition 2.4.4 are essentially the minimal meshes that can lead to robust exponential convergence for the approximation of solutions to (2.1.1). As discussed in Remark 2.4.5 and shown in Fig. 2.4.1, the meshes may be severely distorted in the sense that minimal and maximal angles can be close to 0 or π. For implementational reasons, it is advisable to be able to control the distortion of the elements, in particular the maximal angles. The introduction of *regular admissible meshes* allows for this control. In essence, a regular admissible mesh is an admissible mesh family that is also a normalizable mesh family. The mesh of Fig. 2.4.1 is in fact an example of a regular admissible mesh.

Definition 3.3.10 (regular admissible boundary layer mesh). *Consider a two-parameter family $\mathcal{T}(\kappa, L) = \{(K_i, M_i, \hat{K}_i)\}_{\kappa,L}$, $(\kappa, L) \in (0, \kappa_0] \times \mathbb{N}_0$, of (C_M, γ_M)-*normalizable *meshes (in the sense of Definition 3.3.3). Denote by $h_{x,i}$, $h_{y,i}$ the anisotropic scaling parameters of the element map M_i given by Definition 3.3.3. This family $\mathcal{T}(\kappa, L)$ is said to be* regular admissible *if there are c_i, $i = 1, \dots, 4$, $\sigma \in (0,1)$, and sets Ω_j, $j = 1, \dots, J$, of the form given in Notation 2.3.3 such that each element K_i of the mesh $\mathcal{T}(\kappa, L)$ falls into exactly one of the following three categories:*

(C1) K_i is a boundary layer element, *i.e., for some $j \in \{1, \dots, J\}$ there holds $K_i \subset \mathcal{U}_\kappa(\Gamma_j) \cap \Omega_j \setminus (B_{c_1\kappa}(A_{j-1}) \cup B_{c_1\kappa}(A_j))$ and*

$$M_i(E) \subset \Gamma_j, \qquad E = \{(x, 0) \,|\, 0 < x < 1\},$$
$$h_{x,i} = 1, \qquad h_{y,i} = \kappa.$$

(C2) K_i is a corner layer element, *i.e., for some $j \in \{1, \dots, J\}$, the element K_i satisfies $K_i \subset B_\kappa(A_j) \cap \Omega_j$ or $K_i \subset B_\kappa(A_j) \cap \Omega_{j+1}$. Additionally, denoting $h_i = \operatorname{diam} K_i$, the factors $h_{x,i}$, $h_{y,i}$ satisfy*

$$h_{x,i} = h_{y,i} = h_i.$$

Furthermore, for the element K_i exactly one of the following situations is satisfied: either $A_j \in \overline{K_i}$ together with $h_i \le c_4 \kappa \sigma^L$ or $A_j \notin \overline{K_i}$ together with $c_3 h_i \le \operatorname{dist}(A_j, K_i) \le c_4 h_i$.

(C3) K_i is an interior element, *i.e., $K_i \subset \Omega \setminus \mathcal{U}_{c_2\kappa}(\partial\Omega)$, and the factors $h_{x,i}$, $h_{y,i}$ satisfy*

$$h_{x,i} = 1, \qquad h_{y,i} = \kappa.$$

Analogous to Proposition 3.3.5, we have that regular admissible boundary layer mesh are also admissible boundary layer meshes.

Proposition 3.3.11. *Let $\mathcal{T}(\kappa, L)$ be a regular admissible mesh family in the sense of Definition 3.3.10. Then it is an admissible boundary layer mesh family in the sense of Definition 2.4.4.*

Proof: We start with the most interesting case, that of boundary layer elements. Let K_i be a boundary layer element abutting on Γ_j. As the triangulation $\mathcal{T}$ is assumed to be a normalizable triangulation, the element map M_i can be factored as $M_i = G_i \circ A_i$ for some affine map with

$$A_i' = \begin{pmatrix} h_{x,i} & 0 \\ 0 & h_{y,i} \end{pmatrix},$$

and the mapping G_i satisfies for some C, $\gamma > 0$

$$\|\nabla^p G_i\|_{L^\infty(R_i)} \leq C\gamma^p p! \qquad \forall p \in \mathbb{N}_0, \tag{3.3.6}$$

where the set R_i is $R_i = (0, h_{x,i}) \times (0, h_{y,i})$ in the case of a quadrilateral and $R_i = \{(x,y) \mid 0 < x < h_{x,i}, 0 < y < h_{y,i}(1 - x/h_{x,i})\}$ in the case of a triangle. From the analyticity of G_i and ψ_j, we readily see that for some C, $\gamma > 0$ independent of κ, L there holds

$$\|\nabla^p \left(\psi_j^{-1} \circ G_i\right) \|_{L^\infty(R_i)} \leq C\gamma^p p! \qquad \forall p \in \mathbb{N}_0 .$$

Next, including the assumption that the edge $E = \{(x, 0) \mid 0 < x < 1\}$ is mapped into Γ_j, i.e., $M_i(E) \subset \Gamma_j$, we have that $\psi_j^{-1} \circ G_i$ has the form

$$\psi_j^{-1} \circ G_i(x,y) = (y\mathcal{R}_i(x,y), \Theta_i(x,y)) \qquad \forall (x,y) \in R_i,$$

where the functions $\mathcal{R}_i$, Θ_i are analytic on R_i with

$$\|\nabla^p \mathcal{R}_i\|_{L^\infty(R_i)} + \|\nabla^p \Theta_i\|_{L^\infty(R_i)} \leq C\gamma^p p! \qquad \forall p \in \mathbb{N}_0 .$$

Using now the assumption $h_{x,i} = 1$, $h_{y,i} = \kappa$, we get the representation $s_\kappa^{-1} \circ \psi_j^{-1} \circ M_i(x,y) = s_\kappa^{-1} \circ \psi_j^{-1} \circ G_i \circ A_i(x,y) = (y\mathcal{R}_i(x, h_{y,i}y), \Theta_i(x, h_{y,i}y))$ from which the desired bounds follow. It is easy to see that the conditions on interior elements are satisfied as well. For corner layer elements, we exploit the fact that A_i is affine to get

$$\|\nabla^p \left(G_i \circ A_i\right) \|_{L^\infty(S)} \leq C \|\nabla^p G_i\|_{L^\infty(R_i)} h_i^p .$$

Combining this with $\|\tilde{s}_{h_i}^{-1}\| \leq C h_i^{-1}$, we readily get the desired bounds for $p \geq 1$. It remains to show that $\|\tilde{s}_{j,h_i}^{-1} \circ M_i\| \leq C$. This follows from the fact that $\operatorname{diam} R_i = A_i(S)$ is bounded by $C h_i$ and that hence (3.3.6) implies $\operatorname{diam} G_i(R_i) \leq C h_i$. □

We now turn to the question of constructing admissible boundary layer meshes. A general framework in which this can be accomplished is provided by the notion of *patchwise structured meshes*, which we describe in the ensuing subsection. The meshes constructed in this manner are in fact regular admissible boundary layer meshes.

3.3.2 Patchwise structured meshes

The general idea is to start from a quasi-uniform mesh (the "macro-triangulation" whose elements are called "patches") that resolves the geometry. The actual features such as boundary layer elements and corner layer elements are then defined on the reference element for the appropriate elements and "transported" by the corresponding element map to the computational domain. Rather than starting with a formal definition of this idea, we begin with an example to illustrate the basic idea. Let $\widetilde{\mathcal{T}} = \{(\widetilde{K}_j, \widetilde{M}_j, \hat{K}_j) \,|\, j = 1, \ldots, N\}$ be a fixed triangulation of Ω with N elements. This triangulation is called the macro-triangulation. Next, let $\mathcal{T}_1, \ldots, \mathcal{T}_N$ be N triangulations of the reference elements $\hat{K}_1, \ldots, \hat{K}_N$. These triangulations are supposed to be of the form $\mathcal{T}_j = \{(K_{jk}, M_{jk}, \hat{K}_{jk}) \,|\, k = 1, \ldots, N_j\}$, $j = 1, \ldots, N$. Then the collection $\{\widetilde{M}_j(K_{jk}) \,|\, k \in I(j), j = 1, \ldots, N\}$ forms a partition of Ω satisfying (M1). The corresponding element maps are given by $\widetilde{M}_j \circ M_{jk} : \hat{K}_{jk} \to \widetilde{M}_j(K_{jk})$. We are interested in the case when the collection $\mathcal{T} := \{(\widetilde{M}_j(K_{jk}), \widetilde{M}_j \circ M_{jk}, \hat{K}_{jk})\}$ actually forms a triangulation of Ω in the sense of Definition 2.4.1. In that case, we say that $\widetilde{\mathcal{T}}$ is a macro-triangulation for the triangulation $\mathcal{T}$, and we have the following result.

Lemma 3.3.12. *Let $\widetilde{\mathcal{T}} = \{(\widetilde{K}_j, \widetilde{M}_j, \hat{K}_j), j \in I(\widetilde{\mathcal{T}})\}$ be a triangulation of Ω. For each $j \in I(\widetilde{\mathcal{T}})$, let triangulations $\mathcal{T}_j := \{(K_{jk}, M_{jk}, \hat{K}_{jk}) \,|\, k \in I(\mathcal{T}_j)\}$ of the reference elements $\hat{K}_j$ be given with bilinear/linear element maps, i.e., M_{jk} is an affine map if $\hat{K}_{jk} = T$ and M_{jk} is a bilinear map if $\hat{K}_{jk} = S$. Define the collection of triples $\mathcal{T} := \{(\widetilde{M}_j(K_{jk}), \widetilde{M}_j \circ M_{jk}, \hat{K}_{jk} \,|\, k \in I(\mathcal{T}_j), j \in I(\widetilde{\mathcal{T}})\}$. Then $\mathcal{T}$ satisfies (M1)–(M5) (i.e., it is a triangulation of Ω) if it satisfies (M2).*

Proof: It is clear that $\mathcal{T}$ already satisfies (M1) and (M5). It is therefore enough to see that (M4) follows from (M2). This follows readily from the fact that the restrictions of the element maps M_{jk} to the edges of the elements K_{jk} are linear and that the macro elements maps $\widetilde{M}_j$ satisfy (M4). □

Lemma 3.3.12 allows us to construct in a very convenient way triangulations of Ω: Starting from a fixed, coarse macro-triangulation $\widetilde{\mathcal{T}}$, actual triangulations $\mathcal{T}$ can be generated from triangulations $\mathcal{T}_j$ of the reference elements $\hat{K}_j$. Given the fact that the reference elements $\hat{K}_j$ are only triangles and squares, one would naturally use triangulations of the reference elements that consist of straight triangles and straight quadrilaterals only. Hence, by Lemma 3.3.12, one merely has to check that the resulting collection $\mathcal{T}$ does not have any "hanging nodes". In what follows, this condition is easily checked as we will allow a very limited number of types of subtriangulations. This idea of prescribing a macro-triangulation and a list of possible reference patches is formalized in the following definition:

Definition 3.3.13. *Let $\hat{\mathcal{T}}^S$, $\hat{\mathcal{T}}^T$ be two collections of triangulations of the reference square S and the reference triangle T respectively. Set $\hat{\mathcal{T}}^{ref} := \hat{\mathcal{T}}^T \cup \hat{\mathcal{T}}^S$. Let $\widetilde{\mathcal{T}} = \{(\widetilde{K}_j, \widetilde{M}_j, \hat{K}_j) \,|\, j = 1, \ldots, N\}$ and $\mathcal{T}$ be two triangulations of Ω. We say that $\mathcal{T}$ is of* type *$\hat{\mathcal{T}}^{ref}$ with respect to the macro-triangulation $\widetilde{\mathcal{T}}$ if there exist triangulations $\mathcal{T}_j = \{(\Omega_{jk}, M_{jk}, \hat{K}_{jk})\} \in \hat{\mathcal{T}}^{ref}$, $j = 1, \ldots, N$, such that $\mathcal{T} = \{(\widetilde{M}_j(\Omega_{jk}), \widetilde{M}_j \circ M_{jk}, \hat{K}_{jk}) \,|\, k \in I(\mathcal{T}_j), j = 1, \ldots, N\}$.*

Remark 3.3.14 The concept of a limited number of types of reference patches can have many advantages, both from an implementational point of view and

from an analysis point of view. We do not dwell on the implementational ramifications here. From an analysis point of view the major advantage is that only very few canonical situations can arise which are easy to handle individually. The simplest example of a collection of reference patches $\hat{\mathcal{T}}^{ref}$ is provided by the *trivial reference patch*, i.e., $\mathcal{T}^S := \{(S, \mathrm{Id}, S)\}$, $\mathcal{T}^T := \{(T, \mathrm{Id}, T)\}$, and therefore $\mathcal{T}^{ref} = \{\mathcal{T}^S, \mathcal{T}^T\}$. Obviously, in that case, only $\mathcal{T} = \widetilde{\mathcal{T}}$ is possible. Another simple reference patch is given by the *uniform reference square* $\mathcal{T}^S(M)$, i.e., for a given $M \in \mathbb{N}$, the unit square is uniformly subdivided into M^2 squares (naturally, a similar construction can be done for the reference triangle). One could then choose $\mathcal{T}^{ref} := \mathcal{T}^S(M)$ or even $\mathcal{T}^{ref} := \cup_{M \in \mathbb{N}} \mathcal{T}^S(M)$. More interesting examples are provided by mesh patches that are refined geometrically towards one of the corners.
In our applications below, we only consider (C_M, γ_M)-regular macro triangulations $\widetilde{\mathcal{T}}$; the elements of the reference patches, however, may be highly distorted in order to be able to resolve the boundary and corner layers. ▪

3.3.3 The p-version boundary layer and corner layer patches

We now define the references patches that are the basic building blocks for the patchwise structured boundary layer meshes. For simplicity of notation, we assume that the macro-triangulation $\widetilde{\mathcal{T}}$ consists of quadrilaterals only, i.e., $\hat{K}_j = S$ for all j. This is a purely notational simplification in order to reduce the number of reference patches.
Given $\kappa \in (0, 1/2]$, $\sigma \in (0, 1)$ and $L \in \mathbb{N}_0$ a reference patch can be only one of the following four types:

1. The *trivial* patch: $\check{\mathcal{T}} = \{(S, \mathrm{Id}, S)\}$.
2. The *hp boundary layer* reference patch is of the form

$$\check{\mathcal{T}} = \{(F_1(S), F_1, S), (F_2(S), F_2, S)\}$$

with maps F_1, F_2 given by (cf. Fig. 3.3.2, left)

$$\begin{aligned} F_1 : S &\to S & F_2 : S &\to S \\ (\xi, \eta) &\mapsto (\xi, \kappa\eta), & (\xi, \eta) &\mapsto (\xi, \kappa + (1 - \kappa)\eta). \end{aligned}$$

3. The *hp tensor product corner layer* reference patch with grading factor σ and $L + 1$ layers is given by the simplest triangulation of S consisting of rectangles and triangles that contains the points (cf. Fig. 3.3.2, right)

$$\begin{aligned} &(0,0), (0,1), (1,0), (1,1), (\kappa, 0), (0, \kappa), (1, \kappa), (\kappa, 1), \\ &(\sigma^l \kappa, 0), \quad (0, \sigma^l \kappa), \quad (\sigma^l \kappa\sqrt{2}, \sigma^l \kappa\sqrt{2}), \qquad l = 0, \ldots, L. \end{aligned}$$

4. The *hp mixed corner layer* reference patch with grading factor σ and $L + 1$ layers is given by the simplest triangulation of S consisting of rectangles and triangles that contains the points (cf. Fig. 3.3.3, left)

$$\begin{aligned} &(0,0), (0,1), (1,0), (1,1), (\kappa, 0), (0, \kappa), (1, \kappa), \\ &(\sigma^l \kappa, 0), \quad (0, \sigma^l \kappa), \quad (\sigma^l \kappa\sqrt{2}, \sigma^l \kappa\sqrt{2}), \qquad l = 0, \ldots, L. \end{aligned}$$

5. The *hp mixed corner layer* that is the mirror image with respect to the mid-line $x = 1/2$ of the preceding mixed corner layer patch.
6. The *hp geometric corner layer* reference patch with grading factor σ and $L+1$ layers is given by the simplest triangulation of S consisting of rectangles and triangles that contains the points (cf. Fig. 3.3.3, right)

$$(0,0), (0,1), (1,0), (1,1), (\kappa, 0), (0, \kappa), (1, \kappa),$$
$$(\sigma^l \kappa, 0), \quad (0, \sigma^l \kappa), \quad (\sigma^l \kappa\sqrt{2}, \sigma^l \kappa\sqrt{2}), \qquad l = 0, \ldots, L.$$

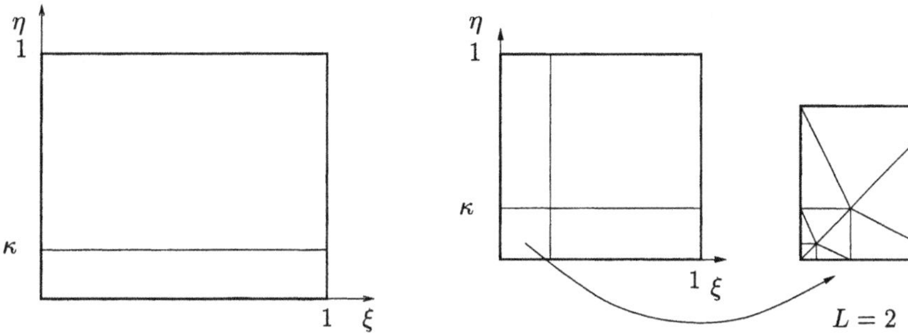

Fig. 3.3.2. Ref. boundary layer patch (left); ref. tensor product patch (right).

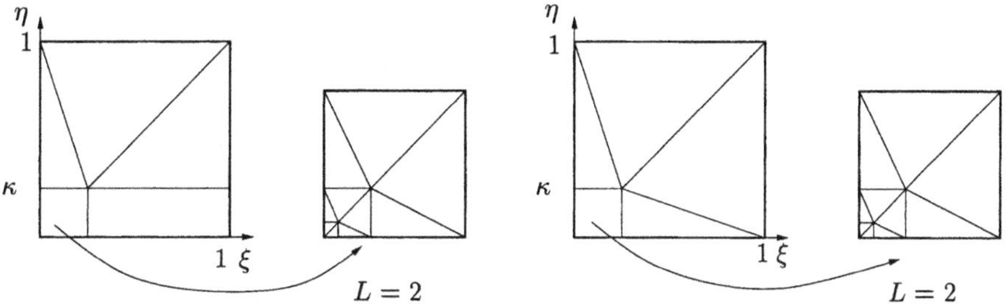

Fig. 3.3.3. Reference mixed patch (left); reference geometric patch (right).

The six types of patches just introduced represent the minimal number of patches required for *hp*-approximation of our model problem. For practical applications of the notion of mesh patches, one would introduce further types. For problems with layers, patches that feature anisotropic geometric refinement towards one (or two) edges of the reference square are useful. We therefore define them at this point:

1. The *anisotropically geometrically* refined patch with $L+1$ layers and grading factor $\sigma \in (0,1)$ is given by the simplest triangulation of S consisting of rectangles that contains the points (cf. Fig. 3.3.4, left)

$$(0, \sigma^i), (1, \sigma^i), \qquad i = 0, 1, \ldots, L.$$

2. The *tensor product anisotropically geometrically* refined patch with $L+1$ layers and grading factor $\sigma \in (0,1)$ is given by the simplest triangulation of S consisting of rectangles that contains the points (cf. Fig. 3.3.4, right)

$$(0,\sigma^i), (\sigma^i, 0), (\sigma^i, \sigma^j), \qquad i,j = 0,1,\ldots,L.$$

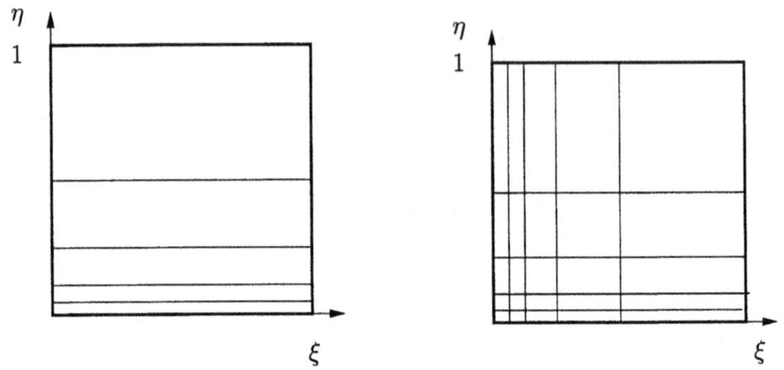

Fig. 3.3.4. Anisotropic and geometric refinement towards one edge (left) and two edges (right).

3.3.4 Boundary layer mesh generation via mesh patches

We are now in position to define meshes via mesh patches. We start by defining the macro-triangulation $\widetilde{\mathcal{T}} = \{(\widetilde{K}_j, \widetilde{M}_j, \hat{K}_j)\}$. It should satisfy the following conditions (cf. Fig. 3.3.5, left):

(MT1) All elements are quadrilaterals, i.e., $\hat{K}_j = S$ for all j.

(MT2) For every $(\widetilde{K}_j, \widetilde{M}_j, S) \in \widetilde{\mathcal{T}}$ exactly one of the following situations can occur:
  a) $\partial\widetilde{K}_j \cap \partial\Omega = \emptyset$;
  b) $\partial\widetilde{K}_j \cap \partial\Omega = \widetilde{M}_j(0,0)$;
  c) $\partial\widetilde{K}_j \cap \partial\Omega = \widetilde{M}_j(E)$ where $E = \{(x,y) \in S \,|\, y = 0\}$;
  d) $\partial\widetilde{K}_j \cap \partial\Omega = \widetilde{M}_j(E)$ where $E = \{(x,y) \in S \,|\, y = 0 \text{ or } x = 0\}$.

(MT3) If A_m is a vertex of Ω at a strictly convex corner then there exists a unique $(\widetilde{K}_j, \widetilde{M}_j, S) \in \widetilde{\mathcal{T}}$ of type (MT2).(a) with $\widetilde{M}_j(0,0) = A_m$; this element is said to be a *tensor-product patch*.

(MT4) If A_m is a vertex of Ω at a concave corner then there are exactly three elements $(\widetilde{K}_j, \widetilde{M}_j, S)$, $(\widetilde{\Omega}_k, \widetilde{M}_k, S)$, $(\widetilde{\Omega}_l, \widetilde{M}_l, S) \in \widetilde{\mathcal{T}}$ satisfying $\widetilde{M}_j(0,0) = \widetilde{M}_k(0,0) = \widetilde{M}_l(1,0) = A_m$ and the following extra conditions: $(\widetilde{K}_j, \widetilde{M}_j, S)$ is of type (MT2).(b) and is said to be a *geometric patch*; the other two elements are of type (MT2).(c) and said to be *mixed patches*.

(MT5) Any element of type (MT2).(c) that does not fall into the category (MT4) (i.e., elements of type (MT2).(c) with $\partial\widetilde{K}_j \cap \cup_m A_m = \emptyset$) is said to be a *boundary layer patch*.

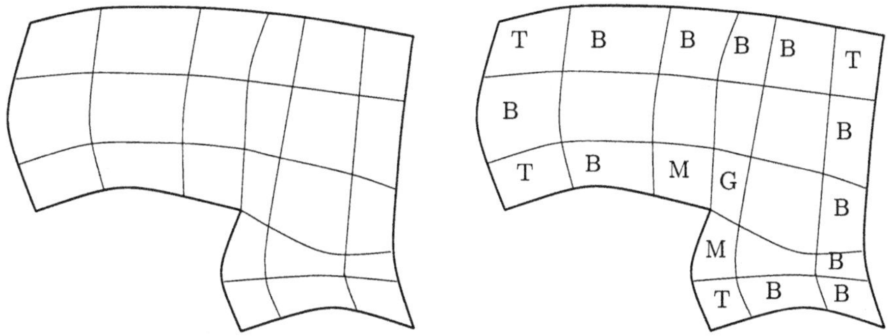

Fig. 3.3.5. Left: Example of a macro-triangulation. Right: B, T,M, G indicate boundary layer, tensor product, mixed, and geometric patches, respectively.

(MT6) Any element of type (MT2).(a) is called a *trivial patch.*

We are now in position to define the meshes $\mathcal{T}(\kappa, L)$ that we wish to use. Let $\widetilde{\mathcal{T}}$ be a fixed macro-triangulation satisfying (MT1)–(MT6) above. For convenience's sake, we also assume that the macro-triangulation $\widetilde{\mathcal{T}}$ is chosen such that the macro-elements abutting on the boundary are sufficiently small so that the decompositions based on asymptotic expansions of Theorem 2.3.4 are valid on whole patches near $\partial\Omega$. Additionally, as the corner layers are only piecewise smooth, we assume that the diagonal of the geometric patch is mapped under the macro-element map $\widetilde{M}_i$ to a curve Γ that can be taken as a curve Γ_j' of Notation 2.3.3 This is formalized in the following two additional assumptions.

(MT7) Let κ_0 be as in (2.4.3). Then the macro-elements $\widetilde{K}_j$ abutting on $\partial\Omega$ are in the κ_0-neighborhood of $\partial\Omega$: $\overline{\widetilde{K}_i} \cap \partial\Omega \neq \emptyset$ implies $\widetilde{K}_i \subset \mathcal{U}_{\kappa_0}(\partial\Omega)$.

(MT8) For geometric patches $(\widetilde{K}_j, \widetilde{M}_j, S)$ at a vertex A_m, we assume that the angles between the image $\widetilde{M}_j(D)$ of the diagonal $D := \{(x, y) \in S \,|\, x = y\}$ and the two boundary components Γ_{m-1}, Γ_m is strictly less than π, i.e., upon setting $T_1 = T$, $T_2 = S \setminus T_1$ the set $\widetilde{M}_j(T_1)$, $\widetilde{M}_j(T_2)$ are completely contained in the subdomain Ω_m or Ω_{m+1}.

For $(\kappa, L) \in (0, 1/2] \times \mathbb{N}_0$ and a fixed macro-triangulation $\widetilde{\mathcal{T}}$ satisfying (MT1)–(MT6) above, the triangulation $\mathcal{T}(\kappa, L)$ is a triangulation of Ω for which $\widetilde{\mathcal{T}}$ is a macro-triangulation where the reference patches are chosen as follows: For elements of type (MT2).(a), the trivial reference patch is chosen, for elements of type (MT5), the boundary layer reference patch is chosen, for elements of type (MT4) the reference tensor product patch is chosen, and in the situation (MT3), the reference geometric patch is chosen for the geometric patch and reference mixed patches are chosen for the two mixed patches (in the notation of (MT3), the reference mixed patch is chosen for the element $(\widetilde{\Omega}_k, \widetilde{M}_k, S)$ and the mirror image of the reference mixed patch for the element $(\widetilde{\Omega}_l, \widetilde{M}_l, S)$). One should note that the conditions (MT2) on the element maps $\widetilde{M}_j$ guarantee that our choices of the reference patches lead to a triangulation of Ω.

Proposition 3.3.15. *Let $\mathcal{T}^{macro}$ be a fixed macro triangulation satisfying the assumptions (MT1)–(MT8) and let $\mathcal{T}(\kappa, L)$ be a two-parameter family of triangulations generated by mesh patches as described above. Then $\mathcal{T}(\kappa, L)$ is a regular admissible boundary layer mesh family in the sense of Definition 3.3.10.*

Proof: The result is obvious as one only has to check the properties for the four types of reference patches (boundary layer, tensor product, mixed, and geometric) on the reference square. □

3.3.5 Properties of the pull-backs to the patches

In the present section, we illustrate that analytic regularity results on the physical domain Ω imply corresponding analytic regularity results on the reference domain $\hat{K}$ if the concept of mesh patches is employed. Hence, in this context, one can perform approximation theory on simple triangulations of the reference domains $\hat{K}$ and then transport those results back to the physical domain with the patch maps. We will not pursue this approach to obtaining approximation theoretical results in the present work as we concentrate on the approximation properties on admissible meshes, which have notably less structure than regular admissible meshes or meshes generated with mesh patches. Nevertheless, we collect some results that show how regularity results on the physical domain can be transferred to the reference domains. Polynomial approximation results on mesh patches can be fairly easily obtained for the operators Π_p^∞ and $\Pi_p^{1,\infty}$ (e.g., Proposition 3.2.25) once regularity assertions on the physical domain Ω have been transferred to the reference elements via the patch maps.

Proposition 3.3.16. *Let $\widetilde{\mathcal{T}} = \{(\widetilde{K}_j, \widetilde{M}_j, \hat{K}_j)\}$ be a (C_M, γ_M)-regular triangulation in the sense of Definition 3.3.1. Then there exists $C > 0$ depending only on the constants C_M, γ_M such that for all $\widetilde{K}_j$ there holds with $h_j = \operatorname{diam}\widetilde{K}_j$ and $u_j := u \circ \widetilde{M}_j$:*

$$\begin{aligned}
\|u\|_{L^\infty(\widetilde{K}_j)} &= \|u_j\|_{L^\infty(\hat{K}_j)} && \forall u \in L^\infty(\widetilde{K}_j),\\
C^{-1}h_j^{-1}\|\nabla u_j\|_{L^\infty(\hat{K}_j)} \le \|\nabla u\|_{L^\infty(\widetilde{K}_j)} &\le C h_j^{-1}\|\nabla u_j\|_{L^\infty(\hat{K}_j)} && \forall u \in W^{1,\infty}(\widetilde{K}_j),\\
C^{-1}h_j^{1/2}\|u_j\|_{L^2(\hat{K}_j)} \le \|u\|_{L^2(\widetilde{K}_j)} &\le C h_j^{1/2}\|u_j\|_{L^2(\hat{K}_j)} && \forall u \in L^2(\widetilde{K}_j),\\
C^{-1}\|\nabla u_j\|_{L^2(\hat{K}_j)} \le \|\nabla u\|_{L^2(\widetilde{K}_j)} &\le C\|\nabla u_j\|_{L^2(\hat{K}_j)} && \forall u \in H^1(\widetilde{K}_j).
\end{aligned}$$

Proof: Follows from the properties of the mapping functions $\widetilde{M}_j$. □

The next lemma shows that, as the maps $\widetilde{M}_j$ provide analytic diffeomorphisms between the reference elements $\hat{K}_j$ and the physical element $\widetilde{K}_j$, regularity results on $\widetilde{K}_j$ are transferred to the reference elements $\hat{K}_j$. The reader will recognize that in Lemma 3.3.17, (ii), the maps s, $\hat{s}$ could be taken as appropriate stretching maps as in the definition of admissible meshes in Definition 2.4.4.

Lemma 3.3.17. *Let $\widetilde{\mathcal{T}} = \{(\widetilde{K}_j, \widetilde{M}_j, \hat{K}_j)\}$ be a (C_M, γ_M)-regular triangulation. Let u be a function defined on $\widetilde{K}_j$ and set $u_j := u \circ \widetilde{M}_j$. Then:*

(i) Let u be analytic on $\overline{\widetilde{K}_j}$ and $\xi \in \overline{\hat{K}_j}$; set $x = \widetilde{M}_j(\xi)$. If for some $C(x)$, $\gamma > 0$

$$|\nabla^p u(x)| \leq C(x)\gamma^p p! \qquad \forall p \in \mathbb{N}_0,$$

then there are C', $K > 0$ depending only on γ and γ_M, C_M such that

$$|\nabla^p u_j(\xi)| \leq C' C(x) K^p p! \qquad \forall p \in \mathbb{N}_0.$$

(ii) Let G, $\hat{G} \subset \mathbb{R}^2$ be two domains, $s : G \to s(G) \subset \overline{\widetilde{K}_j}$, $\hat{s} : \hat{G} \to \hat{s}(\hat{G}) \subset \hat{K}_j$, be two analytic maps with analytic inverses. Assume that there are C, $\gamma > 0$ such that

$$\|\nabla^p (u \circ s)\|_{L^\infty(G)} \leq C\gamma^p p! \qquad \forall p \in \mathbb{N}_0,$$
$$\|\nabla^p (s^{-1} \circ \widetilde{M}_j \circ \hat{s})\|_{L^\infty(\hat{G})} \leq C\gamma^p p! \qquad \forall p \in \mathbb{N}_0.$$

Then there are C', $K > 0$ depending only on C, $\gamma > 0$ such that

$$\|\nabla^p (u_j \circ \hat{s})\|_{L^\infty(\hat{G})} \leq C' K^p p! \qquad \forall p \in \mathbb{N}_0.$$

(iii) Let $G \subset \widetilde{K}_j$ and $u \in H^2_{loc}(G) \cap H^1(G)$. Assume that $\|\Phi \nabla^2 u\|_{L^2(G)} < \infty$ for some weight function $\Phi \in L^\infty(G)$. Set $\hat{G} := \widetilde{M}_j^{-1}(G)$, $\hat{\Phi} := \Phi \circ \widetilde{M}_j$. Then there is $C > 0$ depending only on C_M, γ_M such that

$$\|\hat{\Phi}\nabla^2 u_j\|_{L^2(\hat{G})} \leq C\left[h_j \|\Phi\nabla^2 u\|_{L^2(G)} + \|\Phi\nabla u\|_{L^2(G)}\right].$$

Proof: (i) follows immediately from Lemma 4.3.4 and (ii) is proved similarly. (iii) is a direct consequence of the chain rule. □

Finally, in this connection, it is useful to note Theorems 4.2.20, 6.2.6, which state that for analytic changes of variables (e.g., as provided by the patch maps), a function $u \in \mathcal{B}^2_{\beta,\varepsilon}$ (or $\mathcal{B}^2_{\beta,\varepsilon,\alpha}$) is mapped again to a function in $\mathcal{B}^2_{\beta,\varepsilon}$ (or $\mathcal{B}^2_{\beta,\varepsilon,\alpha'}$ for some appropriate $\alpha' < \alpha$).

3.4 *hp* Approximation on minimal meshes

3.4.1 Regularity on the reference element

In Section 3.2, we presented polynomial approximation results on the reference triangle and square. The assumptions on the function to be approximated were formulated as regularity requirements on the reference element. Therefore, in order to apply these approximation results, we need to know how the element maps transform regularity asserts on Ω to regularity assertions on the reference elements. Analyzing this is the purpose of the present subsection.

We employ notation concerning weighted spaces that will be introduced in more generality in Sections 4.2 and 6.2.1. For the reader's convenience, we briefly compile the relevant notions. The weight functions $\hat{\Phi}_{p,\beta,\varepsilon}$ are defined (see (4.2.2)) as follows:

$$\hat{\Phi}_{p,\beta,\varepsilon}(x) = \left(\min\left\{1, \frac{|x|}{\min\{1, \varepsilon(|p|+1)\}}\right\}\right)^{p+\beta}, \qquad p \in \mathbb{Z}, \quad \beta \in [0,1].$$

For a sector S with apex 0 (see Definition 4.2.1 for the precise definition of a curvilinear sector) the norm $\|\cdot\|_{H^{2,2}_{\beta,\varepsilon}(S)}$ is then defined by

$$\|u\|^2_{H^{2,2}_{\beta,\varepsilon}(S)} = \varepsilon^2\|\hat{\Phi}_{0,\beta,\varepsilon}\nabla^2 u\|^2_{L^2(S)} + \varepsilon\|\nabla u\|^2_{L^2(S)} + \|u\|^2_{L^2(S)}.$$

For constants C_u, $\gamma_u > 0$, the countably normed space $\mathcal{B}^2_{\beta,\varepsilon}(S, C_u, \gamma_u)$ is defined by (4.2.15), viz.,

$$\|u\|_{H^{2,2}_{\beta,\varepsilon}(S)} \le C_u \text{ and } \|\hat{\Phi}_{p,\beta,\varepsilon}\nabla^{p+2}u\|_{L^2(S)} \le C_u\gamma_u^p \max\{p+1, \varepsilon^{-1}\}^{p+2} \quad \forall p \in \mathbb{N}.$$

Finally, exponentially weighted versions of these spaces exist (cf. (6.2.11), (6.2.12)): For $\alpha \ge 0$, we can introduce the weight functions

$$\hat{\Psi}_{p,\beta,\varepsilon,\alpha}(x) := e^{\alpha|x|/\varepsilon}\hat{\Phi}_{p,\beta,\varepsilon}(x).$$

The norm $\|\cdot\|_{H^{2,2}_{\beta,\varepsilon,\alpha}(S)}$ and the countably normed space $\mathcal{B}^2_{\beta,\varepsilon,\alpha}(S, C_u, \gamma_u)$ are then defined analogously with the weights $\hat{\Phi}_{p,\beta,\varepsilon}$ replaced with $\hat{\Psi}_{p,\beta,\varepsilon,\alpha}$. We note in particular that $\|\cdot\|_{H^{2,2}_{\beta,\varepsilon,0}(S)} = \|\cdot\|_{H^{2,2}_{\beta,\varepsilon}(S)}$ and $\mathcal{B}^2_{\beta,\varepsilon,0}(S, C_u, \gamma_u) = \mathcal{B}^2_{\beta,\varepsilon}(S, C_u, \gamma_u)$.

Lemma 3.4.1. *Let S' be a sector with apex 0. Let $K \subset S'$ with $h :=$ diam K. Let A be an affine mapping with $A' = h$. Let finally $M : \hat{K} \to K \subset S'$ be an analytic, invertible mapping satisfying for some C_M, γ_M, $c_{geo} > 0$*

$$\begin{aligned} \|\nabla^p\left(A^{-1}\circ M\right)\|_{L^\infty(\hat{K})} &\le C_M\gamma_M^p p! \qquad \forall p \in \mathbb{N}_0, \\ \|\left((A^{-1}\circ M)'\right)^{-1}\|_{L^\infty(\hat{K})} &\le C_M, \\ c_{geo}^{-1}h \le \operatorname{dist}(K,0) &\le c_{geo}h. \end{aligned}$$

Let $u \in \mathcal{B}^2_{\beta,\varepsilon,\alpha}(S', C_u, \gamma_u)$ for some C_u, $\gamma_u > 0$, $\varepsilon \in (0,1]$, $\alpha \ge 0$. Then there are C, c, $\gamma > 0$ depending only on C_M, γ_M, α, c_{geo}, β such that the mapped function $u \circ M$ satisfies on $\hat{K}$ for all $p \in \mathbb{N}$

$$\|\nabla^p(u\circ M)\|_{L^\infty(\hat{K})} \le CC_u\varepsilon^{-1}\left(h/\varepsilon + (h/\varepsilon)^{1-\beta}\right)e^{ch/\varepsilon}\gamma^p p!. \tag{3.4.1}$$

Remark 3.4.2 The assumptions on the mapping M are those that are typically met for the element maps in *geometric* meshes. ∎

Proof: We set $\tilde{K} := A^{-1}(K)$. Then the assumption $c_{geo}^{-1}h \le \operatorname{dist}(K,0) \le c_{geo}h$ implies that

$$\frac{h}{|x|} \leq c_{geo} \qquad \forall x \in K. \tag{3.4.2}$$

Combining Theorem 4.2.23 for the case $\alpha = 0$ with Theorem 6.2.7 for $\alpha > 0$, we get the existence of C, c, $\gamma > 0$ independent of ε such that

$$\|\hat{\Psi}_{p-1,\beta,\varepsilon,c\alpha} \nabla^p u\|_{L^\infty(K)} \leq C\gamma^p C_u \max\{p+1, \varepsilon^{-1}\}^{p+1} \max\{1, h/\varepsilon\} \qquad \forall p \in \mathbb{N}.$$

Rewriting this last statement using Lemma 4.2.2, we obtain (using the shorthand $r = |x|$ and $\max\{1, h/\varepsilon\} \leq 1 + h/\varepsilon$)

$$|\nabla^p u(x)| \leq C C_u \varepsilon^{-2} (1+h/\varepsilon) \frac{\max\{(p+1)/r, \varepsilon^{-1}\}^{p-1}}{\hat{\Phi}_{0,\beta,\varepsilon}(x)} e^{-c\alpha r/\varepsilon} \gamma^p, \quad p \in \mathbb{N},\ x \in K.$$

Hence, we get for the "stretched" function $\tilde{u}(\xi) := u \circ A$ on the set $\tilde{K}$

$$\begin{aligned} |\nabla^p \tilde{u}(\xi)| &\leq C C_u \varepsilon^{-2} h \frac{1+h/\varepsilon}{\hat{\Phi}_{0,\beta,\varepsilon}(x)} \max\{(p+1)h/r, h/\varepsilon\}^{p-1} \gamma^p e^{-c\alpha r/\varepsilon} \\ &\leq C C_u \varepsilon^{-2} h \frac{1+h/\varepsilon}{\hat{\Phi}_{0,\beta,\varepsilon}(x)} \max\{p+1, h/\varepsilon\}^{p-1} \gamma^p e^{-c\alpha h/\varepsilon} \qquad \forall p \in \mathbb{N}, \end{aligned}$$

where we used (3.4.2) and appropriately adjusted the constants C, γ, c. Furthermore, since

$$\max\{p, h/\varepsilon\}^{p-1} \leq e^p \max\left\{p!, \frac{(h/\varepsilon)}{p!} p!\right\} \leq e^p p! \max\{1, e^{h/\varepsilon}\} \leq e^p p! e^{h/\varepsilon},$$

the function $\tilde{u}$ satisfies for appropriate $c > 0$

$$\begin{aligned} |\nabla^p \tilde{u}(\xi)| &\leq C C_u h \varepsilon^{-2} e^{h/\varepsilon} p! \gamma^p \frac{1+h/\varepsilon}{\hat{\Phi}_{0,\beta,\varepsilon}(x)} e^{-c\alpha h/\varepsilon} \\ &\leq C C_u \varepsilon^{-1} \frac{h}{\varepsilon} \left(1 + \left(\frac{\varepsilon}{h}\right)^\beta\right) \left(1 + \frac{h}{\varepsilon}\right) e^{ch/\varepsilon} p! \gamma^p \quad \forall p \in \mathbb{N}, \quad \xi \in \tilde{K}. \end{aligned}$$

Exploiting now the fact that $\hat{u} := u \circ M = \tilde{u} \circ (A^{-1} \circ M)$ and the assumptions on the map $A^{-1} \circ M$, we get from Corollary 4.3.5 on the reference element $\hat{K}$

$$\|\nabla^p \hat{u}\|_{L^\infty(\hat{K})} \leq C C_u \varepsilon^{-1} \left(\frac{h}{\varepsilon} + \left(\frac{h}{\varepsilon}\right)^{1-\beta}\right) \left(1 + \frac{h}{\varepsilon}\right) e^{ch/\varepsilon} p! \gamma^p \qquad \forall p \in \mathbb{N}.$$

Absorbing now the factor $(1 + h/\varepsilon)$ into the exponential factor $e^{ch/\varepsilon}$ by appropriately adjusting c proves the lemma. □

For the elements abutting on the vertices, we need a different result:

Lemma 3.4.3. *Let $S' \subset \mathbb{R}^2$ be a sector with apex 0 and $K \subset S'$ with $h := \operatorname{diam} K < \infty$. Let $\hat{K} = T$ or $\hat{K} = S$. Let $M : \hat{K} \to K$ be a C^2-diffeomorphism satisfying for some affine map A with $A' = h$ and some constant $C_M > 0$*

$$\|A^{-1} \circ M\|_{C^2(\hat{K})} + \|\left((A^{-1} \circ M)'\right)^{-1}\|_{L^\infty(\hat{K})} \leq C_M. \tag{3.4.3}$$

Then there is $C > 0$ depending only on C_M and $\beta \in (0,1)$ such that the following is true for all $u \in H^{2,2}_{\beta,\varepsilon}(K)$: If $d := \operatorname{dist}(K, 0) > 0$, then

$$\|\nabla^2(u \circ M)\|_{L^2(\hat{K})} \leq C\varepsilon^{-1}\left[\frac{h}{\varepsilon} + \left(\frac{h}{d}\right)^\beta \left(\frac{h}{\varepsilon}\right)^{1-\beta} + \left(\frac{h+d}{\varepsilon}\right)^{1-\beta}\right] U, \tag{3.4.4}$$

where

$$U := \varepsilon^2 \|\hat{\Phi}_{0,\beta,\varepsilon} \nabla^2 u\|_{L^2(K)} + \varepsilon \|\hat{\Phi}_{0,\beta-1,\varepsilon} \nabla u\|_{L^2(K)}.$$

If $d = 0$ and $M(0) = 0$, then

$$\|\,|x|^\beta \nabla^2(u \circ M)\|_{L^2(\hat{K})} \leq C\varepsilon^{-1}\left[\frac{h}{\varepsilon} + \left(\frac{h}{\varepsilon}\right)^{1-\beta}\right] U. \tag{3.4.5}$$

Proof: We start with (3.4.4). From the chain rule and the bounds in (3.4.3), we have upon writing $\hat{u} = u \circ M$, $\tilde{u} = u \circ A$, $\tilde{K} := (A^{-1} \circ M)(\hat{K})$

$$\|\nabla^2 \hat{u}\|_{L^2(\hat{K})} \leq C\left[\|\nabla^2 \tilde{u}\|_{L^2(\tilde{K})} + \|\nabla \tilde{u}\|_{L^2(\tilde{K})}\right] = C\left[h\|\nabla^2 u\|_{L^2(K)} + \|\nabla u\|_{L^2(K)}\right].$$

Inserting

$$1 \leq \frac{\hat{\Phi}_{0,\beta,\varepsilon}(x)}{\inf_{z \in K} \hat{\Phi}_{0,\beta,\varepsilon}(z)} \leq \hat{\Phi}_{0,\beta,\varepsilon}(x) \max\{1, d/\varepsilon\}^{-\beta} \qquad \forall x \in K,$$

$$1 \leq \frac{\hat{\Phi}_{0,\beta-1,\varepsilon}(x)}{\inf_{z \in K} \hat{\Phi}_{0,\beta-1,\varepsilon}(z)} \leq \hat{\Phi}_{0,\beta-1,\varepsilon}(x) \min\{1, (h+d)/\varepsilon\}^{1-\beta} \qquad \forall x \in K,$$

we arrive at

$$\|\nabla^2 \hat{u}\|_{L^2(\hat{K})} \leq C\varepsilon^{-1}\Big[\frac{h}{\varepsilon} \max\{1, d/\varepsilon\}^{-\beta} \varepsilon^2 \|\hat{\Phi}_{0,\beta,\varepsilon} \nabla^2 u\|_{L^2(K)}$$
$$+ \min\{1, (h+d)/\varepsilon\}^{1-\beta} \varepsilon \|\hat{\Phi}_{0,\beta-1,\varepsilon} \nabla u\|_{L^2(K)}\Big];$$

elementary manipulations and the definition of U then imply the bound (3.4.4). We now turn to the proof of (3.4.5). We may assume $A(0) = 0$ and hence

$$\hat{\Phi}_{0,\beta,\varepsilon/h} = \hat{\Phi}_{0,\beta,\varepsilon} \circ A. \tag{3.4.6}$$

Thus

$$\|\hat{\Phi}_{0,\beta,\varepsilon/h}\nabla^2\tilde{u}\|_{L^2(\tilde{K})} = h\|\hat{\Phi}_{0,\beta,\varepsilon}\nabla^2 u\|_{L^2(K)}. \tag{3.4.7}$$

Hence, the chain rule and the facts that $\hat{u} = \tilde{u}\circ(A^{-1}\circ M)$ and $\|A^{-1}\circ M\|_{C^2(\hat{K})} \leq C$ imply

$$\begin{aligned}
\||x|^\beta\nabla^2\hat{u}\|_{L^2(\hat{K})} &\leq C\|\hat{\Phi}_{0,\beta,1}\nabla^2\tilde{u}\|_{L^2(\tilde{K})} + C\|\hat{\Phi}_{0,\beta,1}\nabla\tilde{u}\|_{L^2(\tilde{K})} \\
&\leq C\max\{1,\varepsilon/h\}^\beta\left[\|\hat{\Phi}_{0,\beta,\varepsilon/h}\nabla^2\tilde{u}\|_{L^2(\tilde{K})} + \|\hat{\Phi}_{0,\beta,\varepsilon/h}\nabla\tilde{u}\|_{L^2(\tilde{K})}\right] \\
&\leq C\max\{1,\varepsilon/h\}^\beta\left[h\|\hat{\Phi}_{0,\beta,\varepsilon}\nabla^2 u\|_{L^2(K)} + \|\hat{\Phi}_{0,\beta,\varepsilon}\nabla u\|_{L^2(K)}\right] \\
&\leq C\max\{1,\varepsilon/h\}^\beta\varepsilon^{-1}\left[\frac{h}{\varepsilon}\varepsilon^2\|\hat{\Phi}_{0,\beta,\varepsilon}\nabla^2 u\|_{L^2(K)} + \frac{h}{\varepsilon}\varepsilon\|\hat{\Phi}_{0,\beta-1,\varepsilon}\nabla u\|_{L^2(K)}\right],
\end{aligned}$$

where we used the assumption that $M(0) = 0$, appealed to Lemma 4.2.3 in the second step, and used (3.4.7) in the third one; for the last step, we wrote $\hat{\Phi}_{0,\beta,\varepsilon} = \hat{\Phi}_{0,\beta-1,\varepsilon}\hat{\Phi}_{0,1,\varepsilon}$ and used $\hat{\Phi}_{0,1,\varepsilon}(x) = \min\{1,|x|/\varepsilon\} \leq Ch/\varepsilon$ for all $x \in K$. From this last estimate, the claim of the lemma follows. □

Lemma 3.4.4. *Let u be an analytic function satisfying on a domain $G \subset \mathbb{R}^2$*

$$\|\nabla^p u\|_{L^\infty(G)} \leq C_u\gamma_u^p\max\{p,\varepsilon^{-1}\}^p \qquad \forall p \in \mathbb{N}_0$$

for some C_u, $\gamma_u > 0$, and $\varepsilon \in (0,1]$. Let $M : G' \to G$ be an analytic map satisfying for some affine *map A with $A' = h$, $h \in (0,1]$,*

$$\|\nabla^p(A^{-1}\circ M)\|_{L^\infty(G')} \leq C_M\gamma_M^p p! \qquad \forall p \in \mathbb{N}_0.$$

Then there are constants C, $\gamma > 0$ independent of h, ε such that the function $\hat{u} := u\circ M$ satisfies

$$\|\nabla^p\hat{u}\|_{L^\infty(G')} \leq C\frac{h}{\varepsilon}e^{h/\varepsilon}\gamma^p p! \qquad \forall p \in \mathbb{N}.$$

Proof: We set $\tilde{u} := u\circ A$ and observe that $\hat{u} = \tilde{u}\circ(A^{-1}\circ M)$. Next, we bound for $p \geq 1$ and $h \leq 1$

$$\begin{aligned}
\|\nabla^p\tilde{u}\|_{L^\infty} &\leq C_u\gamma_u^p h^p\max\{p,\varepsilon^{-1}\}^p \leq C_u\frac{h}{\varepsilon}\gamma_u^p\max\{p,h/\varepsilon\}^{p-1} \\
&\leq C_u\frac{h}{\varepsilon}\gamma_u^p e^p\max\{p!,p!\,(h/\varepsilon)^{p-1}/(p-1)!\} \leq C_u\frac{h}{\varepsilon}(e\gamma_u)^p e^{h/\varepsilon}p!.
\end{aligned}$$

Appealing now to Corollary 4.3.5 concludes the argument. □

3.4.2 Approximation on minimal meshes

Our approximation results are based on the operator Π_p^∞ defined on the reference element $\hat{K}$. Theorem 3.2.20 gives approximation results on the reference element $\hat{K}$. In order to "transport" results on the reference element $\hat{K}$ to the physical elements, we use the following lemma.

Lemma 3.4.5. *Let $M : \hat{K} \to K$ be an invertible map satisfying, for some C_M, $\kappa > 0$,*

$$\|M'\|_{L^\infty(\hat{K})} \le C_M, \qquad \|(M')^{-1}\|_{L^\infty(\hat{K})} \le \frac{C_M}{\kappa}.$$

Then there holds for $u := \hat{u} \circ M^{-1}$

$$\begin{aligned}
\|u\|_{L^\infty(K)} &= \|\hat{u}\|_{L^\infty(\hat{K})} & \forall \hat{u} \in L^\infty(\hat{K}),\\
\|\nabla u\|_{L^\infty(K)} &\le \frac{C_M}{\kappa}\|\nabla \hat{u}\|_{L^\infty(\hat{K})} & \forall \hat{u} \in W^{1,\infty}(\hat{K}),\\
\|\nabla \hat{u}\|_{L^\infty(\hat{K})} &\le C_M \|\nabla u\|_{L^\infty(K)} & \forall \hat{u} \in W^{1,\infty}(\hat{K}).
\end{aligned}$$

If the map M satisfies for some affine map A with $A' = h$

$$\|(A^{-1} \circ M)'\|_{L^\infty(\hat{K})} + \| \left((A^{-1} \circ M)'\right)^{-1} \|_{L^\infty(\hat{K})} \le C_M,$$

then we have for some C depending only on C_M

$$\begin{aligned}
\|u\|_{L^\infty(K)} &= \|\hat{u}\|_{L^\infty(\hat{K})} & \forall \hat{u} \in L^\infty(\hat{K}),\\
C^{-1}h^{-1}\|\nabla \hat{u}\|_{L^\infty(\hat{K})} &\le \|\nabla u\|_{L^\infty(K)} \le Ch^{-1}\|\nabla \hat{u}\|_{L^\infty(\hat{K})} & \forall \hat{u} \in W^{1,\infty}(\hat{K}),\\
C^{-1}h\|\hat{u}\|_{L^2(\hat{K})} &\le \|u\|_{L^2(K)} \le Ch\|\hat{u}\|_{L^2(\hat{K})} & \forall \hat{u} \in L^2(\hat{K}),\\
C^{-1}\|\nabla \hat{u}\|_{L^2(\hat{K})} &\le \|\nabla u\|_{L^2(K)} \le C\|\nabla \hat{u}\|_{L^2(\hat{K})} & \forall \hat{u} \in H^1(\hat{K}).
\end{aligned}$$

Proof: We only show the bound involving $\|u\|_{L^2(K)}$ and $\|\hat{u}\|_{L^2(K)}$. Define the set $\tilde{K} := A^{-1}(K)$ and define the function $\tilde{u} := u \circ A$. We note that $\hat{u} = \tilde{u} \circ (A^{-1} \circ M)$. From our assumptions on $A^{-1} \circ M$, we have

$$\|u\|_{L^2(K)} \sim \|\tilde{u}\|_{L^2(\tilde{K})} = h\|u\|_{L^2(\hat{K})};$$

here, the constants hidden in $\sim$ depend only on C_M. The result follows. □

We also need the following result for meshes that are refined geometrically toward a point. It expresses the fact that the element size decreases in geometric progression as the elements approach the point of refinement:

Lemma 3.4.6. *Let $\mathcal{T} = (K_i, M_i, \hat{K}_i)$ be a (C_M, γ_M)-regular triangulation of a domain $\tilde{\Omega}$. Let the mesh be geometrically refined toward a point $A \in \tilde{\Omega} \cup \partial\tilde{\Omega}$ in the sense that for a constant $c_{geo} > 0$ there holds for the choice $h_i = \operatorname{diam} K_i$ of the parameters appearing in Definition 3.3.1*

$$c_{geo}^{-1} h_i \le \operatorname{dist}(K_i, A) \le c_{geo} h_i \qquad \forall K_i \quad \text{with } A \notin \overline{K}_i.$$

Let $\delta > 0$ be given. Then there exists $C > 0$ depending only on C_M, γ_M, c_{geo}, and δ such that

$$\sum_{i: A \notin \overline{K}_i} h_i^{2\delta} \le C(\operatorname{diam} \tilde{\Omega})^{2\delta}.$$

Proof: The (C_M, γ_M)-regularity of the mesh implies the existence of $C > 0$ depending only on C_M, γ_M such that for all elements K_i:

$$C^{-1} h_i^2 \leq |K_i| \leq C h_i^2.$$

Next, we observe that for all i with $A \notin \overline{K}_i$, we have

$$\begin{aligned} c_{geo}^{-1} h_i &\leq \operatorname{dist}(K_i, A) = \inf_{x \in K_i} \operatorname{dist}(x, A) \\ &\leq \sup_{x \in K_i} \operatorname{dist}(x, A) \leq \operatorname{dist}(K_i, A) + h_i \leq (1 + c_{geo}) h_i. \end{aligned}$$

Hence, we can estimate

$$\begin{aligned} \sum_i h_i^{2\delta} &\leq C \sum_i h_i^{2(\delta-1)} \int_{K_i} 1\, dx\, dy \leq C \sum_i \int_{K_i} (\operatorname{dist}(x, A))^{2(\delta-1)}\, dx\, dy \\ &\leq C \int_{\widetilde{\Omega}} (\operatorname{dist}(x, A))^{2(\delta-1)}\, dx\, dy \leq C \int_0^{\operatorname{diam} \widetilde{\Omega}} r^{2(\delta-1)} r\, dr \leq C (\operatorname{diam} \widetilde{\Omega})^{2\delta}, \end{aligned}$$

where the summation is over all i such that $A \notin \overline{K}_i$. □

Lemma 3.4.7. *Let S be a curvilinear sector with apex 0 and let $\mathcal{T} = (K_i, M_i, \hat{K}_i)$ be a (C_M, γ_M)-regular triangulation of $\Omega_0 \subset S$. Assume that the element maps M_i satisfy*

$$\|\nabla^p \left(\tilde{s}_{h_i}^{-1} \circ M_i\right)\|_{L^\infty(\hat{K}_i)} \leq C_M \gamma_M^p p! \qquad \forall p \in \mathbb{N}_0, \tag{3.4.8a}$$

$$\|\left((\tilde{s}_{h_i}^{-1} \circ M_i)'\right)^{-1}\|_{L^\infty(\hat{K}_i)} \leq C_M, \tag{3.4.8b}$$

$$c_{geo}^{-1} h_i \leq \operatorname{dist}(K_i, 0) \leq c_{geo} h_i \quad \text{if } 0 \notin \overline{K}_i, \tag{3.4.8c}$$

where $h_i = \operatorname{diam} K_i$ and $\tilde{s}_h$ is the stretching map $x \mapsto hx$. Set

$$H := \operatorname{diam} \Omega_0, \qquad h_0 := \max_{i: 0 \in \overline{K}_i} h_i$$

and assume $H > \varepsilon$. For a function $u \in \mathcal{B}^2_{\beta,\varepsilon}(S, C_u, \gamma_u)$ and the interpolation operator $\Pi^\infty_{p,\mathcal{T}}$ of (3.3.3), we obtain for the interpolation error

$$e_i = (u - \Pi^\infty_{p,\mathcal{T}} u)|_{K_i}$$

the following bounds:

$$\left\{ \sum_{i: 0 \in \overline{K}_i} p^4 \|e_i\|^2_{L^\infty(K_i)} + \|\nabla e_i\|^2_{L^2(K_i)} \right\}^{1/2} \leq C \frac{C_u}{\varepsilon} p^3 (1 + \ln p) \left[\frac{h_0}{\varepsilon} + \left(\frac{h_0}{\varepsilon}\right)^{1-\beta} \right],$$

$$\left\{ \sum_{i: 0 \notin \overline{K}_i} p^4 \|e_i\|^2_{L^\infty(K_i)} + \|\nabla e_i\|^2_{L^2(K_i)} \right\}^{1/2} \leq C \frac{C_u}{\varepsilon} \left[\frac{H}{\varepsilon} + \left(\frac{H}{\varepsilon}\right)^{1-\beta} \right] e^{cH/\varepsilon} e^{-bp}$$

for some C, b, $c > 0$ that depend only on C_M, γ_M, γ_u, β, and the sector S.

Proof: We start by noting that

$$|K_i| \sim h_i^2, \tag{3.4.9}$$

where the constants hidden in the $\sim$ depend only on C_M. Next, by combining Proposition 3.2.21 with Lemmata 3.4.1, 3.4.3, 3.4.5, we obtain the following error bounds:

$$\begin{aligned}
\|e_i\|_{L^\infty(K_i)} + h_i\|\nabla e_i\|_{L^\infty(K_i)} &\le CC_u F_i e^{-bp} e^{ch_i/\varepsilon} && \text{if } 0 \notin \overline{K}_i, \\
\|e_i\|_{L^\infty(K_i)} + p^{-2}\|\nabla e_i\|_{L^2(K_i)} &\le Cp(1+\ln p) F_i U_i && \text{if } 0 \in \overline{K}_i,
\end{aligned}$$

where

$$\begin{aligned}
F_i &= \varepsilon^{-1}\left[\frac{h_i}{\varepsilon} + \left(\frac{h_i}{\varepsilon}\right)^{1-\beta}\right], \\
U_i &= \left[\varepsilon^2\|\hat{\Phi}_{0,\beta,\varepsilon}\nabla^2 u\|_{L^2(K_i)} + \varepsilon\|\hat{\Phi}_{0,\beta-1,\varepsilon}\nabla u\|_{L^2(K_i)}\right],
\end{aligned}$$

and the constants C, c, $b > 0$ are independent of ε and p. Lemma 3.4.6 implies

$$\sum_{i:0\notin\overline{K}_i} F_i^2 \le C\varepsilon^{-2}\left[(H/\varepsilon)^2 + (H/\varepsilon)^{2(1-\beta)}\right]. \tag{3.4.11}$$

From Corollary 4.2.11 we get

$$\begin{aligned}
\sum_{i:0\notin\overline{K_i}} U_i^2 &\le \varepsilon^4\|\hat{\Phi}_{0,\beta,\varepsilon}\nabla^2 u\|^2_{L^2(S)} + \varepsilon^2\|\hat{\Phi}_{0,\beta-1,\varepsilon}\nabla u\|^2_{L^2(S)} \\
&\le C\|u\|^2_{H^{2,2}_{\beta,\varepsilon}(S)} \le CC_u^2.
\end{aligned} \tag{3.4.12}$$

Bounding $\|\nabla e_i\|_{L^2(K_i)} \le Ch_i\|\nabla e_i\|_{L^\infty(K_i)}$, we obtain using (3.4.11) and (3.4.12)

$$\begin{aligned}
\sum_{i:0\notin\overline{K}_i} \|\nabla e_i\|^2_{L^2(K_i)} &\le CC_u^2 e^{2cH/\varepsilon} e^{-2bp} \sum_{i:0\notin\overline{K}_i} F_i^2 \\
&\le C\varepsilon^{-2} C_u^2 e^{2cH/\varepsilon} e^{-2bp}\left\{(H/\varepsilon)^2 + (H/\varepsilon)^{2(1-\beta)}\right\}, \\
\sum_{i:0\in\overline{K}_i} \|\nabla e_i\|^2_{L^2(K_i)} &\le p^6(1+\ln p)^2\left(\max_{i:0\in\overline{K}_i} F_i^2\right)\sum_{i:0\in\overline{K}_i} U_i^2 \\
&\le CC_u^2\varepsilon^{-2}p^6(1+\ln p)^2\left\{(h_0/\varepsilon)^2 + (h_0/\varepsilon)^{2(1-\beta)}\right\},
\end{aligned}$$

which can be brought to the desired form. The L^∞-bounds $\sum_{i:0\notin\overline{K}_i}\|e_i\|^2_{L^\infty(K_i)}$ and $\sum_{i:0\in\overline{K}_i}\|e_i\|^2_{L^\infty(K_i)}$ are obtained similarly. □

Theorem 3.4.8 (*hp*-interpolant on minimal meshes). *Let Ω be a curvilinear polygon and $\mathcal{T}(\kappa, L)$ be a two-parameter family of admissible meshes in the sense of Definition 2.4.4. Let u be the solution of (1.2.11) with piecewise analytic Dirichlet data g satisfying (1.2.4) and analytic right-hand side f satisfying (1.2.3). Then there are λ_0, C, $b > 0$ independent of ε such that for each $p \in \mathbb{N}$, $\lambda \in (0, \lambda_0)$ there holds on the mesh $\mathcal{T}(\min\{\kappa_0, \lambda p\varepsilon\}, L)$:*

$$\|u - \Pi^{\infty}_{p,\mathcal{T}} u\|_{L^\infty(\Omega)} \leq Cp(1 + \ln p)\left[e^{-b\lambda p} + p^2 e^{-bL}\right], \tag{3.4.13a}$$

$$\|u - \Pi^{\infty}_{p,\mathcal{T}} u\|_{L^2(\Omega)} \leq Cp(1 + \ln p)\left[e^{-b\lambda p} + \varepsilon p^3 e^{-bL}\right], \tag{3.4.13b}$$

$$\lambda p\varepsilon\|\nabla(u - \Pi^{\infty}_{p,\mathcal{T}} u)\|_{L^2(\Omega)} \leq Cp^3(1 + \ln p)\left[e^{-b\lambda p} + \varepsilon p^3 e^{-bL}\right]. \tag{3.4.13c}$$

The interpolation operator $\Pi^{\infty}_{p,\mathcal{T}}$ is defined in (3.3.3). In particular,

$$\Pi^{\infty}_{p,\mathcal{T}} u \in S^p(\mathcal{T}(\min\{\kappa_0, \lambda p\varepsilon\}, L)).$$

Furthermore, on each edge of the mesh $\mathcal{T}(\min\{\kappa_0, \lambda p\varepsilon\}, L)$, the approximant $\Pi^{\infty}_{p,\mathcal{T}} u$ coincides with the Gauss-Lobatto interpolant of u, i.e.,

$$i_{p,\Gamma} u = (\Pi^{\infty}_{p,\mathcal{T}} u)|_\Gamma \qquad \forall \text{ edges } \Gamma \text{ of } \mathcal{T}(\min\{1, \lambda p\varepsilon\}, L). \tag{3.4.14}$$

Proof: For convenience of notation, we introduce the abbreviation

$$C_p := p(1 + \ln p). \tag{3.4.15}$$

The approximation properties of the operator Π^{∞}_p that we utilize are collected in Proposition 3.2.21. These approximation results are formulated on the reference element $\hat{K}_i$, and we therefore frequently appeal to Lemma 3.4.5 to obtain results on the physical elements K_i.
In Section 2.3, we characterized the behavior of the solution u in different ways: Theorem 2.3.1 employs countably normed spaces and Theorem 2.3.4 asymptotic expansions. We use the regularity results in countably normed spaces for the "asymptotic case" $\lambda p\varepsilon \geq \kappa_0$ and the characterization through asymptotic expansions for the "preasymptotic" case $\lambda p\varepsilon < \kappa_0$.

The asymptotic case $\lambda p\varepsilon \geq \kappa_0$.
The key ingredient for obtaining ε-independent bounds is to note that the assumption $\lambda p\varepsilon \geq \kappa_0$ implies

$$\frac{1}{\varepsilon} \leq \frac{1}{\kappa_0}\lambda p, \tag{3.4.16}$$

thus allowing us to replace powers of ε^{-1} by powers of p. We will employ Theorem 2.3.1, which asserts the existence of C, $\gamma > 0$, and $\beta \in [0,1)^J$ independent of ε such that the solution $u \in \mathcal{B}^2_{\beta,\varepsilon}(\Omega, C, \gamma)$.
By Remark 2.4.6 the mesh $\mathcal{T}(\kappa_0, L)$ is a mesh with large elements outside an $O(\kappa_0)$ neighborhood of the vertices and a geometric mesh with $L+1$ layers near the vertices. Thus, in the neighborhoods of the vertices A_j, $j = 1, \ldots, J$, the element maps satisfy the hypotheses (3.4.8) of Lemma 3.4.7. To fix the notation, we concentrate on the approximation on a sector S' with apex A_j.

The estimates of Lemma 3.4.7 give for the error $e_i = u - \Pi^{\infty}_{p,\mathcal{T}} u$

$$\sum_i p^4 \|e_i\|^2_{L^\infty(K_i)} + \|\nabla e_i\|^2_{L^2(K_i)} \leq C\varepsilon^{-2} \left\{ (H/\varepsilon)^2 + (H/\varepsilon)^{2(1-\beta_j)} \right\} e^{cH/\varepsilon} e^{-bp}$$
$$+ C\varepsilon^{-2} p^6 (1+\ln p)^2 \left\{ (\sigma^L/\varepsilon)^2 + (\sigma^L/\varepsilon)^{2(1-\beta_j)} \right\},$$

where we inserted the fact that the element K_i with $A_j \in \overline{K}_i$ satisfy $h_i \leq C\sigma^L$ and wrote $H = \operatorname{diam} S'$. In order to remove the factor $e^{cH/\varepsilon}$, we employ (3.4.16) to bound

$$e^{-bp} e^{cH/\varepsilon} \leq e^{-bp} e^{(c \operatorname{diam} \Omega)\lambda p/\kappa_0} \leq e^{-(b/2)p}$$

if

$$\lambda \leq \lambda_0 \leq \frac{b\kappa_0}{2c \operatorname{diam} \Omega}.$$

Hence, for these choices of λ, we obtain

$$\sum_i p^4 \|e_i\|^2_{L^\infty(K_i)} + \|\nabla e_i\|^2_{L^2(K_i)} \leq C\varepsilon^{-4} e^{-(b/2)p} + \varepsilon^{-4} p^6 (1+\ln p)^2 \sigma^{2(1-\beta_j)L}$$
$$\leq C p^4 (1+\ln p)^2 \left\{ e^{-(b/2)p} + p^6 \sigma^{2(1-\beta_j)L} \right\},$$

where in the last step, we used (3.4.16) to exchange negative powers of ε for powers of p. The bounds (3.4.13a), (3.4.13c) then follow easily. Combining (3.4.13a) with (3.4.16) yields (3.4.13b).

The preasymptotic case $\lambda p \varepsilon < \kappa_0$.
In this regime (which is of course the regime of practical interest), we employ the decomposition of u based on asymptotic expansions. Specifically, we use the decomposition

$$u = w_\varepsilon + \chi^{BL} u^{BL}_\varepsilon + \chi^{CL} u^{CL}_\varepsilon + r_\varepsilon \tag{3.4.17}$$

of (2.3.1) and the regularity assertions of Theorem 2.3.4. We approximate each of the four terms separately.
Before proving approximation results for each of the four terms in (3.4.17), we recall from the definition admissible meshes in Definition 2.4.4 that there are three types of elements: interior elements (collected in the set $\mathcal{T}^{int}$), boundary layer elements (gather in the set $\mathcal{T}^{BL}$), and corner layer elements (combined into the set $\mathcal{T}^{CL}$). The element maps M_i for the three types of elements have the following regularity properties: interior and boundary layer elements satisfy

$$\left. \begin{array}{l} \|\nabla^p M_i\|_{L^\infty(\hat{K}_i)} \leq C\gamma^p p! \quad \forall p \in \mathbb{N}_0 \\ \|(M_i')^{-1}\|_{L^\infty(\hat{K}_i)} \leq \frac{C}{\lambda p \varepsilon} \end{array} \right\} \quad \forall K_i \in \mathcal{T}^{int} \cup \mathcal{T}^{BL}, \tag{3.4.18}$$

and for corner layer elements K_i in the vicinity of vertex A_j we have with $h_i = \operatorname{diam} K_i$

$$\left. \begin{array}{l} \|\nabla^p \left(\tilde{s}^{-1}_{j,h_i} \circ M_i\right)\|_{L^\infty(\hat{K}_i)} \leq C_M \gamma^p_M p! \quad \forall p \in \mathbb{N}_0, \\ \|\left((\tilde{s}^{-1}_{j,h_i} \circ M_i)'\right)^{-1}\|_{L^\infty(\hat{K}_i)} \leq C_M, \end{array} \right\} \quad \forall K_i \in \mathcal{T}^{CL}; \tag{3.4.19a}$$

additionallly, the elements K_i not abutting the vertex A_j satisfy

$$c_{geo}^{-1} h_i \leq \operatorname{dist}(K_i, A_j) \leq c_{geo} h_i. \tag{3.4.19b}$$

1. step: approximation of w_ε. We claim that

$$\|w_\varepsilon - \Pi_{p,\mathcal{T}}^\infty w_\varepsilon\|_{L^\infty(\Omega)} + \lambda p\varepsilon \|\nabla (w_\varepsilon - \Pi_{p,\mathcal{T}}^\infty w_\varepsilon)\|_{L^\infty(\Omega)} \leq C e^{-bp} \tag{3.4.20}$$

for some C, $b > 0$ independent of ε. We prove this by showing

$$\|w_\varepsilon - \Pi_{p,\mathcal{T}}^\infty w_\varepsilon\|_{L^\infty(K_i)} + \lambda p\varepsilon \|\nabla (w_\varepsilon - \Pi_{p,\mathcal{T}}^\infty w_\varepsilon)\|_{L^\infty(K_i)} \leq C e^{-bp} \qquad \forall K_i. \tag{3.4.21}$$

To that end, we define for each element K_i the pull-back $\hat{w}_i := w_\varepsilon \circ M_i$ and consider the cases of $K_i \in \mathcal{T}^{int} \cap \mathcal{T}^{BL}$ and $K_i \in \mathcal{T}^{CL}$ separately.
We recall from Theorem 2.3.4 that w_ε is analytic on $\overline{\Omega}$, i.e.,

$$\|\nabla^n w_\varepsilon\|_{L^\infty(\Omega)} \leq C \gamma^n n! \quad \forall n \in \mathbb{N}_0.$$

For elements $K_i \in \mathcal{T}^{int} \cup \mathcal{T}^{BL}$ we obtain from (3.4.18) and Lemma 3.4.4 (choosing $\varepsilon = 1$ and $A = \mathrm{Id}$ in the statement of Lemma 3.4.4)

$$\|\nabla^n \hat{w}_i\|_{L^\infty(\hat{K}_i)} \leq C \gamma^n n! \qquad \forall n \in \mathbb{N}_0$$

for some C, $\gamma > 0$ independent of ε. Proposition 3.2.21 therefore implies

$$\|\hat{w}_i - \Pi_p^\infty \hat{w}_i\|_{L^\infty(\hat{K}_i)} + \|\nabla (\hat{w}_i - \Pi_p^\infty \hat{w}_i)\|_{L^\infty(\hat{K}_i)} \leq C e^{-bp}$$

for some C, $b > 0$ independent of ε. In view of Lemma 3.4.5 and (3.4.18), this bound implies (3.4.21) for all elements $K_i \in \mathcal{T}^{int} \cup \mathcal{T}^{BL}$.
It remains to see (3.4.21) for $K_i \in \mathcal{T}^{CL}$. Lemma 3.4.4 (choosing $\varepsilon = 1$ in the statement of Lemma 3.4.4) and the regularity property (3.4.19a) yield

$$\|\nabla^n \hat{w}_i\|_{L^\infty(\hat{K}_i)} \leq C h_i \gamma^n n! \qquad \forall n \in \mathbb{N}.$$

Proposition 3.2.21 and Lemma 3.4.5 (using (3.4.19a)) then give

$$h_i^{-1} \|w_\varepsilon - \Pi_{p,\mathcal{T}}^\infty w_\varepsilon\|_{L^\infty(K_i)} + \|\nabla (w_\varepsilon - \Pi_{p,\mathcal{T}}^\infty w_\varepsilon)\|_{L^\infty(K_i)} \leq C e^{-bp}, \tag{3.4.22}$$

which shows (3.4.21), since $\lambda p \varepsilon \leq \kappa_0$.

2. step: approximation of r_ε. We show

$$\|r_\varepsilon - \Pi_{p,\mathcal{T}}^\infty r_\varepsilon\|_{L^\infty(\Omega)} \leq C C_p e^{-b\lambda p}, \tag{3.4.23a}$$

$$\lambda p \varepsilon \|\nabla (r_\varepsilon - \Pi_{p,\mathcal{T}}^\infty r_\varepsilon)\|_{L^2(\Omega)} \leq C C_p p^2 e^{-b\lambda p}. \tag{3.4.23b}$$

From Theorem 2.3.4, we have

$$\|r_\varepsilon\|_{L^\infty(\Omega)} + \|r_\varepsilon\|_{H^{2,2}_{\beta,\varepsilon}(\Omega)} \leq C e^{-\alpha/\varepsilon} \tag{3.4.24}$$

for some suitable C, $\alpha > 0$. The bounds (3.4.23) follow from (3.4.24), the assumption $\lambda p\varepsilon \le \kappa_0$, and the following three claims:

$$\|r_\varepsilon - \Pi^\infty_{p,\mathcal{T}} r_\varepsilon\|_{L^\infty(K_i)} \le C_p \|r_\varepsilon\|_{L^\infty(K_i)} \quad \forall K_i, \tag{3.4.25}$$

$$\|\nabla(r_\varepsilon - \Pi^\infty_{p,\mathcal{T}} r_\varepsilon)\|_{L^2(K_i)} \le \|\nabla r_\varepsilon\|_{L^2(K_i)} + \frac{\sqrt{|K_i|}p^2 C_p}{\lambda p\varepsilon}\|r_\varepsilon\|_{L^\infty(K_i)} \quad \forall K_i \in \mathcal{T}^{int} \cup \mathcal{T}^{BL}, \tag{3.4.26}$$

$$\sum_{i:K_i \in \mathcal{T}^{CL}} \|\nabla(r_\varepsilon - \Pi^\infty_{p,\mathcal{T}} r_\varepsilon)\|^2_{L^2(K_i)} \le C\left(C_p p^2 \varepsilon^{-2} \|r_\varepsilon\|_{H^{2,2}_{\beta,\varepsilon}(\Omega)}\right)^2. \tag{3.4.27}$$

In order to see these claims, we note that taking $v = 0$ in the infimum in Theorem 3.2.20 readily implies (3.4.25).
For (3.4.26), we use the triangle inequality, the assumptions (3.4.18) on the map $(M_i')^{-1}$, and an inverse estimate for polynomials (Lemma 3.2.2) on the reference element to get

$$\begin{aligned}\|\nabla(r_\varepsilon - \Pi^\infty_{p,\mathcal{T}} r_\varepsilon)\|_{L^2(K_i)} &\le \|\nabla r_\varepsilon\|_{L^2(K_i)} + \sqrt{|K_i|}\,\|\nabla \Pi^\infty_{p,\mathcal{T}} r_\varepsilon\|_{L^\infty(K_i)} \\ &\le \|\nabla r_\varepsilon\|_{L^2(K_i)} + C\sqrt{|K_i|}\frac{p^2}{\lambda p\varepsilon}\|\Pi^\infty_{p,\mathcal{T}} r_\varepsilon\|_{L^\infty(K_i)}.\end{aligned}$$

Combining this bound with that of (3.4.25) for $\|\Pi^\infty_{p,\mathcal{T}} r_\varepsilon\|_{L^\infty(K_i)}$ gives (3.4.26).
For (3.4.27), we consider the elements $K_i \in \mathcal{T}^{CL}$ that are in the vicinity of the vertex A_j. We abbreviate

$$R_i := \varepsilon^2 \|\Phi_{0,\beta,\varepsilon} \nabla^2 r_\varepsilon\|_{L^2(K_i)} + \varepsilon \|\Phi_{0,\beta-1,\varepsilon} \nabla r_\varepsilon\|_{L^2(K_i)}$$

and use the fact that

$$\begin{aligned} h_i \sim \operatorname{dist}(A_j, K_i) \quad &\text{and } h_i \le c\lambda p\varepsilon \qquad &&\text{if } A_j \notin \overline{K}_i, \\ &h_i \le c\lambda p\varepsilon\, \sigma^L &&\text{if } A_j \in \overline{K}_i, \end{aligned}$$

to obtain with Lemma 3.4.3 that the pull-back $\hat{r}_\varepsilon := r_\varepsilon \circ M_i$ satisfies

$$\begin{aligned} \|\nabla^2 \hat{r}_\varepsilon\|_{L^2(\hat{K}_i)} &\le C\varepsilon^{-1}\left(\lambda p + (\lambda p)^{1-\beta_j}\right) R_i \qquad &&\text{if } A_j \notin \overline{K}_i, \\ \|\,|x|^{\beta_j} \nabla^2 \hat{r}_\varepsilon\|_{L^2(\hat{K}_i)} &\le C\varepsilon^{-1}\left(\lambda p\,\sigma^L + (\lambda p\,\sigma^L)^{1-\beta_j}\right) R_i &&\text{if } A_j \in \overline{K}_i. \end{aligned}$$

Upon simplifying $\sigma^L \le 1$, $\sigma^{L(1-\beta_j)} \le 1$ and estimating

$$\left(\lambda p + (\lambda p)^{1-\beta_j}\right) \le \varepsilon^{-1}\left(\lambda p\varepsilon + \varepsilon^{\beta_j}(\lambda p\varepsilon)^{1-\beta_j}\right) \le C\varepsilon^{-1}$$

we get from Proposition 3.2.21

$$\|\nabla(\hat{r}_\varepsilon - \Pi^\infty_p \hat{r}_\varepsilon)\|_{L^2(\hat{K}_i)} \le Cp^2 C_p \varepsilon^{-2} R_i.$$

This estimate in turn implies by mapping back to K_i (using Lemma 3.4.5)

$$\|\nabla(\hat{r}_\varepsilon - \Pi^\infty_p \hat{r}_\varepsilon)\|_{L^2(K_i)} \le Cp^2 C_p \varepsilon^{-2} R_i \qquad \forall K_i \in \mathcal{T}^{CL}.$$

Since Corollary 4.2.11 implies $\sum_i R_i^2 \leq C\|r_\varepsilon\|^2_{H^{2,2}_{\beta,\varepsilon}(\Omega)}$, we get the desired bound (3.4.27) by squaring and summing on i.

3. step: approximation of $\chi^{BL} u_\varepsilon^{BL}$. We claim the following estimates:

$$\|\chi^{BL}u_\varepsilon^{BL} - \Pi^\infty_{p,\mathcal{T}}(\chi^{BL}u_\varepsilon^{BL})\|_{L^\infty(K_i)} \leq CC_p e^{-b\lambda p} \qquad \forall K_i \in \mathcal{T}, \quad (3.4.28a)$$

$$\lambda p\varepsilon\|\nabla(\chi^{BL}u_\varepsilon^{BL} - \Pi^\infty_{p,\mathcal{T}}(\chi^{BL}u_\varepsilon^{BL}))\|_{L^\infty(K_i)} \leq CC_p p^2 e^{-b\lambda p} \quad \forall K_i \in \mathcal{T}. \quad (3.4.28b)$$

We consider the approximation error $u_\varepsilon^{BL} - \Pi^\infty_{p,\mathcal{T}} u_\varepsilon^{BL}$ on each of the element types, $\mathcal{T}^{BL}$, $\mathcal{T}^{CL}$, $\mathcal{T}^{int}$ separately. We define $\hat{u}_i^{BL} := u_\varepsilon^{BL} \circ M_i$.
We start with $K_i \in \mathcal{T}^{BL}$. In this case, we note that $\chi^{BL} \equiv 1$ on K_i. Theorem 2.3.4 allows us to bound

$$\|\partial_\rho^r \partial_\theta^s (u_\varepsilon^{BL} \circ \psi_j \circ s_\kappa)\|_{L^\infty} \leq C(\lambda p)^r \gamma^{r+s} s! \leq Ce^{\lambda p}\gamma^{r+s} r! s! \quad \forall (r,s) \in \mathbb{N}_0^2,$$

where we used the bound $(\lambda p)^r = \frac{(\lambda p)^r}{r!} r! \leq e^{\lambda p}/r!$. In view of the assumptions placed on the map $s_\kappa^{-1} \circ \psi_j^{-1} \circ M_i$ (see Definition 2.4.4) and the fact that $\hat{u}_i^{BL} = u_\varepsilon^{BL} \circ M_i = (u_\varepsilon^{BL} \circ \psi \circ s_\kappa) \circ (s_\kappa^{-1} \circ \psi_j^{-1} \circ M_i)$, Lemma 3.4.4 implies

$$\|\nabla^n \hat{u}_i^{BL}\|_{L^\infty(\hat{K}_i)} \leq Ce^{\lambda p}\gamma^n n! \qquad \forall n \in \mathbb{N}_0.$$

Thus, Proposition 3.2.21 implies the existence of C, $b > 0$ independent of ε, p such that

$$\|\hat{u}_\varepsilon^{BL} - \Pi_p^\infty \hat{u}_\varepsilon^{BL}\|_{L^\infty(\hat{K}_i)} + \|\nabla(\hat{u}_\varepsilon^{BL} - \Pi_p^\infty \hat{u}_\varepsilon^{BL})\|_{L^\infty(\hat{K}_i)} \leq Ce^{\lambda p}e^{-bp}.$$

We note that the choice $\lambda_0 \leq b/2$ implies $e^{\lambda p}e^{-bp} \leq e^{-(b/2)p}$. In view of the property (3.4.18) and Lemma 3.4.5, we arrive at

$$\|u_\varepsilon^{BL} - \Pi^\infty_{p,\mathcal{T}} u_\varepsilon^{BL}\|_{L^\infty(K_i)} + \lambda p\varepsilon\|\nabla(u_\varepsilon^{BL} - \Pi^\infty_{p,\mathcal{T}} u_\varepsilon^{BL})\|_{L^\infty(K_i)} \leq e^{-(b/2)p}.$$

Next, we consider corner layer elements $K_i \in \mathcal{T}^{CL}$. Again, we note that $\chi^{CL} \equiv 1$ on K_i. To obtain bounds on the derivatives of the pull-back $\hat{u}_i^{BL}$, we appeal to Lemma 3.4.4 to get

$$\|\nabla^n \hat{u}_i^{BL}\|_{L^\infty(\hat{K}_i)} \leq C\frac{h_i}{\varepsilon}e^{h_i/\varepsilon}\gamma^n n! \leq C\frac{h_i}{\varepsilon}e^{c\lambda p}\gamma^n n! \quad \forall n \in \mathbb{N},$$

where we employed the fact $h_i \leq c\lambda p\varepsilon$ in the second estimate. Hence, using the approximation results of Proposition 3.2.21, we get the existence of C, $b > 0$ independent of ε, p such that

$$\|\hat{u}_i^{BL} - \Pi_p^\infty \hat{u}_i^{BL}\|_{L^\infty(\hat{K}_i)} + \|\nabla(\hat{u}_i^{BL} - \Pi_p^\infty \hat{u}_i^{BL})\|_{L^\infty(\hat{K}_i)} \leq C\frac{h_i}{\varepsilon}e^{c\lambda p}e^{-bp}.$$

Again, the term $e^{c\lambda p}e^{-bp}$ can be replaced with $e^{-(b/2)p}$ for λ_0 sufficiently small. Combining the properties (3.4.18) with Lemma 3.4.5, we obtain (3.4.28).

Finally, for interior elements $K_i \in \mathcal{T}^{int}$, we exploit the fact that $\chi^{BL} u_\varepsilon^{BL}$ is exponentially small due to $\operatorname{dist}(K_i, \partial\Omega) \geq c\lambda p\varepsilon$, viz.,

$$\|\chi^{BL} u_\varepsilon^{BL}\|_{L^\infty(K_i)} + \varepsilon\|\nabla(\chi^{BL} u_\varepsilon^{BL})\|_{L^\infty(K_i)} \leq Ce^{-b\lambda p}.$$

Reasoning as in the proof of (3.4.25) then allows us to bound

$$\|\chi^{BL} u_\varepsilon^{BL} - \Pi_{p,\mathcal{T}}^\infty(\chi^{BL} u_\varepsilon^{BL})\|_{L^\infty(K_i)} \leq CC_p\|\chi^{BL} u_\varepsilon^{BL}\|_{L^\infty(K_i)} \leq CC_p e^{-b\lambda p},$$

and

$$\|\nabla(\chi^{BL} u_\varepsilon^{BL} - \Pi_{p,\mathcal{T}}^\infty(\chi^{BL} u_\varepsilon^{BL}))\|_{L^\infty(K_i)} \leq$$
$$\|\nabla(\chi^{BL} u_\varepsilon^{BL})\|_{L^\infty(K_i)} + \|\nabla(\Pi_{p,\mathcal{T}}^\infty(\chi^{BL} u_\varepsilon^{BL}))\|_{L^\infty(K_i)} \leq C\frac{p^2 C_p}{\lambda p\varepsilon} e^{-b\lambda p}.$$

4. step: approximation of $\chi^{CL} u_\varepsilon^{CL}$. We claim the following bounds:

$$\|\chi^{CL} u_\varepsilon^{CL} - \Pi_{p,\mathcal{T}}^\infty(\chi^{CL} u_\varepsilon^{CL})\|_{L^\infty(K_i)} \leq CC_p e^{-b\lambda p} \quad \text{if } A_j \notin \overline{K}_i \quad (3.4.29a)$$
$$\|\chi^{CL} u_\varepsilon^{CL} - \Pi_{p,\mathcal{T}}^\infty(\chi^{CL} u_\varepsilon^{CL})\|_{L^\infty(K_i)} \leq CC_p\, p e^{-bL} \quad \text{if } A_j \in \overline{K}_i \quad (3.4.29b)$$
$$\lambda p\varepsilon\|\nabla(\chi^{CL} u_\varepsilon^{CL} - \Pi_{p,\mathcal{T}}^\infty(\chi^{CL} u_\varepsilon^{CL}))\|_{L^2(\Omega)} \leq CC_p p^2\left[e^{-b\lambda p} + p^2\varepsilon e^{-bL}\right]. \quad (3.4.29c)$$

We distinguish again the two cases $K_i \in \mathcal{T}^{int} \cup \mathcal{T}^{BL}$ and $K_i \in \mathcal{T}^{CL}$.
For $K_i \in \mathcal{T}^{int} \cup \mathcal{T}^{BL}$, we reason as in the proof of (3.4.28) to exploit that $\chi^{CL} u_\varepsilon^{CL}$ is small away from the vertices and get

$$\|\chi^{CL} u_\varepsilon^{CL} - \Pi_{p,\mathcal{T}}^\infty(\chi^{CL} u_\varepsilon^{CL})\|_{L^\infty(K_i)} \leq CC_p e^{-b\lambda p}, \quad (3.4.30)$$
$$\lambda p\varepsilon\|\nabla(\chi^{CL} u_\varepsilon^{CL} - \Pi_{p,\mathcal{T}}^\infty(\chi^{CL} u_\varepsilon^{CL}))\|_{L^\infty(K_i)} \leq CC_p p^2 e^{-b\lambda p}. \quad (3.4.31)$$

We now turn our attention to elements $K_i \in \mathcal{T}^{CL}$. For simplicity of notation, we restrict our attention to the neighborhood of a single vertex, A_j, say. The line Γ_j' divides the set $\Omega \cap B_{c\lambda p\varepsilon}(A_j)$ into two sectors; our assumptions on admissible boundary layer meshes in Definition 2.4.4 are such that each element $K_i \in \mathcal{T}^{CL}$ is completely contained in one of these two sectors. For simplicity of notation, we fix one of these two sectors and denote it by S' and consider only elements $K_i \in \mathcal{T}^{CL}$ with $K_i \subset S'$.
Theorem 2.3.4 implies that $u_\varepsilon^{CL} \in \mathcal{B}^2_{\beta,\varepsilon,\alpha}(S', C\varepsilon, \gamma)$; in particular, therefore,

$$u_\varepsilon^{CL} \in \mathcal{B}^2_{\beta,\varepsilon}(S', C\varepsilon, \gamma). \quad (3.4.32)$$

Since the element maps M_i of corner layer elements satisfy the hypotheses (3.4.8), Lemma 3.4.7 is applicable, and we obtain for the errors

$$e_i = (u_\varepsilon^{CL} - \Pi_{p,\mathcal{T}}^\infty u_\varepsilon^{CL})|_{K_i},$$

from Lemma 3.4.7 (with $C_u \leq C\varepsilon$, $h_0 \leq C\sigma^L\lambda p\varepsilon$, $H \leq C\lambda p\varepsilon$)

$$\sum_{i:A_j\in\overline{K}_i} p^4\|e_i\|^2_{L^\infty(K_i)} + \|\nabla e_i\|^2_{L^2(K_i)} \le Cp^4C_p^2\left\{(\sigma^L\lambda p)^2 + (\sigma^L\lambda p)^{2(1-\beta_j)}\right\},$$

$$\sum_{i:A_j\notin\overline{K}_i} p^4\|e_i\|^2_{L^\infty(K_i)} + \|\nabla e_i\|^2_{L^2(K_i)} \le C\left\{(\lambda p)^2 + (\lambda p)^{2(1-\beta_j)}\right\}e^{c\lambda p}e^{-bp}$$

Choosing λ_0 sufficiently small allows us to estimate with suitable C, $b>0$

$$\sum_{i:A_j\in\overline{K}_i} p^4\|e_i\|^2_{L^\infty(K_i)} + \|\nabla e_i\|^2_{L^2(K_i)} \le Cp^6C_p^2\sigma^{2(1-\beta_j)L}, \qquad (3.4.33a)$$

$$\sum_{i:A_j\notin\overline{K}_i} p^4\|e_i\|^2_{L^\infty(K_i)} + \|\nabla e_i\|^2_{L^2(K_i)} \le Ce^{-bp}, \qquad (3.4.33b)$$

The bound (3.4.33a) implies (3.4.29b); the estimate (3.4.33b) together with (3.4.30) gives (3.4.29a); combining (3.4.33) with (3.4.31) gives (3.4.29c).
5. step: Conclusion of the proof in the preasymptotic case. In order to complete the proof of the theorem, we combine (3.4.20), (3.4.23a), (3.4.28a), and (3.4.29a), (3.4.29b), to get

$$\|u - \Pi^\infty_{p,\mathcal{T}}u\|_{L^\infty(\Omega)} \le CC_p\left[e^{-b\lambda p} + pe^{-bL}\right],$$

from which the pointwise estimate (3.4.13a) follows.
For the L^2-bound (3.4.13b), we proceed similarly. The only difference is that we exploit the fact that for elements K_i abutting a vertex we have $h_i \le c\lambda p\varepsilon$; thus, the bound (3.4.29b) yields

$$\|\chi^{CL}u_\varepsilon^{CL} - \Pi^\infty_{p,\mathcal{T}}(\chi^{CL}u_\varepsilon^{CL})\|_{L^2(K_i)} \le CC_pp^2\varepsilon e^{-bL} \quad \text{if } A_j\in\overline{K}_i,$$

from which we infer (3.4.13b).
For the H^1-norm bound (3.4.13c), we combine (3.4.20), (3.4.23b), (3.4.28b), and (3.4.29c), to arrive at

$$\lambda p\varepsilon\|\nabla(u - \Pi^\infty_{p,\mathcal{T}}u)\|_{L^2(\Omega)} \le CC_pp^2\left[e^{-b\lambda p} + p\varepsilon e^{-bL}\right].$$

□

Remark 3.4.9 The various powers of p in front of the terms $p^{-b\lambda p}$ and e^{-bL} are largely due to our choice of the convenient *elementwise* interpolant $\Pi^\infty_{p,\mathcal{T}}$. These factors are likely to be suboptimal. A different construction would reduce the powers of p; cf. also Remark 3.3.9. ▪

Remark 3.4.10 For simplicity of exposition (and proof), we assumed that a uniform polynomial degree p is utilized. The arguments presented, however, could be modified to accommodate the use of a reduced polynomial degree in the $L+1$ layers of geometric refinement near the vertices. ▪

Corollary 3.4.11. *Let $\mathcal{T}(\kappa, L)$ be a two-parameter family of* regular *admissible boundary layer meshes (see Definition 3.3.10) or a two-parameter family of meshes generated by mesh patches as in Section 3.3.4. Then the statement of Theorem 3.4.8 holds true.*

Proof: The corollary follows from Theorem 3.4.8 and the fact that by Proposition 3.3.11 regular admissible boundary layer meshes in the sense of Definition 3.3.10 are admissible boundary layer meshes. □

Part II

Regularity in Countably Normed Spaces

4. The Countably Normed Spaces $\mathcal{B}^l_{\beta,\varepsilon}$

4.1 Motivation and outline

4.1.1 Motivation

Chapter 4 and the following Chapter 5 are closely connected. These two chapters describe the regularity properties of the solutions of (1.2.1) in terms of weighted Sobolev spaces. Two kinds of regularity results are presented: Proposition 5.3.2 and Theorem 5.3.8 show that the solution u_ε of (1.2.1) lies in the weighted Sobolev $H^{2,2}_{\beta,\varepsilon}$, the space of H^1 functions whose second derivatives are in a weighted L^2-space. This $H^{2,2}_{\beta,\varepsilon}$ regularity result is then used for a boot-strapping argument to control *all* derivatives of the solution u_ε under the assumption of (piecewise) analyticity of the input data. This control of the growth of the derivatives is cast in the framework of *countably normed space* $\mathcal{B}^l_{\beta,\varepsilon}$ and can be found in Theorem 5.3.8 and Theorem 5.3.10.

The present chapter is preparatory in nature in that the weighted Sobolev space $H^{l,l}_{\beta,\varepsilon}(\Omega)$ and the countably normed spaces $\mathcal{B}^l_{\beta,\varepsilon}$ are defined and their essential properties are proved. The weighted Sobolev spaces $H^{l,l}_{\beta,\varepsilon}$ and the countably normed spaces $\mathcal{B}^l_{\beta,\varepsilon}$ are introduced in such a way that for the case $\varepsilon = 1$ the spaces $H^{l,l}_{\beta,1}$ and $\mathcal{B}^l_{\beta,1}$ coincide with classical weighted Sobolev spaces employed to describe corner singularities for elliptic problems on polygonal domains (see, e.g., [79]) and the countably normed spaces $\mathcal{B}^l_\beta$ introduced by Babuška & Guo in [15, 16]. Our spaces are thus an extension of existing spaces that allow for precise control in terms of the singular perturbation parameter ε.

In order to motivate our weighted Sobolev spaces $H^{l,l}_{\beta,\varepsilon}$, we consider the following model equation:

$$-\varepsilon^2 \Delta u_\varepsilon + u_\varepsilon = f \quad \text{on } \Omega, \qquad u_\varepsilon|_{\partial\Omega} = 0. \tag{4.1.1}$$

It is more convenient for our purposes to write it as

$$-\Delta u_\varepsilon = \varepsilon^{-2}\left[f - u_\varepsilon\right], \qquad u_\varepsilon|_{\partial\Omega} = 0.$$

If Ω has a smooth boundary, then by the classical shift theorem the solution u_ε is in $H^2(\Omega)$ and $\|\nabla^2 u_\varepsilon\|_{L^2(\Omega)} \le C\varepsilon^{-2}\left[\|f\|_{L^2(\Omega)} + \|u_\varepsilon\|_{L^2(\Omega)}\right]$. From this classical shift theorem, we therefore expect $\nabla^2 u$ to be of size $O(\varepsilon^{-2})$. Let us now consider polygonal domains Ω. We will restrict our attention first to a single vertex, i.e., we consider a sector

$$\Omega = S_R(\omega) := \{(r\cos\varphi, r\sin\varphi) \mid 0 < r < R, \quad 0 < \varphi < \omega\} \tag{4.1.2}$$

and fix $R' < R$. Then classical local regularity results give that $u \in H^2_{loc}(S_R(\omega))$. In fact these classical local regularity results (cf. Lemma 5.5.11) allow for the following sharper result: For δ, $r > 0$ we set $\Omega_{\delta,r} := S_r(\omega) \setminus B_\delta(0)$ and have the existence of $C > 0$ depending only on ω and R' such that

$$\|\nabla^2 u_\varepsilon\|_{L^2(\Omega_{2\delta,R'})} \leq C\varepsilon^{-2}\|f - u_\varepsilon\|_{L^2(\Omega_{\delta,R})} + C\left[\delta^{-1}\|\nabla u_\varepsilon\|_{L^2(\Omega_{\delta,R})} + \delta^{-2}\|u_\varepsilon\|_{L^2(\Omega_{\delta,R})}\right].$$

Thus, upon choosing $\delta = \varepsilon$, we obtain

$$\|\nabla^2 u_\varepsilon\|_{L^2(\Omega_{2\varepsilon,R'})} \leq C\varepsilon^{-2}\left[\|f\|_{L^2(\Omega_{\varepsilon,R})} + \varepsilon\|\nabla u_\varepsilon\|_{L^2(\Omega_{\varepsilon,R})} + \|u_\varepsilon\|_{L^2(\Omega_{\varepsilon,R})}\right]. \tag{4.1.3}$$

We remark the similarity of (4.1.3) with the corresponding bound for smooth domains Ω and conclude that the solution u_ε has a weak singularity at the origin in the sense that outside the ball $B_\varepsilon(0)$, the H^2-norm of u_ε can be controlled in the same way as for smooth domains. For the behavior of u_ε in the vicinity of the origin, Proposition 5.3.2 below states the following: For $\beta \in \{\beta \in [0,1) \mid \beta > 1 - \pi/\omega\}$, there is $C > 0$ such that

$$\left\|(|x|/\varepsilon)^\beta \nabla^2 u_\varepsilon\right\|_{L^2(\Omega\cap B_{2\varepsilon}(0))} \leq C\varepsilon^{-2}\left[\left\|(|x|/\varepsilon)^\beta f\right\|_{L^2(\Omega\cap B_{4\varepsilon}(0))} + \varepsilon\|\nabla u_\varepsilon\|_{L^2(\Omega)} + \|u_\varepsilon\|_{L^2(\Omega)}\right]. \tag{4.1.4}$$

Combining (4.1.3), (4.1.4) we see that $\nabla^2 u_\varepsilon$ is in the following *weighted* L^2-space

$$\|\hat{\Phi}_{0,\beta,\varepsilon}\nabla^2 u_\varepsilon\|_{L^2(\Omega\cap B_{R'}(0))} \leq C\varepsilon^{-2}\left[\|\hat{\Phi}_{0,\beta,\varepsilon} f\|_{L^2(\Omega)} + \varepsilon\|\nabla u_\varepsilon\|_{L^2(\Omega)} + \|u_\varepsilon\|_{L^2(\Omega)}\right],$$

where the weight $\hat{\Phi}_{0,\beta,\varepsilon}$ is given by (4.2.2). Essentially, $\hat{\Phi}_{0,\beta,\varepsilon}$ behaves like $(|x|/\varepsilon)^\beta$ in an $O(\varepsilon)$ neighborhood of the origin and reduces to 1 outside an $O(\varepsilon)$ neighborhood of the origin. Multiplying the last estimate by ε^2, we have thus obtained the following estimate:

$$\begin{aligned}\|u_\varepsilon\|_{H^{2,2}_{\beta,\varepsilon}(S_{R'}(\omega))} &:= \varepsilon^2\|\hat{\Phi}_{0,\beta,\varepsilon}\nabla^2 u_\varepsilon\|_{L^2(S_{R'}(\omega))} + \varepsilon\|\nabla u_\varepsilon\|_{L^2(S_{R'}(\omega))} + \|u_\varepsilon\|_{L^2(S_{R'}(\omega))} \\ &\leq C\left[\|\hat{\Phi}_{0,\beta,\varepsilon} f\|_{L^2(\Omega)} + \varepsilon\|\nabla u_\varepsilon\|_{L^2(\Omega)} + \|u_\varepsilon\|_{L^2(\Omega)}\right],\end{aligned}$$

where $C > 0$ is independent of ε and f. The term $\varepsilon\|\nabla u_\varepsilon\|_{L^2(\Omega)} + \|u_\varepsilon\|_{L^2(\Omega)}$ can also be bounded in terms of f with the aid of a Hardy inequality as shown in Theorem 5.3.8 (cf. also the energy estimate (4.1.11) in the proof of Lemma 4.1.1 of the present introduction):

$$\varepsilon\|\nabla u_\varepsilon\|_{L^2(\Omega)} + \|u_\varepsilon\|_{L^2(\Omega)} \leq C\|\hat{\Phi}_{0,\beta,\varepsilon} f\|_{L^2(\Omega)},$$

where again $C > 0$ is independent of ε and f. Thus, we arrive at the desired shift theorem in weighted spaces for sectors:

$$\|u_\varepsilon\|_{H^{2,2}_{\beta,\varepsilon}(S_{R'}(\omega))} \leq C\|\hat{\Phi}_{0,\beta,\varepsilon} f\|_{L^2(\Omega)}. \tag{4.1.5}$$

Here, the constant $C > 0$ is independent of ε. This shift theorem is the motivation for introducing the spaces $H^{l,l}_{\beta,\varepsilon}$ (and, more generally, $H^{l,m}_{\beta,\varepsilon}$) in this chapter. These weighted shift theorems for sectors have natural extensions to polygonal domains. There, one chooses $\beta_j \in [0,1)$ for each vertex A_j, $j = 1\ldots,J$, of the polygon Ω, sets $\beta = (\beta_1,\ldots,\beta_J)$, and defines the weight function $\Phi_{0,\beta,\varepsilon}$ by

$$\Phi_{0,\beta,\varepsilon}(x) := \prod_{j=1}^{J} \hat{\Phi}_{0,\beta_j,\varepsilon}(x - A_j). \tag{4.1.6}$$

With this definition of a weight function $\Phi_{0,\beta,\varepsilon}$, the shift theorem for a single sector of (4.1.5) extends to the case of polygons Ω:

$$\begin{aligned}\|u_\varepsilon\|_{H^{2,2}_{\beta,\varepsilon}(\Omega)} &:= \varepsilon^2\|\Phi_{0,\beta,\varepsilon}\nabla^2 u_\varepsilon\|_{L^2(\Omega)} + \varepsilon\|\nabla u_\varepsilon\|_{L^2(\Omega)} + \|u_\varepsilon\|_{L^2(\Omega)} \\ &\leq C\|\Phi_{0,\beta,\varepsilon} f\|_{L^2(\Omega)},\end{aligned} \tag{4.1.7}$$

provided the components of the vector $\beta \in [0,1)^J$ satisfy $\beta_j > 1 - \pi/\omega_j$, $j = 1,\ldots,J$. If the data of an elliptic equation are sufficiently smooth, then higher order derivatives can be bounded as well. This is done essentially by differentiating the equation and using an elliptic shift theorem. In particular, if the data are analytic, then the solution is analytic as well, [98]. Using the techniques of [98], it was shown in [95] that, if $\partial\Omega$ is a closed analytic curve and f is analytic on $\overline{\Omega}$, then the solution u_ε of (4.1.1) satisfies

$$\|\nabla^{p+2} u_\varepsilon\|_{L^2(\Omega)} \leq CK^{p+2}\max\{p+2, \varepsilon^{-2}\}^{p+2} \qquad \forall p \in \mathbb{N}_0,$$

where the constants C, K are independent of ε but depend on the analytic right-hand side f. In a sector (or, more generally, in a polygon) higher order derivatives of u_ε will of course be in weighted L^2-spaces as we just ascertained for $\nabla^2 u_\varepsilon$, i.e., we expect a result of the following form

$$\|\Phi_{p,\beta,\varepsilon}\nabla^{p+2} u_\varepsilon\|_{L^2(\Omega)} \leq CK^{p+2}\max\{p+2, \varepsilon^{-1}\}^{p+2} \qquad \forall p \in \mathbb{N}_0,$$

where the weight function $\Phi_{p,\beta,\varepsilon}$ depends on the order of the derivative and ε. It is reasonable to expect $\Phi_{p,\beta,\varepsilon}$ to be structurally similar to our definition of $\Phi_{0,\beta,\varepsilon}$ in (4.1.6), i.e., to expect that it can be written as product weight functions $\hat{\Phi}_{p,\beta_j,\varepsilon}$ associated with the vertices A_j of the domain Ω:

$$\Phi_{p,\beta,\varepsilon}(x) = \prod_{j=1}^{J} \hat{\Phi}_{p,\beta_j,\varepsilon}(x - A_j).$$

This is indeed the case. The weight function $\hat{\Phi}_{p,\beta_j,\varepsilon}$ for each vertex has the form (4.2.2). In order to see that the dependence of $\hat{\Phi}_{p,\beta_j,\varepsilon}$ in (4.2.2) on the parameters p, ε, and the location x is a good choice, it is instructive to consider a one-dimensional example with singular right-hand side. While this example exhibit the essential features of two-dimensional problems in polygonal domains, it avoids many of the technical difficulties. It is therefore mostly the proof of the following result that is of interest:

Lemma 4.1.1. *Let $\Omega = (0,1)$ and define for $p \in \mathbb{N}_0$, $\beta \in [0,1)$, $\varepsilon \in (0,1]$ the weight functions $\hat{\Phi}_{p,\beta,\varepsilon}$ by*

$$\hat{\Phi}_{p,\beta,\varepsilon}(x) := \left(\min\left\{1, \frac{x}{\min\{1, \varepsilon(p+1)\}}\right\}\right)^{p+\beta}.$$

Let f be analytic on Ω and satisfy for some C_f, $\gamma_f > 0$, $\beta \in (0,1]$

$$\|\hat{\Phi}_{p,\beta,\varepsilon} f^{(p)}\|_{L^2(\Omega)} \leq C_f \gamma_f^p \max\{p+1, \varepsilon^{-1}\}^p \qquad \forall p \in \mathbb{N}_0.$$

Let u_ε be the solution to

$$-\varepsilon^2 u_\varepsilon'' + u_\varepsilon = f \quad \text{on } \Omega, \qquad u_\varepsilon(0) = u_\varepsilon(1) = 0. \tag{4.1.8}$$

Then there exist constants C_u, $\gamma_u > 0$ depending only on C_f, γ_f such that

$$\|\hat{\Phi}_{p,\beta,\varepsilon} u_\varepsilon^{(p+2)}\|_{L^2(\Omega)} \leq C\gamma^p \max\{p+2, \varepsilon^{-1}\}^{p+2} \qquad \forall p \in \mathbb{N}_0. \tag{4.1.9}$$

Proof: We start with an energy estimate. Defining the energy norm $\|\cdot\|_\varepsilon$ as in (1.2.7) by $\|u\|_\varepsilon^2 = \varepsilon^2\|u'\|^2_{L^2(\Omega)} + \|u\|^2_{L^2(\Omega)}$ we get from the weak formulation of (4.1.8)

$$\|u_\varepsilon\|_\varepsilon^2 = \int_\Omega f u_\varepsilon \, dx \leq \|\hat{\Phi}_{0,\beta,\varepsilon} f\|_{L^2(\Omega)} \|\hat{\Phi}^{-1}_{0,\beta,\varepsilon} u_\varepsilon\|_{L^2(\Omega)}. \tag{4.1.10}$$

Next using $\beta \in [0,1)$ and $u_\varepsilon(0) = 0$ we have from [68, Thm. 327] for a constant $C_H > 0$ independent of $\varepsilon \in (0,1]$ the bound $\int_0^\varepsilon |x^{-1}u_\varepsilon|^2\,dx \leq C_H\|u_\varepsilon\|^2_{H^1(\Omega)}$. This allows us to estimate

$$\begin{aligned}\|\hat{\Phi}^{-1}_{0,\beta,\varepsilon} u_\varepsilon\|^2_{L^2(\Omega)} &\leq \|\hat{\Phi}^{-1}_{0,1,\varepsilon} u_\varepsilon\|^2_{L^2(\Omega)} = \varepsilon^2 \int_0^\varepsilon \left|x^{-1}u_\varepsilon\right|^2 dx + \int_\varepsilon^1 |u_\varepsilon|^2\,dx \\ &\leq C_H\varepsilon^2\|u_\varepsilon\|^2_{H^1(\Omega)} + C_H\|u_\varepsilon\|^2_{L^2(\Omega)} \leq C_H^2\|u_\varepsilon\|^2_\varepsilon,\end{aligned}$$

Inserting this bound into (4.1.10), we obtain the following energy estimate:

$$\|u\|_\varepsilon \leq C_H\|\hat{\Phi}_{0,\beta,\varepsilon} f\|_{L^2(\Omega)} \leq C_H C_f. \tag{4.1.11}$$

The differential equation now gives

$$\begin{aligned}\|\hat{\Phi}_{0,\beta,\varepsilon} u_\varepsilon''\|_{L^2(\Omega)} &\leq \varepsilon^{-2}\|\hat{\Phi}_{0,\beta,\varepsilon}(f - u_\varepsilon)\|_{L^2(\Omega)} \\ &\leq \varepsilon^{-2}\|\hat{\Phi}_{0,\beta,\varepsilon} f\|_{L^2(\Omega)} + \varepsilon^{-2}\|\hat{\Phi}_{0,\beta,\varepsilon} u_\varepsilon\|_{L^2(\Omega)} \\ &\leq C_f(1 + C_H)\max\{1, \varepsilon^{-1}\}^2\end{aligned}$$

from the assumptions on f and the energy estimate (4.1.11). Similarly, we get by differentiating (4.1.8) once

$$\|\hat{\Phi}_{1,\beta,\varepsilon} u_\varepsilon^{(3)}\|_{L^2(\Omega)} \leq \varepsilon^{-2}\|\hat{\Phi}_{1,\beta,\varepsilon}\left(f' - u_\varepsilon'\right)\|_{L^2(\Omega)} \leq C_f(\gamma_f + C_H)\max\{3, \varepsilon^{-1}\}^3.$$

We show (4.1.9) by induction on p. Choosing $C_u := C_f(1 + C_H)$ and $\gamma_u := 2 + \gamma_f + C_H$, we have just proved (4.1.9) for $p = 0$ and $p = 1$. Let us assume that (4.1.9) holds for all $0 \leq p' < p$. Differentiating (4.1.8) $p \geq 2$ times, multiplying by $\hat{\Phi}_{p,\beta,\varepsilon}$ and integrating over Ω, we arrive at

$$\begin{aligned}\|\hat{\Phi}_{p,\beta,\varepsilon} u_\varepsilon^{(p+2)}\|_{L^2(\Omega)} &= \varepsilon^{-2}\|\hat{\Phi}_{p,\beta,\varepsilon}\left(f^{(p)} - u_\varepsilon^{(p)}\right)\|_{L^2(\Omega)} \\ &\leq \varepsilon^{-2}\|\hat{\Phi}_{p,\beta,\varepsilon} f^{(p)}\|_{L^2(\Omega)} + \varepsilon^{-2}\|\hat{\Phi}_{p,\beta,\varepsilon} u_\varepsilon^{(p)}\|_{L^2(\Omega)}. \qquad (4.1.12)\end{aligned}$$

We now claim that for fixed $x \in \Omega$, $\varepsilon \in (0,1]$ the function $p \mapsto \hat{\Phi}_{p,\beta,\varepsilon}(x)$ is decreasing. First, the function $p \mapsto \min\{1, \varepsilon(p+1)\}$ is monotonically increasing and thus

$$\phi : p \mapsto \min\left\{1, \frac{x}{\min\{1, (p+1)\varepsilon\}}\right\}$$

is monotonically decreasing. As $\phi \leq 1$ for all p, we conclude that

$$p \mapsto [\phi(p)]^{p+\beta}$$

is also decreasing. Recognizing this last function as $\hat{\Phi}_{p,\beta,\varepsilon}$ we get that the $p \mapsto \hat{\Phi}_{p,\beta,\varepsilon}(x)$ is indeed decreasing. Thus, we may bound (4.1.12) further by using the induction hypothesis (4.1.9)

$$\begin{aligned}\|\hat{\Phi}_{p,\beta,\varepsilon} u_\varepsilon^{(p+2)}\|_{L^2(\Omega)} &\leq \varepsilon^{-2}\|\hat{\Phi}_{p,\beta,\varepsilon} f^{(p)}\|_{L^2(\Omega)} + \varepsilon^{-2}\|\hat{\Phi}_{p-2,\beta,\varepsilon} u_\varepsilon^{(p)}\|_{L^2(\Omega)} \\ &\leq \varepsilon^{-2} C_f \gamma_f^p \max\{p+1, \varepsilon^{-1}\}^p + \varepsilon^{-2} C_u \gamma_u^p \max\{p, \varepsilon^{-1}\}^p \\ &\leq C_u \gamma_u^{p+2} \max\{p+2, \varepsilon^{-1}\}^{p+2} \gamma_u^{-2}\left[\frac{C_f}{C_u}\left(\frac{\gamma_f}{\gamma_u}\right)^p + 1\right].\end{aligned}$$

Our choice $\gamma_u = 2 + \gamma_f + C_H > 2$ and $C_u = 1 + C_f + C_H$ implies

$$\gamma_u^{-2}\left[\frac{C_f}{C_u}\left(\frac{\gamma_f}{\gamma_u}\right)^p + 1\right] \leq 1$$

so that we indeed obtain the desired bound

$$\|\hat{\Phi}_{p,\beta,\varepsilon} u_\varepsilon^{(p+2)}\|_{L^2(\Omega)} \leq C_u \gamma_u^{p+2} \max\{p+2, \varepsilon^{-1}\}^{p+2}.$$

□

In the notation of this work, Lemma 4.1.1 represents a shift theorem in countably normed spaces: For right-hand sides $f \in \mathcal{B}^0_{\beta,\varepsilon}$, the solution u_ε of (4.1.8) is in the countably normed space $\mathcal{B}^2_{\beta,\varepsilon}$. We will see in Chapter 5 that the analogous result holds in curvilinear polygons.

4.1.2 Outline of Chapter 4

The outline of Chapter 4 is as follows. In Chapter 4 we first define the spaces $H^{m,l}_{\beta,\varepsilon}$ and the countably normed spaces $\mathcal{B}^l_{\beta,\varepsilon}$. For simplicity of exposition, these spaces are defined on sectors in the present chapter. The subscript β is therefore a scalar rather than a vector as suggested above for polygonal domains. In Lemma 4.2.2, we collect some basic properties of the weight functions $\hat{\Phi}_{p,\beta,\varepsilon}$. Of particular relevance for the understanding of the methods of proof employed in this work is the last result of Lemma 4.2.2, (4.2.8). It states that for balls $B_{cR}(x)$ with $|x| = R$, $c \in (0,1)$, the weight function $\hat{\Phi}_{p,\beta,\varepsilon}$ can be bounded above and below (up to a factor K^p with K independent of ε and x) by $\hat{\Phi}_{p,\beta,\varepsilon}(x)$, i.e., by the value of the weight function $\hat{\Phi}_{p,\beta,\varepsilon}$ at the center of the ball. This feature will frequently allow us in local analyses to replace weighted norms by standard Sobolev norms. Section 4.2.1 collects some properties of the spaces $H^{m,l}_{\beta,\varepsilon}$, notably two embedding theorems. The first embedding result, Lemma 4.2.9, shows that the functions from the space $H^{2,2}_{\beta,\varepsilon}$ are continuous up to the boundary. The second embedding result, Lemma 4.2.10, is the key result of Section 4.2.1 where Hardy-type estimates in weighted Sobolev spaces are proved. Lemma 4.2.10 is an important technical tool for the proof of the main result of the ensuing Chapter 5, Theorem 5.3.10 and its variants Propositions 5.4.5, 5.4.8, 5.4.7.
Section 4.2.2 collects properties of the countably normed spaces $\mathcal{B}^l_{\beta,\varepsilon}$. The major result of Section 4.2.2 is Theorem 4.2.20. This result shows that the spaces $\mathcal{B}^l_{\beta,\varepsilon}$ are invariant under analytic changes of variables. This result will prove useful in Chapter 5 in our treatment of curved boundaries: Theorem 4.2.20 allows us to infer regularity results for domains with curved boundaries from those with straight boundaries by mapping arguments. The main idea of the proof of Theorem 4.2.20 is to consider in a first step the change of variables locally and then combine in a second step these local results to a global estimate with the aid of a covering argument. The technical tool for inferring membership in a countably normed space $\mathcal{B}^l_{\beta,\varepsilon}$ from local estimates is provided in Lemma 4.2.17. Local results for changes of variables need to track two parameters: the perturbation parameter ε and the distance to the apex of the sector. Such results are again technically involved and therefore provided in the separate Lemma 4.3.1.
The spaces $\mathcal{B}^l_{\beta,\varepsilon}$ are L^2-based function spaces. It is also of interest to characterize the pointwise behavior of $\mathcal{B}^l_{\beta,\varepsilon}$ functions. This is achieved in Theorem 4.2.23.

4.2 The spaces $H^{m,l}_{\beta,\varepsilon}$ and $\mathcal{B}^l_{\beta,\varepsilon}$ in a Sector

Definition 4.2.1 (sector). *A bounded Lipschitz domain $S \subset \mathbb{R}^2$ is said to be a sector with apex 0 (or simply: a sector) if $0 \in \partial S$. A sector S is called a C^2-curvilinear sector if there are three, mutually disjoint C^2-arcs Γ_i ($i \in \{1,2,3\}$) such that $\partial S = \cup_{i=1}^3 \overline{\Gamma_i}$ and $0 = \overline{\Gamma_1} \cap \overline{\Gamma_2}$. A C^2-curvilinear sector is called an* analytic curvilinear sector *(or simply curvilinear sector) if the three arcs $\overline{\Gamma_i}$ are analytic arcs.*

For $R > 0$ and $\omega \in (0, 2\pi)$ we define straight *sectors $S_R(\omega)$ as the sets*

$$S_R(\omega) = \{(r\cos\varphi, r\sin\varphi) \,|\, 0 < r < R, \quad 0 < \varphi < \omega\}. \tag{4.2.1}$$

For $p \in \mathbb{Z}$ and $\beta \in \mathbb{R}$ we define the weight function $\hat{\Phi}_{p,\beta,\varepsilon}$ by

$$\hat{\Phi}_{p,\beta,\varepsilon}(x) = \left(\min\left\{1, \frac{|x|}{\min\{1, \varepsilon(|p|+1)\}}\right\}\right)^{p+\beta}. \tag{4.2.2}$$

One can check that the weight functions $\hat{\Phi}_{p,\beta,\varepsilon}$ has the following properties:

Lemma 4.2.2 (Properties of $\hat{\Phi}_{p,\beta,\varepsilon}$). *Let $S \in \mathbb{R}^2$ be a sector. There holds for all $p \in \mathbb{N}_0$, $\varepsilon \in (0,1]$, $\beta \in [0,1]$, $l \in \mathbb{N}$, and $x \in S$:*

$$\hat{\Phi}_{p,\beta,\varepsilon}(x) \sim \hat{\Phi}_{p,0,\varepsilon}(x)\hat{\Phi}_{0,\beta,\varepsilon}(x), \tag{4.2.3}$$

$$\hat{\Phi}_{p-l,\beta,\varepsilon}(x) \sim \hat{\Phi}_{p,0,\varepsilon}(x)\hat{\Phi}_{-l,\beta,\varepsilon}(x), \tag{4.2.4}$$

$$\hat{\Phi}_{p-l,\beta,\varepsilon}(x) \sim \hat{\Phi}_{p,\beta-l,\varepsilon}(x), \tag{4.2.5}$$

$$\frac{1}{\hat{\Phi}_{p,\beta,\varepsilon}(x)} \sim 1 + \left(\frac{\min\{1, \varepsilon(p+1)\}}{|x|}\right)^{p+\beta}, \tag{4.2.6}$$

$$\frac{1}{\hat{\Phi}_{p,0,\varepsilon}(x)} \max\{p+1, \varepsilon^{-1}\}^p \sim \max\{(p+1)/|x|, \varepsilon^{-1}\}^p. \tag{4.2.7}$$

Here, the relationship $a \sim b$ means that there exist C, $K > 0$ independent of $p \in \mathbb{N}_0$, $\varepsilon \in (0,1]$, and $x \in S$ such that $C^{-1}K^{-p}a \le b \le CK^p a$.
Furthermore, let $c \in (0,1)$ be given. Then for all balls $B_{c|x|}(x)$ with $x \in S$

$$\min_{z \in B_{c|x|}(x)} \hat{\Phi}_{p,\beta,\varepsilon}(z) \sim \hat{\Phi}_{p,\beta,\varepsilon}(x) \sim \max_{z \in B_{c|x|}(x)} \hat{\Phi}_{p,\beta,\varepsilon}(z). \tag{4.2.8}$$

Here, the constants C, K in the definition of $\sim$ depend additionally on the constant $c \in (0,1)$ but are independent of p, ε, $x \in S$.

Proof: We will only show (4.2.7) to demonstrate the general procedure. If $\operatorname{diam}(S) \ge |x| \ge \min\{1, \varepsilon(p+1)\}$, then (4.2.7) follows easily as $\hat{\Phi}_{p,\beta,\varepsilon}(x) = 1$ and

$$\frac{p+1}{\operatorname{diam}(S)} \le \frac{p+1}{|x|} \le (p+1)\max\left\{1, \frac{1}{\varepsilon(p+1)}\right\} \le \max\{p+1, \varepsilon^{-1}\}.$$

We therefore consider the case $|x| \le \min\{1, \varepsilon(p+1)\}$. We note that this implies

$$\varepsilon^{-1} \le \frac{p+1}{|x|}. \tag{4.2.9}$$

From the definition of the symbol $\sim$, the bound (4.2.7) is proved if we can show the existence of $C > 0$ independent of ε such that for all $p \in \mathbb{N}_0$

$$C^{-1}\frac{\min\{1,\varepsilon(p+1)\}}{|x|}\max\{p+1,\varepsilon^{-1}\} \le \max\{(p+1)/|x|,\varepsilon^{-1}\}$$
$$\le C\frac{\min\{1,\varepsilon(p+1)\}}{|x|}\max\{p+1,\varepsilon^{-1}\}.$$

First, we note that

$$\forall z>0 \qquad \min\{1,z\}\max\{1,z^{-1}\}=1. \tag{4.2.10}$$

This implies readily that

$$\frac{\min\{1,(p+1)\varepsilon\}}{|x|}\max\{p+1,\varepsilon^{-1}\} = \frac{p+1}{|x|}\min\{1,(p+1)\varepsilon\}\max\left\{1,\frac{1}{(p+1)\varepsilon}\right\}$$
$$= \frac{p+1}{|x|} = \max\left\{\frac{p+1}{|x|},\varepsilon^{-1}\right\}.$$

where the last equality follows from (4.2.9). □

Of interest in the following will be monotonicity properties of the weight functions $\hat{\Phi}_{p,\beta,\varepsilon}$ in the arguments β, p, and ε. We have

Lemma 4.2.3 (Monotonicity properties of $\hat{\Phi}_{p,\beta,\varepsilon}$). *Let S be a sector.*

1. *For all fixed $p\in\mathbb{N}$, $\varepsilon>0$, $x\in S$ the function $\beta\mapsto\hat{\Phi}_{p,\beta,\varepsilon}(x)$ is monotonically decreasing on $\mathbb{R}^+_0$.*
2. *For all fixed $p\in\mathbb{N}$, $\beta\in[0,1)$, $x\in S$ the function $\varepsilon\mapsto\hat{\Phi}_{p,\beta,\varepsilon}(x)$ is monotonically decreasing on $\mathbb{R}^+_0$.*
3. *For all fixed $\varepsilon>0$, $\beta>0$, $x\in S$ the function $p\mapsto\hat{\Phi}_{p,\beta,\varepsilon}(x)$ is monotonically decreasing on $\mathbb{N}_0$.*
4. *For all $\beta\in[0,1]$ there are C, $\gamma>0$ independent of ε, $\varepsilon'\in(0,1]$, and $p\in\mathbb{N}_0$ such that*

$$\frac{\hat{\Phi}_{p,\beta,\varepsilon}(x)}{\max\{p+1,\varepsilon^{-1}\}^p} \le C\gamma^p\max\{1,\varepsilon'/\varepsilon\}^\beta\max\{1,\varepsilon/\varepsilon'\}^p\frac{\hat{\Phi}_{p,\beta,\varepsilon'}(x)}{\max\{p+1,(\varepsilon')^{-1}\}^p}.$$

Proof: The first and second assertions of the lemma follow immediately from (4.2.2). The third assertion was already proved in the course of the proof of Lemma 4.1.1. For the fourth assertion, we start by first considering the case $p=0$. We have for $\beta\ge 0$ (writing $r=|x|$)

$$\hat{\Phi}_{0,\beta,\varepsilon}(x) = \min\{1,r/\varepsilon\}^\beta \le \min\{1,r/\varepsilon'\}^\beta\max\{1,\varepsilon'/\varepsilon\}^\beta$$
$$\le \hat{\Phi}_{0,\beta,\varepsilon'}(x)\max\{1,\varepsilon'/\varepsilon\}^\beta. \tag{4.2.11}$$

Next, let us consider the case $\beta=0$. By the definition of $\hat{\Phi}_{p,0,\varepsilon}$ we have

$$\frac{\hat{\Phi}_{p,0,\varepsilon}(x)}{\max\{p+1,\varepsilon^{-1}\}^p} = \left(\frac{\min\{1,(r/m)\}}{\max\{p+1,\varepsilon^{-1}\}}\right)^p, \quad m=\min\{1,(p+1)\varepsilon\}. \tag{4.2.12}$$

Employing the formula

$$\frac{ab}{a+b} \leq \min\{a,b\} \leq 2\frac{ab}{a+b} \qquad \forall a,b>0, \tag{4.2.13}$$

we can bound the expression in parentheses in (4.2.12) as follows:

$$\frac{r}{(p+1)+r\max\{p+1,\varepsilon^{-1}\}} \leq \frac{\min\{1,(r/m)\}}{\max\{p+1,\varepsilon^{-1}\}} \leq \frac{2r}{(p+1)+r\max\{p+1,\varepsilon^{-1}\}}.$$

Next, as

$$\begin{aligned}\max\{p+1,\varepsilon^{-1}\} &= \max\{p+1,(\varepsilon')^{-1}(\varepsilon/\varepsilon')^{-1}\}\\ &\geq \max\{p+1,(\varepsilon')^{-1}\}\min\{1,(\varepsilon/\varepsilon')^{-1}\}\end{aligned}$$

we arrive at

$$\begin{aligned}\frac{\min\{1,(r/m)\}}{\max\{p+1,\varepsilon^{-1}\}} &\leq \frac{2r}{(p+1)+r\max\{p+1,(\varepsilon')^{-1}\}\min\{1,(\varepsilon/\varepsilon')^{-1}\}}\\ &\leq \frac{r}{(p+1)+r\max\{p+1,(\varepsilon')^{-1}\}}\cdot\frac{2}{\min\{1,(\varepsilon/\varepsilon')^{-1}\}}.\end{aligned}$$

Raising both sides to the power p we get

$$\frac{\hat{\Phi}_{p,0,\varepsilon}(x)}{\max\{p+1,\varepsilon^{-1}\}^p} \leq \frac{\hat{\Phi}_{p,0,\varepsilon'}(x)}{\max\{p+1,(\varepsilon')^{-1}\}^p}\max\{1,\varepsilon/\varepsilon'\}^p 2^p,$$

which is the desired bound for the case $\beta = 0$. Appealing now to (4.2.3) of Lemma 4.2.2 allows us to conclude the claim in the desired generality. □

We are now in position to introduce the spaces $H^{m,l}_{\beta,\varepsilon}(S)$ and $\mathcal{B}^l_{\beta,\varepsilon}(S)$. On a sector S, we define for m, $l \in \mathbb{N}_0$, $m \geq l$, and $\beta \in [0,1]$, $\varepsilon \in (0,1]$, the spaces $H^{m,l}_{\beta,\varepsilon}(S)$ as the completion of the space $C^\infty(S)$ under the norms

$$\|u\|^2_{H^{m,l}_{\beta,\varepsilon}(\Omega)} := \sum_{k=0}^{l-1}\varepsilon^{2k}\|\nabla^k u\|^2_{L^2(\Omega)} + \varepsilon^{2l}\sum_{k=l}^{m}\|\hat{\Phi}_{k-l,\beta,\varepsilon}\nabla^k u\|^2_{L^2(\Omega)}, \quad l>0, \tag{4.2.14}$$

$$\|u\|^2_{H^{m,0}_{\beta,\varepsilon}(\Omega)} := \sum_{k=0}^{m}\|\hat{\Phi}_{k,\beta,\varepsilon}\nabla^k u\|^2_{L^2(\Omega)}.$$

For a given sector S and constants C_u, $\gamma_u > 0$, the spaces $\mathcal{B}^l_{\beta,\varepsilon}(S,C_u,\gamma_u)$ are defined as

$$\begin{aligned}\mathcal{B}^l_{\beta,\varepsilon}(S,C_u,\gamma_u) = \{&u \in H^{l,l}_{\beta,\varepsilon}(S)\,|\,\|u\|_{H^{l,l}_{\beta,\varepsilon}(S)} \leq C_u \text{ and}\\ &\|\hat{\Phi}_{k,\beta,\varepsilon}\nabla^{k+l}u\|_{L^2(S)} \leq C_u\gamma_u^k\max\{k+1,\varepsilon^{-1}\}^{k+l} \quad \forall k\in\mathbb{N}_0\}.\end{aligned} \tag{4.2.15}$$

For simplicity of notation, the dependence on the domain S and the constants C_u, γ_u is dropped when no confusion can arise.

Remark 4.2.4 For fixed $\beta \in [0,1)$, the spaces $H^{2,2}_{\beta,\varepsilon}(S)$ are all isomorphic: There holds (cf. Lemma 4.2.7 below)

$$\varepsilon^2 \|u\|_{H^{2,2}_{\beta,1}(S)} \leq \|u\|_{H^{2,2}_{\beta,\varepsilon}(S)} \leq \|u\|_{H^{2,2}_{\beta,1}(S)} \quad \forall u \in H^{2,2}_{\beta,1}(S) \quad \forall \varepsilon \in (0,1]. \quad (4.2.16)$$

Similarly, the spaces $\cup_{C_u>0,\gamma_u>0}\mathcal{B}^l_{\beta,\varepsilon}(S, C_u, \gamma_u)$ are algebraically identical for all $\varepsilon \in (0,1]$. Furthermore, we note that the spaces $H^{m,l}_{\beta,1}(S)$ and $\mathcal{B}^l_{\beta,1}$ coincide with the spaces $H^{m,l}_{\beta}$, $\mathcal{B}^l_{\beta}$ introduced by Babuška and Guo in [15,16]. We will therefore frequently make use of results by them for this special case. ▪

Lemma 4.2.5. *For all $\varepsilon \in (0,1]$ and $\beta \in [0,1)$, there holds the embedding $H^{2,2}_{\beta,\varepsilon}(S) \subset C(\overline{S})$. Moreover, there is $C > 0$ depending only on S and β such that*

$$\|u\|_{C(\overline{S})} \leq C\varepsilon^{-2} \|u\|_{H^{2,2}_{\beta,\varepsilon}(S)} \qquad \forall u \in H^{2,2}_{\beta,\varepsilon}(S).$$

Proof: From [21] we have $\|u\|_{C(\overline{S})} \leq C\|u\|_{H^{2,2}_{\beta,1}(S)}$. The result now follows from (4.2.16). □

We are now interested in analyzing the behavior of the spaces $H^{m,l}_{\beta,\varepsilon}$ and $\mathcal{B}^l_{\beta,\varepsilon}$ under smooth changes of variables. For the purposes of this work, we will limit our attention to the cases $m = l \in \{0,2\}$.

4.2.1 Properties of the spaces $H^{m,l}_{\beta,\varepsilon}(\Omega)$

Lemma 4.2.6. *Let $\hat{S}$, $S \subset \mathbb{R}^2$ be two sectors. Let $g \in C^2(\hat{S}, S)$ be a C^2 diffeomorphism satisfying additionally $g(0) = 0$. Then there exists $C > 0$ depending only on $\hat{S}$ and g such that for every $\varepsilon \in (0,1]$*

$$C^{-1}\|u\|_{H^{l,l}_{\beta,\varepsilon}(S)} \leq \|u \circ g\|_{H^{l,l}_{\beta,\varepsilon}(\hat{S})} \leq C\|u\|_{H^{l,l}_{\beta,\varepsilon}(S)} \quad \forall u \in H^{l,l}_{\beta,\varepsilon}(S), \quad l \in \{0,1,2\}.$$

Proof: As g is a C^2 diffeomorphism, there are c_1, $c_2 > 0$ such that $c_1|x| \leq |g(x)| \leq c_2|x|$. Furthermore, $\|g\|_{C^2(\hat{S})} < \infty$. The upper bounds now follow readily from the chain rule. As g is a C^2 diffeomorphism, the inverse g^{-1} is also a C^2 diffeomorphism and hence the lower bounds can be obtained from the upper bounds by replacing g by g^{-1} (and exchanging $\hat{S}$ and S). □

The monotonicity properties of the weight function $\hat{\Phi}_{0,\beta,\varepsilon}$ of Lemma 4.2.3 imply properties of the norms $\|\cdot\|_{H^{l,l}_{\beta,\varepsilon}}$:

Lemma 4.2.7. *Let S be a sector. Then, for fixed $\beta \in [0,1)$ and $l \in \mathbb{N}_0$*

$$\|u\|_{H^{l,l}_{\beta,\varepsilon}(S)} \leq \|u\|_{H^{l,l}_{\beta,\varepsilon'}(S)} \qquad \forall 0 < \varepsilon \leq \varepsilon' \leq 1 \qquad \forall u \in H^{l,l}_{\beta,1}(S), \quad l > 0,$$

$$\|u\|_{H^{0,0}_{\beta,\varepsilon}(S)} \leq (\varepsilon'/\varepsilon)^{\beta} \|u\|_{H^{0,0}_{\beta,\varepsilon'}(S)} \qquad \forall 0 < \varepsilon \leq \varepsilon' \leq 1 \qquad \forall u \in H^{0,0}_{\beta,1}(S).$$

For fixed $\varepsilon \in (0,1]$ there holds

$$\|u\|_{H^{l,l}_{\beta',\varepsilon}(S)} \leq \|u\|_{H^{l,l}_{\beta,\varepsilon}(S)} \qquad \forall 0 \leq \beta \leq \beta' \leq 1 \qquad \forall u \in H^{l,l}_{\beta,1}(S).$$

Proof: The estimates follow from the monotonicity properties of the weight function $\hat{\Phi}_{0,\beta,\varepsilon}$ of Lemma 4.2.3. In particular, the case of fixed $\beta \in [0,1)$ and $l = 0$ follows from (4.2.11). □

Lemma 4.2.8 ($H^{2,2}_{\beta,\varepsilon}$ compactly embedded in H^1). *Let S be a sector, $\beta \in [0,1)$, $\varepsilon \in (0,1]$. Then $H^{2,2}_{\beta,\varepsilon}(S)$ is compactly embedded in $H^1(S)$. Furthermore, for every $\delta > 0$, there exists $C(\delta) > 0$ independent of $\varepsilon \in (0,1]$ such that*

$$\|u\|_{H^1(\Omega)} \leq \delta \|\hat{\Phi}_{0,\beta,\varepsilon}\nabla^2 u\|_{L^2(S)} + C(\delta)\|u\|_{L^2(S)} \qquad \forall u \in H^{2,2}_{\beta,\varepsilon}(S).$$

Proof: The compactness of the embedding is essentially proved in, e.g., [112] (only the case of a straight polygon is considered there but Lemma 4.2.6 allows us to infer the general case readily). For $\varepsilon = 1$, the compact embedding $H^{2,2}_{\beta,1}(S) \subset\subset H^1(S)$ implies by a standard argument ("Ehrling's Lemma", cf. [129, Thm. 7.3]) that every for every $\delta > 0$ there exists $C(\delta) > 0$ such that

$$\|u\|_{H^1(S)} \leq \delta \|\hat{\Phi}_{0,\beta,1}\nabla^2 u\|_{L^2(S)} + C(\delta)\|u\|_{L^2(S)}.$$

The desired result now follows from the observation that for $\varepsilon \in (0,1]$ there holds $\hat{\Phi}_{0,\beta,\varepsilon} \geq \hat{\Phi}_{0,\beta,1}$ on S. □

We furthermore need a result concerning L^∞ bounds:

Lemma 4.2.9. *Let S be a sector and $\beta \in [0,1)$. There exists $C > 0$ depending only on β and S such that for all $u \in H^{2,2}_{\beta,\varepsilon}(S)$*

$$\|u\|_{L^\infty(S)} \leq C\left[\|\hat{\Phi}_{0,\beta,\varepsilon}\nabla^2 u\|_{L^2(S)} + \|u\|_{L^2(S)}\right].$$

Proof: By [21], we have the embedding $\|u\|_{L^\infty(S)} \leq C\|u\|_{H^{2,2}_{\beta,1}(S)}$. Using the preceding lemma, we therefore have

$$\|u\|_{L^\infty(S)} \leq C\left[\|\hat{\Phi}_{0,\beta,1}\nabla^2 u\|_{L^2(S)} + \|u\|_{L^2(S)}\right].$$

Using again $\hat{\Phi}_{0,\beta,1} \leq \hat{\Phi}_{0,\beta,\varepsilon}$ on S for $\varepsilon \in (0,1]$ gives the desired result. □

Next, we need embedding theorems in the weighted Sobolev spaces $H^{2,2}_{\beta,\varepsilon}(S)$ akin to those studied in [21]:

Lemma 4.2.10 (embedding in weighted spaces). *Let S be a C^2-curvilinear sector, $\beta \in (0,1)$, $l \in \{1,2\}$. Then there is $C > 0$ depending only on S, β, and l such that for every $\varepsilon \in (0,1]$ and every $u \in H^{l,l}_{\beta,\varepsilon}(S)$ there is $\overline{u} \in \mathbb{R}$ such that:*

(i) if $l = 2$ the constant $\overline{u}$ may be taken as $\overline{u} = u(0)$ and there holds

$$\begin{aligned}\|\hat{\Phi}_{0,\beta-2,\varepsilon}(u-\overline{u})\|_{L^2(S\cap B_{2\varepsilon}(0))} &\leq \varepsilon\|\hat{\Phi}_{0,\beta-1,\varepsilon}\nabla u\|_{L^2(S\cap B_{2\varepsilon}(0))}\\ &\leq C\|u\|_{H^{2,2}_{\beta,\varepsilon}(S\cap B_{2\varepsilon}(0))},\end{aligned}$$

$$\|u\|_{L^\infty(S\cap B_{2\varepsilon}(0))} \leq C\varepsilon^{-1}\|u\|_{H^{2,2}_{\beta,\varepsilon}(S\cap B_{2\varepsilon}(0))},$$

$$\|\hat{\Phi}_{0,\beta-2,\varepsilon}u\|_{L^2(S\setminus B_\varepsilon(0))} + \varepsilon\|\hat{\Phi}_{0,\beta-1,\varepsilon}\nabla u\|_{L^2(S\setminus B_\varepsilon(0))} \leq C\|u\|_{H^{2,2}_{\beta,\varepsilon}(S)};$$

(ii) if $l = 1$:

$$\|\hat{\Phi}_{0,\beta-1,\varepsilon} u\|_{L^2(S\cap B_{2\varepsilon}(0))} \le C\|u\|_{H^{1,1}_{\beta,\varepsilon}(S\cap B_{2\varepsilon}(0))},$$

$$\|\hat{\Phi}_{0,\beta-1,\varepsilon} u\|_{L^2(S)} \le C\|u\|_{H^{1,1}_{\beta,\varepsilon}(S)},$$

$$\|\hat{\Phi}_{0,\beta-1,\varepsilon}\,(u-\overline{u})\,\|_{L^2(S\cap B_{2\varepsilon}(0))} \le C\varepsilon\|\hat{\Phi}_{0,\beta,\varepsilon}\nabla u\|_{L^2(S\cap B_{2\varepsilon}(0))} \le C\|u\|_{H^{1,1}_{\beta,\varepsilon}(S)}.$$

Proof: We start with the proof of (i). As $\hat{\Phi}_{0,\beta-2,\varepsilon} = \hat{\Phi}_{0,\beta-1,\varepsilon} = 1$ on $S \setminus B_\varepsilon(0)$ the third assertion of (i) is trivial. We can therefore restrict our attention to the neighborhood $V = B_{2\varepsilon}(0) \cap S$. We start with the following
Assertion: For $\tau > 0$ denote $T_\tau := \{(x,y)\,|\,0 < x < \tau,\quad 0 < y < \tau - x\}$. Then for any $s \in (0,1)$ there is $C > 0$ independent of τ such that for all functions $u \in C(\overline{T_\tau})$ with $\|\,|x|^s\nabla^2 u\|_{L^2(T_\tau)} < \infty$ there holds

$$\begin{aligned}&\|\,|x|^{s-2}\,(u-u(0))\,\|_{L^2(T_\tau)} + \|\,|x|^{s-1}\nabla u\|_{L^2(T_\tau)}\\ &\le C\left[\|\,|x|^s\nabla^2 u\|_{L^2(T_\tau)} + \tau^{s-2}\|u\|_{L^2(T_\tau)}\right].\end{aligned}$$

Proof of the Assertion: By homogeneity, it suffices to consider the case $\tau = 1$. From [21, Lemma 4.4] there exists $C > 0$ such that for the linear interpolant p of u there holds

$$\|\,|x|^{s-2}(u-p)\|_{L^2(T_1)} + \|\,|x|^{s-1}\nabla(u-p)\|_{L^2(T_1)} \le C\|\,|x|^s\nabla^2 u\|_{L^2(T_1)}. \quad (4.2.17)$$

By Sobolev's embedding on $T_1 \setminus T_{1/2}$ we have the existence of $C > 0$ such that

$$\|u\|_{L^\infty(T_1\setminus T_{1/2})} \le C\left[\|\,|x|^s\nabla^2 u\|_{L^2(T_1\setminus T_{1/2})} + \|\,u\|_{L^2(T_1\setminus T_{1/2})}\right]. \quad (4.2.18)$$

Next, we write $p(x) = u(0) + l(x)$ where l is a linear function with $l(0) = 0$ and $\|l\|_{L^\infty(T_1)} + \|\nabla l\|_{L^\infty(T_1)} \le C\|u\|_{L^\infty(T_1\setminus T_{1/2})}$. An application of the reverse triangle inequality in (4.2.17) and (4.2.18) concludes the proof of the assertion.

Let us now consider the neighborhood $V = S \cap B_{2\varepsilon}(0)$. As S is a C^2-curvilinear sector, there are two C^2 curves Γ_1, Γ_2 comprising the boundary of S near the origin 0. Introduce another smooth curve Γ' through 0 (independent of ε) that divides V into two domains V', V'' each having a convex corner at 0. By the smoothness of the curves Γ_1, Γ_2, Γ' there is a C^2 map $F : T_\varepsilon \to V'$ such that $F(0,0) = 0$ and F', $(F^{-1})'$, and F'' can be bounded independently of ε (e.g., the blending map from T_ε to V', [58–60]). From the Assertion and the fact that $F(0) = 0$ we get that the function $\hat{u} = u \circ F$ satisfies

$$\begin{aligned}&\|\,|x|^{s-2}\,(\hat{u}-\hat{u}(0))\,\|_{L^2(T_\varepsilon)} + \|\,|x|^{s-1}\nabla\hat{u}\|_{L^2(T_\varepsilon)}\\ &\le C\left[\|\,|x|^s\nabla^2\hat{u}\|_{L^2(T_\varepsilon)} + \varepsilon^{s-2}\|\hat{u}\|_{L^2(T_\varepsilon)}\right]\\ &\le C\big[\|\,|x|^s(\nabla^2 u)\circ F\|_{L^2(T_\varepsilon)} +\\ &\qquad C\|F''\|_{L^\infty(T_\varepsilon)}\,\|\,|x|^s(\nabla u)\circ F\|_{L^2(T_\varepsilon)} + \varepsilon^{s-2}\|u\circ F\|_{L^2(T_\varepsilon)}\big].\end{aligned}$$

Employing now the fact that F', $(F^{-1})'$, and F'' can be bounded independently of ε, we obtain by transforming this last estimate back to V':

$$\|r^{\beta-2}\left(u-u(0)\right)\|_{L^2(V')}+\|r^{\beta-1}\nabla u\|_{L^2(V')}\leq$$
$$C\left[\|r^{\beta}\nabla^2 u\|_{L^2(V')}+\|r^{\beta}\nabla u\|_{L^2(V')}+\varepsilon^{\beta-2}\|u\|_{L^2(V')}\right],$$

where $r(x)=|x|$. Similarly, we obtain the corresponding bound for V''. Dividing by $\varepsilon^{\beta-2}$ and using that $r^\beta \leq \varepsilon^\beta$ on V we therefore obtain

$$\|\hat{\Phi}_{0,\beta-2,\varepsilon}\left(u-u(0)\right)\|_{L^2(V)}+\varepsilon\|\hat{\Phi}_{0,\beta-1,\varepsilon}\nabla u\|_{L^2(V)}\leq$$
$$C\varepsilon^2\|\Phi_{0,\beta,\varepsilon}\nabla^2 u\|_{L^2(V)}+C\varepsilon\|\nabla u\|_{L^2(V)}+C\|u\|_{L^2(V)}.$$

This last estimate proves $\|\hat{\Phi}_{0,\beta-2,\varepsilon}(u-\overline{u})\|_{L^2(S\cap B_{2\varepsilon}(0))}\leq C\|u\|_{H^{2,2}_{\beta,\varepsilon}(S\cap B_{2\varepsilon}(0))}$ and $\varepsilon\|\hat{\Phi}_{0,\beta,\varepsilon}\nabla u\|_{L^2(S\cap B_{2\varepsilon}(0))}\leq C\|u\|_{H^{2,2}_{\beta,\varepsilon}(S\cap B_{2\varepsilon}(0))}$. We will not explicitly prove the first claim of (i) as its proof is very similar to that of the last claim of (ii), which we prove below.

We now turn to the L^∞-bound. From the embedding theorem Lemma 4.2.9 applied to the triangle T_1 we obtain with scaling arguments

$$\varepsilon^{\beta-1}\|u\|_{L^\infty(T_\varepsilon)}\leq C\left[\|\,|x|^\beta\nabla^2 u\|_{L^2(T_\varepsilon)}+\varepsilon^{\beta-2}\|u\|_{L^2(T_\varepsilon)}\right].$$

Reasoning as above, we conclude

$$\begin{aligned}\varepsilon\|u\|_{L^\infty(V)}&\leq C\left[\varepsilon^2\|\hat{\Phi}_{0,\beta,\varepsilon}\nabla^2 u\|_{L^2(V)}+\varepsilon^2\|\hat{\Phi}_{0,\beta,\varepsilon}\nabla u\|_{L^2(V)}+\|u\|_{L^2(V)}\right]\\&\leq C\|u\|_{H^{2,2}_{\beta,\varepsilon}(V)}.\end{aligned}$$

Part (ii) of the lemma is proved using the same ideas. For the first part estimate of Part (ii), we employ Lemma A.1.7 and a scaling argument to conclude for the triangles T_τ:

$$\|\,|x|^{\beta-1}u\|_{L^2(T_\tau)}\leq C\left[\|\,|x|^\beta\nabla u\|_{L^2(T_\tau)}+\tau^{\beta-1}\|u\|_{L^2(T_\tau)}\right].$$

This implies with arguments as above

$$\|r^{\beta-1}u\|_{L^2(V)}\leq C\left[\|r^\beta\nabla u\|_{L^2(V)}+\varepsilon^{\beta-1}\|u\|_{L^2(V)}\right].$$

Dividing by $\varepsilon^{\beta-1}$ then gives the result.

The second estimate follows by combining the first estimate of Part (ii) with the observation $\Phi_{0,\beta,\varepsilon}(x)=1$ for $x\in S\setminus B_\varepsilon(0)$. The last estimate in Part (ii) is seen as follows: We note that by [21, Lemma 4.3] there is $\overline{u}\in\mathbb{R}$ such that the function $\hat{u}=u\circ F$ satisfies

$$\|\,|x|^{\beta-1}(\hat{u}-\overline{u})\|_{L^2(T_\varepsilon)}\leq C\|\,|x|^\beta\nabla\hat{u}\|_{L^2(T_\varepsilon)}.$$

Transforming back to to V' and dividing by $\varepsilon^{\beta-1}$ yields

$$\varepsilon^{-(\beta-1)}\|\,|x|^{\beta-1}(u-\overline{u})\|_{L^2(T_\varepsilon)}\leq C\varepsilon\varepsilon^{-\beta}\|\,|x|^\beta\nabla u\|_{L^2(T_\varepsilon)}.$$

This proves (ii) of the lemma. □

In the following corollary, we strengthen slightly the assertion of the first part of Lemma 4.2.10.

Corollary 4.2.11. *Under the assumptions of Lemma 4.2.10 there exists $C > 0$ independent of ε such that for all $u \in H^{2,2}_{\beta,\varepsilon}(S)$*

$$\varepsilon\|\Phi_{0,\beta-1,\varepsilon}\nabla u\|_{L^2(S\cap B_{2\varepsilon}(0))} \leq C\left[\varepsilon^2\|\Phi_{0,\beta,\varepsilon}\nabla^2 u\|_{L^2(S\cap B_{2\varepsilon}(0))} + \varepsilon\|\nabla u\|_{L^2(S\cap B_{2\varepsilon}(0))}\right].$$

Proof: Follows from an application of the first estimate of (ii), Lemma 4.2.10 to ∇u. □

We also need the follow variant of Hardy's inequality:

Lemma 4.2.12. *Let $S := S_R(\omega)$ be a straight sector, $u \in H^1(S)$ and $u = 0$ on at least $\overline{\Gamma_1}$ or $\overline{\Gamma_2}$. Then there is $C > 0$ depending only on ω such that for all $R' \in (0, R]$ there holds*

$$\|\frac{1}{|x|}u\|_{L^2(S_{R'}(\omega))} \leq \|\nabla u\|_{L^2(S_{R'}(\omega))}.$$

Proof: By a scaling argument, it suffices to show the result for $R = 1$. The case $R = 1$ is a variant of a standard result, see, e.g., [62, Thm. 1.4.4.3]. □

4.2.2 Properties of the countably normed spaces $\mathcal{B}^l_{\beta,\varepsilon}$

First, we show that the spaces $\mathcal{B}^l_{\beta,\varepsilon}$ are embedded in each other.

Proposition 4.2.13. *Let S be a sector. Then the following holds:*

1. *For each $l \in \mathbb{N}_0$, $\varepsilon > 0$ membership $u \in \mathcal{B}^l_{\beta_0,\varepsilon}(S, C_u, \gamma_u)$ for $\beta_0 \in [0,1)$ implies $u \in \mathcal{B}^l_{\beta,\varepsilon}(S, C_u, \gamma_u)$ for all $\beta \in [\beta_0, 1)$.*
2. *Let $\beta \in [0,1)$, $\varepsilon > 0$, $C_u > 0$, $\gamma_u \geq 1$, $l \in \mathbb{N}_0$. Then $u \in \mathcal{B}^{l+1}_{\beta,\varepsilon}(S, C_u, \gamma_u)$ implies $u \in \mathcal{B}^l_{\beta,\varepsilon}(S, C_u, \gamma_u)$.*
3. *Let $\overline{c} > 0$, $\varepsilon_0 \in (0,1]$, $\beta \in [0,1)$, $l \in \mathbb{N}_0$ be given. Then $u \in \mathcal{B}^l_{\beta,\varepsilon}(S, C_u, \gamma_u)$ implies the existence of C, $\gamma > 0$ independent of $\varepsilon \in (0, \varepsilon_0]$ such that for all $\varepsilon' \in (0,1]$ with $\varepsilon'/\varepsilon \leq \overline{c}$ there holds*

$$\|\hat{\Phi}_{p,\beta,\varepsilon'}\nabla^{p+l}u\|_{L^2(S)} \leq CC_u\gamma^p\left(\frac{\varepsilon'}{\varepsilon}\right)^{l-\beta}\max\{p+1, (\varepsilon')^{-1}\}^{p+l} \qquad \forall p \in \mathbb{N}_0.$$

4. *Under the same assumption as in the preceding statement, we obtain for the special case $l \geq 1$ that $u \in \mathcal{B}^l_{\beta,\varepsilon'}(S, CC_u, \gamma)$ with C, $\gamma > 0$ independent of ε.*

Proof: The first assertion follows immediately from the first assertion of Lemma 4.2.3.
For the second one, we observe that

$$\|u\|_{H^{l,l}_{\beta,\varepsilon}(S)} \leq \|u\|_{H^{l+1,l+1}_{\beta,\varepsilon}(S)} \leq C_u,$$
$$\|\hat{\Phi}_{0,\beta,\varepsilon}\nabla^l u\|_{L^2(S)} \leq \|\nabla^l u\|_{L^2(S)} \leq \varepsilon^{-l}\|u\|_{H^{l+1,l+1}_{\beta,\varepsilon}(S)} \leq C_u\max\{1, \varepsilon^{-1}\}^l.$$

It remains to consider derivatives of order $p \geq 1$. For $p \geq 1$ we have $\hat{\Phi}_{p,\beta,\varepsilon}(x) \leq \hat{\Phi}_{p-1,\beta,\varepsilon}(x)$ by Lemma 4.2.3. Hence, we can write

$$\begin{aligned}\|\hat{\Phi}_{p,\beta,\varepsilon}\nabla^{p+l}u\|_{L^2(S)} &= \|\hat{\Phi}_{p,\beta,\varepsilon}\nabla^{(p-1)+(l+1)}u\|_{L^2(S)} \leq C_u\gamma_u^{p-1}\max\{p,\varepsilon^{-1}\}^{p+l}\\ &\leq C_u\gamma_u^{p-1}\max\{p+1,\varepsilon^{-1}\}^{p+l},\end{aligned}$$

which is the desired bound.
We now turn to the third assertion. From Lemma 4.2.3 we obtain the existence of C, $\gamma > 0$ independent of ε, ε' such that for all $p \in \mathbb{N}_0$

$$\begin{aligned}&\|\hat{\Phi}_{p,\beta,\varepsilon'}\nabla^{p+l}u\|_{L^2(S)} \leq\\ &\quad C\gamma^p\max\{p+1,\varepsilon^{-1}\}^l\max\{1,\varepsilon/\varepsilon'\}^\beta\max\{1,\varepsilon'/\varepsilon\}^p\max\{p+1,(\varepsilon')^{-1}\}^p.\end{aligned}$$

Observing $\varepsilon'/\varepsilon \leq \overline{c}$ allows us to simplify (by adjusting C, γ):

$$\|\hat{\Phi}_{p,\beta,\varepsilon'}\nabla^{p+l}u\|_{L^2(S)} \leq C\gamma^p\max\{p+1,\varepsilon^{-1}\}^l(\varepsilon/\varepsilon')^\beta\max\{p+1,(\varepsilon')^{-1}\}^p \ \forall p \in \mathbb{N}_0.$$

Next, in order to treat the term $\max\{p+1,\varepsilon^{-1}\}$ we bound

$$\begin{aligned}\max\{p+1,\varepsilon^{-1}\} &= \varepsilon^{-1}\max\{(p+1)\varepsilon,1\} \leq \varepsilon^{-1}(p+1)(1+\varepsilon_0)\\ &\leq (1+\varepsilon_0)(p+1)\varepsilon^{-1}\max\{1,(p+1)\varepsilon'\}\\ &\leq (1+\varepsilon_0)(p+1)(\varepsilon'/\varepsilon)\max\{(p+1),(\varepsilon')^{-1}\}.\end{aligned}$$

Hence, we obtain (by again enlarging C, γ) that there holds for all $p \in \mathbb{N}_0$:

$$\|\hat{\Phi}_{p,\beta,\varepsilon'}\nabla^{p+l}u\|_{L^2(S)} \leq C\gamma^p\max\{p+1,(\varepsilon')^{-1}\}^l(\varepsilon/\varepsilon')^{\beta-l}\max\{p+1,(\varepsilon')^{-1}\}^p.$$

For the last assertion, it suffices to note that $\beta \in [0,1)$ and $l \geq 1$. Hence, the factor $(\varepsilon'/\varepsilon)^{l-\beta}$ is bounded. □

Our main goal of this subsection is to show that the spaces $\mathcal{B}^l_{\beta,\varepsilon}$, $l \in \{0,1,2\}$, are invariant under analytic changes of variables. In order to prove that, we start with a variant of Besicovitch's covering theorem:

Lemma 4.2.14 (Besicovitch's covering theorem). *Let S be a sector, $c \in (0,1)$. Then there $N \in \mathbb{N}$ and a family of balls $\mathcal{B} = \{B_i = B_{r_i}(x_i) \,|\, i \in \mathbb{N}\}$ with the following properties:*

1. *$S \subset \cup_i \overline{B_i}$,*
2. *$r_i = c|x_i|$,*
3. *$\forall x \in S$ there holds $\operatorname{card}\{i \in \mathbb{N} \,|\, x \in B_i\} \leq N$.*

Proof: Consider the (uncountable) family $\mathcal{B}' = \cup_{x\in S}B_{c|x|}(x)$ of balls. [136, Thm. 1.3.5] is formulated for families of *closed* balls. Inspection of the proof, however, shows that it also holds for collections of open balls. Hence, by [136, Thm. 1.3.5] there is $N \in \mathbb{N}$ and N countable subfamilies of $\mathcal{B}'$ of the form $\mathcal{B}_i = \{B_{i,j} \,|\, j \in \mathbb{N}\}$, $i = 1,\ldots,N$, with the property that a) $S \subset \cup_{i=1}^N \cup_{j\in\mathbb{N}} B_{ij}$ and b) for each i the balls of the subfamily $\mathcal{B}_i$ are (mutually) disjoint. The lemma follows now by taking as the desired family the union $\cup_{i=1}^N \mathcal{B}_i$. □

Remark 4.2.15 For our purposes, it is actually not important to apply [136, Thm. 1.3.5] with open rather than closed balls. Starting from $\mathcal{B}' = \cup_{x\in S} B_{c|x|}(x_i)$ an application of [136, Thm. 1.3.5] as stated, yields $N \in \mathbb{N}$ and disjointed collections $\mathcal{B}_i = \{\overline{B_{ij}}\,|\, j \in \mathbb{N}\}$ of closed balls with the desired properties. The collection $\mathcal{B} = \{B_{ij}\,|\, i = 1,\ldots,N, j \in \mathbb{N}\}$ now covers $S \setminus A$ where $A = \cup_{i=1}^N \cup_{j\in\mathbb{N}} \partial B_{ij}$ is a set of Lebesgue measure zero. As this covering lemma will be used in order to obtain L^2-bounds on S using L^2-bounds on the balls B_i, the set A of measure zero is irrelevant. ▪

For a sector S, functions belonging to a countably normed space $\mathcal{B}^l_{\beta,\varepsilon}(S, C, \gamma)$ are analytic on S and can be extended analytically across $\partial S \setminus \{0\}$:

Lemma 4.2.16. *Let S be a sector, $u \in \mathcal{B}^l_{\beta,\varepsilon}(S, C_u, \gamma)$ for some $l \in \mathbb{N}_0$, $\beta \in (0,1)$, C_u, $\gamma > 0$. Then there exists a sector $\widetilde{S}$ with $\overline{S} \setminus \{0\} \subset \widetilde{S}$ depending only on S and γ such that u is analytic on $\widetilde{S}$.*

Proof: We start by stressing that the lemma does not allow for an explicit control of the growth of the derivatives of u—merely the analyticity is claimed. It is easy to see that for given $\varepsilon > 0$ there are C_ε, K_ε such that $u \in \mathcal{B}^l_{\beta,1}(S, C_\varepsilon, K_\varepsilon)$. The result now follows from the arguments presented in [14]. □

The following lemma plays a key rôle in the main result of this section, Theorem 4.2.20. In essence, it characterizes the functions from a countably normed spaces $\mathcal{B}^l_{\beta,\varepsilon}$ in terms of their local behavior.

Lemma 4.2.17 (local characterization of countably normed spaces). *Let S be a sector, $l \in \mathbb{N}_0$, $\beta \in (0,1)$, $\varepsilon > 0$. Let $\mathcal{B} = \{B_i\,|\, i \in \mathbb{N}\}$ be a collection of balls $B_i = B_{r_i}(x_i)$ with the following properties:*

1. *there is $c \in (0,1)$ with $r_i = c|x_i|$ for all $i \in \mathbb{N}$;*
2. *there is $N \in \mathbb{N}$ such that $\forall x \in S$ there holds $\operatorname{card}\{i \in \mathbb{N}\,|\, x \in B_i\} \leq N$.*

Let $f \in \mathcal{B}^l_{\beta,\varepsilon}(S, C_f, \gamma_f)$. Then there are C, $K > 0$ independent of ε such that for all $p \in \mathbb{N}_0$, $i \in \mathbb{N}$

$$\hat{\Phi}_{p,\beta,\varepsilon}(x_i)\|\nabla^{p+l} f\|_{L^2(S\cap B_i)} \leq C C(i) K^p \max\{p, \varepsilon^{-1}\}^{p+l}, \tag{4.2.19}$$

$$\sum_{i=1}^{\infty} C^2(i) \leq C_f^2 N \frac{4}{3} < \infty, \tag{4.2.20}$$

where the constants $C(i)$ are given by

$$C^2(i) := \sum_{p=0}^{\infty} \frac{1}{\max\{p+1, \varepsilon^{-1}\}^{2(p+l)}} \frac{1}{(2\gamma_f)^{2p}} \|\hat{\Phi}_{p,\beta,\varepsilon} \nabla^{p+l} f\|^2_{L^2(S\cap B_i)}$$
$$\leq C_f^2 \frac{4}{3} < \infty. \tag{4.2.21}$$

Conversely, let f be analytic on S and assume that there are $\widetilde{C}$, $\gamma > 0$, and $C(i)$ such that f satisfies on the balls B_i:

$$\hat{\Phi}_{p,\beta,\varepsilon}(x_i)\|\nabla^{p+l}f\|_{L^2(S\cap B_i)} \leq C(i)\gamma^p \max\{p+1,\varepsilon^{-1}\}^{p+l} \quad \forall p \in \mathbb{N}_0, \tag{4.2.22a}$$

$$\sum_{i=1}^{\infty} C^2(i) \leq \widetilde{C}^2. \tag{4.2.22b}$$

Then there exist constants C, $K > 0$ depending only on $\gamma > 0$ and the overlap constant N such that

$$\|\hat{\Phi}_{p,\beta,\varepsilon}\nabla^{p+l}f\|_{L^2(S)} \leq C\widetilde{C}K^p \max\{p+1,\varepsilon^{-1}\}^{p+l} \qquad \forall p \in \mathbb{N}_0. \tag{4.2.23}$$

Proof: The bound on $C^2(i)$ in (4.2.21) follows from

$$\begin{aligned} C^2(i) &= \sum_{p=0}^{\infty} \frac{1}{\max\{p+1,\varepsilon^{-1}\}^{2(p+l)}} \frac{1}{(2\gamma_f)^{2p}} \|\hat{\Phi}_{p,\beta,\varepsilon}\nabla^{p+l}f\|^2_{L^2(S\cap B_i)} \\ &\leq \sum_{p=0}^{\infty} \frac{1}{\max\{p+1,\varepsilon^{-1}\}^{2(p+l)}} \frac{1}{(2\gamma_f)^{2p}} \|\hat{\Phi}_{p,\beta,\varepsilon}\nabla^{p+l}f\|^2_{L^2(S)} \leq \sum_{p=0}^{\infty} C_f^2 \frac{1}{2^{2p}} \\ &= C_f^2 \frac{4}{3}. \end{aligned}$$

The bound (4.2.20) is proved similarly using additionally the overlap properties of the sets B_i:

$$\begin{aligned} \sum_{i=1}^{\infty} C^2(i) &= \sum_{p=0}^{\infty} \frac{1}{\max\{p+1,\varepsilon^{-1}\}^{2(p+l)}} \frac{1}{(2\gamma_f)^{2p}} \sum_{i=1}^{\infty} \|\hat{\Phi}_{p,\beta,\varepsilon}\nabla^{p+l}f\|^2_{L^2(S\cap B_i)} \\ &\leq \sum_{p=0}^{\infty} \frac{1}{\max\{p+1,\varepsilon^{-1}\}^{2(p+l)}} \frac{1}{(2\gamma_f)^{2p}} N \|\hat{\Phi}_{p,\beta,\varepsilon}\nabla^{p+l}f\|^2_{L^2(S)} \\ &\leq \sum_{p=0}^{\infty} C_f^2 \frac{1}{2^{2p}} N \leq C_f^2 N \frac{4}{3}. \end{aligned}$$

(4.2.21) implies additionally

$$\|\hat{\Phi}_{p,\beta,\varepsilon}\nabla^{p+l}f\|^2_{L^2(S\cap B_i)} \leq C(i)\max\{p+1,\varepsilon^{-1}\}^{p+l}(2\gamma_f)^p.$$

From Lemma 4.2.2 and our assumptions on the balls B_i, we have $\hat{\Phi}_{p,\beta,\varepsilon}(x) \geq C^{-1}K^{-p}\hat{\Phi}_{p,\beta,\varepsilon}(x_i)$ for all $x \in B_i$ for some C, $K > 0$ independent of ε. Hence, we get

$$\begin{aligned} \hat{\Phi}_{p,\beta,\varepsilon}(x_i)\|\nabla^{p+l}f\|_{L^2(S\cap B_i)} &\leq CK^p\|\hat{\Phi}_{p,\beta,\varepsilon}\nabla^{p+l}f\|_{L^2(S\cap B_i)} \\ &\leq CK^pC(i)\max\{p,\varepsilon^{-1}\}^{p+l}. \end{aligned}$$

The converse statement, i.e., that (4.2.22) implies (4.2.23) follows from (4.2.8) of Lemma 4.2.2 and a summation over the balls B_i. □

For $l > 0$, only derivatives of order greater than or equal to l appear explicitly in Lemma 4.2.17. Local control of derivatives up to order $l-1$ is possible, however. We illustrate this in the following corollary for the case $l = 1$.

Corollary 4.2.18. *Let S be a sector, $l \in \mathbb{N}_0$, $\beta \in (0,1)$, $\varepsilon > 0$. Let $\mathcal{B} = \{B_i \,|\, i \in \mathbb{N}\}$ be a collection of balls $B_i = B_{r_i}(x_i)$ with the following properties:*

1. *there is $c \in (0,1)$ with $r_i = c|x_i|$ for all $i \in \mathbb{N}$;*
2. *there is $N \in \mathbb{N}$ such that $\forall x \in S$ there holds $\operatorname{card}\{i \in \mathbb{N} \,|\, x \in B_i\} \leq N$.*

Let $f \in \mathcal{B}^1_{\beta,\varepsilon}(S, C_f, \gamma_f)$. Then there are C, $K > 0$ independent of ε such that for all $p \in \mathbb{N}_0$, $i \in \mathbb{N}$

$$\hat{\Phi}_{p-1,\beta,\varepsilon}(x_i)\|\nabla^p f\|_{L^2(S\cap B_i)} \leq CC(i)K^p \max\{p, \varepsilon^{-1}\}^p, \tag{4.2.24}$$

$$\sum_{i=1}^{\infty} C^2(i) \leq CC_f^2 < \infty, \tag{4.2.25}$$

where the constants $C(i)$ are given by

$$C^2(i) := \|\hat{\Phi}_{-1,\beta,\varepsilon} f\|^2_{L^2(S\cap B_i)} + \sum_{p=1}^{\infty} \frac{1}{\max\{p+1, \varepsilon^{-1}\}^{2p}} \frac{1}{(2\gamma_f)^{2(p-1)}} \|\hat{\Phi}_{p-1,\beta,\varepsilon}\nabla^p f\|^2_{L^2(S\cap B_i)}.$$

Proof: The bound (4.2.24) follows immediately from the definition of the constants $C(i)$. It remains to see (4.2.25). We express $C(i)$ as $C(i) := C_1(i) + C_2(i)$, where $C_2(i)$ is given by the infinite sum in the definition of $C^2(i)$ and $C_1(i)$ is defined as

$$C_1(i) := \|\hat{\Phi}_{-1,\beta,\varepsilon} f\|^2_{L^2(S\cap B_i)}.$$

From Lemma 4.2.17 we then have

$$\sum_{i\in\mathbb{N}} C_2^2(i) \leq \frac{4}{3} N C_f^2 < \infty.$$

To ascertain the finiteness of $\sum_{i\in\mathbb{N}} C_1^2(i)$, we employ a Hardy inequality of Lemma 4.2.10, (ii):

$$\sum_{i\in\mathbb{N}} \|\hat{\Phi}_{-1,\beta,\varepsilon} f\|^2_{L^2(S\cap B_i)} \leq N\|\hat{\Phi}_{0,\beta-1,\varepsilon} f\|^2_{L^2(S)} \leq NC\|f\|^2_{H^{1,1}_{\beta,\varepsilon}(S)} \leq CNC_f^2.$$

□

Lemma 4.2.19. *Let S be a sector, $c_0 > 0$, $\beta \in (0,1)$, $l \in \{0,1,2\}$, $\varepsilon \in (0,1]$. Let $u \in \mathcal{B}^l_{\beta,\varepsilon}(S, C_u, \gamma_u)$ for some $\varepsilon > 0$, and C_u, $\gamma_u > 0$. Then there are constants C, $\gamma > 0$ independent of ε and C_u and there exists a constant $\overline{u} \in \mathbb{R}$ such that*

$$\|\hat{\Phi}_{p-l,\beta,\varepsilon}\nabla^p(u - \overline{u})\|_{L^2(S\cap B_{c_0\varepsilon}(0))} \leq CC_u\gamma^p \max\{p+1, \varepsilon^{-1}\}^p \qquad \forall p \in \mathbb{N}_0,$$
$$\|\hat{\Phi}_{p-l,\beta,\varepsilon}\nabla^p u\|_{L^2(S\setminus B_{c_0\varepsilon}(0))} \leq CC_u\gamma^p \max\{p+1, \varepsilon^{-1}\}^p \qquad \forall p \in \mathbb{N}_0.$$

Moreover, in the case $l = 2$, the constant $\overline{u}$ may be taken as $\overline{u} = u(0)$ and for $l = 0$ we can take $\overline{u} = 0$.

Proof: The definition of $\mathcal{B}^l_{\beta,\varepsilon}(S, C_u, \gamma_u)$ implies for $p \geq l$:

$$\|\hat{\Phi}_{p-l,\beta,\varepsilon}\nabla^p u\|_{L^2(S)} \leq C_u \gamma_u^{p-l} \max\{p-l+1, \varepsilon^{-1}\}^p \qquad \forall p \geq l.$$

As $\overline{u} = 0$ for $l = 0$, this gives the desired estimates for $p \geq l$. It remains to check the finitely many cases $0 \leq p < l$. On $S \setminus B_{c_0\varepsilon}(0)$, we use the definition (4.2.2) to find (recall: $0 \leq |p-l| \leq l$)

$$\hat{\Phi}_{p-l,\beta,\varepsilon}(x) \sim 1 \qquad \forall p \in \{0, \ldots, l\} \quad \forall x \in S \setminus B_{c_0\varepsilon}(0). \tag{4.2.26}$$

Hence,

$$\|\hat{\Phi}_{p-l,\beta,\varepsilon}\nabla^p u\|_{L^2(S \setminus B_{c_0\varepsilon}(0))} \leq CC_u, \quad p = 0, \ldots, l-1.$$

Let $\overline{u}$ be given by Lemma 4.2.10. Then (with $\overline{u} = u(0)$ in the case $l = 2$)

$$\|\hat{\Phi}_{-2,\beta,\varepsilon}(u - \overline{u})\|_{L^2(S \cap B_{2\varepsilon}(0))} + \varepsilon\|\hat{\Phi}_{-1,\beta,\varepsilon}\nabla u\|_{L^2(S \cap B_{2\varepsilon}(0))} \leq CC_u \quad \text{if } l = 2,$$
$$\|\hat{\Phi}_{-1,\beta,\varepsilon}(u - \overline{u})\|_{L^2(S \cap B_{2\varepsilon}(0))} \leq CC_u \quad \text{if } l = 1,$$

where we exploited that by definition $\|u\|_{H^{l,l}_{\beta,\varepsilon}(S \cap B_{2\varepsilon}(0))} \leq C_u$. If $c_0 \leq 2$, then these estimates imply the desired result. If $c_0 > 2$, then we note that these estimates together with the fact $\hat{\Phi}_{-l,\beta,\varepsilon} \sim 1$ on $S \setminus B_\varepsilon(0)$ (cf. also (4.2.26))

$$|\overline{u}|\varepsilon \sim \|\overline{u}\|_{L^2(S \cap B_{2\varepsilon}(0) \setminus B_\varepsilon(0))} \leq C\left[\|\hat{\Phi}_{-l,\beta,\varepsilon}(u - \overline{u})\|_{L^2(S \cap B_{2\varepsilon}(0))} + \|u\|_{L^2(S)}\right]$$
$$\leq C\|u\|_{H^{l,l}_{\beta,\varepsilon}(S)} \leq CC_u.$$

Exploiting again $\hat{\Phi}_{-l,\beta,\varepsilon} \sim 1$ on $S \setminus B_{2\varepsilon}(0)$, we get from this bound and the triangle inequality

$$\|\hat{\Phi}_{-l,\beta,\varepsilon}(u - \overline{u})\|_{L^2(S \cap B_{c_0\varepsilon}(0) \setminus B_{2\varepsilon}(0))} \leq CC_u \quad \text{for } l \in \{1, 2\}.$$

□

We can now state the main result of this section, namely, the invariance of the countably normed spaces $\mathcal{B}^l_{\beta,\varepsilon}$ under analytic changes of variables:

Theorem 4.2.20 (Invariance of $\mathcal{B}^l_{\beta,\varepsilon}$ under changes of variables). *Let S be a C^2-curvilinear sector and $g : \overline{S \to g(S)} \subset \mathbb{R}^2$ be analytic on $\overline{S}$, $g(0) = 0$, and assume that g^{-1} is analytic on $\overline{g(S)}$. Let C_u, $\gamma_u > 0$, $\beta \in (0,1)$. Then there exist constants C, $\gamma > 0$ depending only on g, S, γ_u, and β, (in particular, they are independent of ε) such that for $l \in \{0,1,2\}$ and $\varepsilon \in (0,1]$*

$$u \in \mathcal{B}^l_{\beta,\varepsilon}(g(S), C_u, \gamma_u) \Longrightarrow u \circ g \in \mathcal{B}^l_{\beta,\varepsilon}(S, CC_u, \gamma).$$

Proof: We start by noting that Lemma 4.2.6 gives

$$\|u \circ g\|_{H^{l,l}_{\beta,\varepsilon}(S)} \leq CC_u, \qquad l \in \{0, 1, 2\}.$$

We may therefore concentrate on the bounds of higher derivatives. To that end, we proceed in two steps and prove the following two estimates separately

$$\|\hat{\Phi}_{p,\beta,\varepsilon}\nabla^{p+l}(u\circ g)\|_{L^2(S_\varepsilon)} \leq CC_u\gamma^p \max\{p,\varepsilon^{-1}\}^{p+l} \quad \forall p\in\mathbb{N}_0, \tag{4.2.27}$$
$$\|\hat{\Phi}_{p,\beta,\varepsilon}\nabla^{p+l}(u\circ g)\|_{L^2(S\setminus S_\varepsilon)} \leq CC_u\gamma^p \max\{p,\varepsilon^{-1}\}^{p+l} \quad \forall p\in\mathbb{N}_0, \tag{4.2.28}$$

where $S_\varepsilon := g^{-1}(g(S)\cap B_\varepsilon(0))$.
Let $\mathcal{B} = \{B_i = B_{r_i}(x_i)\,|\,i\in\mathbb{N}\}$ be a covering of $g(S)$ by balls B_i as given by Lemma 4.2.14. Note that we can choose c is so small that $B_i\cap B_\varepsilon(0)\neq\emptyset$ implies $B_i\subset B_{2\varepsilon}(0)$. Next, we define the index set $I_\varepsilon := \{i\in\mathbb{N}\,|\,B_i\cap B_\varepsilon(0)\neq\emptyset\}$.
We are now in position to prove (4.2.27). From Lemma 4.2.19, we have the existence of $\overline{u}\in\mathbb{R}$ such that

$$\|\hat{\Phi}_{p-l,\beta,\varepsilon}\nabla^p(u-\overline{u})\|_{L^2(g(S)\cap B_{2\varepsilon}(0))} \leq C\gamma^p\max\{p+1,\varepsilon^{-1}\}^p \qquad \forall p\in\mathbb{N}_0.$$

We introduce the shorthand $\tilde{u} := u-\overline{u}$ on $g(S)\cap B_{2\varepsilon}(0)$. Moreover, the constant $\overline{u}=0$ for $l=0$. It suffices therefore to show that similar estimates holds for $\tilde{u}\circ g$. For indices $i\in I_\varepsilon$ we define

$$C^2(i) := \sum_{p\in\mathbb{N}_0}\frac{1}{(2\gamma)^{2p}\max\{p,\varepsilon^{-1}\}^{2p}}\|\hat{\Phi}_{p-l,\beta,\varepsilon}\nabla^p\tilde{u}\|^2_{L^2(B_i\cap g(S))}.$$

We note that

$$\sum_{i\in I_\varepsilon} C^2(i) =: C'C_u < \infty, \tag{4.2.29}$$

where C' depends only on the covering $\mathcal{B}$ (and is independent of ε). This definition of $C(i)$ implies that for each i there holds

$$\|\hat{\Phi}_{p-l,\beta,\varepsilon}\nabla^p\tilde{u}\|_{L^2(B_i\cap g(S))} \leq CC_uC(i)(2\gamma)^p\max\{p,\varepsilon^{-1}\}^p \qquad \forall p\in\mathbb{N}_0.$$

Let us now consider a fixed i. Abbreviating $r=|x_i|$, we get from Lemma 4.2.2 the existence of C, $K>0$ independent of p, ε, and i such that

$$\begin{aligned} C^{-1}K^{-p}\max_{x\in B_i\cap g(S)}\hat{\Phi}_{p-l,\beta,\varepsilon}(x) &\leq \hat{\Phi}_{p-l,\beta,\varepsilon}(x_i) \\ &\leq CK^p\min_{x\in B_i\cap g(S)}\hat{\Phi}_{p-l,\beta,\varepsilon}(x). \end{aligned} \tag{4.2.30}$$

Hence, we have using (4.2.4), (4.2.7) for some C, $\gamma>0$ independent of p, ε, i:

$$\begin{aligned} \hat{\Phi}_{-l,\beta,\varepsilon}(x_i)\|\nabla^p\tilde{u}\|_{L^2(B_i\cap g(S))} &\leq C\gamma^p\hat{\Phi}^{-1}_{p,0,\varepsilon}(x_i)\cdot\hat{\Phi}_{p-l,\beta,\varepsilon}(x_i)\|\nabla^p\tilde{u}\|_{L^2(B_i\cap g(S))} \\ &\leq CC_uC(i)\gamma^p\left(\frac{p!}{r^p}+\max\left\{p,\varepsilon^{-1}\right\}^p\right). \end{aligned}$$

From Lemma 4.3.1 with $f_2=\tilde{u}$, $f_1=1$ we get that there are C, $K>0$ independent of p, ε, and i such that

$$\hat{\Phi}_{-l,\beta,\varepsilon}(x_i)\|\nabla^p\left(\tilde{u}\circ g\right)\|_{L^2(G_i)} \leq CC_uC(i)K^p\left(\frac{p!}{r^p}+\max\left\{p,\varepsilon^{-1}\right\}^p\right), \tag{4.2.31}$$

where we defined $G_i := g^{-1}(B_i \cap g(S))$. Since by (4.2.7) of Lemma 4.2.2 there holds $p!r^{-p} + \max\{p, \varepsilon^{-1}\}^p \leq CK^p \hat{\Phi}^{-1}_{p,0,\varepsilon}(x_i) \max\{p, \varepsilon^{-1}\}^p$, we get

$$\hat{\Phi}_{p-l,\beta,\varepsilon}(x_i) \|\nabla^p(\tilde{u} \circ g)\|_{L^2(G_i)} \leq CC_u C(i) \gamma^p \max\{p, \varepsilon^{-1}\}^p \qquad (4.2.32)$$

for some C, $\gamma > 0$. Next, by the assumptions on g, there exists $C > 0$ such that

$$C^{-1}|x| \leq |g^{-1}(x)| \leq C|x| \qquad \forall x \in g(S).$$

Hence, by the definition of $\hat{\Phi}_{p,\beta,\varepsilon}$, there exist C, $K > 0$ independent of ε, $p \in \mathbb{N}_0$, $x \in g(S)$ such that

$$C^{-1}K^{-p}\hat{\Phi}_{p-l,\beta,\varepsilon}(x) \leq (\hat{\Phi}_{p-l,\beta,\varepsilon} \circ g^{-1})(x) \leq CK^p \hat{\Phi}_{p-l,\beta,\varepsilon}(x) \quad \forall x \in g(S).$$

By combining this estimate with (4.2.30), we get

$$\begin{aligned}\max_{x \in G_i} \hat{\Phi}_{p-l,\beta,\varepsilon}(x) = \max_{x \in B_i \cap g(S)} (\hat{\Phi}_{p-l,\beta,\varepsilon} \circ g^{-1})(x) &\leq CK^p \max_{x \in B_i \cap g(S)} \hat{\Phi}_{p-l,\beta,\varepsilon}(x) \\ &\leq CK^p \hat{\Phi}_{p-l,\beta,\varepsilon}(x_i).\end{aligned}$$

Inserting this in (4.2.32), we get for suitable C, $\gamma > 0$

$$\|\hat{\Phi}_{p-l,\beta,\varepsilon} \nabla^p(\tilde{u} \circ g)\|_{L^2(G_i)} \leq CC_u C(i) \gamma^p \max\{p, \varepsilon^{-1}\}^p \qquad (4.2.33)$$

As $\cup_i G_i \supset g^{-1}(g(S) \cap B_\varepsilon(0)) = S_\varepsilon$, we get the desired result (4.2.27) by squaring (4.2.33), summing on $i \in I_\varepsilon$ and using (4.2.29).
The estimate (4.2.28) is proved completely analogously: We consider the index set

$$\tilde{I}_\varepsilon := \{i \in \mathbb{N} \,|\, B_i \cap g(S) \setminus B_\varepsilon(0) \neq \emptyset\};$$

since $\mathcal{B}$ is a covering of $g(S)$, we get

$$\cup_{i \in \tilde{I}_\varepsilon} g^{-1}(B_i \cap g(S)) \supset S \setminus S_\varepsilon.$$

Furthermore, we may assue that the parameter $c \in (0,1)$ of the covering may be chosen so small that $B_i \subset g(S) \setminus B_{\varepsilon/2}(0)$ for all $i \in \tilde{I}_\varepsilon$. Additionally, we get from Lemma 4.2.19 the existence of C, $\gamma > 0$ such that

$$\|\hat{\Phi}_{p-l,\beta,\varepsilon} \nabla^p u\|_{L^2(g(S) \setminus B_{\varepsilon/2}(0))} \leq C\gamma^p \max\{p+1, \varepsilon^{-1}\}^p \qquad \forall p \in \mathbb{N}_0.$$

We may then prove (4.2.28) in the same way as (4.2.27) by taking $\tilde{u} = u$. □

Let us consider $l = 2$ in Theorem 4.2.20. In the above proof, we essentially assumed that $\|u\|_{L^2(g(S))}$ and $\varepsilon\|\nabla u\|_{L^2(g(S))}$ are of the same size. We will see later on that this is not always the case. Inspection of the proof of Theorem 4.2.20 allows us to refine the results as follows.

Corollary 4.2.21. *Let the sector S and the map g satisfy the same hypotheses as in Theorem 4.2.20. Assume that $u \in \mathcal{B}^2_{\beta,\varepsilon}(g(S))$ satisfies the following estimates:*

$$\|u\|_{L^2(g(S))} \leq C_0, \qquad \varepsilon\|\nabla u\|_{L^2(g(S))} \leq C_1,$$
$$\|\hat{\Phi}_{p,\beta,\varepsilon}\nabla^{p+2}u\|_{L^2(g(S))} \leq C_1\gamma^p \max\{p+1,\varepsilon^{-1}\}^{p+2} \qquad \forall p \in \mathbb{N}_0.$$

Then there are C, $K > 0$ independent of C_0, C_1, and $\varepsilon \in (0,1]$ such that the function $u \circ g$ satisfies

$$\|u \circ g\|_{L^2(S)} \leq CC_0, \qquad \varepsilon\|\nabla(u \circ g)\|_{L^2(S)} \leq CC_1, \tag{4.2.34}$$
$$\|\hat{\Phi}_{p,\beta,\varepsilon}\nabla^{p+2}(u \circ g)\|_{L^2(S)} \leq CC_1K^p \max\{p+1,\varepsilon^{-1}\}^{p+2} \quad \forall p \in \mathbb{N}_0. \tag{4.2.35}$$

Proof: (4.2.34) follows easily from the chain rule. In order to see (4.2.35), it is more convenient to consider ∇u. We first claim that there are C, $K > 0$ depending only on g and γ such that

$$\|\hat{\Phi}_{p-1,\beta,\varepsilon}\nabla^p\nabla u\|_{L^2(g(S))} \leq CC_1\gamma^p\varepsilon^{-1} \max\{p+1,\varepsilon^{-1}\}^p \qquad \forall p \in \mathbb{N}_0. \tag{4.2.36}$$

For $p \in \mathbb{N}$ this follows directly from the assumptions on u. For $p = 0$, we note that on $g(S) \setminus B_\varepsilon(0)$ we have $\hat{\Phi}_{-1,\beta,\varepsilon} = 1$; it suffices therefore to show that

$$\|\hat{\Phi}_{-1,\beta,\varepsilon}\nabla u\|_{L^2(g(S)\cap B_\varepsilon(0))} \leq CC_1\varepsilon^{-1},$$

which follows from Corollary 4.2.11.
With the formula $\nabla(u \circ g) = (g' \cdot (\nabla u) \circ g)$ we see that we can proceed verbatim as in the proof of Theorem 4.2.20 for the case $l = 1$ (using Lemma 4.3.1 with $f_1 = \nabla u$, $f_2 = g'$) to conclude that

$$\|\hat{\Phi}_{p-1,\beta,\varepsilon}\nabla^p\nabla(u \circ g)\|_{L^2(S)} \leq CC_1\gamma^p\varepsilon^{-1} \max\{p+1,\varepsilon^{-1}\}^p \qquad \forall p \in \mathbb{N}_0.$$

This last estimate is readily brought to the desired form. □

In the following, we will be particularly interested in the case $l = 2$. For elements of the spaces $\mathcal{B}^2_{\beta,\varepsilon}$ we have the following pointwise characterization.

Proposition 4.2.22. *Let $S = S_R(\omega)$ be a straight sector, $l \in \{0,1,2\}$, $\varepsilon \in (0,1]$, and let $u \in \mathcal{B}^l_{\beta,\varepsilon}(S, C_u, \gamma_u)$ for some C_u, $\gamma_u > 0$, and $\beta \in (0,1)$. Then for every $R' \in (0,R)$ there are C, $\gamma > 0$ independent of ε such that*

$$\left\|\frac{\hat{\Phi}_{p-l+1,\beta,\varepsilon}}{\max\{1,|x|/\varepsilon\}}\nabla^p u\right\|_{L^\infty(S_{R'}(\omega))} \leq C\gamma^p \max\{p+1,\varepsilon^{-1}\}^{p+1} \qquad \forall p \in \mathbb{N}.$$

This estimate is also valid for $p = 0$ if either $l = 0$ or $l = 2$ together with $u(0) = 0$.

Proof: For given R', let $\rho_0 < 1$ be such that $R'\rho_0 < R$. We cover $S_{R'}(\omega)$ by balls $B_i = B_{r_i}(x_i)$ with the following properties: $r_i = \rho_0|x_i|$ and $B_i \cap S_R(\omega)$ is either a full ball or a half ball, i.e., either B_i is completely contained in $S_R(\omega)$ or x_i is on one of the two straight sides of $S_R(\omega)$. For each $i \in \mathbb{N}$ let F_i be the affine map from the reference domain $\widehat{B}$ to $B_i \cap S_R(\omega)$, where $\widehat{B} = B_1(0)$ if $B_i \cap S_R(\omega)$ is a full ball and $\widehat{B} = B_1(0) \cap \{y > 0\}$ if $B_i \cap S_R(\omega)$ is a half-ball. From $\rho_0 < 1$ we have that for every B_i there holds $B_i \cap S \subset S'_1 := S_{(1+\rho_0)\varepsilon}$ or $B_i \cap S \subset S'_2 := S \setminus S_{(1-\rho_0)\varepsilon}$. By Lemma 4.2.19 there is $\overline{u} \in \mathbb{R}$ such that the functions $\widetilde{u}_1 := u - \overline{u}$ and the function $\widetilde{u}_2 := u$ satisfy on S'_1, S'_2:

$$\|\hat{\Phi}_{p-l,\beta,\varepsilon}\nabla^p\widetilde{u}_1\|_{L^2(S'_1)} \leq CC_u\gamma^p \max\{p+1,\varepsilon^{-1}\}^p \qquad \forall p \in \mathbb{N}_0,$$
$$\|\hat{\Phi}_{p-l,\beta,\varepsilon}\nabla^p\widetilde{u}_2\|_{L^2(S'_2)} \leq CC_u\gamma^p \max\{p+1,\varepsilon^{-1}\}^p \qquad \forall p \in \mathbb{N}_0$$

For each $i \in \mathbb{N}$ consider the function $\widehat{u} := \widetilde{u} \circ F_i$. Then there holds for some C, $\gamma > 0$ independent of ε:

$$\hat{\Phi}_{p-l,\beta,\varepsilon}(x_i)\|\nabla^p\widehat{u}\|_{L^2(\widehat{B})} \leq C\gamma^p \max\{p+1,\varepsilon^{-1}\}^p r_i^{-1+p} \qquad \forall p \in \mathbb{N}_0.$$

Next, from the Sobolev embedding $L^\infty(\widehat{B}) \subset H^2(\widehat{B})$ and the norm equivalence $\|\cdot\|_{H^2(\widehat{B})} \sim \|\cdot\|_{L^2(\widehat{B})} + \|\nabla^2\cdot\|_{L^2(\widehat{B})}$, we infer

$$\hat{\Phi}_{p-l,\beta,\varepsilon}(x_i)\|\nabla^p\widehat{u}\|_{L^\infty(\widehat{B})} \leq C\gamma^p \max\{p+1,\varepsilon^{-1}\}^p r_i^{-1+p}\left[1 + r_i^2\max\{p+1,\varepsilon^{-1}\}^2\right].$$

Mapping back to $B_i \cap S_R(\omega)$ yields

$$\begin{aligned}
&\hat{\Phi}_{p-l,\beta,\varepsilon}(x_i)\|\nabla^p\widetilde{u}\|_{L^\infty(B_i\cap S)} \\
&\leq C\gamma^p \max\{p+1,\varepsilon^{-1}\}^p r_i^{-1}\left[1 + r_i^2\max\{p+1,\varepsilon^{-1}\}^2\right] \\
&\leq C(\gamma')^p \max\{p+1,\varepsilon^{-1}\}^{p+1}\left[\frac{1}{r_i\max\{1,\varepsilon^{-1}\}} + 1 + r_i\max\{1,\varepsilon^{-1}\}\right] \\
&\leq C(\gamma')^p \max\{p+1,\varepsilon^{-1}\}^{p+1}\left(\frac{\varepsilon}{r_i} + \frac{r_i}{\varepsilon}\right).
\end{aligned}$$

Next, we estimate

$$\begin{aligned}
\frac{r_i}{\varepsilon} + \frac{\varepsilon}{r_i} &\leq 2\max\{1,r_i/\varepsilon\}\max\{1,\varepsilon/r_i\} = 2\frac{\max\{1,r_i/\varepsilon\}}{\min\{1,r_i/\varepsilon\}} \\
&\leq C\frac{\max\{1,r_i/\varepsilon\}}{\hat{\Phi}_{1,0,\varepsilon}(x_i)}
\end{aligned}$$

and therefore conclude

$$\left\|\frac{\hat{\Phi}_{p-l+1,\beta,\varepsilon}}{\max\{1,|x|/\varepsilon\}}\nabla^p\widetilde{u}\right\|_{L^\infty(S'_i)} \leq C\gamma^p \max\{p+1,\varepsilon^{-1}\}^{p+1} \qquad \forall p \in \mathbb{N}_0, \quad i \in \{1,2\}.$$

For $l = 0$ we have $\overline{u} = 0$ so that $\widetilde{u} = u$ and this implies the desired result. For $l \in \{1,2\}$ we note that $\widetilde{u}$ and u differ by a constant so that we have the result for $p \geq 1$. It remains to analyze $p = 0$ and $l = 2$. We then have $\overline{u} = u(0) = 0$ by Lemma 4.2.19 so that we have again $\widetilde{u} = u$ and hence the result. □

Theorem 4.2.23 (pointwise estimates for $\mathcal{B}^l_{\beta,\varepsilon}$-functions). *Let S be an analytic sector, $l \in \{0,1,2\}$, $u \in \mathcal{B}^l_{\beta,\varepsilon}(S, C_u, \gamma_u)$ for some C_u, $\gamma_u > 0$ and $\beta \in (0,1)$, $\varepsilon > 0$. Then for every neighborhood $\mathcal{U}$ of Γ_3 there are there are C, $\gamma > 0$ independent of ε such that on $S' := S \setminus \mathcal{U}$*

$$\left\| \frac{\hat{\Phi}_{p-l+1,\beta,\varepsilon}}{\max\{1, |x|/\varepsilon\}} \nabla^p u \right\|_{L^\infty(S')} \leq C\gamma^p \max\{(p+1), \varepsilon^{-1}\}^{p+1} \qquad \forall p \in \mathbb{N}.$$

This result is also valid for $p = 0$ if either $l = 0$ or $l = 2$ together with $u(0) = 0$.

Proof: Theorem 4.2.23 follows easily from Proposition 4.2.22 as follows. For simplicity, we assume that there are R, $\omega > 0$, and an analytic mapping $\Lambda : \overline{S} \to \overline{S_R(\omega)}$ that is analytic on the closure S and whose inverse is also analytic $S_R(\omega)$ (such a mapping can, for example, be constructed using blending maps, [58–60]). Otherwise, the argument below has to be carried out in a piecewise fashion. By Theorem 4.2.20, the function $u \circ \Lambda$ is in the space $\mathcal{B}^l_{\beta,\varepsilon}(S_R(\omega), C, \gamma)$ for some C, $\gamma > 0$ independent of ε. Taking R' appropriately ensures that $S_{R'}(\omega) \supset \Lambda(S \setminus \mathcal{U})$. Next, we employ Proposition 4.2.22 to conclude that $u \circ \Lambda$ satisfies for all $p \in \mathbb{N}$

$$\left\| \frac{\hat{\Phi}_{p-l+1,\beta,\varepsilon}}{\max\{1, |x|/\varepsilon\}} \nabla^p (u \circ \Lambda) \right\|_{L^\infty(S_{R'}(\omega))} \leq C\gamma^p \max\{p+1, \varepsilon^{-1}\}^{p+1}. \qquad (4.2.37)$$

Using Lemma 4.2.2, we get that $u \circ \Lambda$ satisfies equivalently

$$|\nabla^p \nabla (u \circ \Lambda)(x)| \leq C\gamma^p \frac{1}{\hat{\Phi}_{1-l,\beta,\varepsilon}(x)} \max\{1, |x|/\varepsilon\} \varepsilon^{-1} \max\{(p+1)/|x|, \varepsilon^{-1}\}^{p+1}$$

for all $x \in S_{R'}(\omega)$ and $p \in \mathbb{N}_0$. Hence, applying Lemma 4.3.3 to $(\nabla u) \circ \Lambda \cdot \Lambda'$ we conclude that for all $x \in S \setminus \mathcal{U}$ and $p \in \mathbb{N}_0$

$$|\nabla^p \nabla u(x)| \leq C\gamma^p \max\{1, |x|/\varepsilon\} \frac{1}{\hat{\Phi}_{1-l,\beta,\varepsilon}(x)} \varepsilon^{-1} \max\{(p+1)/|x|, \varepsilon^{-1}\}^{p+1}.$$

Using again Lemma 4.2.2, we get for all $x \in S \setminus \mathcal{U}$ and $p \in \mathbb{N}_0$

$$|\nabla^{p+1} u(x)| \leq C\gamma^p \max\{1, |x|/\varepsilon\} \frac{1}{\hat{\Phi}_{1-l,\beta,\varepsilon}(x)} \frac{\max\{p+1, \varepsilon^{-1}\}^{p+1}}{\hat{\Phi}_{p+1,0,\varepsilon}(x)},$$

from which we conclude

$$\left\| \frac{\hat{\Phi}_{p-l+1,\beta,\varepsilon}}{\max\{1, |x|/\varepsilon\}} \nabla^p u \right\|_{L^\infty(S \setminus \mathcal{U})} \leq C\gamma^p \max\{(p+1), \varepsilon^{-1}\}^{p+1} \qquad \forall p \in \mathbb{N}.$$

Since $S' = S \setminus \mathcal{U}$, this is the desired estimate for $p \in \mathbb{N}$. We note that (4.2.37) is also valid for $p = 0$ if $l = 0$ or if $l = 2$ together with $u(0) = 0$. The theorem is now proved in full generality. □

4.3 Local changes of variables for analytic functions

Lemma 4.3.1. *Let $G, G_1 \subset \mathbb{R}^2$ be bounded open sets. Assume that $g = (g_1, g_2) : \overline{G_1} \to \mathbb{R}^2$ is analytic and injective on $\overline{G_1}$, $\det g' \neq 0$ and that it satisfies $g(G_1) \subset G$. Let $f_1 : \overline{G_1} \to \mathbb{C}$, $f_2 : \overline{G} \to \mathbb{C}$ be analytic and assume that f_2 satisfies for some ε, C_f, $\gamma_f > 0$, $r \leq 1$,*

$$\|\nabla^p f_2\|_{L^2(G)} \leq C_f \gamma^p \left[\frac{p!}{r^p} + \max\{p, \varepsilon^{-1}\}^p\right].$$

Then there are constants C, $K > 0$ depending only on γ and g, f_1 such that

$$\|\nabla^p(f_1 \cdot (f_2 \circ g))\|_{L^2(G_1)} \leq C_f C K^p \left[\frac{p!}{r^p} + \max\{p, \varepsilon^{-1}\}^p\right].$$

Proof: First, it is more convenient to formulate the assumption on f_2 as

$$\|\nabla^p f_2\|_{L^2(G)} \leq C_f \gamma^p \frac{p!}{r^p} \max\{1, \frac{r}{(p+1)\varepsilon}\}^p \qquad \forall p \in \mathbb{N}_0.$$

Here, the constants C_f, $\gamma > 0$ may be different from those in the statement of the lemma but are independent of r, ε.
The growth conditions on the derivatives of f_2 imply that f_2 can be extended to a holomorphic function (also denoted f_2) on $\tilde{G} \subset \mathbb{C} \times \mathbb{C}$ with $\overline{G} \subset \tilde{G}$ and $\tilde{G}$ independent of $\varepsilon > 0$ ($\tilde{G}$ depends only on the ratio γ/r). First, we claim that there are δ_0, γ', and $C > 0$ depending only on f_1, γ and C_f such that

$$\|f_2(\cdot + z_1(\cdot), \cdot + z_2(\cdot))\|_{L^2(G)} \leq C e^{\gamma' \delta r/\varepsilon} \tag{4.3.38}$$

for all continuous functions z_1, $z_2 : G \to \mathbb{C}$ with $\|z_i\|_{L^\infty(G)} \leq \delta r \leq r\delta_0$, $i = 1, 2$. As f_2 is holomorphic on $\tilde{G}$, there is $\delta_0 > 0$ such that for all $(x, y) \in \overline{G}$ the power series expansion of f_2 about (x, y) converges on a ball of radius $2\delta_0 r$. For functions z_1, z_2 with $\|z_i\|_{L^\infty(G)} \leq \delta r \leq \delta_0 r$ we obtain:

$$\begin{aligned} |f_2(x + z_1(x,y), y + z_2(x,y))| &= \left|\sum_{\alpha \in \mathbb{N}_0^2} \frac{1}{\alpha!} D^\alpha f(x,y)(z_1, z_2)^\alpha\right| \\ &\leq \sum_{\alpha \in \mathbb{N}_0^2} \frac{1}{\alpha!} |D^\alpha f_2(x,y)| \, (r\delta)^{|\alpha|}. \end{aligned}$$

Therefore we get

$$\|f_2(\cdot + z_1(\cdot), \cdot + z_2(\cdot))\|_{L^2(G)} \le \sum_{\alpha\in\mathbb{N}_0^2} \frac{1}{\alpha!}\|D^\alpha f_2\|_{L^2(G)}(r\delta)^{|\alpha|}$$

$$\le \sum_{p=0}^{\infty}\sum_{|\alpha|=p}\left((p!)^{1/2}(\alpha!)^{-1/2}\|D^\alpha f_2\|_{L^2(G)}\right)\left((\alpha!)^{-1/2}p!^{-1/2}(r\delta)^p\right)$$

$$\le \sum_{p=0}^{\infty}\|\nabla^p f_2\|_{L^2(G)}\Big(\sum_{|\alpha|=p}\frac{1}{\alpha!p!}(r\delta)^{2p}\Big)^{1/2} = \sum_{p=0}^{\infty}\|\nabla^p f_2\|_{L^2(G)}\frac{1}{p!}2^{p/2}(r\delta)^p$$

$$\le C_f \sum_{0\le p\le r/\varepsilon}\frac{1}{p!}\frac{p!}{(p+1)^p}\left(\sqrt{2}\gamma\varepsilon^{-1}r\delta\right)^p + C_f\sum_{p>r/\varepsilon}\frac{p!}{p!}\gamma^p 2^{p/2}\delta^p$$

$$\le C_f e^{\sqrt{2}\gamma\delta/\varepsilon} + C\sum_{p>r/\varepsilon}\left(e\sqrt{2}\gamma\delta\right)^p \le C_f e^{\sqrt{2}\gamma\delta/\varepsilon} + \frac{1}{1-\sqrt{2}\gamma\delta_0} \le Ce^{\sqrt{2}\gamma\delta/\varepsilon},$$

where we made made the tacit assumption that δ_0 is so small that $e\sqrt{2}\gamma\delta_0 < 1$ for the second sum to be finite. This proves (4.3.38).
Since g is analytic on $\overline{G}_1$ it has a holomorphic extension (also denoted g) to $\tilde{G}_1 \subset \mathbb{C}\times\mathbb{C}$. Thus, there are η, $\delta_0' > 0$ such that for $i \in \{1,2\}$ and all $(x,y) \in \overline{G}_1$

$$|g_i(x+z_1, y+z_2) - g_i(x,y)| \le \eta\delta \quad \forall z_1, z_2 \in \mathbb{C} \text{ s.t. } |z_1|, |z_2| \le \delta \le \delta_0'. \quad (4.3.39)$$

Furthermore, since f_1 is assumed analytic on $\overline{G_1}$, we may suppose that the set $\tilde{G}_1$ is such that f_1 is analytic on $\tilde{G}_1$ and satisfies $\|f_1\|_{L^\infty(\tilde{G}_1)} \le C$. For any $0 < \delta \le \min(\delta_0', \delta_0/\eta)$ we obtain by Cauchy's integral theorem for derivatives for every $(x,y) \in G_1$ and every $\alpha = (\alpha_1, \alpha_2) \in \mathbb{N}_0^2$ (note that $r \le 1$)

$$D^\alpha\left(f_1\cdot(f_2\circ g)\right)(x,y) = \frac{-\alpha!}{4\pi^2}\int_{|z_1|=\delta r}\int_{|z_2|=\delta r}\frac{(f_1\cdot(f_2\circ g))(x+z_1, y+z_2)}{(-z_1)^{\alpha_1+1}(-z_2)^{\alpha_2+1}}\,dz_1 dz_2.$$

Hence, we can bound

$$|D^\alpha\left(f_1\cdot(f_2\circ g)\right)(x,y)|^2 \le \frac{\alpha!^2}{4\pi^2(\delta r)^{2|\alpha|+2}}\times$$
$$\int_{|z_1|=\delta r}\int_{|z_2|=\delta r}\left|f\left(g_1(x+z_1,y+z_2), g_2(x+z_1,y+z_2)\right)\right|^2 |dz_1|\,|dz_2|.$$

By (4.3.39), we can write

$$g_1(x+z_1, y+z_2) = g_1(x,y) + \zeta_1, \qquad g_2(x+z_1,y+z_2) = g_2(x,y) + \zeta_2$$

where ζ_1, ζ_2 are smooth functions of x, y, z_1, z_2, and $|\zeta_i| \le \eta\delta$, $i = 1,2$. Integrating over G_1, we obtain after the smooth change of variables $g(x,y) = (x', y')$ and denoting by ζ_1', ζ_2' the functions corresponding to ζ_1, ζ_2 after this change of variables

$$|D^\alpha(f_1\cdot(f_2\circ g))(x,y)|^2_{L^2(G_1)} \le$$
$$c_2\frac{(\alpha!)^2}{4\pi^2(\delta r)^{2|\alpha|+2}}\int_{|z_1|=\delta r}\int_{|z_2|=\delta r}\int_G |f(x'+\zeta_1', y'+\zeta_2')|^2\, dx'dy'|dz_1|\,|dz_2|.$$

As $|\zeta_1'|$, $|\zeta_2'| \le \eta\delta r$ uniformly in $(x',y') \in G$, $|z_1|$, $|z_2| \le \delta r$, the estimate (4.3.38) yields

$$\|D^\alpha(f_1 \cdot (f_2 \circ g))\|_{L^2(G_1)} \le C \frac{\alpha!}{(r\delta)^{|\alpha|}} e^{\gamma' \eta\delta r/\varepsilon} \qquad \forall 0 < \delta \le \min(\delta_0', \delta_0/\eta).$$

In order to extract from this estimate the claim of the lemma, we choose

$$\delta := \min\left\{\frac{(p+1)\varepsilon}{r}, \overline{\delta}\right\}, \qquad \overline{\delta} := \min(\delta_0', \delta_0/\eta).$$

This choice of δ implies with $|\alpha| = p$:

$$\begin{aligned}\|D^\alpha(f_1 \cdot (f_2 \circ g))\|_{L^2(G_1)} &\le C\frac{p!}{r^p} \max\{\overline{\delta}^{-1}, \frac{r}{(p+1)\varepsilon}\}^p e^{\gamma'\eta r \min\{(p+1)\varepsilon, \overline{\delta}\}/\varepsilon} \\ &\le \frac{p!}{r^p}\tilde{\gamma}^p \max\left\{1, \frac{r}{(p+1)\varepsilon}\right\}^p e^{\gamma'\eta(p+1)}\end{aligned}$$

for some appropriate $\tilde{\gamma} > 0$ independent of ε, r. This completes the proof. □

Remark 4.3.2 The conditions on f_1 be relaxed in the following way: It suffices that f_1 satisfies a bound of the form $|\nabla^p f_1(x)| \le C(\gamma/r)^p p!$ for all $p \in \mathbb{N}_0$. ▪

Lemma 4.3.3. *Let $G \subset \mathbb{R}^2$ and $G' \subset \mathbb{R}^n$ $(n \in \mathbb{N})$ be two bounded domains. Let $\Lambda : G' \to G$, $v : G' \to \mathbb{C}$ be* analytic *on $\overline{G'}$ and $u : G \to \mathbb{C}$ be analytic on G and satisfy*

$$|\nabla^p u(x)| \le C(x) K^p \max\{(p+1)/r, \varepsilon^{-1}\}^p \qquad \forall p \in \mathbb{N}_0, \quad x \in G$$

for some K, r, $\varepsilon > 0$ and a function $C : G \to \mathbb{R}^+$. Then there exist C', $\gamma > 0$ depending only on K, Λ, and v such that

$$|\nabla^p (v \cdot (u \circ \Lambda))(z)| \le C'\gamma^p \max\{(p+1)/r, \varepsilon^{-1}\}^p C(\Lambda(z)) \qquad \forall p \in \mathbb{N}_0, \quad z \in G'.$$

Proof: The proof is very similar to that of Lemma 4.3.1. It is worth pointing out, however, that the mapping Λ is not required to be invertible–a condition that was necessary in the proof of Lemma 4.3.1. □

We now address the special case $r = 1$, $\varepsilon = 1$. There, the dependence of the various constants can be tracked more easily.

Lemma 4.3.4. *Let $G \subset \mathbb{R}^2$ and $G' \subset \mathbb{R}^n$ $(n \in \mathbb{N})$ be two bounded domains. Let $\Lambda : G' \to G$, $v : G' \to \mathbb{C}$ be* analytic *on $\overline{G'}$ and $u : G \to \mathbb{C}$ be analytic on G and satisfy for some γ_u and a function $C : G \to \mathbb{R}^+$*

$$|\nabla^p u(x)| \le C(x)\gamma_u^p p! \qquad \forall p \in \mathbb{N}_0, \quad x \in G.$$

Then there are C', $\gamma > 0$ depending only on v and Λ such that

$$|\nabla^p (v \cdot (u \circ \Lambda))(z)| \le C'(\gamma(1+\gamma_u))^p p!(\Lambda(z)) \qquad \forall p \in \mathbb{N}_0, \quad z \in G'.$$

Proof: The proof is very similar to that of Lemma 4.3.1. There only difference is that one has to track the dependence on the constant γ_u. □

As another special case of Lemma 4.3.3, we have the following corollary, in which we stipulate only control over ∇u.

Corollary 4.3.5. *Let $G \subset \mathbb{R}^2$ and $G' \subset \mathbb{R}^n$ ($n \in \mathbb{N}$) be two bounded domains. Let $\Lambda : G' \to G$, and $u : G \to \mathbb{C}$ be analytic on G and satisfy*

$$\|\nabla^p \Lambda\|_{L^\infty(G')} \leq C_\Lambda \gamma_\Lambda^p p! \qquad \forall p \in \mathbb{N}_0,$$
$$|\nabla^p u(x)| \leq C(x) K^p \max\{(p+1)/r, \varepsilon^{-1}\}^p \qquad \forall p \in \mathbb{N}, \quad x \in G$$

for some C_Λ, γ_Λ, K, r, $\varepsilon > 0$, and a function $C : G \to \mathbb{R}^+$. Then there are C', $\gamma > 0$ depending only on K and the constants C_Λ, γ_Λ such that

$$|\nabla^p (u \circ \Lambda)) (z)| \leq C' \gamma^p \max\{(p+1)/r, \varepsilon^{-1}\}^p C(\Lambda(z)) \qquad \forall p \in \mathbb{N}, \quad z \in G'.$$

Proof: It is clear that in Lemma 4.3.3, the dependence of all constants on the mapping functions Λ is in fact through the constants C_Λ, γ_Λ that determine its growth of the derivatives. In order to apply Lemma 4.3.3, we note that the function $\tilde{u} := \nabla u$ satisfies

$$|\nabla^p \tilde{u}(x)| \leq C \max\{r^{-1}, \varepsilon^{-1}\} C(x) \tilde{K}^p \max\{(p+1)/r, \varepsilon^{-1}\}^p \qquad \forall p \in \mathbb{N}_0, \quad x \in G$$

for some appropriate $\tilde{K}$. Next, we observe that $\nabla(u \circ \Lambda) = \Lambda' \cdot (\nabla u) \circ \Lambda = \Lambda' \cdot (\tilde{u} \circ \Lambda)$. Hence, applying Lemma 4.3.3 we get

$$|\nabla^{p+1}(u \circ \Lambda)(z)| \leq C \gamma^p C(\Lambda(z)) \max\{r^{-1}, \varepsilon^{-1}\} \max\{(p+1)/r, \varepsilon^{-1}\}^p$$

for all $p \in \mathbb{N}_0$ and $z \in G'$, which is the desired bound. □

In the above results involving analytic changes of variables, we always assumed that the transformation is analytic up to the boundary. The important case of a change of variables from polar coordinates to Cartesian coordinates is therefore not covered here, but we refer to Section 6.3 where this issue is addressed.

5. Regularity Theory in Countably Normed Spaces

5.1 Motivation and outline

5.1.1 Motivation

In this chapter we prove shift theorems in countably normed spaces. The most important example is Theorem 5.3.10 where we consider the Dirichlet problem (1.2.1). Analogous results for other kinds of boundary conditions, i.e., Neumann problems and transmission problems, are proved in Propositions 5.4.5, 5.4.8, 5.4.7. Such shift theorems have the following structure: If the right-hand side $f \in \mathcal{B}^0_{\beta,\mathcal{E}}$, then the solution u_ε of (1.2.1) is in the countably normed space $\mathcal{B}^2_{\beta,\mathcal{E}}$. Here, $\beta = (\beta_1, \ldots, \beta_J) \in [0,1)^J$ is a vector of numbers associated with the vertices A_j of the curvilinear polygon and $\mathcal{E}$ is the smallest characteristic length scale of solution u_ε. In the ensuing two subsections, we will motivate our notion of smallest characteristic length scale $\mathcal{E}$ and then outline the key steps of the proof of this shift theorem in countably normed spaces.

Smallest characteristic length scale. The characteristic length scale depends on the size of ε relative to the coefficients A, b, and c in (1.2.1). To see this, let us consider a one-dimensional example:

$$-\varepsilon^2 u'' + bu' + cu = f \quad \text{on } (0,1), \qquad u(0) = u(1) = 0.$$

For constant $b \in \mathbb{R}$, $c \geq 0$, the solution of this problem is given by $u = u_{part} + \alpha_1 e^{\lambda_1 x} + \alpha_2 e^{\lambda_2 x}$, where u_{part} is a particular solution of the equation and λ_1, λ_2 are given by

$$\lambda_{1,2} = \frac{b}{\varepsilon^2} \pm \sqrt{\left(\frac{b}{\varepsilon^2}\right)^2 + 4\frac{c}{\varepsilon^2}};$$

here, α_1, α_2 are suitable constants determined by the choice of u_{part} and the boundary conditions. Ignoring for the moment the size of the coefficients α_1, α_2, we see that the growth of the derivatives of $\alpha_1 e^{\lambda_1 x} + \alpha_2 e^{\lambda_2 x}$ is controlled by the size of λ_1 and λ_2. Defining now $\mathcal{E}$ by

$$\mathcal{E}^{-1} := \frac{|b|}{\varepsilon^2} + \frac{\sqrt{c}}{\varepsilon} + 1, \tag{5.1.1}$$

it is not difficult to see that

$$\mathcal{E}^{-1} \leq \max\{|\lambda_1|, |\lambda_2|\} + 1 \leq 2\mathcal{E}^{-1}.$$

Thus, we expect the p-th derivatives of u to be of size $\mathcal{E}^{-p}$. This qualitative consideration neglects of course the contribution of the constants α_1, α_2, and the particular solution u_{part}. In general, we expect—in analogy to the result of Lemma 4.1.1—that the derivatives of the solution u can be controlled by expressions of the form

$$C\gamma^p \max\{p+2, \mathcal{E}^{-1}\}^{p+2}. \tag{5.1.2}$$

We note that in the example of Lemma 4.1.1, i.e., $b \equiv 0$, $c \equiv 1$, we have $\mathcal{E} = \varepsilon$, and (5.1.2) is indeed the estimate obtained in Lemma 4.1.1. The other case of interest is $b = 1$ together with $c = O(1)$. Then, $\mathcal{E} = \varepsilon^2$. Thus, for $c = O(1)$ the cases $b = 0$ and $b = O(1)$ have two different length scales. We remark in passing that our heuristic claims on being able to control the growth of the derivatives of solutions of these one-dimensional singularly perturbed problems by expressions of the form (5.1.2) were rigorously established in [92, 96]. In the present chapter, we show that for solutions of (1.2.1) in curvilinear polygons, the growth of the derivatives can indeed be characterized by the characteristic length scale $\mathcal{E}$ defined in (5.3.8), which is the analog of (5.1.1) in two-dimensional problems with variable coefficients.

Outline of the proof of Theorem 5.3.10. At the heart of the proof is Proposition 5.3.4, where a result similar to that of Theorem 5.3.10 is shown for a single straight sector, $S_R(\omega)$. The general case of curved boundaries then follows easily by a mapping argument and Theorem 4.2.20.
Let us outline the main ideas of the proof of Proposition 5.3.4 for the simple model equation already discussed in the introduction to Chapter 4:

$$-\varepsilon^2 \Delta u_\varepsilon + u_\varepsilon = f \quad \text{on } S_R(\omega), \tag{5.1.3a}$$

$$u_\varepsilon = 0 \quad \text{on the two sides } \Gamma_1,\, \Gamma_2 \text{ of } S_R(\omega). \tag{5.1.3b}$$

We note that the smallest characteristic length scale is $\mathcal{E} \sim \varepsilon$. Let $R' < R$ be fixed and $\beta \in \{\beta \in (0,1) \,|\, \beta > 1-\pi/\omega\}$. Then Proposition 5.3.4 states that, if the right-hand side $f \in \mathcal{B}^0_{\beta,\varepsilon}(S_R(\omega), C_f, \gamma_f)$, then for any solution u_ε of (5.1.3) there are C_u, $\gamma_u > 0$ such that $u_\varepsilon \in \mathcal{B}^2_{\beta,\varepsilon}(S_{R'}(\omega), C_u, \gamma_u)$; furthermore, the constant C_u has the from

$$C_u = C_{geo}\left[C''_f + \varepsilon\|\nabla u_\varepsilon\|_{L^2(S_R(\omega))} + \|u_\varepsilon\|_{L^2(S_R(\omega))}\right],$$

where the constant C''_f depends only on the right-hand side f and is independent of ε and u_ε; the constant C_{geo} depends only on R, R', ω, and β.
The proof of this result proceeds in several steps.

1. The first ingredient of the proof are *local* regularity results on balls B_r or half-balls G_r (near the boundary) as in [98]. Let u_ε solve $-\varepsilon^2 \Delta u_\varepsilon + u_\varepsilon = f$ on a ball B_r, where the right-hand side f satisfies

$$\|\nabla^p f\|_{L^2(B_r)} \leq C'_f \gamma^p \max\{(p+1)/r, \varepsilon^{-1}\}^p \qquad \forall p \in \mathbb{N}_0 \tag{5.1.4}$$

for some C'_f, $\gamma > 0$. The local regularity result of Proposition 5.5.1 then states that there exist C, $\gamma_u > 0$ depending only on γ such that

$$r^{p+2}\|\nabla^{p+2}u_\varepsilon\|_{L^2(B_{r/2})} \le C\gamma_u^p \max\{p+2, r/\varepsilon\}^{p+2}\,\mathrm{Loc}(u_\varepsilon), \quad \text{where}$$
$$\mathrm{Loc}(u_\varepsilon) := \min\{1, r/\varepsilon\}\varepsilon\|\nabla u_\varepsilon\|_{L^2(B_r)} + \min\{1, r/\varepsilon\}^2\left[\|u_\varepsilon\|_{L^2(B_r)} + C'_f\right].$$

A completely analogous result, Proposition 5.5.2, holds on half-balls G_r if homogeneous boundary conditions are imposed on u_ε on the straight part of ∂G_r. For notational convenience in this outline of the proof of Proposition 5.3.4, however, we will ignore the technical complications introduced at the boundary of $S_R(\omega)$.

2. The domain $S_{R'}(\omega)$ is covered by balls $B_{r_i/2}(x_i)$ that have the following key properties: The balls $B_{r_i}(x_i)$ have finite overlap and the radii r_i are (essentially) proportional to $|x_i|$, i.e., $r_i = c|x_i|$ for some fixed $c \in (0,1)$. In order to avoid the regularity study near $\Gamma_3 = \{(R\cos\varphi, R\sin\varphi) \,|\, 0 < \varphi < \omega\}$, we assume that for some $R'' \in (R', R)$ the balls $B_{r_i}(x_i)$ are all contained in $B_{R''}(0)$.
The characterization of functions f from the space $\mathcal{B}^0_{\beta,\varepsilon}(S_R(\omega), C_f, \gamma_f)$ that we provided in Lemma 4.2.17 allows us to bound C'_f, $\gamma > 0$ in (5.1.4): The constant γ of (5.1.4) depends only on c and γ_f for each ball $B_{r_i}(x_i)$, and the constant C'_f can be bounded by

$$C'_f \le \frac{C(i)}{\hat{\Phi}_{0,\beta,\varepsilon}(x_i)}$$

for some $C(i) > 0$ satisfying

$$\sum_{i=1}^{\infty} C^2(i) = C''_f < \infty;$$

the constant $C''_f > 0$ depends only on C_f, γ_f, and the covering by balls.

3. As the balls $B_{r_i/2}$ cover $S_{R'}(\omega)$, we merely sum up the local regularity results on balls (or half-balls near the boundary). Here, it is important to note that, due to our choice $r_i \sim |x_i|$, the weight functions $\hat{\Phi}_{p,\beta,\varepsilon}$ satisfy (4.2.8). Using additional properties of the weight function from Lemma 4.2.2, we obtain

$$\|\hat{\Phi}_{p,\beta,\varepsilon}\nabla^{p+2}u_\varepsilon\|^2_{L^2(S_{R'}(\omega))} \le CK^p \sum_{i=1}^{\infty} \hat{\Phi}^2_{p,\beta,\varepsilon}(x_i)\|\nabla^{p+2}u_\varepsilon\|^2_{L^2(B_{r_i/2}(x_i))}$$
$$\le CK^p \max\{p+2, \varepsilon^{-1}\}^{2(p+2)} \times$$
$$\sum_{i=1}^{\infty}\left\{\varepsilon^2\|\hat{\Phi}_{-1,\beta,\varepsilon}\nabla u_\varepsilon\|^2_{L^2(B_{r_i}(x_i))} + \|\hat{\Phi}_{0,\beta,\varepsilon}u_\varepsilon\|^2_{L^2(B_{r_i}(x_i))} + C^2(i)\right\}.$$

4. Next, we employ the finite overlap property of the balls $B_{r_i}(x_i)$ to bound the infinite sum by

$$\sum_{i=1}^{\infty} \left\{ \varepsilon^2 \|\hat{\Phi}_{-1,\beta,\varepsilon} \nabla u_\varepsilon\|^2_{L^2(B_{r_i}(x_i))} + \|\hat{\Phi}_{0,\beta,\varepsilon} u_\varepsilon\|^2_{L^2(B_{r_i}(x_i))} + C^2(i) \right\}$$
$$\leq \varepsilon \|\hat{\Phi}_{-1,\beta,\varepsilon} \nabla u_\varepsilon\|^2_{L^2(S_{R''}(\omega))} + \|\hat{\Phi}_{0,\beta,\varepsilon} u_\varepsilon\|^2_{L^2(S_{R''}(\omega))} + C''_f.$$

Using Lemma 4.2.10 we can control the term $\|\hat{\Phi}_{-1,\beta,\varepsilon} \nabla u_\varepsilon\|^2_{L^2(S_{R''}(\omega))}$ to get

$$\|\hat{\Phi}_{p,\beta,\varepsilon} \nabla^{p+2} u_\varepsilon\|_{L^2(S_{R'}(\omega))} \leq C K^p \max\{p+2, \varepsilon^{-1}\}^{(p+2)} \times$$
$$\left[\|u_\varepsilon\|_{H^{2,2}_{\beta,\varepsilon}(S_{R''}(\omega))} + C''_f \right].$$

5. It remains to replace the term $\|u_\varepsilon\|_{H^{2,2}_{\beta,\varepsilon}(S_{R''}(\omega))}$ by expressions involving the right-hand side f and the energy $\varepsilon\|\nabla u_\varepsilon\|_{L^2(S_R(\omega))} + \|u_\varepsilon\|_{L^2(S_R(\omega))}$. This is done with the aid of the shift theorem in Proposition 5.3.2. It states that a solution u_ε of (5.1.3) satisfies

$$\|u_\varepsilon\|_{H^{2,2}_{\beta,\varepsilon}(S_{R''}(\omega))} \leq$$
$$C \left[\|\hat{\Phi}_{0,\beta,\varepsilon} f\|_{L^2(S_R(\omega))} + \varepsilon \|\nabla u_\varepsilon\|_{L^2(S_R(\omega))} + \|u_\varepsilon\|_{L^2(S_R(\omega))} \right]$$

provided that $\beta > 0$ satisfies additionally $\beta \in (1 - \pi/\omega, 1)$. Inserting this bound in the previous one gives the desired result.

5.1.2 Outline of Chapter 5

The outline of this chapter is as follows. We start with a brief review of the analytic regularity results of Babuška & Guo for the case $\varepsilon = 1$ in Section 5.2. These results are taken from [15] and are phrased in our notation. In [15], the case of Laplace's equation in a straight sector was analyzed. The extension to problems with variable coefficients was done with a perturbation argument. The main result of this chapter is Theorem 5.3.10, which states a shift theorem in countably normed spaces for equation (1.2.1). Two cases of interest are discussed separately in Corollary 5.3.12 for the case $b \equiv 0$ and in Corollary 5.3.13 for the case $|b| > 0$ on Ω. The reason for discussing them separately is that the smallest characteristic length scale is different in these two cases: For the case $b \equiv 0$ in Corollary 5.3.12 the length scale is $O(\varepsilon)$ whereas for the case $|b| > 0$ in Corollary 5.3.13, the length scale is $O(\varepsilon^2)$. We remark at this point that Theorem 5.3.10 represents a slight improvement over the results of Babuška & Guo in [14] in the context of curved boundaries: There, for the case $\varepsilon = 1$ solutions u_1 of (1.2.1) were shown to be in the space $\mathcal{C}^2_\beta$ rather than in $\mathcal{B}^2_\beta$ (see the discussion following Proposition 5.2.4). The key to our improvement is the invariance of $\mathcal{B}^l_{\beta,\varepsilon}$ functions under analytic changes of variables ascertained in Theorem 4.2.20.

Theorem 5.3.10 is proved using analytic regularity results on sectors which are then combined to obtain results on (curvilinear) polygons. The analysis on sectors is done in Section 5.3.1. This is done in two steps. The first step consists in Proposition 5.3.2, which shows that the solution of (1.2.1) satisfies a shift theorem in weighted Sobolev spaces: If the right-hand side f is in $H^{0,0}_{\beta,\mathcal{E}}$, then the solution u of (1.2.1) is in the space $H^{2,2}_{\beta,\mathcal{E}}$. In the second step this regularity result is extended to countably normed spaces: If f is in $\mathcal{B}^0_{\beta,\mathcal{E}}$, then the solution u is in $\mathcal{B}^2_{\beta,\mathcal{E}}$ (Proposition 5.3.4). These results are formulated for straight sectors but can readily be extended to curvilinear sectors with the aid of Theorem 4.2.20. Theorem 5.3.10 is formulated for a Dirichlet problem. In Propositions 5.4.5, 5.4.8, 5.4.7 we formulate analogous results for a variety of other boundary conditions. The main technical tool for obtaining the analytic regularity results in sectors (Proposition 5.3.4) are local regularity results for the solutions of (1.2.1). These local regularity results are provided in Section 5.5. These results are obtained with the techniques of [98]. The novel feature here is that the dependence on the parameter ε and the diameter of the ball R is explicit. We prove four types of local regularity results: interior regularity on balls of radius R, regularity on half-balls of radius R with homogeneous Dirichlet conditions, regularity results for the Neumann problem, and finally regularity results for transmission problems. These last local regularity results could easily be used to extend Theorem 5.3.10 to problems with piecewise analytic data A, b, c, and f. In fact, our procedure in Chapter 6 shows how this can be done.

5.2 Analytic regularity results of Babuška and Guo

Many elliptic problems arising in practice have piecewise analytic input data such as the coefficients of the differential equations, the right-hand side, and the geometry of the domain. Thus, the solution is in general piecewise analytic as well, [86,98]. From an approximation point of view this suggests that exponential rates of convergence could be possible with spectral methods. This exponential convergence of spectral methods is not obvious for problems of the form (1.2.1) as analyticity of the solution is not given up to the boundary: The solution has singularities at the vertices A_j as we discussed in Section 1.4.1. Nevertheless, as pointed out in Section 1.4.3, exponential convergence of the hp-FEM is possible in this situation, if the increase of the polynomial degree is combined with an appropriate mesh refinement toward the singularities. For a rigorous proof of the exponential rate of convergence of this scheme, it is essential to control all derivatives of the solution with bounds that are explicit in their dependence on the location in the domain. A framework for controlling such bounds was developed by Babuška and Guo with the notion of countably normed spaces, which coincide with the spaces $\mathcal{B}^l_{\beta,1}$ introduced in Chapter 4 for the special case $\varepsilon = 1$. In this framework, Babuška and Guo proved that solutions of (linear) elliptic problems of second order are elements of countably normed spaces $\mathcal{B}^2_{\beta,1}$ as we ascertained in Proposition 1.4.2. The original proof of Proposition 1.4.2 was accomplished by Babuška and Guo by induction on the order of the derivative. The start of the induction argument was an $H^{2,2}_{\beta,1}$ regularity result. Such regularity results in weighted Sobolev spaces are intimately linked to Proposi-

tion 1.4.1 and go back to a seminal paper by Kondrat'ev, [76,77]. In the present section, we briefly highlight the key results from [14–17]. We will formulate their results for the Dirichlet problem although analogous results were obtained for the Neumann problem and problems with mixed boundary conditions (see also Sections 5.4).
We start with the $H^{2,2}_{\beta,1}$ regularity result for the Laplacian in a straight sector:

Proposition 5.2.1. *Let $S_R(\omega)$ be a sector, $\omega \in (0,2\pi)$, and let $R' < R$. Then for $\beta \in [0,1) \cap (1-\pi/\omega, 1)$ there exists $C > 0$ depending only on ω, β, R, R' such that the solution $u \in H^1_0(S_R(\omega))$ of $-\Delta u = f$ with $f \in H^{0,0}_{\beta,1}(S_R(\omega))$ is in $H^2_{loc}(S_R(\omega))$ and satisfies*

$$\|u\|_{H^1(S_R(\omega))} + \|u\|_{H^{2,2}_{\beta,1}(S_{R'}(\omega))} \leq C\|f\|_{H^{0,0}_{\beta,1}(S_R(\omega))}. \tag{5.2.1}$$

Perturbation arguments allow for an extension to the case of variable coefficients:

Proposition 5.2.2. *Let $0 < R' < R$, $\omega \in (0,2\pi)$, and $A \in C^1(\overline{S_R(\omega)}, \mathbb{S}^2_>)$ with $0 < \lambda_{min} \leq A$ on $S_R(\omega)$ be given. Then there are $\beta \in [0,1)$, which depends only on ω and $A(0)$, and $C > 0$ such that for all $f \in H^{0,0}_{\beta,1}(S_R(\omega))$ the solution u of*

$$-\nabla \cdot (A\nabla u) = f \quad \text{on } S_R(\omega), \quad u = 0 \quad \text{on } \partial S_R(\omega),$$

is in $H^2_{loc}(S_R(\omega))$ and satisfies

$$\|u\|_{H^{2,2}_{\beta,1}(S_{R'}(\omega))} \leq C\|f\|_{H^{0,0}_{\beta,1}(S_R(\omega))}.$$

Proof: First, from Lemma 5.3.7 (with $\varepsilon = 1$ there), we get $\left(H^{0,0}_{\beta,1}(S_R(\omega))\right)' \subset H^{-1}(S_R(\omega))$ so that for every $\beta \in [0,1)$ there exists $C_\beta > 0$ such that

$$\|u\|_{H^1(S_R(\omega))} \leq C\|f\|_{H^{0,0}_{\beta,1}(S_R(\omega))}. \tag{5.2.2}$$

Next, by standard elliptic regularity theory, the solution $u \in H^2_{loc}(S_R(\omega))$. We then fix $R'' \in (R', R)$ and claim the existence of $C > 0$ independent of f such that

$$\|\,|x|\nabla^2 u\|_{L^2(S_{R'}(\omega))} \leq C\left[\|\,|x|f\|_{L^2(S_{R''}(\omega))} + \|\nabla u\|_{L^2(S_{R''}(\omega))}\right]. \tag{5.2.3}$$

To see (5.2.3), we employ elliptic regularity locally in the following way. We choose a covering (see, e.g., Lemma 5.3.1)) of $S_{R'}(\omega)$ by balls $\mathcal{B} = \{B_i \,|\, i \in \mathbb{N}\}$ with the following properties:

1. $B_i = B_{r_i}(x_i)$ with $r_i = c|x_i|$ for suitable $c \in (0,1)$;
2. $B_i \cap S_{R''}(\omega)$ is either completely contained in $S_{R''}(\omega)$ or a half ball;
3. $\operatorname{card}\{i \,|\, x \in B_i\} \leq N$ for all $x \in S_{R''}(\omega)$ for some fixed $N \in \mathbb{N}$;
4. the "stretched" balls $\widehat{B}_i := B_{r_i/2}(x_i)$ also form a covering of $S_{R'}(\omega)$.

By elliptic regularity (cf. (5.5.22) of Lemma 5.5.12 for the case of balls $B_i \subset S_{R''}(\omega)$ and (5.5.27) of Lemma 5.5.15 for the case of $B_i \cap S_{R''}(\omega)$ being a half ball) we have for a constant $C > 0$ independent of i

$$r_i^2 \|\nabla^2 u\|_{L^2(\widehat{B}_i \cap S_{R''}(\omega))} \le C \left[r_i^2 \|f\|_{L^2(B_i \cap S_{R''}(\omega))} + r_i \|\nabla u\|_{L^2(B_i \cap S_{R''}(\omega))} \right].$$

Dividing by r_i and using $r_i \le C \inf_{x \in B_i} |x| \le C \sup_{x \in B_i} |x| \le C r_i$, we get

$$\| \, |x| \nabla^2 u\|_{L^2(\widehat{B}_i \cap S_R(\omega))} \le C \left[\| \, |x| f\|_{L^2(B_i \cap S_{R''}(\omega))} + \|\nabla u\|_{L^2(B_i \cap S_{R''}(\omega))} \right].$$

Squaring this last estimate and summing on $i \in \mathbb{N}$ gives with the overlap properties of the balls

$$\| \, |x| \nabla^2 u\|_{L^2(S_{R'}(\omega))} \le C \left[\| \, |x| f\|_{L^2(S_{R''}(\omega))} + \|\nabla u\|_{L^2(S_{R''}(\omega))} \right],$$

which is (5.2.3).

Next, it is easy to see with affine changes of variables that for every fixed $\tilde{A} \in \mathbb{S}^2_>$, there exist $\beta \in [0,1)$ and $C > 0$ such that for every $\tilde{f} \in H^{0,0}_{\beta,1}(S_R(\omega))$, the $H^1_0(S_R(\omega))$-solution to $-\nabla \cdot \left(\tilde{A} \nabla u \right) = \tilde{f}$ satisfies

$$\|u\|_{H^{2,2}_{\beta,1}(S_{R'}(\omega))} \le C \|\tilde{f}\|_{H^{0,0}_{\beta,1}(S_{R''}(\omega))}. \tag{5.2.4}$$

This puts us in position to prove Proposition 5.2.2 by a perturbation argument. Denoting $\tilde{A} := A(0)$, we calculate

$$-\nabla \cdot \left(\tilde{A} \nabla u \right) = \tilde{f} \quad \text{a.e. on } S_R(\omega), \tag{5.2.5}$$

where

$$\tilde{f} = f - \left(\nabla \cdot (\tilde{A} - A) \right) \nabla u - (\tilde{A} - A) : \nabla^2 u.$$

In view of $A \in C^1(\overline{S_R(\omega)})$ there is $C > 0$ with $|A(x) - \tilde{A}| \le C|x|$. Hence, we get by combining (5.2.4) with (5.2.3) and exploiting $\beta < 1$

$$\begin{aligned}
\|u\|_{H^{2,2}_{\beta,1}(S_{R'}(\omega))} &\le C \|\tilde{f}\|_{H^{0,0}_{\beta,1}(S_{R''}(\omega))} \\
&\le C \left[\|f\|_{H^{0,0}_{\beta,1}(S_{R''}(\omega))} + \|\nabla u\|_{L^2(S_{R''}(\omega))} + \| \, |x| \nabla^2 u\|_{L^2(S_{R''}(\omega))} \right] \\
&\le C \|f\|_{H^{0,0}_{\beta,1}(S_R(\omega))}.
\end{aligned}$$

□

Remark 5.2.3 Note that in the proof of Proposition 5.2.2, the assumption that A be Lipschitz is exploited in an essential way—jumping coefficients need a different treatment. In that case, we need the analog of Proposition 5.2.1 for the case of a differential operator with piecewise constant coefficients. Such a result is provided in Proposition A.2.1. Most of the details for this case are worked out in Chapter 6 below. ▪

By induction on the order of the derivatives, akin to the way Morrey proceeds in [98] (see also Section 5.5 ahead), one can obtain bounds on all derivatives of the solutions of elliptic equations with analytic coefficients. Prototypical is the following result (a proof can be found below in Theorem 5.3.10).

Proposition 5.2.4. *Let $S_R(\omega)$ be a sector, $R' < R$, $A \in \mathcal{A}(\overline{S_R(\omega)}, \mathbb{S}^2_>)$ with $0 < \lambda_{min} \leq A$ on $S_R(\omega)$ for some fixed λ_{min}. Then there exists $\beta \in [0,1)$, which depends only on ω, $A(0)$, such that for $f \in \mathcal{B}^0_{\beta,1}(S_R(\omega), C_f, \gamma_f)$ (with C_f, $\gamma_f > 0$) the solution $u \in H^1_0(S_R(\omega))$ of $-\nabla \cdot (A\nabla u) = f$ on $S_R(\omega)$ satisfies $u \in \mathcal{B}^2_{\beta,1}(S_{R'}(\omega), C_u, \gamma_u)$ for some C_u, γ_u.*

The case of curvilinear polygons was addressed by Babuška and Guo in [14]. In order to characterize the solutions in curvilinear sectors S, they introduced the countably normed spaces

$$\mathcal{C}^2_\beta(S, C, \gamma) := \{u \in H^{2,2}_{\beta,1} \mid \|\hat{\Phi}_{p-1,\beta,1}\nabla^p u\|_{L^\infty(S)} \leq Cp!\gamma^p \quad \forall p \in \mathbb{N}\}.$$

In [14], the following regularity result was shown:

Proposition 5.2.5. *Let S be a curvilinear sector, $A \in \mathcal{A}(\overline{S}, \mathbb{S}^2_>)$ with $0 < \lambda_{min} \leq A$ on $S_R(\omega)$. Then there exists $\beta \in [0,1)$ with the following property: For any neighborhood $\mathcal{U}$ of Γ_3 and $f \in \mathcal{B}^0_{\beta,1}(S, C_f, \gamma_f)$ (with C_f, $\gamma_f > 0$), the solution $u \in H^1_0(S)$ of $-\nabla \cdot (A\nabla u) = f$ on S satisfies $u \in \mathcal{C}^2_\beta(S \setminus \mathcal{U}, C, \gamma)$ for some C, $\gamma > 0$.*

In [14] it is also shown that $\mathcal{B}^2_{\beta,1} \subset \mathcal{C}^2_\beta \subset \mathcal{B}^2_{\beta+\delta,1}$ for all $\delta > 0$. Hence, Proposition 5.2.5 implies that the solution $u \in \mathcal{B}^2_{\beta+\delta,1}$ for all $\delta > 0$. This leads the authors to raise in [14] the question whether the solution u is in fact in $\mathcal{B}^2_{\beta,1}$ for curvilinear polygons as well. Theorem 5.3.10 below answers this question in the affirmative. The key ingredient for this assertion is Theorem 4.2.20, which states that the spaces $\mathcal{B}^l_{\beta,\varepsilon}$ are invariant under analytic changes of variables. This invariance allows us to restrict our attention to the case of straight sectors and then infer the case of curvilinear polygons by a mapping argument.

Remark 5.2.6 Proposition 5.2.4 is a form of an elliptic shift theorem in countably normed spaces $\mathcal{B}^l_{\beta,1}$, which is a class of analytic functions. Other forms of elliptic shift theorems exist in the literature for classes of functions larger than those of analytic functions. We mention here in particular the notion of Gevrey regularity, for which we refer the reader to [30] and the reference there. ▪

5.3 Analytic regularity: Dirichlet problems

For the sake of definiteness, we consider in this section the case of homogeneous Dirichlet boundary conditions. The similar cases of of homogeneous Neumann boundary conditions or transmission conditions are treated in Section 5.4.

5.3.1 Analytic regularity in sectors

In this subsection, we prove analytic regularity results on straight sectors. We start with the following covering lemma.

Lemma 5.3.1 (Covering). *Let $0 < R' < R$, $\omega \in (0, 2\pi)$. Then there exist constants $c \in (0,1)$, $N \in \mathbb{N}$ and a covering $\mathcal{B} = \{B_i \,|\, i \in \mathbb{N}\}$ of $S_{R'}(\omega)$ by balls B_i with the following properties:*

1. *$\mathcal{B}$ covers $S_{R'}(\omega)$, i.e., $S_{R'}(\omega) \subset \cup_i B_i$;*
2. *the balls have the form $B_i = B_{r_i}(x_i)$ with $r_i = c|x_i|$;*
3. *the balls B_i satisfy a finite overlap condition, i.e., there exists $N \in \mathbb{N}$ such that $\operatorname{card}\{i \in \mathbb{N} \,|\, x \in B_i\} \leq N \ \forall x \in S_R(\omega)$;*
4. *the sets $D_i := B_i \cap S_R(\omega)$ satisfy the following dichotomy: either D_i is a ball (i.e., $D_i = B_i$) or D_i is a half-ball (i.e., the center x_i is on one of the straight parts of $\partial S_{R'}(\omega)$ and $|x_i| + r_i < R$);*
5. *the "stretched" balls $\widehat{B}_i := B_{cr_i/2}(x_i)$ also form a covering of $S_{R'}(\omega)$.*

Proof: The existence of such coverings is geometrically clear. In order to ensure that the sets D_i indeed satisfy the above dichotomy, one has to choose the constant c sufficiently small in dependence on the ratio of R to R'. □

On sectors $S_R(\omega)$, we are interested in the regularity of solutions u to the following equation

$$-\varepsilon^2 \nabla \cdot (A(x)\nabla u) + b(x) \cdot \nabla u + c(x)u = f(x) \quad \text{on } S_R(\omega), \tag{5.3.6a}$$
$$u = 0 \quad \text{on } \partial S_R(\omega) \setminus \partial B_R(0). \tag{5.3.6b}$$

We assume that $\varepsilon \in (0,1]$ and that the coefficients of (5.3.6) are analytic; that is, $A \in \mathcal{A}(S_R(\omega), \mathbb{S}^2_>)$, $b \in \mathcal{A}(S_R(\omega))$, $c \in \mathcal{A}(S_R(\omega))$ satisfy, for some C_b, γ_b, C_c, $\gamma_c \geq 0$,

$$\|\nabla^p A\|_{L^\infty(S_R(\omega))} \leq C_A \gamma_A^p p! \qquad \forall p \in \mathbb{N}_0, \tag{5.3.7a}$$
$$0 < \lambda_{min} \leq A \qquad \text{on } S_R(\omega), \tag{5.3.7b}$$
$$\|\nabla^p b\|_{L^\infty(S_R(\omega))} \leq C_b \gamma_b^p p! \qquad \forall p \in \mathbb{N}_0, \tag{5.3.7c}$$
$$\|\nabla^p c\|_{L^\infty(S_R(\omega))} \leq C_c \gamma_c^p p! \qquad \forall p \in \mathbb{N}_0. \tag{5.3.7d}$$

Next, we define for $\varepsilon > 0$ the *relative diffusivity* $\mathcal{E}$ as

$$\mathcal{E}^{-1} := \frac{C_b}{\varepsilon^2} + \frac{\sqrt{C_c}}{\varepsilon} + 1. \tag{5.3.8}$$

Note that ε and $\mathcal{E}$ satisfy trivially the relationships

$$\frac{C_b}{\varepsilon^2} \leq \mathcal{E}^{-1}, \qquad \frac{C_c}{\varepsilon^2} \leq \mathcal{E}^{-2}. \tag{5.3.9}$$

H^1 solutions u of (5.3.6) are in fact in the space $H^{2,2}_{\beta,\mathcal{E}}(S_R(\omega))$ for some $\beta \in (0,1)$:

Proposition 5.3.2 (Shift theorem). *Assume that the coefficients A, b, c satisfy (5.3.7). Then there exists $\beta \in [0,1)$, which depends only on ω and $A(0)$, and there exists $C > 0$ such that each solution $u \in H^1(S_R(\omega))$ of (5.3.6) with right-hand side $f \in H^{0,0}_{\beta,\mathcal{E}}(S_R(\omega))$ satisfies*

$$\|\hat{\Phi}_{0,\beta,\mathcal{E}}\nabla^2 u\|_{L^2(S_{R'}(\omega))} \leq$$
$$C\mathcal{E}^{-2}\left[(\mathcal{E}/\varepsilon)^2\|\hat{\Phi}_{0,\beta,\mathcal{E}}f\|_{L^2(S_R(\omega))} + \mathcal{E}\|\nabla u\|_{L^2(S_R(\omega))} + C_c\left(\mathcal{E}/\varepsilon\right)^2\|u\|_{L^2(S_R(\omega))}\right].$$

Remark 5.3.3 If $A(0) = \mathrm{Id}$, then we may choose $\beta \in (1-\pi/\omega, 1) \cap [0,1)$. ▪

Proof: In this proof, we will use the shorthand S_r for the sectors $S_r(\omega)$. Let $\mathcal{B} = \{B_i \,|\, i \in \mathbb{N}\}$ be the covering of $S_{R'}$ given by Lemma 5.3.1. Recall that the balls B_i have radii $c|x_i|$, that the corresponding "stretched" balls $\widehat{B}_i$ have radii $c/2|x_i|$, and that the sets $B_i \cap S_R$ are either full balls or half-balls. Define the index set $I_\mathcal{E} := \{i \in \mathbb{N} \,|\, B_i \cap S_\mathcal{E} \neq \emptyset\}$. Elementary geometric considerations show

$$i \in I_\mathcal{E} \Longrightarrow |x_i| < (1-c)^{-1}\mathcal{E}, \qquad i \in \mathbb{N}\setminus I_\mathcal{E} \Longrightarrow |x_i| \geq (1-c)^{-1}\mathcal{E}. \quad (5.3.10)$$

It readily follows that the stretched balls $\widehat{B}_i$ satisfy

$$\bigcup_{i\in I_\mathcal{E}} \widehat{B}_i \subset S_{(1+c/2)(1-c)^{-1}\mathcal{E}}, \qquad \bigcup_{i\in\mathbb{N}\setminus I_\mathcal{E}} \widehat{B}_i \subset \mathbb{R}^2 \setminus S_{(1-c/2)(1-c)^{-1}\mathcal{E}}.$$

Setting $h := \min\{R', (1+c/2)(1-c)^{-1}\mathcal{E}\}$, we infer from $\cup_{i\in\mathbb{N}}\widehat{B}_i \supset S_{R'}$ that

$$\bigcup_{i\in\mathbb{N}\setminus I_\mathcal{E}} \widehat{B}_i \supset S_{R'} \setminus S_{(1+c/2)(1-c)^{-1}\mathcal{E}} =: S_{R'} \setminus S_h, \quad (5.3.11)$$

Introducing $H := \min\{R, 2h\}$, we readily ascertain the existence of $C > 0$ independent of $\mathcal{E}$ such that

$$C^{-1} \leq \min_{x\in B_i} \hat{\Phi}_{0,\beta,\mathcal{E}}(x) \leq 1 \qquad \forall i \in \mathbb{N}\setminus I_\mathcal{E}, \quad (5.3.12)$$
$$C^{-1}\mathcal{E} \leq |x_i| \qquad \forall i \in \mathbb{N}\setminus I_\mathcal{E}, \quad (5.3.13)$$
$$|x| \leq C\mathcal{E} \qquad \forall x \in S_H, \quad (5.3.14)$$
$$C^{-1}\left(\frac{|x|}{\mathcal{E}}\right)^\beta \leq \hat{\Phi}_{0,\beta,\mathcal{E}}(x) \leq C\left(\frac{|x|}{\mathcal{E}}\right)^\beta \qquad \forall x \in S_H. \quad (5.3.15)$$

We estimate the L^2 norm of $\hat{\Phi}_{0,\beta,\mathcal{E}}\nabla^2 u$ in two steps by first estimating the L^2 norm over S_h and then over $S_{R'} \setminus S_h \subset \cup_{i\in\mathbb{N}\setminus I_\mathcal{E}}\widehat{B}_i$. In order to estimate the L^2 norm over S_h, let χ be a smooth cut-off function satisfying $\chi \equiv 1$ on $B_h(0)$, $\chi \equiv 0$ on $\mathbb{R}^2 \setminus B_H(0)$, and $\|\nabla^j\chi\|_{L^\infty(\mathbb{R}^2)} \leq C\mathcal{E}^{-j}$, $j \in \{0,1,2\}$. Next, we calculate that $u\chi$ satisfies on S_R

$$-\nabla\cdot(A\nabla(u\chi)) = \chi\varepsilon^{-2}[f - b\cdot\nabla u - cu] - 2\nabla\chi\cdot(A\nabla u) - u\nabla\cdot(A\nabla\chi),$$
$$u\chi = 0 \quad \text{on } \partial S_R,$$

where we used the fact that u satisfies (5.3.6). Applying Proposition 5.2.2 yields the existence of $\beta \in [0,1)$, $C > 0$ such that

$$\|r^\beta \nabla^2 (u\chi)\|_{L^2(S_R)} \leq C\|r^\beta \nabla \cdot (A\nabla(\chi u))\|_{L^2(S_R)}.$$

Dividing both sides by $\mathcal{E}^\beta$, using (5.3.15), and the support properties of χ yields

$$\|\hat{\Phi}_{0,\beta,\mathcal{E}} \nabla^2 u\|_{L^2(S_h)} \leq C\|\hat{\Phi}_{0,\beta,\mathcal{E}} \nabla \cdot (A\nabla(\chi u))\|_{L^2(S_H)}.$$

Expanding the right hand-side and observing (5.3.9), (5.3.14) together with $\hat{\Phi}_{0,\beta,\mathcal{E}} \leq 1$ gives

$$\begin{aligned}
\|\hat{\Phi}_{0,\beta,\mathcal{E}} \nabla^2 u\|_{L^2(S_h)} &\leq C\Big[\varepsilon^{-2}\|\hat{\Phi}_{0,\beta,\mathcal{E}} f\|_{L^2(S_H)} + C_b\varepsilon^{-2}\|\nabla u\|_{L^2(S_H)} \\
&\quad + C_c\varepsilon^{-2}\|u\|_{L^2(S_H)} + \mathcal{E}^{-1}\|\nabla u\|_{L^2(S_H)} + \mathcal{E}^{-2}\|u\|_{L^2(S_H)}\Big] \\
&\leq C\Big[\varepsilon^{-2}\|\hat{\Phi}_{0,\beta,\mathcal{E}} f\|_{L^2(S_H)} + \mathcal{E}^{-1}\|\nabla u\|_{L^2(S_H)} \qquad (5.3.16)\\
&\quad + C_c(\mathcal{E}/\varepsilon)^2\mathcal{E}^{-2}\|u\|_{L^2(S_H)} + \mathcal{E}^{-2}\|u\|_{L^2(S_H)}\Big].
\end{aligned}$$

Bounding further $|x| \leq C\mathcal{E}$ for $x \in S_H$, we get in view of Lemma 4.2.12

$$\mathcal{E}^{-2}\|u\|_{L^2(S_H)} \leq C\mathcal{E}^{-1}\|\frac{1}{|x|}u\|_{L^2(S_H)} \leq C\mathcal{E}^{-1}\|\nabla u\|_{L^2(S_H)}.$$

Inserting this in (5.3.16) finally yields

$$\begin{aligned}
&\|\hat{\Phi}_{0,\beta,\mathcal{E}} \nabla^2 u\|_{L^2(S_h)} \leq \qquad (5.3.17)\\
&\quad C\left[\varepsilon^{-2}\|\hat{\Phi}_{0,\beta,\mathcal{E}} f\|_{L^2(S_H)} + \mathcal{E}^{-1}\|\nabla u\|_{L^2(S_H)} + \mathcal{E}^{-2}C_c(\mathcal{E}/\varepsilon)^2\|u\|_{L^2(S_H)}\right].
\end{aligned}$$

We now turn to estimating $\|\Phi_{0,\beta,\mathcal{E}}\nabla u^2\|_{L^2(S_{R'}\setminus S_h)}$. For each ball B_i, u satisfies $-\nabla\cdot(A\nabla u) = \varepsilon^{-2}[f - b\nabla u - cu]$ on $B_i \cap S_R$. As in the proof of Proposition 5.2.2, we appeal to elliptic regularity, viz., we use (5.5.22) of Lemma 5.5.12 for the case of balls B_i completely contained in $S_R(\omega)$ and (5.5.27) of Lemma 5.5.15 for the case of $B_i \cap S_R(\omega)$ being a half ball to bound with a $C > 0$ independent of i

$$r_i^2\|\nabla^2 u\|_{L^2(\widehat{B}_i\cap S_R)} \leq C\left[r_i^2\varepsilon^{-2}\|f - b\nabla u - cu\|_{L^2(B_i\cap S_R)} + r_i\|\nabla u\|_{L^2(B_i\cap S_R)}\right].$$

For $i \in \mathbb{N} \setminus I_{\mathcal{E}}$, we may employ (5.3.12), (5.3.13) and additionally use (5.3.9) to get for all $i \in \mathbb{N} \setminus I_{\mathcal{E}}$:

$$\begin{aligned}
&\|\hat{\Phi}_{0,\beta,\mathcal{E}} \nabla^2 u\|_{L^2(\widehat{B}_i\cap S_R)} \leq \qquad (5.3.18)\\
&C\big[\varepsilon^{-2}\|\hat{\Phi}_{0,\beta,\mathcal{E}} f\|_{L^2(B_i\cap S_R)} + \mathcal{E}^{-1}\|\nabla u\|_{L^2(B_i\cap S_R)} + \mathcal{E}^{-2}C_c(\mathcal{E}/\varepsilon)^2\|u\|_{L^2(B_i\cap S_R)}\big].
\end{aligned}$$

Squaring this last estimate and summing on $i \in \mathbb{N} \setminus I_{\mathcal{E}}$ gives in view of (5.3.11) and the fact that the sets B_i satisfy an overlap condition:

$$\|\hat{\Phi}_{0,\beta,\mathcal{E}}\nabla^2 u\|_{L^2(S_{R'}\setminus S_h)} \le$$
$$C\left[\varepsilon^{-2}\|\hat{\Phi}_{0,\beta,\mathcal{E}} f\|_{L^2(S_R)} + \mathcal{E}^{-1}\|\nabla u\|_{L^2(S_R)} + \mathcal{E}^{-2}C_c(\mathcal{E}/\varepsilon)^2\|u\|_{L^2(S_R)}\right]. \quad (5.3.19)$$

Combining (5.3.17), (5.3.19) and observing that $\varepsilon^{-2} = (\mathcal{E}/\varepsilon)^2\mathcal{E}^{-2}$ concludes the proof. □

We can now formulate an analytic regularity result for the solution u of (5.3.6) for right-hand sides f from the countably normed space $\mathcal{B}^0_{\beta,\mathcal{E}}$:

Proposition 5.3.4. *Let $R > 0$, $\omega \in (0, 2\pi)$, and coefficients A, b, c satisfying (5.3.7) be given. Then there exists $\beta \in [0,1)$ with the following properties: For $\varepsilon \in (0,1]$, let $\mathcal{E}$ be given by (5.3.8). Let $u \in H^1(S_R(\omega))$ be a solution of (5.3.6) for a right-hand side $f \in \mathcal{B}^0_{\beta,\mathcal{E}}(S_R(\omega), C_f, \gamma_f)$ for some C_f, γ_f. Then for every $R' \in (0,R)$ there are C, $K > 0$ independent of ε and C_f such that for all $p \in \mathbb{N}_0$*

$$\|\hat{\Phi}_{p,\beta,\mathcal{E}}\nabla^{p+2}u\|_{L^2(S_{R'}(\omega))} \le$$
$$CK^p \max\{p, \mathcal{E}^{-1}\}^{p+2}\left[(\mathcal{E}/\varepsilon)^2 C_f + \mathcal{E}\|\nabla u\|_{L^2(S_R(\omega))} + C_c(\mathcal{E}/\varepsilon)^2\|u\|_{L^2(S_R(\omega))}\right].$$

Remark 5.3.5 We note that the coercivity condition (1.2.2e) is not explicitly required in Propositions 5.3.2, 5.3.4. These propositions are regularity assertions for H^1-solution, whose existence is part of the assumptions. Closely connected with this fact is the observation that the data A, b, c, f, and ε need not be real-valued: The data b, c, f, and ε may be complex-valued in which case Propositions 5.3.2, 5.3.4 still hold provided that ε in the definition of $\mathcal{E}$ and in the statement of the propositions is replaced with $|\varepsilon|$.
If $A(0) = \mathrm{Id}$, then any $\beta \in (0,1) \cap (1 - \pi/\omega, 1)$ may be chosen. We excluded the case $\beta = 0$ for technical convenience: Inspection of the proof shows that the $\beta = 0$ requires control of $\|\,|x|^{-1}\nabla u\|_{L^2(S_R(\omega)\cap B_{\mathcal{E}}(0))}$ by additional arguments. ▪

Proof: Let $R'' \in (R', R)$. We use the shorthand notation $S = S_{R''}(\omega)$, $S' = S_{R'}(\omega)$. Let $\mathcal{B} = \{B_i \,|\, i \in \mathbb{N}\}$ be the covering of S by balls $B_i = B_{r_i}(x_i)$, $r_i = c|x_i|$, given by Lemma 5.3.1. Denote by $\widehat{B}_i$ the "stretched balls" $\widehat{B}_i = B_{r_i/2}(x_i)$ as introduced in Lemma 5.3.1. An application of Lemma 4.2.17 with the covering $\mathcal{B}$ yields the existence of C, $K > 0$ independent of ε and i such that with the numbers $C(i)$ as given by Lemma 4.2.17 we have

$$\sum_{i\in\mathbb{N}} C^2(i) \le CC_f^2 \quad (5.3.20)$$

together with

$$\begin{aligned}
\|\nabla^p f\|_{L^2(S\cap B_i)} &= \frac{1}{\hat{\Phi}_{p,\beta,\mathcal{E}}(x_i)}\hat{\Phi}_{p,\beta,\mathcal{E}}(x_i)\|\nabla^p f\|_{L^2(S\cap B_i)} \qquad (5.3.21)\\
&\le \frac{1}{\hat{\Phi}_{p,\beta,\mathcal{E}}(x_i)} CK^p C(i)\max\{p+1,\mathcal{E}^{-1}\}^p\\
&\le \frac{1}{\hat{\Phi}_{0,\beta,\mathcal{E}}(x_i)}\frac{1}{\hat{\Phi}_{p,0,\mathcal{E}}(x_i)} CK^p C(i)\max\{p+1,\mathcal{E}^{-1}\}^p\\
&\le \frac{1}{\hat{\Phi}_{0,\beta,\mathcal{E}}(x_i)} CK^p C(i)\max\{(p+1)/|x_i|,\mathcal{E}^{-1}\}^p,
\end{aligned}$$

where, in the last estimate, we appealed to (4.2.6) of Lemma 4.2.2. We note that for all i, the set $B_i \cap S$ is either a full ball or a half ball in which case u vanishes on the straight part of $\partial(B_i \cap S)$. Hence Propositions 5.5.1, 5.5.2 yield the existence of C, $K > 0$ independent of ε such that

$$\left(\frac{r_i}{2}\right)^{p+2}\|\nabla^{p+2}u\|_{L^2(\widehat{B}_i\cap S)} \le C_u(i)K^{p+2}\max\{(p+3), r_i/\mathcal{E}\}^{p+2} \qquad \forall p\in\mathbb{N}_0,$$

where we abbreviate

$$\begin{aligned}
C_u(i) &:= \min\{1, r_i/\mathcal{E}\}\mathcal{E}\|\nabla u\|_{L^2(B_i\cap S)}\\
&\quad + \min\{1, r_i/\mathcal{E}\}^2(\mathcal{E}/\varepsilon)^2\left[C_c\|u\|_{L^2(B_i\cap S)} + \frac{C(i)}{\hat{\Phi}_{0,\beta,\mathcal{E}}(x_i)}\right].
\end{aligned}$$

We bound with (4.2.8) of Lemma 4.2.2

$$\begin{aligned}
&\|\hat{\Phi}_{p,\beta,\mathcal{E}}\nabla^{p+2}\|^2_{L^2(S')} \le CK^p\sum_{i=1}^{\infty}\hat{\Phi}^2_{p,\beta,\mathcal{E}}(x_i)\|\nabla^{p+2}u\|^2_{L^2(\widehat{B}_i\cap S')}\\
&\le CK^{2(p+2)}\sum_{i=1}^{\infty}\Phi^2_{p,\beta,\mathcal{E}}(x_i)r_i^{-2(p+2)}\max\{p+3, r_i/\mathcal{E}\}^{2(p+2)}C_u^2(i).
\end{aligned}$$

We estimate further using (4.2.7) of Lemma 4.2.2

$$\begin{aligned}
&\Phi^2_{p,\beta,\mathcal{E}}(x_i)r_i^{-2(p+2)}\max\{p+3, r_i/\mathcal{E}\}^{2(p+2)}\\
&\le CK^{2(p+2)}\hat{\Phi}^2_{0,\beta,\mathcal{E}}(x_i)\hat{\Phi}^2_{p,0,\mathcal{E}}(x_i)\max\{(p+1)/r_i,\mathcal{E}^{-1}\}^{2p}\max\{(p+1)/r_i,\mathcal{E}^{-1}\}^4\\
&\le K^{2p}\max\{p+1,\mathcal{E}^{-1}\}^{2p}\hat{\Phi}^2_{0,\beta,\mathcal{E}}(x_i)\max\{(p+1)/r_i,\mathcal{E}^{-1}\}^4,\\
&\le K^{2p}\max\{p+1,\mathcal{E}^{-1}\}^{2p}\hat{\Phi}^2_{0,\beta,\mathcal{E}}(x_i)\max\{1/r_i,\mathcal{E}^{-1}\}^4.
\end{aligned}$$

Hence, we obtain

$$\begin{aligned}
&\|\hat{\Phi}_{p,\beta,\mathcal{E}}\nabla^{p+2}u\|^2_{L^2(S')} \le\\
&\quad CK^{2p}\max\{p+1,\mathcal{E}^{-1}\}^{2p}\sum_{i=1}^{\infty}\hat{\Phi}^2_{0,\beta,\mathcal{E}}(x_i)\max\{1/r_i,\mathcal{E}^{-1}\}^4C_u^2(i). \qquad (5.3.22)
\end{aligned}$$

Recognizing $\min\{1, r_i/\mathcal{E}\} \leq C\hat{\Phi}_{1,0,\mathcal{E}}(x_i)$ and using $\hat{\Phi}_{2,\beta,\mathcal{E}} \leq \hat{\Phi}_{1,\beta,\mathcal{E}}$, $\hat{\Phi}_{2,0,\mathcal{E}} \leq (\hat{\Phi}_{1,0,\mathcal{E}})^2$, we can estimate

$$\hat{\Phi}_{0,\beta,\mathcal{E}}(x_i)C_u(i) \leq C\Big(\hat{\Phi}_{1,\beta,\mathcal{E}}(x_i)\mathcal{E}\|\nabla u\|_{L^2(B_i\cap S)} + C_c(\mathcal{E}/\varepsilon)^2\hat{\Phi}_{1,\beta,\mathcal{E}}(x_i)\|u\|_{L^2(B_i\cap S)} + (\mathcal{E}/\varepsilon)^2 C(i)\hat{\Phi}^2_{1,0,\mathcal{E}}(x_i)\Big).$$

Upon setting

$$W_{i,\beta} := \max\{1/r_i, \mathcal{E}^{-1}\}^2\hat{\Phi}_{1,\beta,\mathcal{E}}(x_i),$$
$$W_{i,0} := \max\{1/r_i, \mathcal{E}^{-1}\}^2\hat{\Phi}^2_{1,0,\mathcal{E}}(x_i),$$

we see that we have

$$\sum_{i=1}^{\infty}\hat{\Phi}^2_{0,\beta,\mathcal{E}}(x_i)\max\{1/r_i,\mathcal{E}^{-1}\}^4C_u^2(i) \leq \mathcal{E}^2\sum_{i=1}^{\infty}W^2_{i,\beta}\|\nabla u\|^2_{L^2(B_i\cap S)} + C_c^2(\mathcal{E}/\varepsilon)^4\sum_{i=1}^{\infty}W^2_{i,\beta}\|u\|^2_{L^2(B_i\cap S)} + (\mathcal{E}/\varepsilon)^4\sum_{i=1}^{\infty}W^2_{i,0}C^2(i).$$

Estimating these three sums with the aid of Lemma 5.3.6, and inserting the result into (5.3.22) gives

$$\|\hat{\Phi}_{p,\beta,\mathcal{E}}\nabla^{p+2}u\|^2_{L^2(S')} \leq CK^{2p}\max\{p+1,\mathcal{E}^{-1}\}^{2p}\times \Big\{\|\hat{\Phi}_{0,\beta,\mathcal{E}}\nabla^2 u\|^2_{L^2(S)} + \mathcal{E}^{-2}\|\nabla u\|^2_{L^2(S)} + C_c^2(\mathcal{E}/\varepsilon)^4\mathcal{E}^{-2}\|\nabla u\|^2_{L^2(S)} + C_c^2(\mathcal{E}/\varepsilon)^4\mathcal{E}^{-4}\|u\|^2_{L^2(S)} + C_f^2(\mathcal{E}/\varepsilon)^4\mathcal{E}^{-4}\Big\}.$$

Upon simplifying $C_c(\mathcal{E}/\varepsilon)^2 \leq 1$ and using Proposition 5.3.2 to bound the term $\|\hat{\Phi}_{0,\beta,\mathcal{E}}\nabla^2 u\|_{L^2(S)}$ (note that $S = S_{R''}(\omega)$ with $R'' < R$) we arrive at

$$\|\hat{\Phi}_{p,\beta,\mathcal{E}}\nabla^{p+2}u\|^2_{L^2(S')} \leq CK^{2p}\max\{p+1,\mathcal{E}^{-1}\}^{2p}\mathcal{E}^{-4}\times \Big\{C_f^2(\mathcal{E}/\varepsilon)^2 + \mathcal{E}^2\|\nabla u\|^2_{L^2(S)} + C_c^2(\mathcal{E}/\varepsilon)^2\|u\|^2_{L^2(S)}\Big\}.$$

Bounding $\max\{p+1,\mathcal{E}^{-1}\}^{2p}\mathcal{E}^{-4} \leq \max\{p+1,\mathcal{E}^{-1}\}^{2(p+2)}$ finishes the proof. □

Lemma 5.3.6. *Let $R > R' > 0$, $\omega \in (0, 2\pi)$, $\beta \in (0,1)$. Let $\mathcal{B} = \{B_i \,|\, i \in \mathbb{N}\}$, $B_i = B_{c|x_i|}(x_i)$, be the covering of $S_{R'}(\omega)$ given by Lemma 5.3.1. Then there exists $C > 0$ such that for every $\mathcal{E} \in (0,1]$ and every $u \in H^{2,2}_{\beta,\mathcal{E}}(S_R(\omega))$*

$$\sum_{i\in\mathbb{N}}W^2_{i,\beta}\|\nabla u\|^2_{L^2(B_i\cap S_R(\omega))} \leq \Big\{\mathcal{E}^{-2}\|\hat{\Phi}_{0,\beta,\mathcal{E}}\nabla^2 u\|^2_{L^2(S_R(\omega))} + \mathcal{E}^{-4}\|\hat{\Phi}_{0,\beta,\mathcal{E}}\nabla u\|^2_{L^2(S_R(\omega))}\Big\},$$

$$\sum_{i\in\mathbb{N}}W^2_{i,\beta}\hat{\Phi}^2_{1,0,\mathcal{E}}(x_i)\|\nabla u\|^2_{L^2(B_i\cap S_R(\omega))} \leq C\mathcal{E}^{-4}\|\hat{\Phi}_{0,\beta,\mathcal{E}}\nabla u\|^2_{L^2(S_R(\omega))},$$

$$\sum_{i\in\mathbb{N}}W^2_{i,\beta}\|u\|^2_{L^2(B_i\cap S_R(\omega))} \leq C\Big\{\mathcal{E}^{-2}\|\hat{\Phi}_{0,\beta,\mathcal{E}}\nabla u\|^2_{L^2(S_R(\omega))} + \mathcal{E}^{-4}\|u\|^2_{L^2(S_R(\omega))}\Big\},$$

where we set

$$W_{i,\beta} := \max\{1/|x_i|, \mathcal{E}^{-1}\}^2 \,\hat{\Phi}_{1,\beta,\mathcal{E}}(x_i).$$

Additionally, if numbers $C(i)$ *satisfy* $\sum_{i\in\mathbb{N}} C^2(i) \leq C_f^2 < \infty$*, then, upon setting* $W_{i,0} := \max\{1/|x_i|, \mathcal{E}^{-1}\}^2 \hat{\Phi}^2_{1,0,\mathcal{E}}(x_i)$*,*

$$\sum_{i\in\mathbb{N}} W_{i,0}^2 C^2(i) \leq C\mathcal{E}^{-4} C_f^2.$$

Proof: We abbreviate $S := S_R(\omega)$. With Lemma 4.2.2, we get

$$W_{i,\beta} = \max\{1/|x_i|, \mathcal{E}^{-1}\}^2 \hat{\Phi}_{1,\beta,\mathcal{E}}(x_i) \leq C\mathcal{E}^{-2}\frac{\hat{\Phi}_{1,\beta,\mathcal{E}}(x_i)}{\hat{\Phi}_{2,\beta,\mathcal{E}}(x_i)} \leq C\mathcal{E}^{-2}\hat{\Phi}_{-1,\beta,\mathcal{E}}(x_i).$$

Hence, we get exploiting (4.2.8) of Lemma 4.2.2

$$W_{i,\beta}^2 \|\nabla u\|^2_{L^2(B_i\cap S)} \leq C\mathcal{E}^{-4}\|\hat{\Phi}_{-1,\beta,\mathcal{E}}\nabla u\|^2_{L^2(B_i\cap S)}.$$

On summing over i, we obtain using the overlap properties of the covering $\mathcal{B}$

$$\sum_{i\in\mathbb{N}} W_{i,\beta}^2 \|\nabla u\|^2_{L^2(B_i\cap S)} \leq C\mathcal{E}^{-4}\|\hat{\Phi}_{-1,\beta,\mathcal{E}}\nabla u\|^2_{L^2(S)}.$$

Writing $\|\hat{\Phi}_{-1,\beta,\mathcal{E}}\nabla u\|^2_{L^2(S)} = \|\hat{\Phi}_{-1,\beta,\mathcal{E}}\nabla u\|^2_{L^2(S\cap B_{2\mathcal{E}}(0))} + \|\hat{\Phi}_{-1,\beta,\mathcal{E}}\nabla u\|^2_{L^2(S\setminus B_{2\mathcal{E}}(0))}$ and using Corollary 4.2.11 for the first integral and the fact $\hat{\Phi}_{-1,\beta,\mathcal{E}}(x) = 1$ for $x \in S\setminus B_{2\mathcal{E}}(0)$ we get

$$\sum_{i\in\mathbb{N}} W_{i,\beta}^2 \|\nabla u\|^2_{L^2(B_i\cap S)} \leq C\mathcal{E}^{-2}\|\hat{\Phi}_{0,\beta,\mathcal{E}}\nabla^2 u\|^2_{L^2(S\cap B_{2\mathcal{E}}(0))} + C\mathcal{E}^{-4}\|\nabla u\|^2_{L^2(S)},$$

from which the desired bound for the sum of gradients follows. The second estimate follows easily since $W_{i,\beta}\hat{\Phi}_{1,0,\mathcal{E}}(x_i) \leq C\mathcal{E}^{-2}\hat{\Phi}_{0,\beta,\mathcal{E}}(x_i)$.
For the third estimate, we proceed completely analogously to arrive at

$$\sum_{i\in\mathbb{N}} W_{i,\beta}^2 \|u\|^2_{L^2(B_i\cap S)} \leq C\mathcal{E}^{-4}\|\hat{\Phi}_{-1,\beta,\mathcal{E}} u\|^2_{L^2(S)}.$$

Writing $\|\hat{\Phi}_{-1,\beta,\mathcal{E}} u\|^2_{L^2(S)} = \|\hat{\Phi}_{-1,\beta,\mathcal{E}} u\|^2_{L^2(S\cap B_{2\mathcal{E}}(0))} + \|\hat{\Phi}_{-1,\beta,\mathcal{E}} u\|^2_{L^2(S\setminus B_{2\mathcal{E}}(0))}$ and using Part (ii) of Lemma 4.2.10 for the first integral and the fact $\hat{\Phi}_{-1,\beta,\mathcal{E}}(x) = 1$ for $x \in S\setminus B_{2\mathcal{E}}(0)$ we get

$$\sum_{i\in\mathbb{N}} W_{i,\beta}^2 \|u\|^2_{L^2(B_i\cap S)} \leq C\mathcal{E}^{-2}\|\hat{\Phi}_{0,\beta,\mathcal{E}}\nabla u\|^2_{L^2(S\cap B_{2\mathcal{E}}(0))} + C\mathcal{E}^{-4}\|u\|^2_{L^2(S)},$$

which is the desired bound.
The last claim of the lemma follows from the fact that by Lemma 4.2.2 $\max\{1/|x_i|, \mathcal{E}^{-1}\}^4 \hat{\Phi}^4_{1,0,\mathcal{E}}(x_i) \leq C\mathcal{E}^{-4}$. □

5.3.2 Regularity in curvilinear polygons

In the present subsection Ω denotes a fixed curvilinear polygon as defined in Section 1.2. For $\beta \in [0,1)^J$ and $\varepsilon > 0$ we can define the spaces $H^{l,m}_{\beta,\varepsilon}(\Omega)$ and the countably normed spaces $\mathcal{B}^l_{\beta,\varepsilon}(\Omega, C, \gamma)$ analogously to the way they are defined on sectors as the completions of the space $C^\infty(\Omega)$ under weighted norms with weight functions $\Phi_{p,\beta,\varepsilon}$ instead of $\hat{\Phi}_{p,\beta,\varepsilon}$, where

$$\Phi_{p,\beta,\varepsilon}(x) := \Pi_{j=1}^J \widehat{\Phi}_{p,\beta_j,\varepsilon}(x - A_j). \tag{5.3.23}$$

Our variational setting is based on the energy spaces $H^1_{0,\varepsilon}(\Omega)$: For $\varepsilon > 0$, let $H^1_{0,\varepsilon}(\Omega)$ be the space $H^1_0(\Omega)$ equipped with the energy norm $\|\cdot\|_\varepsilon$ given by (1.2.7), i.e., $\|u\|^2_\varepsilon := \varepsilon^2\|\nabla u\|^2_{L^2(\Omega)} + \|u\|^2_{L^2(\Omega)}$. We have

Lemma 5.3.7. *Let Ω be a curvilinear polygon. Then for all $\varepsilon > 0$ and all $\beta \in [0,1]^J$ there holds $H^{0,0}_{\beta,\varepsilon}(\Omega) \subset \left(H^1_{0,\varepsilon}(\Omega)\right)'$. Furthermore, there exists $C > 0$ independent of $\varepsilon \in (0,1]$ such that*

$$\|f\|_{(H^1_{0,\varepsilon}(\Omega))'} \leq C\|f\|_{H^{0,0}_{1,\varepsilon}(\Omega)} \qquad \forall f \in H^{0,0}_{1,\varepsilon}(\Omega).$$

Proof: From (4.2.6) of Lemma 4.2.2 we get the existence of $C > 0$ independent of ε such that

$$\Phi^{-1}_{0,1,\varepsilon}(x) \leq C\left(1 + \frac{\varepsilon}{\operatorname{dist}(x, \{A_j \,|\, 1 \leq j \leq J\})}\right) \leq C\left(1 + \frac{\varepsilon}{r}\right), \qquad r = \operatorname{dist}(x, \partial\Omega).$$

We then calculate for $v \in H^1_{0,\varepsilon}(\Omega)$

$$\begin{aligned}\left|\int_\Omega fv\,dx\right| &\leq \|\Phi_{0,1,\varepsilon} f\|_{L^2(\Omega)}\|\Phi^{-1}_{0,1,\varepsilon} v\|_{L^2(\Omega)}\\ &\leq C\|\Phi_{0,1,\varepsilon} f\|_{L^2(\Omega)}\left(\|v\|_{L^2(\Omega)} + \varepsilon\|\frac{1}{r}v\|_{L^2(\Omega)}\right)\\ &\leq C\|\Phi_{0,1,\varepsilon} f\|_{L^2(\Omega)}\|v\|_\varepsilon,\end{aligned}$$

where, in the last step we employed the embedding in weighted Sobolev spaces $\|\frac{1}{r}v\|_{L^2(\Omega)} \leq C\|\nabla u\|_{L^2(\Omega)}$ of Lemma 4.2.12. □

We now consider the following boundary value problem:

$$-\varepsilon^2\nabla\cdot(A(x)\nabla u) + b(x)\cdot\nabla u + cu = f \quad \text{on } \Omega, \tag{5.3.24a}$$

$$u = 0 \quad \text{on } \partial\Omega. \tag{5.3.24b}$$

The coefficients A, b, c are assumed to satisfy: $A \in W^{1,\infty}(\Omega, \mathbb{S}^2_>)$, $b \in C^0(\Omega)$ and b piecewise in $W^{1,\infty}$, $c \in L^\infty(\Omega)$. Furthermore, we assume that

$$0 < \lambda_{\min} \leq A(x) \qquad \text{on } \Omega, \tag{5.3.25a}$$

$$\|b\|_{L^\infty(\Omega)} \leq C_b, \tag{5.3.25b}$$

$$\|c\|_{L^\infty(\Omega)} \leq C_c, \tag{5.3.25c}$$

$$0 < \mu \leq c - \frac{1}{2}\nabla\cdot b \qquad \text{a.e. on } \Omega. \tag{5.3.25d}$$

As in the case of solutions on sectors, is convenient to introduce the notion of the *relative diffusivity* $\mathcal{E}$ via

$$\mathcal{E}^{-1} := \frac{C_b}{\varepsilon^2} + \frac{C_c}{\varepsilon} + 1.$$

The coercivity statement of Lemma 1.2.2 implies that the linear form B_ε is coercive and continuous on the spaces $H^1_{0,\varepsilon}(\Omega) \times H^1_{0,\varepsilon}(\Omega)$. Hence, the variational formulation

$$\text{Find } u_\varepsilon \in H^1_{0,\varepsilon}(\Omega) \text{ s.t. } B_\varepsilon(u_\varepsilon, v) = F(v) \qquad \forall v \in H^1_{0,\varepsilon}(\Omega)$$

has a unique solution for any right-hand side $F \in \left(H^1_{0,\varepsilon}(\Omega)\right)'$. In particular, as in the proof of Lemma 1.2.2, we get the existence of $C > 0$ independent of ε such that

$$\|u\|_\varepsilon \leq C\|F\|_{\left(H^1_{0,\varepsilon}(\Omega)\right)'}. \tag{5.3.26}$$

This observation allows us to formulate the following theorem.

Theorem 5.3.8 (shift theorem in curvilinear polygons). *Let Ω be a curvilinear polygon, A, b, c satisfy (5.3.25). Let $\mathcal{E}$ be defined by (5.3.8). Then there exists $C > 0$ independent of $\varepsilon \in (0,1]$ such that for each $f \in H^{0,0}_{1,\varepsilon}(\Omega)$ the (weak) solution u of (5.3.24) satisfies*

$$\|u\|_\varepsilon \leq C\|f\|_{H^{0,0}_{1,\varepsilon}(\Omega)}.$$

Furthermore, if the data A, b, c satisfy additionally (1.2.2a), (1.2.2b), (1.2.2c), then there exists $\beta \in [0,1)^J$ such that for $f \in H^{0,0}_{\beta,\mathcal{E}}(\Omega)$ the solution u satisfies $u \in H^2_{loc}(\Omega)$, and there exists $C > 0$ independent of $\varepsilon \in (0,1]$ and f such that

$$\|\Phi_{0,\beta,\mathcal{E}}\nabla^2 u\|_{L^2(\Omega)} \leq$$
$$C\mathcal{E}^{-2}\left[(\mathcal{E}/\varepsilon)^2\|f\|_{H^{0,0}_{\beta,\mathcal{E}}(\Omega)} + (\mathcal{E}/\varepsilon)\,\varepsilon\|\nabla u\|_{L^2(\Omega)} + C_c(\mathcal{E}/\varepsilon)^2\|u\|_{L^2(\Omega)}\right].$$

Proof: The first assertion of the theorem follows readily from the coercivity statement of Lemma 1.2.2 and (5.3.26). For polygons, the second assertion follows from Proposition 5.3.2. For the general case of curvilinear polygons, we require additionally a mapping argument. These arguments are provided by Corollary 4.2.21. Strictly speaking, Corollary 4.2.21 is formulated for analytic functions; inspection of the proof, however, shows that an analogous result holds in the present $H^{2,2}_{\beta,\mathcal{E}}$-setting. □

Remark 5.3.9 In typical applications, one has $\mathcal{E} \leq C\varepsilon$ (cf. the definition of $\mathcal{E}$ in (5.3.8)). Lemma 4.2.3 then implies the existence of $C > 0$ independent of ε and $\beta \in [0,1]^J$ such that

$$\Phi_{0,1,\varepsilon} \leq \Phi_{0,\beta,\varepsilon} \leq C\Phi_{0,\beta,\mathcal{E}}.$$

Theorem 5.3.8 then implies that for $f \in H^{0,0}_{\beta,\mathcal{E}}(\Omega)$ we have

$$\|\Phi_{0,\beta,\mathcal{E}}\nabla^2 u\|_{L^2(\Omega)} \le C\mathcal{E}^{-2}(\mathcal{E}/\varepsilon)\|f\|_{H^{0,0}_{\beta,\mathcal{E}}(\Omega)}.$$

■

If the data A, b, c, f are analytic on Ω, then the variational solution u of (5.3.24) is analytic on Ω as well. We assume that the data $A \in \mathcal{A}(\Omega, \mathbb{S}^2_>)$, $b \in \mathcal{A}(\Omega, \mathbb{R}^2)$, $c \in \mathcal{A}(\Omega)$, satisfy (5.3.25) and (1.2.2). Then we can then formulate

Theorem 5.3.10 (analytic regularity in curvilinear polygons). *Let Ω be a curvilinear polygon, and let A, b, c satisfy (1.2.2). Then there exists $\beta \in [0,1)^J$ depending only on Ω and A such that the following holds: For each $f \in \mathcal{B}^0_{\beta,\mathcal{E}}(\Omega, C_f, \gamma_f)$ the solution u of (5.3.24) is analytic on Ω, and there are C, $K > 0$ depending only on γ_f, Ω, and the constants of (1.2.2) such that $\forall p \in \mathbb{N}_0$*

$$\|u\|_\varepsilon \le CC_f,$$
$$\|\Phi_{p,\beta,\mathcal{E}}\nabla^{p+2}u\|_{L^2(\Omega)} \le CK^{p+2}\max\{p,\mathcal{E}^{-1}\}^{p+2} \times \left[C_f(\mathcal{E}/\varepsilon)^2 + (\mathcal{E}/\varepsilon)\varepsilon\|\nabla u\|_{L^2(\Omega)} + C_c(\mathcal{E}/\varepsilon)^2\|u\|_{L^2(\Omega)}\right].$$

Proof: The proof follows readily by combining Theorem 5.3.8, Proposition 5.3.4, and mapping arguments provided by Corollary 4.2.21. □

Remark 5.3.11 Several comment concerning Theorem 5.3.10 are in order. Firstly, piecewise analyticity of the data A, b, c, f is sufficient. Then, of course, the solution u is piecewise analytic and the assertion concerning the growth of the derivatives of u has to be understood in a piecewise sense. Secondly, as observed in Remark 5.3.5, it is not necessary to require the coercivity condition (1.2.2e) and that the coefficients b, c, and ε be real-valued for the weighted H^2 bound in Theorem 5.3.8 and the second bound in the statement of Theorem 5.3.10 to hold. These bounds also hold if b, c, f, and ε are complex-valued provided that ε is replaced with $|\varepsilon|$ in these statements. ■

Theorem 5.3.10 covers two particular cases of interest: the case of a reaction-diffusion equation where $C_b = 0$ and $C_c = O(1)$ on the one hand and the case of a convection-diffusion equation where $C_b = O(1)$. In the former case, $\mathcal{E} = O(\varepsilon)$ whereas in the latter case $\mathcal{E} = O(\varepsilon^2)$. For future reference, we collect this observation in the following two corollaries:

Corollary 5.3.12 (reaction-diffusion). *Let Ω be a curvilinear polygon, A, b, c satisfy (1.2.2) and assume that $b \equiv 0$, $g \equiv 0$. Then there exists $\beta \in [0,1)^J$ depending only on Ω and A such that for every $f \in \mathcal{B}^0_{\beta,\varepsilon}(\Omega, C_f, \gamma_f)$ there exist C, $K > 0$ depending only on Ω, γ_f, and the constants of (1.2.2) such that the solution u of (1.2.11) is in $\mathcal{B}^2_{\beta,\varepsilon}(\Omega, CC_f, K)$.*

Corollary 5.3.13 (convection-diffusion). *Let Ω be a curvilinear polygon, A, b, c satisfy (1.2.2), $g \equiv 0$, and assume that $C_b \geq \underline{c} > 0$. Then there exists $\beta \in [0,1)^J$ depending only on Ω and A such that for every $f \in \mathcal{B}^0_{\beta,\varepsilon^2}(\Omega, C_f, \gamma_f)$ there exist C, $K > 0$ depending only on Ω, γ_f, $\underline{c}$, and the constants of (1.2.2) such that the solution u of (1.2.1) satisfies:*

$$\|u\|_{L^2(\Omega)} + \varepsilon\|\nabla u\|_{L^2(\Omega)} \leq CC_f,$$
$$\|\Phi_{p,\beta,\varepsilon^2}\nabla^{p+2}u\|_{L^2(\Omega)} \leq CC_f K^{p+2}\max\{p,\varepsilon^{-2}\}^{p+2}\varepsilon \qquad \forall p \in \mathbb{N}_0.$$

So far, we assumed homogeneous boundary conditions. The regularity theory can be extended to piecewise analytic (Dirichlet) boundary data.

Theorem 5.3.14 (non-homogeneous Dirichlet conditions). *Let Ω be a curvilinear polygon, let the data A, b, c satisfy (1.2.2), and let $g \in C^0(\partial\Omega)$ be piecewise analytic, i.e., g satisfies (1.2.4). Let $\mathcal{E}$ be defined by (5.3.8). Then there exists $\beta \in [0,1)^J$ such that for each $f \in \mathcal{B}^0_{\beta,\mathcal{E}}(\Omega, C_f, \gamma_f)$ the solution u of (1.2.1) is analytic on Ω and there exist C, $K > 0$ depending only on Ω, f, g, and the constants of (1.2.2) such that*

$$\|u\|_\varepsilon \leq C,$$
$$\|\hat{\Phi}_{p,\beta,\mathcal{E}}\nabla^{p+2}u\|_{L^2(\Omega)} \leq CK^{p+2}\max\{p+1,\mathcal{E}^{-1}\}^{p+2}\left[(\mathcal{E}/\varepsilon) + (\mathcal{E}/\varepsilon)^2\right].$$

Proof: Introduce the auxiliary function u_0 as the solution of the Dirichlet problem: $-\nabla \cdot (A(x)\nabla u) = 0$ on Ω and $u_0 = g$ on $\partial\Omega$. Then, for some $\beta \in [0,1)^J$ we have from [15, 16] that $u_0 \in \mathcal{B}^2_{\beta,1}(\Omega, C_g, \gamma_g)$ with C_g, $\gamma_g > 0$ depending only on Ω and g. From Proposition 4.2.13, we immediately get that $u \in \mathcal{B}^2_{\beta,\mathcal{E}}(\Omega, C, \gamma)$ for some C, $\gamma > 0$ independent of ε. Additionally Proposition 4.2.13 implies

$$\|\Phi_{p,\beta,\mathcal{E}}\nabla^{p+2}u_0\|_{L^2(\Omega)} \leq C\gamma^p\max\{p+1,\mathcal{E}^{-1}\}^{p+2}\mathcal{E}^{2-\beta_{max}} \qquad \forall p \in \mathbb{N}_0,$$

where $\beta_{max} = \max_{j=1,\dots,J}\beta_j < 1$. Hence, u_0 satisfies the desired estimates since $\varepsilon \leq 1$. It remains to see that the function $\widetilde{u} := u - u_0$ satisfies them as well. The function $\widetilde{u}$ solves

$$-\varepsilon^2\nabla \cdot (A(x)\nabla\widetilde{u}) + b(x)\cdot\nabla\widetilde{u} + c(x)\widetilde{u} = f + b(x)\cdot\nabla u_0 + c(x)u_0 \qquad \text{on } \Omega,$$
$$\widetilde{u} = 0 \quad \text{on } \partial\Omega.$$

It suffices to check that $\widetilde{f} := b(x)\cdot\nabla u_0 + c(x)u_0 \in \mathcal{B}^0_{\beta,\mathcal{E}}(\Omega, C, \gamma)$ for some C, γ independent of ε. As $u_0 \in \mathcal{B}^2_{\beta,1}$, we get that $\widetilde{f} \in \mathcal{B}^1_{\beta,1}$. From the last assertion of Proposition 4.2.13, we infer that $\widetilde{f} \in \mathcal{B}^1_{\beta,\mathcal{E}}(\Omega, C, \gamma)$ for some constants C, $\gamma > 0$ independent of ε. Appealing now to the second statement of Proposition 4.2.13 allows us to conclude $\widetilde{f} \in \mathcal{B}^0_{\beta,\mathcal{E}}(\Omega, C, \gamma)$ for some constants C, $\gamma > 0$ independent of ε. The result now follows from Theorem 5.3.10. □

Remark 5.3.15 The Dirichlet data g are piecewise analytic. This is, of course, not the weakest possible assumption. The appropriate setting for the boundary data are trace spaces $\mathcal{B}^{3/2}_{\beta,\varepsilon}$, defined as the traces of the elements of the spaces $\mathcal{B}^2_{\beta,\varepsilon}$. The question of an "intrinsic" characterization of the spaces $\mathcal{B}^{3/2}_{\beta,\varepsilon}$ then arises. For the case $\varepsilon = 1$, such intrinsic characterizations are given in [16]. ▪

5.4 Neumann and transmission problems

We discussed in the preceding section the specific case of Dirichlet boundary conditions. The techniques employed, however, are also amenable to treating other kinds of boundary conditions. In the present section, we state the regularity assertions for Neumann boundary conditions and for transmission conditions.

5.4.1 Neumann and Robin corners

On sectors $S_R(\omega)$, we are interested in the regularity of solutions u to the following equation

$$-\varepsilon^2\nabla\cdot(A(x)\nabla u)+b(x)\cdot\nabla u+c(x)u=f(x)\quad\text{on } S_R(\omega), \tag{5.4.27a}$$
$$\varepsilon^2\partial_{n_A}u=\varepsilon(G_{1,0}+G_{2,0}u)\quad\text{on } \Gamma_0, \tag{5.4.27b}$$
$$\varepsilon^2\partial_{n_A}u=\varepsilon(G_{1,\omega}+G_{2,\omega}u)\quad\text{on } \Gamma_\omega, \tag{5.4.27c}$$

where we defined the lateral boundary $\Gamma_0\cup\Gamma_\omega$ by

$$\Gamma_0=\{(r,0)\,|\,0<r<R\},\qquad \Gamma_\omega=\{(r\cos\omega,r\sin\omega)\,|\,0<r<R\}. \tag{5.4.28}$$

In some cases, it will be convenient to combine Γ_0 and Γ_ω into the lateral boundary Γ_{lat} given by

$$\Gamma_{lat}:=\Gamma_0\cup\Gamma_\omega=\partial S_R(\omega)\setminus\partial B_R(0). \tag{5.4.29}$$

In order to formulate conditions on the functions $G_{1,0}$, $G_{2,0}$, $G_{1,\omega}$, $G_{2,\omega}$, it will be convenient to introduce *conical neighborhoods* $S_R^{0,\delta}(\omega)$, $S_R^{\omega,\delta}(\omega)$ of Γ_0, Γ_ω by

$$S_R^{0,\delta}(\omega):=\{(r\cos\varphi,r\sin\varphi)\,|\,0<r<R,\quad 0<\varphi<\delta\} \tag{5.4.30a}$$
$$S_R^{\omega,\delta}(\omega):=\{(r\cos\varphi,r\sin\varphi)\,|\,0<r<R,\quad \omega-\delta<\varphi<\omega\}. \tag{5.4.30b}$$

We note that these conical neighborhoods are sectors in the sense of Definition 4.2.1. The minimal assumptions on the functions $G_{1,0}$, $G_{2,0}$, $G_{1,\omega}$, $G_{2,\omega}$, that we consider are that $G_{2,0}$, $G_{2,\omega}$ are Lipschitz functions on the lines Γ_0, Γ_ω, respectively, and that, for some $\delta>0$, the functions $G_{1,0}$, $G_{1,\omega}$ are traces on the lines Γ_0, Γ_ω of functions (again denoted $G_{1,0}$, $G_{1,\omega}$) $G_{1,0}\in H^{1,1}_{\beta,\mathcal{E}}(S_R^{0,\delta}(\omega))$, $G_{1,\omega}\in H^{1,1}_{\beta,\mathcal{E}}(S_R^{\omega,\delta}(\omega))$.
We note the following lemma, which states that the functions $G_{1,0}$, $G_{1,\omega}$, $G_{2,0}$, $G_{2,\omega}$ may be viewed as the traces of functions defined on $S_R(\omega)$:

Lemma 5.4.1. *Let $R>0$, $\omega\in(0,2\pi)$, $\beta\in(0,1)$, $\delta>0$. Let $\chi\in C^\infty(\mathbb{R})$ with $\chi(0)=1$ and* $\operatorname{supp}\chi\subset B_{\omega/2}(0)\cap B_\delta(0)$. *Then there exists $C>0$ such that for every $\mathcal{E}\in(0,1]$ and for functions $G_{2,0}$, $G_{2,\omega}\in W^{1,\infty}((0,R))$ and $G_{1,0}\in S_R^{0,\delta}(\omega)$, $G_{1,\omega}\in S_R^{\omega,\delta}(\omega)$, the functions G_1, G_2 defined in polar coordinates by*

$$G_1(r\cos\varphi,r\sin\varphi):=G_{1,0}(r\cos\varphi,r\sin\varphi)\chi(\varphi)+G_{1,\omega}(r\cos\varphi,r\sin\varphi)\chi(\varphi-\omega),$$
$$G_2(r\cos\varphi,r\sin\varphi):=G_{2,0}(r)\chi(\varphi)+G_{2,\omega}(r)\chi(\varphi-\omega)$$

satisfy

$$\|G_2\|_{L^\infty(S_R(\omega))} + \| \, |x|\nabla G_2\|_{L^\infty(S_R(\omega))} \le C \sum_{k\in\{0,\omega\}} \|G_{2,k}\|_{W^{1,\infty}((0,R))},$$

$$\|G_1\|_{H^{1,1}_{\beta,\varepsilon}(S_R(\omega))} \le C \sum_{k\in\{0,\omega\}} \|G_{1,k}\|_{H^{1,1}_{\beta,\varepsilon}(S_R^{k,\delta}(\omega))}.$$

Proof: The estimate for G_2 is straight forward. The estimate for G_1 follows from Part (ii) of Lemma 4.2.10. □

We start with a lemma that is analogous to case of the Dirichlet problem, treated in Proposition 5.2.2:

Lemma 5.4.2. *Let $R > R' > 0$, $\omega \in (0, 2\pi)$, and $A \in C^1(\overline{S_R(\omega)}, \mathbb{S}^2_>)$ with $0 < \lambda_{min} \le A$ on $S_R(\omega)$ be given. Then there exist $\beta \in [0,1)$, which depends only on ω and $A(0)$, and a constant $C > 0$ with the following properties: For each $f \in H^{0,0}_{\beta,1}(S_R(\omega))$, $G \in H^{1,1}_{\beta,1}(S_R(\omega))$, the solutions u_i, $i \in \{1,2\}$, of the problems*

$$-\nabla \cdot (A\nabla u_i) = f \quad \text{on } S_R(\omega)$$

with $u_1 = u_2 = 0$ on $\partial S_R(\omega) \setminus \Gamma_{lat}$ and lateral boundary conditions

$$\begin{aligned} \partial_{n_A} u_1 &= G \quad \text{on } \Gamma_0 \cup \Gamma_\omega, \\ \partial_{n_A} u_2 &= G \quad \text{on } \Gamma_0 \text{ and } u_2 = 0 \text{ on } \Gamma_\omega, \end{aligned}$$

satisfy $u_i \in H^2_{loc}(S_R(\omega))$ and

$$\|u_i\|_{H^{2,2}_{\beta,1}(S_{R'}(\omega))} \le C \left[\|f\|_{H^{0,0}_{\beta,1}(S_R(\omega))} + \|G\|_{H^{1,1}_{\beta,1}(S_R(\omega))} \right].$$

Proof: The case $A = \mathrm{Id}$ is proved in [15]. The case of variable A is handled by the same type of perturbation argument as in the proof of Proposition 5.2.2. We will therefore merely outline the main ingredients and use the notation of Proposition 5.2.2 concerning the covering $\mathcal{B}$ by balls. For definiteness' sake, we will consider the Neumann problem, i.e., the regularity of u_1. In the remainder of the proof, we will simply write u instead of u_1.
By [15, Lemma 3.1], for every every $\beta \in [0,1)$ there exists $C_\beta > 0$ such that

$$\|u\|_{H^1(S_R(\omega))} \le C \left[\|f\|_{H^{0,0}_{\beta,1}(S_R(\omega))} + \|G\|_{H^{1,1}_{\beta,1}(S_R(\omega))} \right].$$

By elliptic regularity, viz., (5.5.22) of Lemma 5.5.12 for the case of balls B_i completely contained in $S_R(\omega)$ and Lemma 5.5.26 for the case of $B_i \cap S_R(\omega)$ being a half-ball yields

$$\begin{aligned} r_i^2 \|\nabla^2 u\|_{L^2(\widehat{B}_i \cap S_{R'}(\omega))} \le C \Big(& r_i^2 \|f\|_{L^2(B_i \cap S_{R''}(\omega))} + r_i \|\nabla u\|_{L^2(B_i \cap S_{R''}(\omega))} \\ & + r_i \|G\|_{L^2(B_i \cap S_{R''}(\omega))} + r_i^2 \|\nabla G\|_{L^2(B_i \cap S_{R''}(\omega))} \Big). \end{aligned}$$

Hence, by squaring and summing over all $i \in \mathbb{N}$ and using the overlap properties of the covering, we get

$$\| \, |x|\nabla^2 u\|_{L^2(S_{R'}(\omega))} \leq$$
$$C\Big(\| \, |x| f\|_{L^2(S_{R''}(\omega))} + \|G\|_{L^2(S_{R''}(\omega))} + \| \, |x|\nabla G\|_{L^2(S_{R''}(\omega))} + \|\nabla u\|_{L^2(S_{R''}(\omega))}\Big).$$

Denoting $\tilde{A} := A(0)$, we see that u satisfies the following problem:

$$-\nabla \cdot \left(\tilde{A}\nabla u\right) = \tilde{f} \quad \text{a.e. on } S_R(\omega)$$
$$\partial_{n_{\tilde{A}}} u = \tilde{G} \quad \text{on } \Gamma_0 \cup \Gamma_\omega,$$

where

$$\tilde{f} = f - \left(\nabla \cdot (\tilde{A} - A)\right) \cdot \nabla u - (\tilde{A} - A) : \nabla^2 u,$$
$$\tilde{G} = G - \tilde{\mathbf{n}} \cdot (\tilde{A} - A)\nabla u.$$

The function $\tilde{\mathbf{n}}$ is best expressed in polar coordinates as

$$\tilde{\mathbf{n}}(r, \varphi) = (\varphi/\omega)\mathbf{n}_\omega + (1 - \varphi/\omega)\mathbf{n}_0,$$

where $\mathbf{n}_0$ and $\mathbf{n}_\omega$ are the outer normal vectors on lines Γ_0, Γ_ω. Since $A \in C^1(\overline{S_R(\omega)})$, we have

$$|\tilde{\mathbf{n}} \cdot (\tilde{A} - A)| \leq C|x|, \qquad |\nabla(\tilde{\mathbf{n}} \cdot (\tilde{A} - A))| \leq C \qquad \forall x \in S_R(\omega).$$

These estimates allow us to conclude with our bounds on $\|\nabla u\|_{L^2(S_R(\omega))}$ and $\| \, |x|\nabla^2 u\|_{L^2(S_{R'}(\omega))}$ that

$$\|\tilde{f}\|_{H^{0,0}_{\beta,1}(S_{R''}(\omega))} + \|\tilde{G}\|_{H^{1,1}_{\beta,1}(S_{R''}(\omega))} \leq C\left[\|f\|_{H^{0,0}_{\beta,1}(S_R(\omega))} + \|G\|_{H^{1,1}_{\beta,1}(S_R(\omega))}\right].$$

The argument is completed by observing that [15] ascertains the existence of $\beta \in [0, 1)$ such that

$$\|u\|_{H^{2,2}_{\beta,1}(S_{R'}(\omega))} \leq C\left[\|\tilde{f}\|_{H^{0,0}_{\beta,1}(S_{R''}(\omega))} + \|\tilde{G}\|_{H^{1,1}_{\beta,1}(S_{R''}(\omega))} + \|u\|_{H^1(S_{R'}(\omega))}\right].$$

□

Remark 5.4.3 The term $\|G\|_{H^{1,1}_{\beta,1}(S_R(\omega))}$ can be rewritten as follows: By mapping arguments, the Hardy inequality Lemma A.1.7 allows us to conclude, for arbitrary fixed $R_0 \in (0, R)$, the existence of C depending only on R_0, R, $\beta \in (0, 1)$, and ω such that

$$\|G\|_{H^{1,1}_{\beta,1}(S_R(\omega))} \leq C\left[\|\hat{\Phi}_{0,\beta,1}\nabla G\|_{L^2(S_R(\omega))} + \|G\|_{L^2(S_R(\omega)\setminus S_{R_0}(\omega))}\right]. \quad (5.4.31)$$

■

Proposition 5.4.4. *Let $0 < R' < R$, $\omega \in (0, 2\pi)$, $\delta > 0$. Assume that the coefficients A, b, c satisfy (5.3.7). Define $\mathcal{E}$ by (5.3.8). Then there exists $\beta \in [0, 1)$ depending only on ω and $A(0)$, and there exists $C > 0$ independent of $\mathcal{E}$ such that each solution $u \in H^1(S_R(\omega))$ of (5.4.27) with right-hand side $f \in H^{0,0}_{\beta,\mathcal{E}}(S_R(\omega))$, and boundary data $G_{1,0} \in H^{1,1}_{\beta,\mathcal{E}}(S_R^{0,\delta}(\omega))$, $G_{1,\omega} \in H^{1,1}_{\beta,\mathcal{E}}(S_R^{\omega,\delta}(\omega))$, $G_{2,0}$, $G_{2,\omega} \in W^{1,\infty}((0, R))$ satisfies*

$$\|\hat{\Phi}_{0,\beta,\varepsilon}\nabla^2 u\|_{L^2(S_{R'}(\omega))} \leq C\mathcal{E}^{-2}\Big[(\mathcal{E}/\varepsilon)^2\|f\|_{H^{0,0}_{\beta,\mathcal{E}}(S_R(\omega))} + (\mathcal{E}/\varepsilon)C_{G_1}$$
$$+ \mathcal{E}\|\nabla u\|_{L^2(S_R(\omega))} + C_c(\mathcal{E}/\varepsilon)^2\|u\|_{L^2(S_R(\omega))} + C_{G_2}(\mathcal{E}/\varepsilon)\|u\|_{H^{1,1}_{\beta,\mathcal{E}}(S_R(\omega))}\Big],$$

where

$$C_{G_1} := \|G_{1,0}\|_{H^{1,1}_{\beta,\mathcal{E}}(S_R^{0,\delta}(\omega))} + \|G_{1,\omega}\|_{H^{1,1}_{\beta,\mathcal{E}}(S_R^{\omega,\delta}(\omega))},$$
$$C_{G_2} := \|G_{2,0}\|_{W^{1,\infty}((0,R))} + \|G_{2,\omega}\|_{W^{1,\infty}((0,R))}.$$

Proof: For simplicity of notation, we introduce the functions G_1, G_2 as in Lemma 5.4.1 such that $G_{i,0} = G_i|_{\Gamma_0}$, $G_{i,\omega} = G_i|_{\Gamma_\omega}$ and

$$\|G_1\|_{H^{1,1}_{\beta,\mathcal{E}}(S_R(\omega))} \leq CC_{G_1}, \qquad \|G_2\|_{L^\infty(S_R(\omega))} + \|\,|x|\nabla G_2\|_{L^\infty(S_R(\omega))} \leq CC_{G_2},$$

for a constant $C > 0$ that depends only on δ, $\beta \in (0, 1)$, R, ω. We may therefore formulate the proof for function G_1, G_2 that are defined on $S_R(\omega)$ and have these regularity properties.

The proof follows very closely that of Proposition 5.3.2. Since we will use the same covering $\mathcal{B}$ by balls, the same values of h, H, and the same abbreviations, we refer the reader to the proof of Proposition 5.3.2 for these notions.
Let χ be a cut-off function satisfying $\chi \equiv 1$ on $B_h(0)$, $\chi \equiv 0$ on $S \setminus S_H$, $\|\nabla^j\chi\|_{L^\infty(S)} \leq C\mathcal{E}^{-j}$, $j \in \{0, 1, 2\}$, $\partial_{n_A}\chi = 0$ on $\partial S_R(\omega) \setminus \partial B_R(0)$ as given by Lemma A.1.2. We then observe that χu satisfies

$$-\nabla \cdot (A\nabla(u\chi)) = \tilde{f} := \chi\varepsilon^{-2}\left[f - b \cdot \nabla u - cu\right] - 2\nabla\chi \cdot (A\nabla u) - u\nabla \cdot (A\nabla\chi),$$
$$\partial_{n_A}(u\chi) = \tilde{G}_1 + \tilde{G}_2 := \varepsilon^{-1}(\chi G_1 + G_2(\chi u)) \quad \text{on } \partial S_R.$$

Lemma 5.4.2 gives the existence of $\beta \in [0, 1)$ and $C > 0$ such that

$$\|r^\beta\nabla^2(u\chi)\|_{L^2(S_R)} \leq C\left[\|\tilde{f}\|_{H^{0,0}_{\beta,1}(S_R)} + \|\tilde{G}_1 + \tilde{G}_2\|_{H^{1,1}_{\beta,1}(S_R)}\right].$$

In view of the fact that $\tilde{G}$ is supported by $B_H(0)$ with $H < R$, we may apply (5.4.31) to arrive at

$$\|r^\beta\nabla^2(u\chi)\|_{L^2(S_R)} \leq C\left[\|r^\beta\tilde{f}\|_{L^2(S_R)} + \|r^\beta\nabla\tilde{G}_1\|_{L^2(S_R)} + \|r^\beta\nabla\tilde{G}_2\|_{L^2(S_R)}\right].$$

Since $\|r^\beta\nabla\tilde{G}_2\|_{L^2(S_R)} \leq C\varepsilon^{-1}C_{G_2}\|(\chi u)\|_{H^{1,1}_{\beta,1}(S_R)} \leq C\varepsilon^{-1}C_{G_2}\|r^\beta\nabla(\chi u)\|_{L^2(S_R)}$ by Lemma A.1.7 and the support properties of χ, we arrive at

$$\|r^\beta \nabla^2(u\chi)\|_{L^2(S_R)} \le$$
$$C\left[\|r^\beta \tilde{f}\|_{L^2(S_R)} + \varepsilon^{-1}\|r^\beta \nabla(\chi G_1)\|_{L^2(S_R)} + \varepsilon^{-1} C_{G_2}\|r^\beta \nabla(\chi u)\|_{L^2(S_R)}\right].$$

Dividing both sides by $\mathcal{E}^\beta$ and exploiting the support properties of χ gives

$$\|\hat{\Phi}_{0,\beta,\mathcal{E}}\nabla^2 u\|_{L^2(S_h)} \le C\Big[\|\hat{\Phi}_{0,\beta,\mathcal{E}}\tilde{f}\|_{L^2(S_H)}$$
$$+ \varepsilon^{-1}\|\hat{\Phi}_{0,\beta,\mathcal{E}}\nabla(\chi G_1)\|_{L^2(S_H)} + \frac{C_{G_2}}{\varepsilon}\|\hat{\Phi}_{0,\beta,\mathcal{E}}\nabla(\chi u)\|_{L^2(S_H)}\Big].$$

Expanding the term involving $\tilde{f}$ as in the proof of Proposition 5.3.2 and expanding the terms involving χG_1, χu gives

$$\|\hat{\Phi}_{0,\beta,\mathcal{E}}\nabla^2 u\|_{L^2(S_h)} \le C\mathcal{E}^{-2}\Big\{ \tag{5.4.32}$$
$$(\mathcal{E}/\varepsilon)^2\|f\|_{H^{0,0}_{\beta,\mathcal{E}}(S_H)} + \mathcal{E}\|\nabla u\|_{L^2(S_H)} + C_c(\mathcal{E}/\varepsilon)^2\|u\|_{L^2(S_H)} + \|u\|_{L^2(S_H)}$$
$$+ (\mathcal{E}/\varepsilon)\|G_1\|_{H^{1,1}_{\beta,\mathcal{E}}(S_H)} + C_{G_2}(\mathcal{E}/\varepsilon)\|u\|_{H^{1,1}_{\beta,\mathcal{E}}(S_H)}\Big\},$$

which has, with the exception of the term $\|u\|_{L^2(S_H)}$, the desired from. In order to remove this term, we take $\overline{u}$ as the average of u over S_H and note that $u - \overline{u}$ satisfies

$$-\varepsilon^2\nabla\cdot(A\nabla(u-\overline{u})) = f - b\cdot\nabla(u-\overline{u}) - c(u-\overline{u}) - c\overline{u}, \qquad \text{on } S_R$$
$$\varepsilon^2\partial_{n_A}(u-\overline{u}) = \varepsilon(G_1 + G_2(u-\overline{u})) + \varepsilon G_2\overline{u} \qquad \text{on } \Gamma_{lat}.$$

Hence, we may apply (5.4.32) to this problem with G_1 replaced with $G_1 + G_2\overline{u}$ and f replaced with $f + c\overline{u}$; employing then the Poincaré inequality $\|u-\overline{u}\|_{L^2(S_H)} \le C\mathcal{E}\|\nabla u\|_{L^2(S_H)}$, the bound $\|\overline{u}\|_{L^2(S_H)} \le \|u\|_{L^2(S_H)}$, and Lemma 4.2.10 allows us to remove the term $\|u\|_{L^2(S_H)}$ in (5.4.32).

Paralleling the proof of Proposition 5.3.2, we now turn to the bound on $S_{R'}\setminus S_h$. By (5.5.22) for balls B_i with $B_i \subset S_R$ and by (5.5.35) for balls with $x_i \in \Gamma_{lat}$ we have for all $i \in \mathbb{N}\setminus I_\mathcal{E}$,

$$r_i^2\|\nabla^2 u\|_{L^2(\hat{B}_i\cap S_R)} \le C\Big\{\varepsilon^{-2}r_i^2\|f - b\cdot\nabla u - cu\|_{L^2(B_i\cap S_R)} + r_i\|\nabla u\|_{L^2(B_i\cap S_R)}$$
$$+ r_i\varepsilon^{-1}\|G_1\|_{L^2(B_i\cap S_R)} + r_i^2\varepsilon^{-1}\|\nabla G_1\|_{L^2(B_i\cap S_R)}$$
$$+ C_{G_2}r_i\varepsilon^{-1}\|u\|_{L^2(B_i\cap S_R)} + C_{G_2}r_i^2\varepsilon^{-1}\|\nabla u\|_{L^2(B_i\cap S_R)}\Big\}.$$

Dividing this estimate by r_i^2, summing over all $i \in \mathbb{N}\setminus I_\mathcal{E}$ and exploiting the overlap properties of the covering $\mathcal{B}$ together with $r_i \ge C\mathcal{E}$ and gives

$$\|\nabla^2 u\|_{L^2(S_{R'}\setminus S_h)} \le C\mathcal{E}^{-2}\Big\{(\mathcal{E}/\varepsilon)^2\|f\|_{H^{0,0}_{\beta,\mathcal{E}}(S_R)} + \mathcal{E}\|\nabla u\|_{L^2(S_R)}$$
$$+ C_c(\mathcal{E}/\varepsilon)^2\|u\|_{L^2(S_R)} + (\mathcal{E}/\varepsilon)\|G_1\|_{H^{1,1}_{\beta,\mathcal{E}}(S_R)} + C_{G_2}(\mathcal{E}/\varepsilon)\|\nabla u\|_{H^{1,1}_{\beta,\mathcal{E}}(S_R)}\Big\}.$$

□

Proposition 5.4.5. *Let* $0 < R' < R$, $\omega \in (0, 2\pi)$, $\delta > 0$. *Assume that the coefficients* A, b, c *satisfy (5.3.7). Let* $\mathcal{E}$ *be given by (5.3.8). Then there exists* $\beta \in [0,1)$ *depending only on* ω *and* $A(0)$, *and there exist* C, $K > 0$ *independent of* $\mathcal{E}$, C_f, C_{G_1}, C_{G_2} *such that each solution* $u \in H^1(S_R(\omega))$ *of (5.4.27) with right-hand side* $f \in \mathcal{B}^0_{\beta,\mathcal{E}}(S_R(\omega), C_f, \gamma_f)$, *and boundary data* $G_{1,0} \in \mathcal{B}^1_{\beta,\mathcal{E}}(S_R^{0,\delta}(\omega), C_{G_1}, \gamma_{G_2})$, $G_{1,\omega} \in \mathcal{B}^1_{\beta,\mathcal{E}}(S_R^{\omega,\delta}(\omega), C_{G_1}, \gamma_{G_1})$, *and* $G_{2,0}$, $G_{1,\omega}$ *satisfying*

$$\|D^n G_{2,0}\|_{L^\infty((0,R))} + \|D^n G_{2,\omega}\|_{L^\infty((0,R))} \leq C_{G_2} \gamma_{G_2}^n n! \qquad \forall n \in \mathbb{N}_0,$$

satisfies for all $p \in \mathbb{N}_0$

$$\begin{aligned}\|\hat{\Phi}_{p,\beta,\mathcal{E}} \nabla^{p+2} u\|_{L^2(S_{R'}(\omega))} \leq{}& C K^{p+2} \max\{p+1, \mathcal{E}^{-1}\}^{p+2} \times \\ & \Big\{ (\mathcal{E}/\varepsilon)^2 C_f + \mathcal{E}\|\nabla u\|_{L^2(S_R(\omega))} + C_c (\mathcal{E}/\varepsilon)^2 \|u\|_{L^2(S_R(\omega))} \\ & + (\mathcal{E}/\varepsilon) C_{G_1} + (\mathcal{E}/\varepsilon) C_{G_2} \|u\|_{H^{1,1}_{\beta,\mathcal{E}}(S_R(\omega))} \Big\}.\end{aligned}$$

Proof: The proof is very similar to that of Proposition 5.3.4; we will therefore merely point out the appropriate modifications. First, we note that the assumptions on $G_{2,0}$, $G_{2,\omega}$ imply that they can be extended analytically to functions (again denoted $G_{2,0}$, $G_{2,\omega}$) with

$$\|\nabla^n G_{2,0}\|_{L^\infty(\mathcal{U}_\delta(\Gamma_0))} + \|\nabla^n G_{2,\omega}\|_{L^\infty(\mathcal{U}_\delta(\Gamma_\omega))} \leq C C_{G_2} \gamma^n n! \quad \forall n \in \mathbb{N}_0,$$

for appropriately chosen C, γ and $\delta > 0$ sufficiently small, where the neighborhoods $\mathcal{U}_\delta(\Gamma_0)$, $\mathcal{U}_\delta(\Gamma_\omega)$ are given by $\mathcal{U}_\delta(\Gamma_0) = \cup_{x \in \Gamma_0} B_\delta(x)$, $\mathcal{U}_\delta(\Gamma_\omega) = \cup_{x \in \Gamma_\omega} B_\delta(x)$. We will assume that the covering $\mathcal{B}$ that is chosen in Proposition 5.3.4 is such that the half-balls $B_i \cap S_R(\omega)$ are contained in $S_R^{0,\delta}(\omega) \cup S_R^{\omega,\delta}(\omega)$.
We split the index set $\mathbb{N}$ into

$$\mathbb{N} = I_{bdy} \cup I_{int}, \qquad I_{int} := \{i \in \mathbb{N} \,|\, B_i \subset S_R(\omega)\}, \quad I_{bdy} := \mathbb{N} \setminus I_{int}.$$

The treatment of the balls B_i, $i \in I_{int}$, is completely analogous to that in the proof of Proposition 5.3.4. We will therefore restrict our attention to the half balls $B_i \cap S_R(\omega)$, $i \in I_{bdy}$.
Reasoning as in the proof of Proposition 5.3.4 (cf. (5.3.21)), the application of Lemma 4.2.17 applied to the function $f \in \mathcal{B}^0_{\beta,\mathcal{E}}(S_R(\omega), C_f, \gamma_f)$ with the covering $\mathcal{B}$ yields the existence of C, $K > 0$ independent of $\mathcal{E}$ and i such that with the numbers $C_f(i)$ as given by Lemma 4.2.17 we have

$$\|\nabla^p f\|_{L^2(S \cap B_i)} \leq \frac{1}{\hat{\Phi}_{0,\beta,\mathcal{E}}(x_i)} C K^p C_f(i) \max\{(p+1)/|x_i|, \mathcal{E}^{-1}\}^p. \tag{5.4.33}$$

$$\sum_{i \in I_{bdy}} C_f^2(i) \leq C C_f^2 \tag{5.4.34}$$

Completely analogously, we can exploit with Corollary 4.2.18 the fact that $G_{1,k} \in \mathcal{B}^1_{\beta,\mathcal{E}}(S_R^{k,\delta}(\omega), C_{G_1}, \gamma_{G_1})$, $k \in \{0, \omega\}$, to get the existence of numbers $C_G(i)$ as given by Corollary 4.2.18 such that for $k \in \{0, \omega\}$

$$\|\nabla^p G_{2,k}\|_{L^2(S\cap B_i)} \leq C \frac{K^p C_G(i)}{\hat{\Phi}_{-1,\beta,\mathcal{E}}(x_i)} \max\{(p+1)/|x_i|, \mathcal{E}^{-1}\}^p, \quad (5.4.35)$$

$$\sum_{i\in I_{bdy}} C_G^2(i) \leq C C_{G_1}^2 \quad (5.4.36)$$

As in the proof of Proposition 5.3.4 we now employ a local analytic regularity result, namely, Proposition 5.5.3, to get

$$\left(\frac{r_i}{2}\right)^{p+2} \|\nabla^{p+2}u\|_{L^2(\widehat{B}_i\cap S)} \leq C_u(i) K^{p+2} \max\{(p+3), r_i/\mathcal{E}\}^{p+2} \qquad \forall p \in \mathbb{N}_0,$$

where we abbreviate

$$\begin{aligned} C_u(i) := {} & \min\{1, r_i/\mathcal{E}\}\mathcal{E}\|\nabla u\|_{L^2(B_i\cap S)} + C_c(\mathcal{E}/\varepsilon)^2 \min\{1, r_i/\mathcal{E}\}^2 \|u\|_{L^2(B_i\cap S)} \\ & + C C_f(i) \frac{1}{\hat{\Phi}_{0,\beta,\mathcal{E}}(x_i)} \min\{1, r_i/\mathcal{E}\}^2 (\mathcal{E}/\varepsilon)^2 \\ & + C C_G(i) \frac{1}{\hat{\Phi}_{-1,\beta,\mathcal{E}}(x_i)} \min\{1, r_i/\mathcal{E}\}(\mathcal{E}/\varepsilon) \\ & + C C_{G_2}(\mathcal{E}/\varepsilon) \min\{1, r_i/\mathcal{E}\} \left(\min\{1, r_i/\mathcal{E}\}\mathcal{E}\|\nabla u\|_{L^2(B_i\cap S)} + \|u\|_{L^2(B_i\cap S)}\right). \end{aligned}$$

Reasoning as in the proof of Proposition 5.3.4, we arrive at

$$\begin{aligned} & \|\hat{\Phi}_{p,\beta,\mathcal{E}} \nabla^{p+2} u\|^2_{L^2(S')} \leq \quad (5.4.37) \\ & C K^{2p} \max\{p+1, \mathcal{E}^{-1}\}^{2p} \sum_{i=1}^{\infty} \hat{\Phi}^2_{0,\beta,\mathcal{E}}(x_i) \max\{1/r_i, \mathcal{E}^{-1}\}^4 C_u^2(i). \end{aligned}$$

Using $\min\{1, r_i/\mathcal{E}\} \leq C\hat{\Phi}_{1,0,\mathcal{E}}(x_i)$, we further bound

$$\begin{aligned} \hat{\Phi}_{0,\beta,\mathcal{E}}(x_i) C_u(i) \leq {} & \hat{\Phi}_{1,\beta,\mathcal{E}}(x_i)\mathcal{E}\|\nabla u\|_{L^2(B_i\cap S)} + C_c(\mathcal{E}/\varepsilon)^2 \hat{\Phi}_{1,\beta,\mathcal{E}}(x_i)\|u\|_{L^2(B_i\cap S)} \\ & + C_f(i)(\mathcal{E}/\varepsilon)^2 \hat{\Phi}^2_{1,0,\mathcal{E}}(x_i) + C_G(i)(\mathcal{E}/\varepsilon)\hat{\Phi}^2_{1,0,\mathcal{E}}(x_i) \\ & + C_{G_2}(\mathcal{E}/\varepsilon)\left[\hat{\Phi}_{2,\beta,\mathcal{E}}(x_i)\mathcal{E}\|\nabla u\|_{L^2(B_i\cap S)} + \hat{\Phi}_{1,\beta,\mathcal{E}}\|u\|_{L^2(B_i\cap S)}\right]. \end{aligned}$$

Inserting this in (5.4.37), appealing to Lemma 5.3.6, and employing Proposition 5.4.4 yields

$$\begin{aligned} \|\hat{\Phi}_{p,\beta,\mathcal{E}} \nabla^{p+2} u\|_{L^2(S')} \leq {} & C K^p \max\{p+1, \mathcal{E}^{-1}\}^p \mathcal{E}^{-2} \Big\{\mathcal{E}\|\nabla u\|_{L^2(S)} \\ & + C_c(\mathcal{E}/\varepsilon)^2 \|u\|_{L^2(S)} + C_f(\mathcal{E}/\varepsilon)^2 + C_{G_1}(\mathcal{E}/\varepsilon) + C_{G_2}(\mathcal{E}/\varepsilon)\|u\|_{H^{1,1}_{\beta,\mathcal{E}}(S)}\Big\}. \end{aligned}$$

□

Remark 5.4.6 Proposition 5.4.5 considers the case of straight sectors. However, since by Theorem 4.2.20 the spaces $\mathcal{B}^l_{\beta,\mathcal{E}}$ are invariant under analytic changes of

variables, the case of case of a curvilinear sector can be directly inferred from Proposition 5.4.5 by a mapping argument.
As in Remark 5.3.5, it is not essential that the data f, b, c, and ε be real. In the case of complex ε, it suffices to replace ε with $|\varepsilon|$ in all estimates.
The proof shows that the result could be slightly sharpened: Some of the norms $\|u\|_{L^2(S_R(\omega))}$, $\|\nabla u\|_{L^2(S_R(\omega))}$, can be replaced with weighted norms. ▪

5.4.2 Mixed corners

The case of corners with mixed boundary conditions is treated in exactly the same way as the cases of Dirichlet or Neumann boundary conditions. We will therefore merely state the results and use some of the notation already introduced at the outset of Section 5.4.1.
We consider

$$-\varepsilon^2\nabla\cdot(A(x)\nabla u)+b(x)\cdot\nabla u+c(x)u=f(x) \quad \text{on } S_R(\omega), \tag{5.4.38a}$$
$$\varepsilon^2\partial_{n_A}u=\varepsilon(G_1+G_2u) \quad \text{on } \Gamma_0, \tag{5.4.38b}$$
$$u=0 \quad \text{on } \Gamma_\omega. \tag{5.4.38c}$$

Then the following result holds:

Proposition 5.4.7. *Let $0<R'<R$, $\omega\in(0,2\pi)$, $\delta>0$. Assume that the coefficients A, b, c satisfy (5.3.7). Let $\mathcal{E}$ be given by (5.3.8). Then there exist $\beta\in[0,1)$, which depends only on ω and $A(0)$, and $C>0$, which is independent of $\mathcal{E}$, such that each solution $u\in H^1(S_R(\omega))$ of (5.4.38) with right-hand side $f\in H^{0,0}_{\beta,\varepsilon}(S_R(\omega))$ and boundary data $G_1\in H^{1,1}_{\beta,\varepsilon}(S^{0,\delta}_R(\omega))$, $G_2\in W^{1,\infty}((0,R))$ satisfies*

$$\|\hat\Phi_{0,\beta,\varepsilon}\nabla^2u\|_{L^2(S_{R'}(\omega))}\le C\mathcal{E}^{-2}\Big\{\mathcal{E}\|\nabla u\|_{L^2(S_R(\omega))}+C_c(\mathcal{E}/\varepsilon)^2\|u\|_{L^2(S_R(\omega))}$$
$$+(\mathcal{E}/\varepsilon)^2\|\hat\Phi_{0,\beta,\varepsilon}f\|_{L^2(S_R(\omega))}+(\mathcal{E}/\varepsilon)\|G_1\|_{H^{1,1}_{\beta,\varepsilon}(S_R(\omega))}+C_{G_2}\|u\|_{H^{1,1}_{\beta,\varepsilon}(S_R(\omega))}\Big\},$$

where we set $C_{G_2}:=\|G_2\|_{W^{1,\infty}((0,R))}$.
If furthermore $f\in\mathcal{B}^0_{\beta,\varepsilon}(S_R(\omega),C_f,\gamma_f)$, $G_1\in\mathcal{B}^1_{\beta,\varepsilon}(S^{0,\delta}_R(\omega),C_{G_1},\gamma_{G_1})$, and G_2 satisfies

$$\|D^nG_2\|_{L^\infty((0,R))}\le C_{G_2}\gamma^n n! \quad \forall n\in\mathbb{N}_0,$$

then there exist C, $K>0$ independent of $\mathcal{E}$, C_f, C_{G_1}, C_{G_2} such that $\forall p\in\mathbb{N}_0$

$$\|\hat\Phi_{p,\beta,\varepsilon}\nabla^{p+2}u\|_{L^2(S_{R'}(\omega))}\le CK^p\max\{p,\mathcal{E}^{-1}\}^{p+2}\Big\{\mathcal{E}\|\nabla u\|_{L^2(S_R(\omega))}$$
$$+C_c(\mathcal{E}/\varepsilon)^2\|\nabla u\|_{L^2(S_R(\omega))}+C_f(\mathcal{E}/\varepsilon)^2+C_{G_1}(\mathcal{E}/\varepsilon)+C_{G_2}\|u\|_{H^{1,1}_{\beta,\varepsilon}(S_R(\omega))}\Big\}.$$

Proof: The proof is completely analogous to that of Propositions 5.4.4, 5.4.5. □

5.4.3 Transmission problems

To fix the notation, we assume that a sector $S_R(\omega) = \{(r\cos\varphi, r\sin\varphi)\,|\,0 < r < R,\ 0 < \varphi < \omega\}$ is divided into S_R^+, S_R^- by a line $\Gamma_{\omega'}$ passing through the origin. Specifically, we set for an $\omega' \in (0,\omega)$

$$\begin{aligned}
S_R^+ &:= \{(r\cos\varphi, r\sin\varphi)\,|\,0 < r < R,\ \omega' < \varphi < \omega\},\\
S_R^- &:= \{(r\cos\varphi, r\sin\varphi)\,|\,0 < r < R,\ 0 < \varphi < \omega'\},\\
\Gamma_0 &:= \{(r,0)\,|\,0 < r < R\}, \quad \Gamma_\omega := \{(r\cos\omega, r\sin\omega)\,|\,0 < r < R\},\\
\Gamma_{\omega'} &:= \{(r\cos\omega', r\sin\omega')\,|\,0 < r < R\},\\
S_R^{\omega',\delta}(\omega) &:= \{(r\cos\varphi\,|\,r\sin\varphi)\,|\,0 < r < R, \quad \omega' < \varphi < \omega' + \delta\}.
\end{aligned}$$

We consider H^1-solutions to the following differential equation:

$$-\varepsilon^2\nabla\cdot(A(x)\nabla u) + b(x)\cdot\nabla u + c(x)u = f(x) \quad \text{on } S_R^+ \cup S_R^-, \tag{5.4.39a}$$

$$[\varepsilon^2\partial_{n_A} u] = \varepsilon\,(G_{1,\omega'} + G_{2,\omega'}u) \qquad \text{on } \Gamma_{\omega'}, \tag{5.4.39b}$$

where the expression $[\varepsilon^2\partial_{n_A} u]$ denotes the jump across Γ. Concerning boundary conditions on the lateral sides Γ_0, Γ_ω, we assume any of the following three types:

$$u = 0 \qquad \text{on } \Gamma_0 \cup \Gamma_\omega \qquad \textbf{or} \tag{5.4.40a}$$

$$\begin{cases} \varepsilon^2\partial_{n_A} u = \varepsilon(G_{1,0} + G_{2,0}u) \text{ on } \Gamma_0\\ \varepsilon^2\partial_{n_A} u = \varepsilon(G_{1,\omega} + G_{2,\omega}u) \text{ on } \Gamma_\omega \end{cases} \qquad \textbf{or} \tag{5.4.40b}$$

$$\partial_{n_A} u = \varepsilon(G_{1,0} + G_{2,0}u) \text{ on } \Gamma_0 \quad \text{and} \quad u = 0 \text{ on } \Gamma_\omega. \tag{5.4.40c}$$

We assume $\varepsilon \in (0,1]$ and that the data $A \in \mathcal{A}(S_R^+ \cup S_R^-, \mathbb{S}^2_>)$, $b \in \mathcal{A}(S_R^+ \cup S_R^-)$, $c \in \mathcal{A}(S_R^+ \cup S_R^-)$ satisfy, for some C_A, C_b, γ_A, γ_b, C_c, $\gamma_c \geq 0$,

$$\|\nabla^p A\|_{L^\infty(S_R^+\cup S_R^-)} \leq C_A\gamma_A^p p! \qquad \forall p \in \mathbb{N}_0, \tag{5.4.41a}$$

$$0 < \lambda_{min} \leq A \qquad \text{on } S_R^+ \cup S_R^-, \tag{5.4.41b}$$

$$\|\nabla^p b\|_{L^\infty(S_R^+\cup S_R^-)} \leq C_b\gamma_b^p p! \qquad \forall p \in \mathbb{N}_0, \tag{5.4.41c}$$

$$\|\nabla^p c\|_{L^\infty(S_R^+\cup S_R^-)} \leq C_c\gamma_c^p p! \qquad \forall p \in \mathbb{N}_0. \tag{5.4.41d}$$

The data $G_{2,k}$, $k \in \{0,\omega',\omega\}$, are assumed analytic on Γ_k, $k \in \{0,\omega',\omega\}$, respectively; i.e., there exist C_{G_2}, $\gamma_{G_2} > 0$ such that $G_{2,k}$, $k \in \{0,\omega',\omega\}$, as functions parametrized by arclength satisfy

$$\sum_{k\in\{0,\omega',\omega\}} \|D^n G_{2,k}\|_{L^\infty((0,R))} \leq C_{G_2}\gamma_{G_2}^n n! \quad \forall n \in \mathbb{N}_0. \tag{5.4.41e}$$

In order to treat the different types of lateral boundary conditions in a unified way, the following proposition *assumes* that a solution u to (5.4.39) is piecewise in $H^{2,2}_{\beta,\varepsilon}$.

Proposition 5.4.8. *Assume (5.4.41) and let the relative diffusivity $\mathcal{E}$ be defined by (5.3.8). Assume that for a $\beta \in (0,1)$ a function u satisfies*

1. $u \in H^1(S_R(\omega)) \cap H^{2,2}_{\beta,\mathcal{E}}(S_R^+ \cup S_R^-)$;
2. *u solves (5.4.39) together with any of the three lateral boundary conditions (5.4.40);*
3. $f \in \mathcal{B}^0_{\beta,\mathcal{E}}(S_R^+ \cup S_R^-, C_f, \gamma_f)$ *and* $G_{1,k} \in \mathcal{B}^1_{\beta,\mathcal{E}}(S_R^{k,\delta}(\omega), C_{G_1}, \gamma_{G_1})$, $k \in \{0, \omega', \omega\}$.

Then $u \in \mathcal{A}(S_R^+ \cup S_R^-)$ and for every $R' < R$ there exist C, $K > 0$ independent of $\mathcal{E}$, C_f, C_{G_2}, C_{G_1} such that for all $p \in \mathbb{N}_0$

$$\|\hat{\Phi}_{p,\beta,\mathcal{E}} \nabla^{p+2} u\|_{L^2(S_{R'}^+ \cup S_{R'}^-)} \leq C K^{p+2} \max\{p+1, \mathcal{E}^{-1}\}^{p+2} \times$$
$$\Big[\mathcal{E}^2 \|\hat{\Phi}_{0,\beta,\mathcal{E}} \nabla^2 u\|_{L^2(S_R^+ \cup S_R^-)} + \mathcal{E} \|\hat{\Phi}_{0,\beta,\mathcal{E}} \nabla u\|_{L^2(S_R(\omega))} + (\mathcal{E}/\varepsilon)^2 C_c \|u\|_{L^2(S_R(\omega))}$$
$$+ (\mathcal{E}/\varepsilon)^2 C_f + (\mathcal{E}/\varepsilon) C_{G_1} + (\mathcal{E}/\varepsilon) C_{G_2} \|u\|_{H^{1,1}_{\beta,\mathcal{E}}(S_R(\omega))}\Big].$$

Proof: The proof is very similar to that of Proposition 5.4.5 and therefore omitted. For the treatment of the balls near the interface Γ, we employ the local regularity result Proposition 5.5.4. □

Remark 5.4.9 *Mutatis mutandis*, the Remarks 5.3.5, 5.4.6 apply: The coefficients b, c may be complex and curvilinear sectors can be treated with mapping arguments provided by Theorem 4.2.20. ▪

5.5 Local regularity

The aim of the present section is the proof of four local analytic regularity assertions, Propositions 5.5.1–5.5.4. The first one is concerned with regularity on balls (thus leading to interior regularity results); the second and third ones on half balls (thus leading to regularity results at the boundary); the fourth one with regularity results on balls where the data are assumed to be only piecewise analytic.

We start by introducing the following notation: For $n \geq 2$ and $R > 0$ we denote by $B_R := B_R(0) \subset \mathbb{R}^n$ the ball of radius R with center in the origin; B_R^+ and B_R^- denote the "upper" and "lower" parts:

$$\begin{aligned} B_R &:= B_R(0), & \Gamma_R &:= \{x \in B_R \,|\, x_n = 0\} \\ B_R^+ &:= \{x \in B_R \,|\, x_n > 0\}, & B_R^- &:= \{x \in B_R \,|\, x_n < 0\}. \end{aligned} \tag{5.5.1}$$

The intersection of B_R and the hyper-plane $x_n = 0$ is denoted Γ_R: $\Gamma_R := B_R \cap \{x_n = 0\}$. For problems defined on the half ball B_R^+ or on the two half balls $B_R^+ \cup B_R^-$, the direction normal to Γ_R is special. Hence, we will denote the variable x_n by y. In this context, we will denote by ∇_x differentiation with respect to the tangential variables $x_1, \ldots, x_{n-1}$. The local analytic regularity results are based on the following notation:

$$
\begin{aligned}
[p] &= \max(1,p),\\
N_{R,p}(v) &:= \frac{1}{[p]!}\sup_{R/2\le r<R}(R-r)^{2+p}\|\nabla^{p+2}v\|_{L^2(B_r)}, \quad p\in\mathbb{N}_0\cup\{-2,-1\},\\
N'_{R,p,q}(v) &= \frac{1}{[p+q]!}\sup_{R/2\le r<R}(R-r)^{p+q+2}\|\partial_y^{q+2}\nabla_x^p v\|_{L^2(B_r^+)}, \quad p\ge 0, q\ge -2,\\
N'^{,\pm}_{R,p,q}(v) &= \frac{1}{[p+q]!}\sup_{R/2\le r<R}(R-r)^{p+q+2}\|\partial_y^{q+2}\nabla_x^p v\|_{L^2(B_r^+\cup B_r^-)}, \ p\ge 0, q\ge -2,
\end{aligned}
$$

With these notions, we can formulate the following four local analytic regularity results. Their proofs are technical and relegated to the ensuing subsections.

Proposition 5.5.1 (Interior analytic regularity). *Let $R\in(0,1]$ and $B_R\subset\mathbb{R}^n$ be a ball of radius R. Let $A\in\mathcal{A}(B_R,\mathbb{S}^n_>)$, $b\in\mathcal{A}(B_R,\mathbb{R}^n)$, $c\in\mathcal{A}(B_R)$ satisfy*

$$
\begin{aligned}
\lambda_{min}\le A \quad \text{on } B_R, \qquad & \|\nabla^p A\|_{L^\infty(B_R)}\le C_A\gamma_A^p p!, \qquad \forall p\in\mathbb{N}_0, && (5.5.2a)\\
\|\nabla^p b\|_{L^\infty(B_R)}\le C_b\gamma_b^p p!, \qquad & \|\nabla^p c\|_{L^\infty(B_R)}\le C_c\gamma_c^p p!. \qquad \forall p\in\mathbb{N}_0. && (5.5.2b)
\end{aligned}
$$

Let $\varepsilon\in(0,1]$, and define the relative diffusivity *$\mathcal{E}$ by*

$$
\mathcal{E}^{-1}:=\frac{C_b}{\varepsilon^2}+\frac{\sqrt{C_c}}{\varepsilon}+1. \tag{5.5.3}
$$

Let $f\in\mathcal{A}(B_R)$ satisfy for all $p\in\mathbb{N}_0$

$$
\|\nabla^p f\|_{L^2(B_R)}\le C_f\gamma_f^p\max\{p/R,\mathcal{E}^{-1}\}^p \tag{5.5.4}
$$

and let u be a solution of

$$
-\varepsilon^2\nabla\cdot(A\nabla u)+b\cdot\nabla u+cu=f \qquad \text{on } B_R.
$$

Then for some $K>0$ depending only on the constants of (5.5.2) and γ_f

$$
\begin{aligned}
N_{R,p}(u) &\le C_u K^{p+2}\frac{\max\{(p+3),R/\mathcal{E}\}^{p+2}}{[p]!} \qquad \forall p\in\mathbb{N}_0\cup\{-1\}, && (5.5.5)\\
C_u &:= \min\{1,R/\mathcal{E}\}\mathcal{E}\|\nabla u\|_{L^2(B_R)}\\
&\quad+(\mathcal{E}/\varepsilon)^2\min\{1,R/\mathcal{E}\}^2\left[C_c\|u\|_{L^2(B_R)}+C_f\right]. && (5.5.6)
\end{aligned}
$$

Proposition 5.5.2 (Dirichlet conditions). *Let $R\in(0,1]$ and $B_R^+\subset\mathbb{R}^n$ be a semi-ball of radius R. Let $A\in\mathcal{A}(B_R^+,\mathbb{S}^n_>)$, $b\in\mathcal{A}(B_R^+,\mathbb{R}^n)$, $c\in\mathcal{A}(B_R^+)$ satisfy*

$$
\begin{aligned}
\lambda_{min}\le A \quad \text{on } B_R^+, \qquad & \|\nabla^p A\|_{L^\infty(B_R^+)}\le C_A\gamma_A^p p!, \qquad \forall p\in\mathbb{N}_0, && (5.5.7a)\\
\|\nabla^p b\|_{L^\infty(B_R^+)}\le C_b\gamma_b^p p!, \qquad & \|\nabla^p c\|_{L^\infty(B_R^+)}\le C_c\gamma_c^p p! \qquad \forall p\in\mathbb{N}_0. && (5.5.7b)
\end{aligned}
$$

Let $\varepsilon\in(0,1]$, define the relative diffusivity *$\mathcal{E}$ by (5.5.3). Let $f\in\mathcal{A}(B_R^+)$ satisfy*

$$
\|\nabla^p f\|_{L^2(B_R^+)}\le C_f\gamma_f^p\max\{p/R,\mathcal{E}^{-1}\}^p \quad \forall p\in\mathbb{N}_0, \tag{5.5.8}
$$

and let u be a solution of

$$-\varepsilon^2\nabla\cdot(A\nabla u)+b\cdot\nabla u+cu=f \quad \text{on } B_R^+, \qquad u=0 \quad \text{on } \partial B_R^+ .$$

Then for some K_1, $K_2>0$ *depending only on the constants of (5.5.7) and* γ_f *there holds for all* $p\in\mathbb{N}_0$, $q\in\mathbb{N}_0\cup\{-2,-1\}$ *such that* $p+q\neq -2$

$$N'_{R,p,q}(u) \le C_u K_1^{p+2}K_2^{q+2}\frac{\max\{(p+q+3),R/\mathcal{E}\}^{p+q+2}}{[p+q]!}, \tag{5.5.9}$$

$$\begin{aligned} C_u &:= \min\{1,R/\mathcal{E}\}\mathcal{E}\|\nabla u\|_{L^2(B_R^+)} \\ &\quad + (\mathcal{E}/\varepsilon)^2\min\{1,R/\mathcal{E}\}^2\left[C_c\|u\|_{L^2(B_R^+)}+C_f\right]. \end{aligned} \tag{5.5.10}$$

Proposition 5.5.3 (Neumann conditions). *Assume the same hypotheses on* A, b, c, ε, $\mathcal{E}$ *as in Proposition 5.5.2. Let* $f\in\mathcal{A}(B_R^+)$, G_1, $G_2\in\mathcal{A}(B_R^+)$ *satisfy*

$$\|\nabla^p f\|_{L^2(B_R^+)} \le C_f\gamma_f^p\max\{p/R,\mathcal{E}^{-1}\}^p \qquad \forall p\in\mathbb{N}_0, \tag{5.5.11}$$

$$\|\nabla^p G_1\|_{L^2(B_R^+)} \le C_{G_1}\gamma_{G_1}^p\max\{p/R,\mathcal{E}^{-1}\}^p \qquad \forall p\in\mathbb{N}_0, \tag{5.5.12}$$

$$\|\nabla^p G_2\|_{L^\infty(B_R^+)} \le C_{G_2}\gamma_{G_2}^p p! \qquad \forall p\in\mathbb{N}_0, \tag{5.5.13}$$

and let $u\in H^1(B_{R^+})$ *solve*

$$-\varepsilon^2\nabla\cdot(A\nabla u)+b\cdot\nabla u+cu=f \quad \text{on } B_{R^+}, \qquad \varepsilon^2\partial_{n_A}u=\varepsilon\,(G_1+G_2u) \quad \text{on } \Gamma_R.$$

Then for some K_1, $K_2>0$ *depending only on the constants of (5.5.7) and* γ_f, γ_{G_1}, γ_{G_2}, *there holds for all* $p\in\mathbb{N}_0$, $q\in\mathbb{N}_0\cup\{-2,-1\}$ *such that* $p+q\neq -2$

$$N'_{R,p,q}(u) \le C_u K_1^{p+2}K_2^{q+2}\frac{\max\{(p+q+3),R/\mathcal{E}\}^{p+q+2}}{[p+q]!}, \tag{5.5.14}$$

$$\begin{aligned} C_u &:= \min\{1,R/\mathcal{E}\}\mathcal{E}\|\nabla u\|_{L^2(B_R^+)} \qquad (5.5.15)\\ &\quad + (\mathcal{E}/\varepsilon)^2\min\{1,R/\mathcal{E}\}^2\left[C_c\|u\|_{L^2(B_R^+)}+C_f\right] \\ &\quad + C_{G_1}\min\{1,R/\mathcal{E}\}(\mathcal{E}/\varepsilon) \\ &\quad + C_{G_2}(\mathcal{E}/\varepsilon)\min\{1,R/\mathcal{E}\}\left[\min\{1,R/\mathcal{E}\}\mathcal{E}\|\nabla u\|_{L^2(B_R^+)}+\|u\|_{L^2(B_R^+)}\right]. \end{aligned}$$

Proposition 5.5.4 (Transmission problem). *Let* $R\in(0,1]$, $A\in\mathcal{A}(B_R^+\cup B_R^-,\mathbb{S}^n_>)$, $b\in\mathcal{A}(B_R^+\cup B_R^-,\mathbb{R}^n)$, $c\in\mathcal{A}(B_R^+\cup B_R^-)$ *satisfy*

$$\lambda_{min}\le A \quad \text{on } B_R^+\cup B_R^-, \qquad \|\nabla^p A\|_{L^\infty(B_R^+\cup B_R^-)}\le C_A\gamma_A^p p! \ \forall p\in\mathbb{N}_0, \tag{5.5.16a}$$

$$\|\nabla^p b\|_{L^\infty(B_R^+\cup B_R^-)}\le C_b\gamma_b^p p!, \qquad \|\nabla^p c\|_{L^\infty(B_R^+\cup B_R^-)}\le C_c\gamma_c^p p! \quad \forall p\in\mathbb{N}_0. \tag{5.5.16b}$$

Let $\varepsilon\in(0,1]$, *and define the* relative diffusivity $\mathcal{E}$ *by (5.5.3). Let* $f\in\mathcal{A}(B_R^+\cup B_R^-)$, G_1, $G_2\in\mathcal{A}(B_R^+)$ *satisfy*

$$\|\nabla^p f\|_{L^2(B_R^+\cup B_R^-)} \le C_f\gamma_f^p\max\{p/R,\mathcal{E}^{-1}\}^p \qquad \forall p\in\mathbb{N}_0, \tag{5.5.17}$$

$$\|\nabla^p G_1\|_{L^2(B_R^+)} \le C_{G_1}\gamma_{G_1}^p\max\{p/R,\mathcal{E}^{-1}\}^p \qquad \forall p\in\mathbb{N}_0, \tag{5.5.18}$$

$$\|\nabla^p G_2\|_{L^\infty(B_R^+)} \le C_{G_2}\gamma_{G_2}^p p! \qquad \forall p\in\mathbb{N}_0. \tag{5.5.19}$$

Let $u \in H^1(B_R)$ be a solution of the transmission problem:

$$-\varepsilon^2\nabla\cdot(A\nabla u)+b\cdot\nabla u+cu = f \quad \text{on } B_R, \qquad [\varepsilon^2\partial_{n_A} u] = \varepsilon\,(G_1 + G_2 u) \quad \text{on } \Gamma_R.$$

Then for some K_1, $K_2 > 0$ depending only on the constants of (5.5.16) and γ_f, γ_{G_1}, γ_{G_2}, there holds for all $p \in \mathbb{N}_0$, $q \in \mathbb{N}_0 \cup \{-2,-1\}$ such that $p+q \neq -2$

$$N'^{,\pm}_{R,p,q}(u) \le C_u K_1^{p+2} K_2^{q+2} \frac{\max\{(p+q+3), R/\mathcal{E}\}^{p+q+2}}{[p+q]!}, \tag{5.5.20}$$

$$\begin{aligned} C_u := {} & \min\{1, R/\mathcal{E}\}\mathcal{E}\|\nabla u\|_{L^2(B_R)} \\ & + \min\{1,R/\mathcal{E}\}^2(\mathcal{E}/\varepsilon)^2 \left[C_f + C_c\|u\|_{L^2(B_R)}\right] \\ & + C_{G_1}\min\{1,R/\mathcal{E}\}(\mathcal{E}/\varepsilon) \\ & + C_{G_2}(\mathcal{E}/\varepsilon)\min\{1,R/\mathcal{E}\}\left[\min\{1,R/\mathcal{E}\}\mathcal{E}\|\nabla u\|_{L^2(B_R)} + \|u\|_{L^2(B_R)}\right]. \end{aligned} \tag{5.5.21}$$

5.5.1 Preliminaries: local H^2-regularity

Lemma 5.5.5 (H^2-regularity). *Let $B_R := B_R(0)$. Let $f \in L^2(B_R)$, $A \in C^1(\overline{B_R}, \mathbb{S}^n_>)$ with $0 < \lambda_{min} \le A(x)$ on B_R. Then there exists $C_I > 0$ depending only on λ_{min}, $\|A\|_{L^\infty(B_R)}$, and $R\|A'\|_{L^\infty(B_R)}$ such that the weak solution u of*

$$-\nabla\cdot(A\nabla u) = f \quad \text{on } B_R, \qquad u = 0 \quad \text{on } \partial B_R$$

satisfies

$$\|\nabla^2 u\|_{L^2(B_R)} \le C_I\|f\|_{L^2(B_R)}.$$

Remark 5.5.6 The proof shows that the constant C_I depends only on lower bounds on λ_{min} and upper bounds on $\|A\|_{L^\infty(B_R)}$, $R\|A'\|_{L^\infty(B_R)}$. Furthermore, it is worth stressing that it is independent of $R \le 1$. ▪

Proof: Introduce the affine map $F : B_1 \to B_R$ given by $F(x) = Rx$. Upon setting $\hat{u} = u \circ F$, $\hat{A} = A \circ F$, $\hat{f} = f \circ F$, we get from Lemma A.1.1 that $\hat{u}$ solves

$$-\nabla\cdot(\hat{A}\nabla\hat{u}) = R^2\hat{f} \quad \text{on } B_1, \qquad \hat{u} = 0 \quad \text{on } \partial B_1.$$

Standard regularity theory (see, e.g., [49, 56]) gives the existence of $C_I > 0$ depending only on $\lambda_{min}(\hat{A}) = \lambda_{min}$, $\|\hat{A}\|_{L^\infty(B_1)} = \|A\|_{L^\infty(B_R)}$, $\|\hat{A}'\|_{L^\infty(B_1)} = R\|A'\|_{L^\infty(B_R)}$ such that $\|\nabla^2\hat{u}\|_{L^2(B_1)} \le C_I R^2\|\hat{f}\|_{L^2(B_1)}$. Transforming back to B_R yields the desired result. □

Analogously, we have a *priori* estimates on semi balls:

Lemma 5.5.7 (H^2-regularity, Dirichlet conditions). *Let $f \in L^2(B_R^+)$, $A \in C^1(\overline{B_R^+}, \mathbb{S}^n_>)$ with $0 < \lambda_{min} \le A$ on B_R^+. Then there exists $C_B > 0$ depending only on λ_{min}, $\|A\|_{L^\infty(B_R^+)}$, and $R\|A'\|_{L^\infty(B_R^+)}$ such that the weak solution u of*

$$-\nabla\cdot(A\nabla u) = f \quad \text{on } B_R^+, \qquad u = 0 \quad \text{on } \Gamma_R$$

satisfies

$$\|\nabla^2 u\|_{L^2(B_R^+)} \le C_B\|f\|_{L^2(B_R^+)}.$$

Finally, we need a local regularity result for transmission problems (see also Lemma 5.5.22 below for a version on balls B_R with $R < 1$).

Lemma 5.5.8 (H^2-regularity, transmission conditions). *Let $f \in L^2(B_1)$, $g \in H^{1/2}(\Gamma_1)$ (with $\Gamma_1 = \{x \in B_1 \,|\, x_n = 0\}$), $A \in L^\infty(B_1, \mathbb{S}^n_>)$ with $0 < \lambda_{min} \leq A$ on B_1. Assume that $A|_{B_1^+} \in C^1(\overline{B_1^+}, \mathbb{S}^n_>)$, $A|_{B_1^-} \in C^1(\overline{B_1^-}, \mathbb{S}^n_>)$. Then for every $R \in (0,1)$ there exists $C_R > 0$ depending on R, λ_{min}, $\|A\|_{L^\infty(B_1)}$, and $\|A'\|_{L^\infty(B_1^+ \cup B_1^-)}$ such that the weak solution $u \in H_0^1(B_1)$ of*

$$-\nabla \cdot (A \nabla u) = f \quad \text{on } B_1, \qquad [\partial_{n_A} u] = g \quad \text{on } \Gamma_1$$

satisfies

$$\|\nabla^2 u\|_{L^2(B_R^+ \cup B_R^-)} \leq C_R \left[\|f\|_{L^2(B_1)} + \|g\|_{H^{1/2}(\Gamma_1)} \right].$$

Proof: First, we note that the assumptions give the *a priori* bound

$$\|u\|_{H^1(B_1)} \leq C \left[\|f\|_{H^{-1}(B_1)} + \|g\|_{H^{-1/2}(\Gamma_1)} \right] \leq C \left[\|f\|_{L^2(B_1)} + \|g\|_{H^{1/2}(\Gamma_1)} \right].$$

Standard local regularity results also imply that $u \in H^2_{loc}(B_1^+ \cup B_1^-)$, and it merely remains to obtain the bound. This is proved with the method of tangential differential quotients of Nirenberg (see, e.g., [49, 62]) in a manner completely similar to that of local regularity for the Neumann problem (note that $R < 1$ so that our bounds are really just assertions at flats parts of the boundary of B_1^+, B_1^-). In particular, one checks that in this procedure, only tangential derivatives of A are needed to obtain bounds for the tangential derivatives of ∇u. Then, the differential equation is employed to get a bound on $\partial^2_{x_n} u$ on $B_R^+ \cup B_R^-$. □

Lemma 5.5.9 (H^2-regularity, Neumann conditions). *Let $f \in L^2(B_1^+)$, $g \in H^{1/2}(\Gamma_1)$ (with $\Gamma_1 = \{x \in B_1 \,|\, x_n = 0\}$), $A \in C^1(\overline{B_1^+})$ with $0 < \lambda_{min} \leq A$ on B_1^+. Then for every $R \in (0,1)$ there exists $C_R > 0$ depending on R, λ_{min}, $\|A\|_{L^\infty(B_1^+)}$, and $\|A'\|_{L^\infty(B_1^+)}$ such that the weak solution $u \in H_0^1(B_1^+)$ of*

$$-\nabla \cdot (A \nabla u) = f \quad \text{on } B_1^+, \qquad \partial_{n_A} u = g \quad \text{on } \Gamma_1$$

satisfies

$$\|\nabla^2 u\|_{L^2(B_R^+)} \leq C_R \left[\|f\|_{L^2(B_1^+)} + \|g\|_{H^{1/2}(\Gamma_1)} \right].$$

Proof: The proof is very similar to that of Lemma 5.5.8. □

Lemma 5.5.10. *Let $G \subset \mathbb{R}^n$ be open, $A \in C^{p+1}(G, \mathbb{M}^n)$, $u \in C^{p+2}(G)$. Then there holds on G*

$$\left\{ \sum_{|\alpha|=p} \frac{|\alpha|!}{\alpha!} \left| D^\alpha (\nabla \cdot (A \nabla u)) - \nabla \cdot (A D^\alpha \nabla u) \right|^2 \right\}^{1/2} \leq \sum_{q=1}^{p+1} \binom{p+1}{q} |\nabla^q A| \, |\nabla^{p+2-q} u|.$$

Proof: We start by noting that

$$\nabla \cdot (A\nabla u) = (\nabla \cdot A)\nabla u + A : \nabla^2 u,$$

where $\nabla \cdot A = (\sum_i \partial_i a_{ij})_{j=1}^n$, $\nabla^2 u = (\partial_i \partial_j u)_{i,j=1}^n$. Hence,

$$\begin{aligned} &D^\alpha(\nabla \cdot (A\nabla u)) - \nabla \cdot (AD^\alpha \nabla u) = \\ &\quad \left[D^\alpha((\nabla \cdot A)\nabla u) - (\nabla \cdot A)D^\alpha \nabla u\right] + \left[D^\alpha(A : \nabla^2 u) + A : (D^\alpha \nabla^2 u)\right]. \end{aligned}$$

Applying Lemma A.1.4 we arrive at

$$\begin{aligned} &\left\{\sum_{|\alpha|=p} \frac{|\alpha|!}{\alpha!} \left|D^\alpha(\nabla \cdot (A\nabla u)) - \nabla \cdot (AD^\alpha \nabla u)\right|^2\right\}^{1/2} \\ &\le \left\{\sum_{|\alpha|=p} \frac{|\alpha|!}{\alpha!} \left|D^\alpha((\nabla \cdot A)\nabla u) - (\nabla \cdot A)D^\alpha \nabla u\right|^2\right\}^{1/2} \\ &\qquad + \left\{\sum_{|\alpha|=p} \frac{|\alpha|!}{\alpha!} \left|D^\alpha(A : \nabla^2 u) - A : D^\alpha \nabla^2 u\right|^2\right\}^{1/2} \\ &\le \sum_{q=1}^{p} \binom{p}{q} |\nabla^{q+1} A|\,|\nabla^{p+1-q} u| + \sum_{q=1}^{p} \binom{p}{q} |\nabla^q A|\,|\nabla^{p+2-q} u| \\ &= |\nabla^{p+1} A|\,|\nabla u| + \binom{p}{1} |\nabla A|\,|\nabla^{p+1} u| + \sum_{q=2}^{p} \binom{p+1}{q} |\nabla^q A|\,|\nabla^{p+2-q} u| \\ &\le \sum_{q=1}^{p+1} \binom{p+1}{q} |\nabla^q A|\,|\nabla^{p+2-q} u|. \end{aligned}$$

□

5.5.2 Interior regularity: Proof of Proposition 5.5.1

We introduce the additional notation

$$M_{R,p}(v) := \frac{1}{p!} \sup_{R/2 \le r < R} (R-r)^{2+p} \|\nabla^p v\|_{L^2(B_r)}, \qquad p \in \mathbb{N}_0.$$

Lemma 5.5.11 (interior regularity). *Let $A \in C^1(\overline{B_R}, \mathbb{S}^n_>)$ with $0 < \lambda_{min} \le A(x)$ on B_R. Then there exists $C > 0$ depending only on λ_{min}, $\|A\|_{L^\infty(B_R)}$, $R\|A'\|_{L^\infty(B_R)}$ such that any solution u of*

$$-\nabla \cdot (A\nabla u) = f \in L^2(B_R) \quad \text{on } B_R$$

satisfies for all r, $\delta > 0$ with $r + \delta < R$,

$$\|\nabla^2 u\|_{L^2(B_r)} \le C\left[\|f\|_{L^2(B_{r+\delta})} + \delta^{-1}\|\nabla u\|_{L^2(B_{r+\delta})} + \delta^{-2}\|u\|_{L^2(B_{r+\delta})}\right].$$

Proof: The proof is essentially the same as that of [98, Lemma 5.7.1]. The key ingredient is to employ a smooth cut-off function χ being identically one on $B_{R-\delta}$, vanishing on $B_R \setminus B_{R-\delta/2}$, satisfying $|\nabla^j \chi| \le C\delta^{-j}$, $j \in \{0,1,2\}$ and then to consider the function $u\chi$. □

The following lemma generalizes this result to higher order derivatives. It should be noted that control of the L^2-norm of u is not required—control of derivatives of u suffices.

Lemma 5.5.12 (interior regularity). *Let $p \in \mathbb{N}_0$, $A \in C^{p+1}(\overline{B_R}, \mathbb{S}^n_>)$ with $0 < \lambda_{min} \le A$ on B_R, $f \in H^p(B_R)$. Then there exists $C_I' > 0$ depending only on λ_{min}, $\|A\|_{L^\infty(B_R)}$, $R\|A'\|_{L^\infty(B_R)}$, and n such that any solution u of*

$$-\nabla \cdot (A\nabla u) = f \qquad \text{on } B_R$$

satisfies for all $p \in \mathbb{N}_0$

$$N_{R,p}(u) \le C_I'\Big[M_{R,p}(f) + \sum_{q=1}^{p+1} \binom{p+1}{q} \left(\frac{R}{2}\right)^q \|\nabla^q A\|_{L^\infty(B_R)} \frac{[p-q]!}{p!} N_{R,p-q}(u) + N_{R,p-1}(u) + N_{R,p-2}(u)\Big].$$

For $p = 0$ we have the sharper bound

$$N_{R,0}(u) \le C_I' \left[\left(\frac{R}{2}\right)^2 \|f\|_{L^2(B_R)} + \frac{R}{2}\|\nabla u\|_{L^2(B_R)}\right]. \tag{5.5.22}$$

Proof: Let $p \in \mathbb{N}_0$. For all $\alpha \in \mathbb{N}_0^n$ with $|\alpha| = p$ the function $D^\alpha u$ satisfies

$$-\nabla \cdot (A\nabla D^\alpha u) = D^\alpha f + D^\alpha(\nabla \cdot (A\nabla u)) - \nabla \cdot \big(AD^\alpha \nabla u\big) \qquad \text{on } B_R.$$

By Lemma 5.5.5 we therefore get for r, $\delta > 0$ with $r + \delta < R$

$$\begin{aligned}\|\nabla^2 D^\alpha u\|^2_{L^2(B_r)} \le C\Big[&\|D^\alpha f\|^2_{L^2(B_{r+\delta})} \\ &+ \|D^\alpha(\nabla \cdot (A\nabla u)) - \nabla \cdot (D^\alpha \nabla u)\|^2_{L^2(B_{r+\delta})} \\ &+ \delta^{-2}\|D^\alpha \nabla u\|^2_{L^2(B_{r+\delta})} + \delta^{-4}\|D^\alpha u\|^2_{L^2(B_{r+\delta})}\Big].\end{aligned}$$

Multiplying by $\frac{|\alpha|!}{\alpha!}$ and summing on α with $|\alpha| = p$ we get

$$\begin{aligned}\|\nabla^{p+2} u\|^2_{L^2(B_r)} \le C\Big[&\|\nabla^p f\|^2_{L^2(B_{r+\delta})} \\ &+ \sum_{|\alpha|=p} \frac{|\alpha|!}{\alpha!}\|D^\alpha(\nabla \cdot (A\nabla u)) - \nabla \cdot (D^\alpha \nabla u)\|^2_{L^2(B_{r+\delta})} \\ &+ \delta^{-2}\|\nabla^{p+1} u\|^2_{L^2(B_{r+\delta})} + \delta^{-4}\|\nabla^p u\|^2_{L^2(B_{r+\delta})}\Big].\end{aligned}$$

Together with Lemma 5.5.10, we then arrive at

$$\begin{aligned}||\nabla^{p+2}u||_{L^2(B_{r+\delta})} \le C\Big[&||\nabla^p f||_{L^2(B_{r+\delta})}\\ &+\sum_{q=1}^{p+1}\binom{p+1}{q}||\nabla^q A||_{L^\infty(B_R)}||\nabla^{p+2-q}u||_{L^2(B_{r+\delta})}\\ &+\delta^{-1}||\nabla^{p+1}u||_{L^2(B_{r+\delta})}+\delta^{-2}||\nabla^p u||_{L^2(B_{r+\delta})}\Big]\end{aligned}$$

with $\delta > 0$ still at our disposal. From the definition of the quantities $M_{R,p}(f)$, $N_{R,p}(u)$ we infer

$$\begin{aligned}||\nabla^p f||_{L^2(B_{r+\delta})} &\le (R-r-\delta)^{-(p+2)}p!M_{R,p}(f),\\ ||\nabla^{p+2-q}u||_{L^2(B_{r+\delta})} &\le (R-r-\delta)^{-(p+2-q)}[p-q]!N_{R,p-q}(u),\\ ||\nabla^{p+1}u||_{L^2(B_{r+\delta})} &\le (R-r-\delta)^{-(p+1)}[p-1]!N_{R,p-1}(u),\\ ||\nabla^{p}u||_{L^2(B_{r+\delta})} &\le (R-r-\delta)^{-p}[p-2]!N_{R,p-2}(u).\end{aligned}$$

Choosing now

$$\delta := \frac{R-r}{p+2}, \qquad R-r-\delta = (R-r)\frac{p+1}{p+2}, \tag{5.5.23}$$

we arrive at

$$\begin{aligned}\frac{1}{p!}(R-r)^{p+2}||\nabla^{p+2}u||_{L^2(B_r)} \le C\Big[&\left\{\left(\frac{p+2}{p+1}\right)^{p+2}\right\}M_{R,p}(f)\\ &+\sum_{q=1}^{p+1}\binom{p+1}{q}||\nabla^q A||_{L^\infty(B_R)}\frac{[p-q]!}{p!}(R-r)^q\left\{\left(\frac{p+2}{p+1}\right)^{p+2-q}\right\}N_{R,p-q}(u)\\ &+\left\{\frac{(p+2)[p-1]!}{p!}\left(\frac{p+2}{p+1}\right)^{p+1}\right\}N_{R,p-1}(u)\\ &+\left\{\frac{(p+2)^2[p-2]!}{p!}\left(\frac{p+2}{p+1}\right)^{p}\right\}N_{R,p-2}(u)\Big].\end{aligned}$$

It is easy to see that the terms in curly braces can be bounded uniformly in p. Taking the supremum over $r \in (R/2, R)$ and bounding $(R-r)^q \le (R/2)^q$, we get the desired result.
We now turn to the special case $p = 0$. Replacing the function u by $u - \overline{u}$, where $\overline{u}$ is the average of u over B_R, we see that the same reasoning as above allows us to get with a Poincaré inequality the desired bound (5.5.22) if we appropriately adjust the constant C_I'. □

Lemma 5.5.13. *Let $b \in \mathcal{A}(B_R, \mathbb{R}^n)$, c, $u \in \mathcal{A}(B_R)$ and assume that b, c satisfy the estimates (5.5.2b). Then*

$$M_{R,p}(cu) \leq C_c \sum_{q=0}^{p} \left(\gamma_c \frac{R}{2}\right)^q \left(\frac{R}{2}\right)^2 \frac{[p-q-2]!}{(p-q)!} N_{R,p-q-2}(u)$$
$$= C_c \sum_{q=0}^{p-1} \left(\gamma_c \frac{R}{2}\right)^q \left(\frac{R}{2}\right)^2 \frac{[p-q-2]!}{(p-q)!} N_{R,p-q-2}(u)$$
$$+ C_c \left(\gamma_c \frac{R}{2}\right)^p \left(\frac{R}{2}\right)^2 N_{R,-2}(u), \tag{5.5.24}$$
$$M_{R,p}(b \cdot \nabla u) \leq C_b \sum_{q=0}^{p} \left(\gamma_b \frac{R}{2}\right)^q \frac{R}{2} \frac{[p-q-1]!}{(p-q)!} N_{R,p-q-1}(u). \tag{5.5.25}$$

Proof: We will only prove the second estimate. By Lemma A.1.3 we have

$$M_{R,p}(b \cdot \nabla u) \leq \frac{1}{p!} \sup_{R/2 \leq r < R} (R-r)^{p+2} \|\nabla^p (b \cdot \nabla u)\|_{L^2(B_r)}$$
$$\leq \frac{1}{p!} \sup_{R/2 \leq r < R} (R-r)^{p+2} \sum_{q=0}^{p} \binom{p}{q} \| |\nabla^{p-q} \nabla u| \, |\nabla^q b| \|_{L^2(B_r)}$$
$$\leq C_b \sum_{q=0}^{p} \binom{p}{q} \frac{q!}{p!} \gamma_b^q \sup_{R/2 \leq r < R} (R-r)^{p+2} \|\nabla^{p-q+1} u\|_{L^2(B_r)}.$$

The desired bound (5.5.25) is now obtained by observing

$$\sup_{R/2 \leq r < R} (R-r)^{p+2} \|\nabla^{p-q+1} u\|_{L^2(B_r)}$$
$$\leq \left(\frac{R}{2}\right)^{q+1} \sup_{R/2 \leq r < R} (R-r)^{(p-q-1)+2} \|\nabla^{(p-q-1)+2} u\|_{L^2(B_r)}$$
$$= \left(\frac{R}{2}\right)^{q+1} [p-q-1]! N_{R,p-q-1}(u).$$

□

Proof of Proposition 5.5.1: We start by choosing

$$K > \max\{1, \gamma_f/2, \gamma_b R/2, \gamma_c R/2, \gamma_A R/2\}$$

such that for C_I' of Lemma 5.5.12

$$C_I' \Big[\frac{1}{4} K^{-2} + K^{-1} \frac{1/2}{1 - \gamma_b R/(2K)} + K^{-2} \frac{1/4}{1 - \gamma_c R/(2K)}$$
$$+ K^{-1} + K^{-2} + K^{-1} \frac{C_A \gamma_A R/2}{1 - \gamma_A R/(2K)} \Big] \leq 1. \tag{5.5.26}$$

We first note that the claim holds true for $p = -1$ since $K \geq 1$ and

$$
\begin{aligned}
N_{R,-1}(u) &\leq \frac{R}{2}\|\nabla u\|_{L^2(B_R)} \leq \frac{1}{2}\frac{R}{\mathcal{E}}\mathcal{E}\|\nabla u\|_{L^2(B_R)} \\
&\leq \frac{1}{2}\max\{1, R/\mathcal{E}\}\min\{1, R/\mathcal{E}\}\,\mathcal{E}\|\nabla u\|_{L^2(B_R)} \leq \frac{1}{2}\max\{1, R/\mathcal{E}\}\,C_u.
\end{aligned}
$$

We now show that the claim is also true for $p = 0$. From (5.5.22) and the differential equation in the form $-\nabla \cdot (A\nabla u) = \varepsilon^{-2}[f - b\cdot\nabla u - cu]$ we get

$$
N_{R,0}(u) \leq
$$

$$
C_I'\left[\left(\frac{R}{2\varepsilon}\right)^2\|f\|_{L^2(B_R)} + \frac{C_b}{\varepsilon^2}\frac{R^2}{4}\|\nabla u\|_{L^2(B_R)} + \frac{C_c}{\varepsilon^2}\frac{R^2}{4}\|u\|_{L^2(B_R)} + \frac{R}{2}\|\nabla u\|_{L^2(B_R)}\right].
$$

From (5.3.9) and $R/\mathcal{E} = \min\{1, R/\mathcal{E}\}\max\{1, R/\mathcal{E}\}$ we get

$$
\begin{aligned}
\frac{R^4}{4\varepsilon^2} &\leq \frac{1}{4}(R/\mathcal{E})^2(\mathcal{E}/\varepsilon)^2 \leq \frac{1}{4}\min\{1, R/\mathcal{E}\}^2(\mathcal{E}/\varepsilon)^2\max\{1, R/\mathcal{E}\}^2, \\
\frac{C_b}{\varepsilon^2}\frac{R^2}{4} &\leq \frac{1}{4}\mathcal{E}^{-1}R^2 \leq \frac{1}{4}\mathcal{E}\min\{1, R/\mathcal{E}\}^2\max\{1, R/\mathcal{E}\}^2, \\
R &\leq \min\{1, R/\mathcal{E}\}\max\{1, R/\mathcal{E}\}\mathcal{E}.
\end{aligned}
$$

Inserting these bounds into the estimate for $N_{R,0}(u)$ gives

$$
N_{R,0}(u) \leq \max\{1, R/\mathcal{E}\}^2 C_I' C_u \leq K^2\max\{1, R/\mathcal{E}\}^2 C_u\, K^{-2}C_I'.
$$

By our choice of K, we have $K^{-2}C_I' \leq 1$ so that the desired bound for $p = 0$ holds.

Let us next proceed by induction on p and assume that (5.5.5) holds for all $-1 \leq p' < p$. With C_I' of Lemma 5.5.12 we have

$$
\begin{aligned}
N_{R,p}(u) \leq C_I'\Big[&\varepsilon^{-2}M_{R,p}(f - b\cdot\nabla u - cu) + N_{R,p-1}(u) + N_{R,p-2}(u) \\
&+ \sum_{q=1}^{p+1}\binom{p+1}{q}\left(\frac{R}{2}\right)^q\|\nabla^q A\|_{L^\infty(B_R)}\frac{[p-q]!}{p!}N_{R,p-q}(u)\Big].
\end{aligned}
$$

Next, $M_{R,p}(f - b\cdot\nabla u - cu) \leq M_{R,p}(f) + M_{R,p}(b\cdot\nabla u) + M_{R,p}(cu)$. We first turn our attention to $M_{R,p}(f)$. We have

$$
\begin{aligned}
\varepsilon^{-2}M_{R,p}(f) &\leq \varepsilon^{-2}\frac{1}{p!}C_f\gamma_f^p\left(\frac{R}{2}\right)^{p+2}\max\{p/R, \mathcal{E}^{-1}\}^p \\
&\leq \frac{R^2}{\mathcal{E}^2}\frac{\mathcal{E}^2}{\varepsilon^2}\frac{1}{p!}\frac{C_f}{4}\left(\frac{\gamma_f}{2}\right)^p\max\{p+3, R/\mathcal{E}\}^p.
\end{aligned}
$$

As $R^2/\mathcal{E}^2 \leq \max\{p+3, R/\mathcal{E}\}^2\min\{1, R/\mathcal{E}\}^2$ we infer

$$
\begin{aligned}
\varepsilon^{-2}M_{R,p}(f) &\leq \frac{\max\{p+3, R/\mathcal{E}\}^{p+2}}{p!}\cdot\frac{C_f}{4}\left(\frac{\gamma_f}{2}\right)^p\min\{1, R/\mathcal{E}\}^2(\mathcal{E}/\varepsilon)^2 \\
&\leq C_u K^{p+2}\frac{\max\{p+3, R/\mathcal{E}\}^{p+2}}{p!}\frac{1}{4}K^{-2}\left(\frac{\gamma_f}{2K}\right)^p.
\end{aligned}
$$

We now turn to $\varepsilon^{-2}M_{R,p}(b\cdot\nabla u)$. From

$$\begin{aligned}\frac{[p-q-1]!}{(p-q)!}N_{R,p-q-1}(u) &\le \frac{1}{p!}\frac{p!}{(p-q)!}[p-q-1]!N_{R,p-q-1}(u)\\ &\le \frac{1}{p!}p^q C_u K^{p-q+1}\max\{p+3,R/\mathcal{E}\}^{p-q+1}\\ &\le \frac{1}{p!}C_u K^{p-q+1}\max\{p+3,R/\mathcal{E}\}^{p+1},\end{aligned}$$

we infer from Lemma 5.5.13 and the observation $\varepsilon^{-2}C_b \le \mathcal{E}^{-1}$:

$$\begin{aligned}\varepsilon^{-2}M_{R,p}(b\cdot\nabla u) &\le \frac{R}{2}C_b\varepsilon^{-2}\sum_{q=0}^{p}\left(\frac{\gamma_b R}{2}\right)^q C_u K^{p-q+1}\frac{\max\{p+3,R/\mathcal{E}\}^{p+1}}{p!}\\ &\le C_u K^{p+2}\frac{\max\{p+3,R/\mathcal{E}\}^{p+2}}{p!}\frac{1}{2}K^{-1}\sum_{q=0}^{p}\left(\frac{\gamma_b R}{2K}\right)^q.\end{aligned}$$

In order to treat the term $M_{R,p}(cu)$, we first note that

$$\begin{aligned}\varepsilon^{-2}C_c R^2 N_{R,-2}(u) &= (\mathcal{E}/\varepsilon)^2 C_c (R/\mathcal{E})^2\|u\|_{L^2(B_R)}\\ &\le C_c(\mathcal{E}/\varepsilon)^2\min\{1,R/\mathcal{E}\}^2\max\{1,R/\mathcal{E}\}^2\|u\|_{L^2(B_R)}\\ &\le C_u\frac{p!}{p!}\max\{1,R/\mathcal{E}\}^2 \le C_u\frac{\max\{p+3,R/\mathcal{E}\}^{p+2}}{p!}.\end{aligned}$$

Hence, observing that $\varepsilon^{-2}C_c \le \mathcal{E}^{-2}$ we can bound $M_{R,p}(cu)$ using Lemma 5.5.13 as follows:

$$\begin{aligned}\varepsilon^{-2}M_{R,p}(cu) \le& \left(\frac{R}{2}\right)^2 C_c\varepsilon^{-2}\sum_{q=0}^{p-1}\left(\frac{\gamma_c R}{2}\right)^q C_u K^{p-q}\frac{\max\{p+3,R/\mathcal{E}\}^p}{p!}\\ &+\left(\frac{\gamma_c R}{2K}\right)^p\frac{1}{4K^2}C_u K^{p+2}\frac{\max\{p+3,R/\mathcal{E}\}^{p+2}}{p!}\\ \le& C_u K^{p+2}\frac{\max\{p+3,R/\mathcal{E}\}^{p+2}}{p!}\frac{1}{4}K^{-2}\sum_{q=0}^{p}\left(\frac{\gamma_c R}{2K}\right)^q.\end{aligned}$$

Estimating

$$\begin{aligned}\frac{1}{p!}\frac{p!}{[p-1]!}\max\{p+2,R/\mathcal{E}\}^{p+1} &\le \frac{\max\{p+3,R/\mathcal{E}\}^{p+2}}{p!},\\ \frac{1}{p!}\frac{p!}{[p-2]!}\max\{p+1,R/\mathcal{E}\}^{p} &\le \frac{\max\{p+3,R/\mathcal{E}\}^{p+2}}{p!},\end{aligned}$$

we can bound

$$N_{R,p-1}(u)+N_{R,p-2}(u) \le C_u K^{p+2}\frac{\max\{p+3,R/\mathcal{E}\}^{p+2}}{p!}\left[K^{-1}+K^{-2}\right].$$

Finally, bounding $\binom{p+1}{q} \leq \frac{(p+1)^q}{q!}$ we derive

$$\sum_{q=1}^{p+1} \binom{p+1}{q} \left(\frac{R}{2}\right)^q \|\nabla^q A\|_{L^\infty(B_R)} \frac{[p-q]!}{p!} N_{R,p-q}(u) \leq$$
$$C_u C_A \sum_{q=1}^{p+1} \left(\frac{\gamma_A R}{2}\right)^q K^{p-q+2}(p+1)^q \frac{\max\{p-q+3, R/\mathcal{E}\}^{p-q+2}}{p!} \leq$$
$$C_u K^{p+2} \frac{\max\{p+3, R/\mathcal{E}\}^{p+2}}{p!} C_A \sum_{q=1}^{p+1} \left(\frac{\gamma_A R}{2K}\right)^q.$$

This allows us to conclude

$$N_{R,p}(u) \leq C_u K^{p+2} \frac{\max\{p+3, R/\mathcal{E}\}^{p+2}}{p!} C_I' \left[\frac{1}{4} K^{-2} \left(\frac{\gamma_f}{2K}\right)^p \right.$$
$$\left. + K^{-1} \frac{1/2}{1-\gamma_b R/(2K)} + K^{-2} \frac{1/4}{1-\gamma_c R/(2K)} + K^{-1} + K^{-2} + \frac{C_A \gamma_A R/(2K)}{1-\gamma_A R/(2K)}\right].$$

As the expression in brackets is bounded by 1 uniformly in p by our choice of K in (5.5.26), the induction argument is completed. □

5.5.3 Regularity at the boundary: Proof of Proposition 5.5.2

We will consider half-balls at the boundary and show analytic regularity on such half-balls. This will be done in two steps: First, the growth of the tangential derivatives (x denotes the tangential variables; derivatives of u with respect to x are denoted in this section by $\nabla_x u$) is controlled. In a second step, the derivatives of u in the normal direction (denoted y) are controlled. Most of the arguments are quite similar to analogous ones of Subsection 5.5.2; proofs will therefore only be provided for selected steps.

We start by introducing the following auxiliary notation suitable for controlling tangential derivatives:

$$M'_{R,p}(v) = \frac{1}{p!} \sup_{R/2 \leq r < R} (R-r)^{p+2} \|\nabla_x^p v\|_{L^2(B_r^+)},$$
$$N'_{R,p}(v) = \begin{cases} \frac{1}{p!} \sup_{R/2 \leq r < R} (R-r)^{p+2} \|\nabla^2 \nabla_x^p v\|_{L^2(B_r^+)} & \text{if } p \geq 0 \\ \sup_{R/2 \leq r < R} (R-r)^{p+2} \|\nabla^{2+p} v\|_{L^2(B_r^+)} & \text{if } p = -2, -1, \end{cases}$$
$$\tilde{M}_{R,p}(v) = \frac{1}{p!} \sup_{R/2 \leq r < R} (R-r)^{p+2} \|\nabla^p v\|_{L^2(B_r^+)}.$$

Control of tangential derivatives. Analogous to Lemma 5.5.10 is the following result.

Lemma 5.5.14. *Let $G \subset \mathbb{R}^n$ be open, $A \in C^{p+1}(G, \mathbb{M}^n)$, $u \in C^{p+2}(G)$. Denote by D_x^α differentiation with respect to the variables x. Then on G*

$$\left\{ \sum_{|\alpha|=p} \frac{|\alpha|!}{\alpha!} \left| D_x^\alpha (\nabla \cdot (A \nabla u)) - \nabla \cdot \left(A D_x^\alpha \nabla u \right) \right|^2 \right\}^{1/2} \leq \sum_{q=1}^{p+1} \binom{p+1}{q} |\nabla^q A| \, |\nabla_x^{p-q} \nabla^2 u|.$$

The analog of Lemma 5.5.12 is the following lemma.

Lemma 5.5.15. *Let $p \in \mathbb{N}_0$, $A \in C^{p+1}(\overline{B_R^+}, \mathbb{S}^n_>)$ with $0 < \lambda_{min} \leq A$ on B_R^+, $f \in H^p(B_R^+)$. Then there exists $C'_B > 0$ depending only on λ_{min}, $\|A\|_{L^\infty(B_R^+)}$, $R\|A'\|_{L^\infty(B_R^+)}$, and n such that for all solutions u of*

$$-\nabla \cdot (A \nabla u) = f \quad \text{on } B_R^+, \qquad u = 0 \quad \text{on } \Gamma_R$$

there holds

$$N'_{R,p}(u) \leq C'_B \Big[M'_{R,p}(f) + \sum_{q=1}^{p+1} \binom{p+1}{q} \left(\frac{R}{2} \right)^q \|\nabla^q A\|_{L^\infty(B_R^+)} \frac{[p-q]!}{p!} N'_{R,p-q}(u)$$
$$+ N'_{R,p-1}(u) + N'_{R,p-2}(u) \Big].$$

For $p = 0$, we have the sharper bound

$$N'_{R,0}(u) \leq C \left[\left(\frac{R}{2} \right)^2 \|f\|_{L^2(B_R^+)} + \frac{R}{2} \|\nabla u\|_{L^2(B_R^+)} \right]. \tag{5.5.27}$$

Proposition 5.5.16 (control of tangential derivatives). *Let $R \in (0,1]$ and Let $A \in \mathcal{A}(B_R^+, \mathbb{S}^n_>)$, $b \in \mathcal{A}(B_R^+, \mathbb{R}^n)$, $c \in \mathcal{A}(B_R^+)$, $f \in \mathcal{A}(B_R^+)$ satisfy (5.5.7), (5.5.8). Let $\varepsilon \in (0,1]$ and let $\mathcal{E}$ be defined by (5.5.3). Finally let u be a solution of*

$$-\varepsilon^2 \nabla \cdot (A \nabla u) + b \cdot \nabla u + cu = f \quad \text{on } B_R^+, \qquad u = 0 \quad \text{on } \Gamma_R.$$

Then for some $K_1 > 0$ depending only on the constants of (5.5.7) and γ_f of (5.5.8) there holds

$$N'_{R,p}(u) \leq C_u K_1^{p+2} \frac{\max\{(p+3), R/\mathcal{E}\}^{p+2}}{[p]!}, \qquad p \geq -1, \tag{5.5.28}$$
$$C_u := \min\{1, R/\mathcal{E}\} \mathcal{E} \|\nabla u\|_{L^2(B_R^+)}$$
$$+ (\mathcal{E}/\varepsilon)^2 \min\{1, R/\mathcal{E}\}^2 \left[C_c \|u\|_{L^2(B_R^+)} + C_f \right]. \tag{5.5.29}$$

Proof: The proof proceeds by induction analogous to the way Proposition 5.5.1 is proved. The cases $p = -1$ and $p = 0$ are shown, as in the proof of Proposition 5.5.1, by inspection and Lemma 5.5.15. The induction argument then parallels that of Proposition 5.5.1 using Lemma 5.5.15. □

Control of normal derivatives. We now turn to bounding the normal derivatives. This is done using the fact that u satisfies a differential equation. We stress the fact that boundary conditions are immaterial for this step. We start with the following lemma.

Lemma 5.5.17. *Let* $R > 0$, $A \in C^\infty(B_R^+, \mathbb{M}^n)$, $u \in C^\infty(B_R^+)$, $A_{nn} \equiv 0$ *on* B_R^+. *Then there holds*

$$\frac{1}{[p+q]!} \sup_{R/2 \le r < R} (R-r)^{p+q+2} \|\nabla_x^p \partial_{x_n}^q (A : \nabla^2 u)\|_{L^2(B_r^+)} \le$$
$$\sum_{r=0}^{p} \sum_{s=0}^{q} \binom{p}{r} \binom{q}{s} \|\partial_{x_n}^s \nabla_x^r A\|_{L^\infty(B_R^+)} \left(\frac{R}{2}\right)^{p+s} \times$$
$$\frac{[p-r+q-s]!}{[p+q]!} \left[N'_{R,p+1-r,q-1-s}(u) + N'_{R,p+2-r,q-s-2}(u)\right].$$

Proof: As $A_{nn} = 0$, we have $A : \nabla^2 u = \sum_{(i,j) \neq (n,n)} A_{ij} \partial_i \partial_j u$. By Lemma A.1.5 therefore on B_R^+ for p, $q \in \mathbb{N}_0$

$$|\nabla_x^p \partial_{x_n}^q A : \nabla^2 u| \le \sum_{(i,j)\neq(n,n)} \sum_{r=0}^{q} \sum_{s=0}^{q} \binom{p}{r} \binom{q}{s} |\nabla_x^r \partial_{x_n}^s A_{ij}| |\nabla_x^{p-r} \partial_{x_n}^{q-s} \partial_i \partial_j u|$$
$$\le \sum_{r=0}^{q} \sum_{s=0}^{q} \binom{p}{r} \binom{q}{s} \Big| \sum_{(i,j)\neq(n,n)} |\nabla_x^r \partial_{x_n}^s A_{ij}|^2 \Big|^{1/2} \Big| \sum_{(i,j)\neq(n,n)} |\nabla_x^{p-r} \partial_{x_n}^{q-s} \partial_i \partial_j u|^2 \Big|^{1/2}$$
$$\le \sum_{r=0}^{q} \sum_{s=0}^{q} \binom{p}{r} \binom{q}{s} |\nabla_x^r \partial_{x_n}^s A| \left[|\nabla_x^{p-r+2} \partial_{x_n}^{q-s} u| + |\nabla_x^{p-r+1} \partial_{x_n}^{q-s+1} u|\right].$$

From the definition of $N'_{p,q,R}(u)$ we infer

$$\sup_{R/2 \le r < R} (R-r)^{p+q+2} \|\nabla_x^{p-r+2} \partial_{x_n}^{q-s} u\|_{L^2(B_r^+)}$$
$$\le \left(\frac{R}{2}\right)^{r+s} \sup_{R/2 \le r < R} (R-r)^{(p-r+2)+(q-2-s)+2} \|\nabla_x^{p-r+2} \partial_{x_n}^{(q-2-s)+2} u\|_{L^2(B_r^+)}$$
$$\le \left(\frac{R}{2}\right)^{r+s} [p-r+q-s]! N'_{R,p-r+2,q-s-2}(u)$$

and also

$$\sup_{R/2 \le r < R} (R-r)^{p+q+2} \|\nabla_x^{p-r+1} \partial_{x_n}^{q-s+1} u\|_{L^2(B_r^+)}$$
$$\le \left(\frac{R}{2}\right)^{r+s} [p-r+q-s]! N'_{R,p-r+1,q-s-1}(u).$$

The result now follows easily. □

We obtain by similar reasoning a result of the following form:

Lemma 5.5.18. *Let $R > 0$, $b \in C^\infty(B_R^+, \mathbb{R}^n)$, $c \in C^\infty(B_R^+)$, $u \in C^\infty(B_R^+)$. Then there holds*

$$\frac{1}{[p+q]!} \sup_{R/2 \le r < R} (R-r)^{p+q+2} \|\nabla_x^p \partial_{x_n}^q (b \cdot \nabla u)\|_{L^2(B_r^+)} \le$$
$$\sum_{r=0}^{p} \sum_{s=0}^{q} \binom{p}{r} \binom{q}{s} \|\partial_{x_n}^s \nabla_x^r b\|_{L^\infty(B_R^+)} \left(\frac{R}{2}\right)^{r+s+1} \times$$
$$\frac{[p-r+q-s-1]!}{[p+q]!} \left[N'_{R,p-r,q-1-s}(u) + N'_{R,p+1-r,q-s-2}(u)\right],$$

and

$$\frac{1}{[p+q]!} \sup_{R/2 \le r < R} (R-r)^{p+q+2} \|\nabla_x^p \partial_{x_n}^q (cu)\|_{L^2(B_r^+)} \le$$
$$\sum_{r=0}^{p} \sum_{s=0}^{q} \binom{p}{r} \binom{q}{s} \|\partial_{x_n}^s \nabla_x^r c\|_{L^\infty(B_R^+)} \left(\frac{R}{2}\right)^{r+s+2} \frac{[p-r+q-s-2]!}{[p+q]!} N'_{R,p-r,q-2-s}(u)$$
$$= \sum_{\substack{0 \le r \le p \\ 0 \le s \le q \\ (r,s) \ne (p,q)}} \binom{p}{r} \binom{q}{s} \|\partial_{x_n}^s \nabla_x^r c\|_{L^\infty(B_R^+)} \left(\frac{R}{2}\right)^{r+s+2} \frac{[p-r+q-s-2]!}{[p+q]!} N'_{R,p-r,q-2-s}(u)$$
$$+ \frac{1}{[p+q]!} \|\partial_{x_n}^q \nabla_x^p c\|_{L^\infty(B_R^+)} \left(\frac{R}{2}\right)^{p+q+2} N_{R,0,-2}(u).$$

Lemma 5.5.19. *Let $G \subset \mathbb{R}^n$ be open, $a \in \mathcal{A}(G)$, $a \ge \lambda > 0$ on G and let a satisfy for some C_A, $\gamma_A > 0$*

$$\|\nabla^p a\|_{L^\infty(G)} \le C_A \gamma_A^p p! \qquad \forall p \in \mathbb{N}_0 .$$

Then there are C'_A, $\gamma'_A > 0$ depending only on C_A, γ_A, and λ (in particular, they are independent of G) such that the function $\widetilde{a} := 1/a$ satisfies

$$\|\nabla^p \widetilde{a}\|_{L^\infty(G)} \le C'_A (\gamma'_A)^p p! \qquad \forall p \in \mathbb{N}_0 .$$

Proof: Follows from Cauchy's integral representation of derivatives. □

Lemma 5.5.20. *Let $G \subset \mathbb{R}^n$ be open and let a, $b \in \mathcal{A}(G)$ satisfy*

$$\|\nabla^p a\|_{L^\infty(G)} \le C_A \gamma_A^p p!, \qquad \|\nabla^p b\|_{L^2(G)} \le C_B \gamma_B^p p! \qquad \forall p \in \mathbb{N}_0 .$$

Then the product ab satisfies

$$\|\nabla^p (ab)\|_{L^2(G)} \le C_A C_B (\gamma_A + \gamma_B)^p p! \qquad \forall p \in \mathbb{N}_0 .$$

Proof: By Lemma A.1.3, we have on G

$$|\nabla^p(ab)(x)| \le \sum_{q=0}^{p} \binom{p}{q} |\nabla^q a(x)|\,|\nabla^q b(x)| \le \sum_{q=0}^{p} \binom{p}{q} \|\nabla^q a\|_{L^\infty(G)}\,|\nabla^{p-q} b(x)|.$$

Hence,

$$\begin{aligned}\|\nabla^p(ab)\|_{L^2(G)} &\le C_A \sum_{q=0}^{p} \frac{p!}{(p-q)!}\gamma_A^q \|\nabla^{p-q} b\|_{L^2(G)}\\ &\le C_A C_B \sum_{q=0}^{p} p!\gamma_A^q \gamma_B^{p-q} \le C_A C_B p! (\gamma_A + \gamma_B)^p.\end{aligned}$$

□

Proof of Proposition 5.5.2: We will choose K_1, $K_2 > 1$ sufficiently large below (however, both constants will only depend on the constants of (5.5.7) and γ_f of (5.5.8)). As we choose $K_2 \ge 1$ we see that the claim (5.5.9) holds true for $q = 0$ by Proposition 5.5.16 and the observation that $N'_{R,p,0} \le N'_{R,p}$ for all $p \in \mathbb{N}_0$. For $q = -2$ and $q = -1$, we calculate directly for $p \in \mathbb{N}_0$:

$$\begin{aligned}N'_{R,p,-2}(u) &\le \frac{1}{[p-2]!} \sup_{R/2\le r<R} (R-r)^p \|\nabla_x^p u\|_{L^2(B_r^+)} \le N'_{R,p-2}(u),\\ N'_{R,p,-1}(u) &\le \frac{1}{[p-1]!} \sup_{R/2\le r<R} (R-r)^p \|\partial_{x_n} \nabla_x^p u\|_{L^2(B_r^+)} \le N'_{R,p-1}(u).\end{aligned}$$

We now proceed by induction on q. Let us assume that (5.5.9) holds for all $-2 \le q' < q$ and all $p \in \mathbb{N}_0$ with $p + q \ge -1$. Upon denoting by $\hat{A}$ the $n \times n$ matrix given by $\hat{A}_{ij} = A_{ij}$ for $(i,j) \ne (n,n)$ and $\hat{A}_{nn} = 0$ we note that u satisfies

$$-A_{nn}\partial_n^2 u = \varepsilon^{-2}\left[f - b\cdot\nabla u - cu\right] + (\nabla\cdot A)\nabla u + \hat{A} : \nabla^2 u.$$

Thus, after divided both sides by A_{nn}, the function u satisfies

$$-\partial_n^2 u = \varepsilon^{-2}\left[\widetilde{f} - \widetilde{b}\cdot\nabla u - \widetilde{c}u\right] + \widetilde{A}\nabla u + B : \nabla^2 u, \tag{5.5.30}$$

where, by Lemmata 5.5.19, 5.5.20, 4.3.1 there are C', $\gamma > 0$ depending only on λ_{min}, C_A, γ_A, γ_f, γ_b, γ_c such that for all $p \in \mathbb{N}_0$

$$\begin{aligned}&\|\nabla^p \widetilde{A}\|_{L^\infty(B_R^+)} \le C'\gamma^p p!,\\ &\|\nabla^p B\|_{L^\infty(B_R^+)} \le C'\gamma^p p!, \qquad B_{nn} = 0,\\ &\|\nabla^p \widetilde{f}\|_{L^2(B_R^+)} \le C_f C'\gamma^p \max\{p/R, \mathcal{E}^{-1}\}^p,\\ &\|\nabla^p \widetilde{b}\|_{L^\infty(B_R^+)} \le C_b C'\gamma^p p!, \qquad \|\nabla^p \widetilde{c}\|_{L^\infty(B_R^+)} \le C_c C'\gamma^p p!.\end{aligned}$$

We now use (5.5.30) in order to bound

$$N'_{R,p,q}(u) = \frac{1}{[p+q]!} \sup_{R/2\le r<R} (R-r)^{p+q+2} \|\nabla_x^p \partial_{x_n}^{q+2} u\|_{L^2(B_r^+)}.$$

We start with the contribution from $\widetilde{f}$: For $r \in [R/2, R)$ we have

$$
\begin{aligned}
&(R-r)^{p+q+2}\varepsilon^{-2}\|\nabla_x^p \partial_{x_n}^q \widetilde{f}\|_{L^2(B_r^+)} \\
&\leq \varepsilon^{-2}\left(\frac{R}{2}\right)^{p+q+2} C_f C' \gamma^{p+q} \max\{(p+q)/R, \mathcal{E}^{-1}\}^{p+q} \\
&\leq \frac{C'}{4} C_f \left(\frac{\gamma}{2}\right)^{p+q} \left(\frac{R}{\mathcal{E}}\right)^2 \left(\frac{\mathcal{E}}{\varepsilon}\right)^2 \max\{(p+q), R/\mathcal{E}\}^{p+q}.
\end{aligned}
$$

Bounding $(R/\mathcal{E})^2 \leq \min\{1, R/\mathcal{E}\}^2 \max\{p+q+3, R/\mathcal{E}\}^2$ we arrive at

$$
\begin{aligned}
&\frac{1}{[p+q]!}(R-r)^{p+q+2}\varepsilon^{-2}\|\nabla_x^p \partial_{x_n}^q \widetilde{f}\|_{L^2(B_r^+)} \\
&\leq \frac{C'}{4} C_f \left(\frac{\gamma}{2}\right)^{p+q} \min\{1, R/\mathcal{E}\}^2 \left(\frac{\mathcal{E}}{\varepsilon}\right)^2 \frac{\max\{(p+q), R/\mathcal{E}\}^{p+q+2}}{[p+q]!} \\
&\leq C_u K_1^{p+2} K_2^{q+2} \frac{\max\{(p+q), R/\mathcal{E}\}^{p+q+2}}{[p+q]!} \left[K_1^{-2} K_2^{-2} \frac{C'}{4}\left(\frac{\gamma}{2K_1}\right)^p \left(\frac{\gamma}{2K_2}\right)^q\right]
\end{aligned}
$$

with C_u of (5.5.10). We now turn to the term involving $\widetilde{b} \cdot \nabla u$, which we handle using Lemma 5.5.18. By the induction hypothesis

$$
\begin{aligned}
&[p-r+q-s-1]!\left[N'_{R,p-r,q-1-s}(u) + N'_{R,p+1-r,q-s-2}(u)\right] \leq \\
&\quad C_u K_1^{p-r+2} K_2^{q-s+2} \max\{p+q+3, R/\mathcal{E}\}^{p+q-r-s+1}\left[K_2^{-1} + K_1 K_2^{-2}\right].
\end{aligned}
$$

Next, we observe that

$$
\begin{aligned}
\varepsilon^{-2}\binom{p}{r}\binom{q}{s}\|\nabla_x^r \partial_{x_n}^s \widetilde{b}\|_{L^\infty(B_R^+)} &\leq C' C_b \varepsilon^{-2}\binom{p}{r}\binom{q}{s}(r+s)!\gamma^{r+s} \\
&\leq C'\mathcal{E}^{-1} p^r q^s (2\gamma)^{r+s}.
\end{aligned}
$$

Thus, by Lemma 5.5.18

$$
\begin{aligned}
&\frac{1}{[p+q]!}\varepsilon^{-2} \sup_{R/2 \leq r < R} (R-r)^{p+q+2}\|\nabla_x^p \partial_{x_n}^q (\widetilde{b}\cdot\nabla u)\|_{L^2(B_r^+)} \\
&\leq \frac{1}{[p+q]!}\frac{C'}{2} C_u\left[K_2^{-1} + K_1 K_2^{-2}\right] \times \\
&\qquad \sum_{r=0}^{p}\sum_{s=0}^{q} \frac{R}{\mathcal{E}}(\gamma R)^{r+s} p^r q^s K_1^{p-r+2} K_2^{q-s+2} \max\{p+q+3, R/\mathcal{E}\}^{p+q-r-s+1} \\
&\leq C_u K_1^{p+2} K_2^{q+2} \frac{\max\{p+q+3, R/\mathcal{E}\}^{p+q+2}}{[p+q]!} \frac{C'}{2}\left[K_2^{-1} + K_1 K_2^{-2}\right] \times \\
&\qquad \sum_{r=0}^{p}\sum_{s=0}^{q}\left(\frac{\gamma R}{K_1}\right)^r \left(\frac{\gamma R}{K_2}\right)^s.
\end{aligned}
$$

We now turn to the terms involving $\widetilde{c}u$. First, we note that $N'_{0,-2}(u) = \|u\|_{L^2(B_R^+)}$; thus,

$$\varepsilon^{-2}\frac{1}{[p+q]!}\|\partial^q_{x_n}\nabla^p_x\widetilde{c}\|_{L^\infty(B^+_R)}\left(\frac{R}{2}\right)^{p+q+2} N_{R,0,-2}(u)$$
$$\le\frac{1}{4}C'\left(\frac{\gamma R}{2}\right)^{p+q}\frac{[p+q]!}{[p+q]!}(R/\mathcal{E})^2C_c(\mathcal{E}/\varepsilon)^2\|u\|_{L^2(B^+_R)}$$
$$\le\frac{1}{4}C'\left(\frac{\gamma R}{2}\right)^{p+q}C_c(\mathcal{E}/\varepsilon)^2\min\{1,R/\mathcal{E}\}^2\max\{1,R/\mathcal{E}\}^2\|u\|_{L^2(B^+_R)}$$
$$\le\frac{1}{4}C'\left(\frac{\gamma R}{2}\right)^{p+q}\frac{\max\{p+q+3,R/\mathcal{E}\}^{p+q+2}}{[p+q]!}C_u.$$

This estimate implies for the term involving $\widetilde{c}u$:

$$\frac{1}{[p+q]!}\varepsilon^{-2}\sup_{R/2\le r<R}(R-r)^{p+q+2}\|\nabla^p_x\partial^q_{x_n}(\widetilde{c}u)\|_{L^2(B^+_r)}\le$$
$$C_uK_1^{p+2}K_2^{q+2}\frac{\max\{p+q+3,R/\mathcal{E}\}^{p+q+2}}{[p+q]!}\frac{C'}{4}K_2^{-2}\sum_{r=0}^{p}\sum_{s=0}^{q}\left(\frac{\gamma R}{K_1}\right)^r\left(\frac{\gamma R}{K_2}\right)^s.$$

It remains to bound the terms involving $\widetilde{A}\nabla u$ and $B:\nabla^2 u$. Reasoning analogously as in the treatment of the term involving $\widetilde{b}\cdot\nabla u$, we obtain

$$\frac{1}{[p+q]!}\sup_{R/2\le r<R}(R-r)^{p+q+2}\|\nabla^p_x\partial^q_{x_n}(\widetilde{A}\nabla u)\|_{L^2(B^+_r)}\le C_uK_1^{p+2}K_2^{q+2}.$$

Finally, for $B:\nabla^2 u$ we apply Lemma 5.5.17. First, we note that by the induction hypothesis

$$[p-r+q-s]!\left[N'_{R,p-r+1,q-1-s}(u)+N'_{R,p+2-r,q-s-2}(u)\right]\le$$
$$C_uK_1^{p-r+2}K_2^{q-s+2}\max\{p+q+3,R/\mathcal{E}\}^{p+q-r-s+2}\left[K_1K_2^{-1}+K_1^2K_2^{-2}\right].$$

Inserting this estimate in the bound of Lemma 5.5.17 applied to $B:\nabla^2 u$ we get

$$\frac{1}{[p+q]!}\sup_{R/2\le r<R}(R-r)^{p+q+2}\|\nabla^p_x\partial^q_{x_n}(B:\nabla^2 u)\|_{L^2(B^+_r)}$$
$$\le C_uK_1^{p+2}K_2^{q+2}\frac{\max\{p+q+3,R/\mathcal{E}\}^{p+q+2}}{[p+q]!}C'\left[K_1K_2^{-1}+K_1^2K_2^{-2}\right]\times$$
$$\sum_{r=0}^{p}\sum_{s=0}^{q}\left(\frac{\gamma R}{K_1}\right)^r\left(\frac{\gamma R}{K_2}\right)^s.$$

Next, we assume that $K_1>\max\{1,\gamma/2,\gamma R\}$, $K_2>\max\{1,\gamma/2,\gamma R\}$, so that all geometric sums arising in the above estimates are convergent series. Hence, by combing all the above estimates and observing that $\max\{p+q+3,R/\mathcal{E}\}^{p+q+1}\le\max\{p+q+3,R/\mathcal{E}\}^{p+q+2}$, we get

$$N'_{R,p,q}(u) \le C_u K_1^{p+2} K_2^{q+2} \frac{\max\{p+q+3, R/\mathcal{E}\}^{p+q+2}}{[p+q]!} \times$$
$$\Big[\frac{C'}{4} K_1^{-2} K_2^{-2} + \frac{C'}{(1-\gamma R/K_1)(1-\gamma R/K_2)} \Big\{ \frac{1}{2}(K_2^{-1} + K_1 K_2^{-2})$$
$$+ \frac{1}{4} K_2^{-2} + \frac{R}{2}[K_2^{-1} + K_1 K_2^{-2}] + [K_1 K_2^{-2} + K_1^2 K_2^{-2}] \Big\}\Big].$$

Choose now $K_1 > \max\{1, \gamma/2, \gamma R\}$ such that it is greater than the constant K_1 given by Proposition 5.5.16. Next, choose $K_2 > \max\{1, \gamma/2, \gamma R\}$ such that the expression in brackets is bounded by one. This allows us to complete the induction argument. □

5.5.4 Regularity of transmission problems: Proof of Prop. 5.5.4

The proof of Proposition 5.5.4 is very similar to that of Proposition 5.5.2. One proceeds in two steps by first estimating the growth of the derivatives in the tangential direction and then controlling the derivatives in the normal direction. In order to do that, we need the analog of Lemma 5.5.11. To infer such a result from Lemma 5.5.22, we need the following lemma.

Lemma 5.5.21. *Let $A \in L^\infty(B_1, \mathbb{S}^n_>)$ satisfy $0 < \lambda_{min} \le A$ on B_1. Assume that the restrictions $A^+ := A|_{B_1^+}$, $A^- := A|_{B_1^-}$ satisfy $A^+ \in C^1(\overline{B_1^+}, \mathbb{S}^n_>)$, $A^- \in C^1(\overline{B_1^-}, \mathbb{S}^n_>)$. Then there exists $C > 0$ depending only on λ_{min}, $\|A\|_{C^1(B_1^+)} + \|A\|_{C^1(B_1^-)}$ such that for every $\delta \in (0,1)$ we can find a piecewise smooth cut-off function $\chi_\delta \in W^{1,\infty}(\mathbb{R}^n)$, $\chi_\delta \equiv 1$ on $B_{1-\delta}(0)$, $\chi \equiv 0$ on $\mathbb{R}^2 \setminus B_{1-\delta/2}(0)$ with the following properties:*

1. *$\partial_{n_{A^+}} \chi_\delta = 0$ and $\partial_{n_{A^-}} \chi_\delta = 0$ on $x_n = 0$; in particular, therefore, $[\partial_{n_A} \chi_\delta] = 0$ on $x_n = 0$;*
2. *$\|\nabla^j \chi_\delta\|_{L^\infty(B_1^+ \cup B_1^-)} \le C\delta^{-j}$ for $j \in \{0,1,2\}$.*

Proof: For simplicity of exposition, we construct χ_δ in the two-dimensional case $n = 2$ only. We construct χ_δ in polar coordinates (r, φ). Furthermore, we will only construct χ_δ for the right half-plane (i.e., $\varphi > 0$); the general case follows easily.
Let χ be a smooth cut-off function satisfying $\chi \equiv 1$ for $r \in (0, 1-\delta)$, $\chi \equiv 0$ for $r > 1 - \delta/2$ and $\|\chi^{(j)}\|_{L^\infty(\mathbb{R})} \le C\delta^{-j}$, $j \in \{0,1,2\}$ for some $C > 0$ independent of δ. We compute for the co-normal derivatives on the line $\varphi = 0$. To that end, we write $\nabla \chi_\delta$ as

$$\nabla \chi_\delta = (\mathbf{n} \cdot \nabla \chi_\delta)\mathbf{n} + (\mathbf{t} \cdot \nabla \chi_\delta)\mathbf{t} = -\frac{\partial_\varphi \chi_\delta}{r}\mathbf{n} + (\partial_r \chi_\delta)\,\mathbf{t},$$

where $\mathbf{n}$ is the outer normal vector on $\varphi = 0$ and $\mathbf{t}$ is the (normalized) tangential vector. Inserting this decomposition in the definition of $\partial^{\chi}_{n_{A^+}}\delta$ gives

$$\partial_{n_{A^+}}\chi_\delta = -(\mathbf{n}^\top A^+\mathbf{n})\frac{\partial_\varphi\chi_\delta}{r} + (\mathbf{n}^T A^+\mathbf{t})\,\partial_r\chi_\delta =: a^+(r)\frac{\partial_\varphi\chi_\delta}{r} + b^+(r)\partial_r\chi_\delta(r),$$

where the functions a^+, b^+ are smooth. Furthermore, from $\mathbf{n}^T A^+\mathbf{n} \geq \lambda_{min} > 0$ we get that $|a^+|$ is bounded from below by λ_{min}. We now set for $\varphi \in (0,\pi/2)$:

$$\chi_\delta(r,\varphi) := \chi(r) - \frac{rb^+(r)}{a^+(r)}\chi'(r)\,\varphi\,\rho(\varphi/\delta),$$

where ρ is a smooth cutoff function satisfying $\rho \equiv 1$ in a (fixed) neighborhood of $\varphi = 0$ and $\rho \equiv 0$ on $\mathbb{R}\backslash(-\pi/4,\pi/4)$. We see that χ_δ satisfies $\chi_\delta \equiv 1$ for $r \leq 1-\delta$, $\chi_\delta = 0$ for $r > 1-\delta/2$ and that χ_δ is smooth (for $\varphi \in [0,\pi/2]$). Furthermore, the conditions on $\nabla^j\chi_\delta$ are satisfied because

$$\sup_{\varphi\in[0,\pi/2]} |\partial_\varphi^j(\varphi\rho(\varphi/\delta))| \leq C\delta^{-j+1}, \qquad j \in \{0,1,2\}$$

in view of the support properties of ρ. Inserting χ_δ in $\partial_{n_A^+}\chi_\delta$ for $\varphi = 0$, we see that $\partial_{n_A^+}\chi_\delta = 0$. This completes the construction for $\varphi \in (0,\pi/2)$. For $\varphi \in (-\pi/2,0)$, the construction is completely analogous (note that choosing the same "base" function χ guarantees the continuity across $\varphi = 0$). □

We can now prove the analog of Lemma 5.5.11.

Lemma 5.5.22. *Let $R \leq 1$, $A \in L^\infty(B_R,\mathbb{S}^n_>)$ with $0 < \lambda_{min} \leq A$, whose restrictions $A^+ := A|_{B_R^+}$, $A^- := A|_{B_R^-}$ satisfy $A^+ \in C^1(\overline{B_R^+},\mathbb{S}^n_>)$, $A^- \in C^1(\overline{B_R^-},\mathbb{S}^n_>)$. Let $f \in L^2(B_R)$, $G^+ \in H^1(B_R^+)$. Let $u \in H^1(B_R)$ be a solution of the transmission problem*

$$-\nabla\cdot(A\nabla u) = f \quad \text{on } B_R, \qquad [\partial_{n_A} u] = G^+ \quad \text{on } x_n = 0.$$

Then there exists $C > 0$ depending only on λ_{min}, $\|A\|_{L^\infty(B_R)}$, $R\|A'\|_{L^\infty(B_R^+)} + R\|A'\|_{L^\infty(B_R^-)}$, and n such that for all r, $\delta > 0$ with $r+\delta < R$ there holds

$$\|\nabla^2 u\|_{L^2(B_r^+\cup B_r^-)} \leq C\Big[\|f\|_{L^2(B_{r+\delta})} + \|\nabla G^+\|_{L^2(B_{r+\delta}^+)} + \delta^{-1}\|G^+\|_{L^2(B_{r+\delta}^+)} + \delta^{-1}\|\nabla u\|_{L^2(B_{r+\delta})} + \delta^{-2}\|u\|_{L^2(B_{r+\delta})}\Big].$$

Proof: Let $F : x \mapsto Rx$ be the affine mapping transforming the unit ball B_1 to the B_R. From Lemma A.1.1, we see that the transformed function $\hat{u} := u \circ F$ satisfies the following transmission problem

$$-\nabla\cdot(\hat{A}\nabla\hat{u}) = R^2\hat{f} \quad \text{on } B_1, \qquad \partial_{n_{\hat{A}}}\hat{u} = R\hat{G} \quad \text{on } x_n = 0,$$

where we set $\hat{f} := f\circ F$, $\hat{G}^+ := G^+\circ F$, $\hat{A} := A\circ F$. We note in passing that $\|\hat{A}\|_{L^\infty(B_1)} = \|A\|_{L^\infty(B_R)}$, $\|\hat{A}'\|_{L^\infty(B_1^+\cup B_1^-)} = R\|A'\|_{L^\infty(B_R^+\cup B_R^-)}$. Let $\delta' \in (0,1)$

and let χ be the cut-off function given by Lemma 5.5.21. A calculation then shows that the function $\chi\hat{u}$ satisfies the following transmission problem on B_1:

$$-\nabla\cdot(\hat{A}\nabla(\chi\hat{u})) = R^2\chi\hat{f} - (\hat{A}\nabla\hat{u})\cdot\nabla\chi - \left(\nabla\cdot(\hat{A}\nabla\chi)\right)\hat{u} - (\hat{A}\nabla\chi)\cdot\nabla\hat{u} \quad \text{on } B_1,$$
$$[\partial_{n_{\hat{A}}}\chi\hat{u}] = R\chi\hat{G}^+ \quad \text{on } x_n = 0.$$

Furthermore, as $\chi\hat{u}$ and the right-hand side vanish in a neighborhood of ∂B_1, this equation in fact holds on B_2:

$$-\nabla\cdot(\hat{A}\nabla(\chi\hat{u})) = \tilde{f} := R^2\chi\hat{f} - (\hat{A}\nabla\hat{u})\cdot\nabla\chi - \left(\nabla\cdot(\hat{A}\nabla\chi)\right)\hat{u} - (\hat{A}\nabla\chi)\cdot\nabla\hat{u}$$
$$[\partial_{n_{\hat{A}}}\chi\hat{u}] = \tilde{G} := R\chi\hat{G} \qquad \text{on } x_n = 0.$$

(We implicitly extended $\hat{A}$ to B_2 appropriately). Thus, the interior regularity assertion of Lemma 5.5.8 is applicable (with B_2 taking the place of B_1) and yields with the trace theorem

$$\begin{aligned}\|\nabla^2\hat{u}\|_{L^2(B^+_{1-\delta'}\cup B^-_{1-\delta'})} &\leq \|\nabla^2(\chi\hat{u})\|_{L^2(B_1^+\cup B_1^-)} \leq C\left[\|\tilde{f}\|_{L^2(B_2)} + \|\tilde{G}\|_{H^{1/2}(\Gamma_2)}\right]\\ &\leq C\Big[R^2\|\hat{f}\|_{L^2(B_{1-\delta'/2})} + R\|\nabla\hat{G}^+\|_{L^2(B^+_{1-\delta'/2})} + R\delta'^{-1}\|\hat{G}^+\|_{L^2(B^+_{1-\delta'/2})}\\ &\quad + \delta'^{-1}\|\nabla\hat{u}\|_{L^2(B_{1-\delta'/2})} + \delta'^{-2}\|\hat{u}\|_{L^2(B_{1-\delta'/2})}\Big].\end{aligned}$$

Scaling back to the original domain B_R gives the desired bounds for $r := R(1-\delta')$, $\delta := R\delta'/2$. □

In order to prove Proposition 5.5.4, we introduce the following notation, analogous to the notation used previously for the regularity assertions in the interior and at the boundary. We define for $p \in \mathbb{N}_0 \cup \{-2,-1\}$

$$N'^{,\pm}_{R,p}(v) = \begin{cases} \dfrac{1}{p!}\sup\limits_{R/2\leq r<R}(R-r)^{p+2}\|\nabla^2\nabla^p_x v\|_{L^2(B_r^+\cup B_r^-)} & \text{if } p \geq 0\\ \sup\limits_{R/2\leq r<R}(R-r)^{p+2}\|\nabla^{2+p}v\|_{L^2(B_r^+\cup B_r^-)} & \text{if } p = -2,-1,\end{cases}$$

and for all $p \in \mathbb{N}_0$ we introduce

$$H_{R,p}(v) := \frac{1}{[p-1]!}\sup_{R/2\leq r<R}(R-r)^{p+1}\left[\|\nabla^p_x v\|_{L^2(B_r^+)} + \frac{R-r}{[p]}\|\nabla^p_x\nabla v\|_{L^2(B_r^+)}\right],$$
$$M'^{,\pm}_{R,p}(f) := \frac{1}{p!}\sup_{R/2<r<R}(R-r)^{p+2}\|\nabla^p_x f\|_{L^2(B_r^+\cup B_r^-)}.$$

The analog of Lemma 5.5.15 is the following lemma.

Lemma 5.5.23. *Let $p \in \mathbb{N}_0$, $A \in L^\infty(B_R, \mathbb{S}^n_>)$ with $0 < \lambda_{min} \le A$, whose restrictions $A^+ := A|_{B_R^+}$, $A^- := A|_{B_R^-}$ satisfy $A^+ \in C^{p+1}(\overline{B_R^+}, \mathbb{S}^n_>)$, $A^- \in C^{p+1}(\overline{B_R^-}, \mathbb{S}^n_>)$. Assume $f \in H^p(B_R^+ \cup B_R^-)$. Let $G \in H^{p+1}(B_R^+)$. Then there exists $C'_B > 0$ depending only on λ_{min}, $\|A\|_{L^\infty(B_R)}$, $R\|A'\|_{L^\infty(B_R^+ \cup B_R^-)}$, and n such that any solution $u \in H^1(B_R)$ of the transmission problem*

$$-\nabla \cdot (A\nabla u) = f \quad \text{on } B_R, \qquad [\partial_{n_A} u] = G \quad \text{on } \Gamma_R$$

satisfies

$$N'^{,\pm}_{R,p}(u) \le C'_B \Big[M'^{,\pm}_{R,p}(f) + H_{R,p}(G) + \sum_{q=1}^{p+1} \binom{p+1}{q} \Big\{ \left(\frac{R}{2}\right)^q \|\nabla^q A\|_{L^\infty(B_R^+ \cup B_R^+)} + \left(\frac{R}{2}\right)^{q-1} q \|\nabla^{q-1} A\|_{L^\infty(B_R^+ \cup B_R^+)} \Big\} \times \frac{[p-q]!}{p!} N'^{,\pm}_{R,p-q}(u) + N'^{,\pm}_{R,p-1}(u) + N'^{,\pm}_{R,p-2}(u) \Big].$$

For $p = 0$, we have the sharper bound

$$N'^{,\pm}_{R,0}(u) \le C'_B \left[R^2 \|f\|_{L^2(B_R)} + R\|G\|_{L^2(B_R^+)} + R^2 \|\nabla G\|_{L^2(B_R^+)} + R\|\nabla u\|_{L^2(B_R)} \right]. \tag{5.5.31}$$

Proof: For brevity of notation, we set $B'_r := B_r^+ \cup B_r^-$.
One proceeds as in the proof of Lemma 5.5.15 by differentiating the differential equation in the tangential direction. Let $p \in \mathbb{N}_0$ and $\alpha \in \mathbb{N}_0^{n-1}$ with $|\alpha| = p$ be given. Then $D^\alpha u$ solves the following transmission problem:

$$\begin{aligned} -\nabla \cdot (A\nabla D^\alpha u) &= \tilde f_\alpha := D^\alpha f + D^\alpha(\nabla \cdot (A\nabla u)) - \nabla \cdot (AD^\alpha \nabla u) \qquad \text{on } B_R, \\ [\partial_{n_A} D^\alpha u] &= \tilde G_\alpha + \tilde G^+_\alpha + \tilde G^-_\alpha \\ &:= D^\alpha G + \{ D^\alpha(\mathbf{n}^T A^+ \nabla u^+) - \mathbf{n}^T A^+ D^\alpha \nabla u^+ \} \\ &\quad - \{ D^\alpha(\mathbf{n}^T A^- \nabla u^-) - \mathbf{n}^T A^- D^\alpha \nabla u^- \} \qquad \text{on } \Gamma_R, \end{aligned}$$

where we used the notation u^+, u^- to denote the restriction of u to B_R^+, B_R^-, respectively. From Lemma 5.5.22, we therefore have the existence of $C > 0$ such that for all r, $\delta > 0$ with $r + \delta < R$:

$$\begin{aligned} \|\nabla^2 D^\alpha u\|^2_{L^2(B'_r)} \le C \Big[& \|\tilde f_\alpha\|^2_{L^2(B'_{r+\delta})} + \delta^{-2} \|\tilde G_\alpha\|^2_{L^2(B^+_{r+\delta})} + \|\nabla \tilde G_\alpha\|^2_{L^2(B^+_{r+\delta})} \\ & + \|\nabla \tilde G^+_\alpha\|^2_{L^2(B^+_{r+\delta})} + \delta^{-2}\|\tilde G^+_\alpha\|^2_{L^2(B^+_{r+\delta})} + \|\nabla \tilde G^-_\alpha\|^2_{L^2(B^-_{r+\delta})} + \delta^{-2}\|\tilde G^-_\alpha\|^2_{L^2(B^-_{r+\delta})} \\ & + \delta^{-2} \|D^\alpha \nabla u\|^2_{L^2(B^+_{r+\delta} \cup B^-_{r+\delta})} + \delta^{-4} \|D^\alpha u\|^2_{L^2(B'_{r+\delta})} \Big]. \end{aligned}$$

Recalling that $\|\nabla_x^p \nabla^2 u\|^2_{L^2(B'_r)} = \sum_{|\alpha|=p} \frac{|\alpha|!}{\alpha!} \|D^\alpha \nabla^2 u\|^2_{L^2(B'_r)}$, we see that we have to bound $\sum_{|\alpha|=p} \frac{|\alpha|!}{\alpha!} \|\tilde f_\alpha\|^2_{L^2(B'_{r+\delta})}$ etc. We have

$$\sum_{|\alpha|=p} \frac{|\alpha|!}{\alpha!} \|D^\alpha f\|^2_{L^2(B'_{r+\delta})} = \|\nabla^p_x f\|^2_{L^2(B'_{r+\delta})},$$

$$\sum_{|\alpha|=p} \frac{|\alpha|!}{\alpha!} \left\{\delta^{-2}\|D^\alpha \nabla u\|^2_{L^2(B'_{r+\delta})} + \delta^{-4}\|D^\alpha u\|^2_{L^2(B'_{r+\delta})}\right\}$$
$$= \delta^{-2}\|\nabla^p_x \nabla u\|^2_{L^2(B'_{r+\delta})} + \delta^{-4}\|\nabla^p_x u\|^2_{L^2(B'_{r+\delta})},$$

$$\sum_{|\alpha|=p} \frac{|\alpha|!}{\alpha!} \left\{\|D^\alpha \nabla G\|^2_{L^2(B^+_{r+\delta})} + \delta^{-2}\|D^\alpha G\|^2_{L^2(B^+_{r+\delta})}\right\}$$
$$= \|\nabla^p_x \nabla G\|^2_{L^2(B^+_{r+\delta})} + \delta^{-2}\|\nabla^p_x G\|^2_{L^2(B^+_{r+\delta})}.$$

From Lemma A.1.4, we obtain (as in the proofs of Lemmata 5.5.12, 5.5.10), with the abbreviation $\|\nabla^q A\|_\infty := \|\nabla^q A\|_{L^\infty(B'_R)}$:

$$\sum_{|\alpha|=p} \frac{|\alpha|!}{\alpha!} \|D^\alpha(\nabla\cdot(A\nabla u) - \nabla\cdot(AD^\alpha\nabla u))\|^2_{L^2(B'_{r+\delta})} \le$$
$$C\left\{\sum_{q=1}^{p}\binom{p}{q}\left[\|\nabla^{q+1}A\|_\infty\|\nabla^{p-q}_x\nabla u\|_{L^2(B'_{r+\delta})} + \|\nabla^q A\|_\infty\|\nabla^{p-q}_x\nabla^2 u\|_{L^2(B'_{r+\delta})}\right]\right\}^2,$$

and analogously (see Lemma A.1.4) for the terms involving $\tilde{G}^+_\alpha$:

$$\sum_{|\alpha|=p} \frac{|\alpha|!}{\alpha!} \|\nabla\tilde{G}^+_\alpha\|^2_{L^2(B^+_{r+\delta})} \le$$
$$C\left\{\sum_{q=1}^{p}\binom{p}{q}\left[\|\nabla^{q+1}A\|_\infty\|\nabla^{p-q}_x\nabla u\|_{L^2(B^+_{r+\delta})} + \|\nabla^q A\|_\infty\|\nabla^{p-q}_x\nabla^2 u\|_{L^2(B^+_{r+\delta})}\right]\right\}^2$$

and

$$\sum_{|\alpha|=p} \frac{|\alpha|!}{\alpha!} \|\tilde{G}^+_\alpha\|^2_{L^2(B^+_{r+\delta})} \le C\left\{\sum_{q=1}^{p}\binom{p}{q}\|\nabla^q A\|_\infty\|\nabla^{p-q}_x\nabla u\|_{L^2(B^+_{r+\delta})}\right\}^2.$$

Completely analogous bounds hold for $\tilde{G}^-_\alpha$, $\nabla\tilde{G}^-_\alpha$. Combining the bounds obtained so far, we have the existence of $C > 0$ (independent of δ) such that

$$\|\nabla^p_x\nabla^2 u\|_{L^2(B'_r)} \le C\Big\{\|\nabla^p_x f\|_{L^2(B'_{r+\delta})} + \|\nabla^p_x\nabla G\|_{L^2(B^+_{r+\delta})} + \delta^{-1}\|\nabla^p_x G\|_{L^2(B^+_{r+\delta})}$$
$$+ \delta^{-1}\|\nabla^p_x\nabla u\|_{L^2(B'_{r+\delta})} + \delta^{-2}\|\nabla^p_x u\|_{L^2(B'_{r+\delta})}$$
$$+ \delta^{-1}\sum_{q=1}^{p}\binom{p}{q}\|\nabla^q A\|_\infty\|\nabla^{p-q}_x\nabla u\|_{L^2(B^+_{r+\delta})}$$
$$+ \sum_{q=1}^{p}\binom{p}{q}\left[\|\nabla^{q+1}A\|_\infty\|\nabla^{p-q}_x\nabla u\|_{L^2(B^+_{r+\delta})} + \|\nabla^q A\|_\infty\|\nabla^{p-q}_x\nabla^2 u\|_{L^2(B^+_{r+\delta})}\right]\Big\}.$$

The remainder of the proof is now similar to that of Lemma 5.5.12. We use the definition of the quantities $M'^{,\pm}_{R,p}$, $H_{R,p}$, $N'^{,\pm}_{R,p}$ and choose δ as in (5.5.23) to obtain with similar arguments as given there:

$$\frac{1}{p!}(R-r)^{p+2}\|\nabla_x^p \nabla^2 u\|_{L^2(B'_r)} \leq$$
$$C\Big\{M'^{,\pm}_{R,p}(f) + H_{R,p}(G) + N'^{,\pm}_{R,p-1}(u) + N'^{,\pm}_{R,p-2}(u)$$
$$+ (p+1)\sum_{q=1}^{p}\binom{p}{q}\|\nabla^q A\|_\infty \left(\frac{R}{2}\right)^q \frac{[p-1-q]!}{p!} N'^{,\pm}_{R,p-q-1}(u)$$
$$+ \sum_{q=1}^{p+1}\binom{p+1}{q}\|\nabla^q A\|_\infty \left(\frac{R}{2}\right)^q \frac{[p-q]!}{p!} N'^{,\pm}_{R,p-q}(u)\Big\}.$$

The claim of the lemma now follows by some further manipulations of this last expression and observing that taking the supremum over r yields $N'^{,\pm}_{R,p}(u)$ on the left-hand side.
We finally turn to (5.5.31), i.e., the case $p = 0$. This is done with a Poincaré argument. The assertion for general p yields for some $C > 0$

$$N'^{,\pm}_{R,0}(u) \leq \tag{5.5.32}$$
$$C\big[R^2\|f\|_{L^2(B_R)} + R\|G\|_{L^2(B_R)} + R^2\|\nabla G\|_{L^2(B_R)} + R\|\nabla u\|_{L^2(B_R)} + \|u\|_{L^2(B_R)}\big].$$

Let $\overline{u}$ be the average of u over B_R. We note that $u - \overline{u}$ satisfies the same differential equation as u; thus, (5.5.32) also holds with u replaced with $u - \overline{u}$. Using the Poincaré estimate $\|u-\overline{u}\|_{L^2(B_R)} \leq CR\|\nabla u\|_{L^2(B_R)}$ then gives the desired (5.5.31) after suitably adjusting the constant C'_B. □

Lemma 5.5.24. *Let G_2 satisfy (5.5.19) and assume $u \in \mathcal{A}(B_R^+ \cup B_R^-)$. Then*

$$\varepsilon^{-1}H_{R,p}(G_2 u) \leq C_{G_2}\frac{1}{[p]!}\frac{1}{2}\frac{\mathcal{E}}{\varepsilon}\min\{1, R/\mathcal{E}\}\times$$
$$\Big\{\max\{p+1, R/\mathcal{E}\}^2 \sum_{q=0}^{p}\binom{p}{q}\left[\left(\frac{\gamma_{G_2}R}{2}\right)^q + \left(\frac{\gamma_{G_2}R}{2}\right)^{q+1}\right] q![p-q-2]! N'^{,\pm}_{R,p-q-2}(u)$$
$$+ \max\{p+1, R/\mathcal{E}\}\sum_{q=0}^{p}\binom{p}{q}\left(\frac{\gamma_{G_2}R}{2}\right)^q [p-q-1]! N'^{,\pm}_{R,p-q-1}(u)\Big\}.$$

Proof: The definition of $H_{R,p}$ consists of two terms, which we will estimate separately. We start with estimating

$$H' := \varepsilon^{-1}\frac{1}{[p-1]!}\sup_{R/2<r<R}(R-r)^{p+1}\|\nabla_x^p(G_2 u)\|_{L^2(B_r^+)}.$$

Using Leibniz' formula and the bounds (5.5.19), we get

$$H' \le \varepsilon^{-1}\frac{[p]}{[p]!}C_{G_2}\frac{R}{2}\sum_{q=0}^{p}\binom{p}{q}\gamma_{G_2}^{q}q!\left(\frac{R}{2}\right)^{q}\sup_{R/2<r<R}(R-r)^{p-q}\|\nabla_x^{p-q}u\|_{L^2(B_r^+)}$$
$$\le \frac{C_{G_2}}{2}\frac{\mathcal{E}}{\varepsilon}[p]\frac{R}{\mathcal{E}}\sum_{q=0}^{p}\binom{p}{q}q!\left(\frac{R\gamma_{G_2}}{2}\right)^{q}[p-q-2]!N'^{,\pm}_{R,p-q-2}(u).$$

Estimating further

$$[p]\frac{R}{\mathcal{E}} = [p]\min\{1,R/\mathcal{E}\}\max\{1,R/\mathcal{E}\} \le \min\{1,R/\mathcal{E}\}\max\{p+1,R/\mathcal{E}\}^2,$$

we obtain

$$H' \le \frac{C_{G_2}}{2}\frac{\mathcal{E}}{\varepsilon}\frac{1}{[p]!}\max\{p+1,R/\mathcal{E}\}^2\min\{1,R/\mathcal{E}\}\times$$
$$\sum_{q=0}^{p}\binom{p}{q}q!\left(\frac{R\gamma_{G_2}}{2}\right)^{q}[p-q-2]!N'^{,\pm}_{R,p-q-2}(u).$$

In a similar way, we estimate

$$H'' := \varepsilon^{-1}\frac{1}{[p]!}\sup_{R/2<r<R}(R-r)^{p+2}\|\nabla_x^p\nabla(G_2u)\|_{L^2(B_r^+)},$$

namely, we bound

$$|\nabla_x^p\nabla(G_2u)| \le \sum_{q=0}^{p}\binom{p}{q}\left(|\nabla_x^q\nabla G_2|\,|\nabla_x^{p-q}u| + |\nabla_x^qG_2|\,|\nabla_x^{p-q}\nabla u|\right)$$

and estimate $\sum_{q=0}^{p}\varepsilon^{-1}\frac{1}{[p]!}\binom{p}{q}(S_{1,p,q}+S_{2,p,q})$, where

$$S_{1,p,q} := \sup_{R/2<r<R}(R-r)^{p+2}\|\,|\nabla_x^q\nabla G_2|\,|\nabla_x^{p-q}u|\,\|_{L^2(B_r^+)},$$
$$S_{2,p,q} := \sup_{R/2<r<R}(R-r)^{p+2}\|\,|\nabla_x^qG_2|\,|\nabla_x^{p-q}\nabla u|\,\|_{L^2(B_r^+)}.$$

For $S_{1,p,q}$, we obtain

$$S_{1,p,q} \le C_{G_2}\gamma_{G_2}^{q+1}(q+1)!\sup_{R/2<r<R}(R-r)^{p+2}\|\nabla_x^{p-q}u\|_{L^2(B_r^+)}$$
$$\le C_{G_2}\gamma_{G_2}^{q+1}(q+1)!\left(\frac{R}{2}\right)^{q+2}[p-q-2]!N'^{,\pm}_{p-q-2}(u)$$
$$\le \frac{C_{G_2}}{2}R(p+1)q!\left(\frac{R\gamma_{G_2}}{2}\right)^{q+1}[p-q-2]!N'^{,\pm}_{p-q-2}(u).$$

For $S_{2,p,q}$ we bound

$$S_{2,p,q} \le C_{G_2}\gamma_{G_2}^{q}q!\sup_{R/2<r<R}(R-r)^{p+2}\|\nabla_x^{p-q}\nabla u\|_{L^2(B_r^+)}$$
$$\le C_{G_2}\gamma_{G_2}^{q}q!\left(\frac{R}{2}\right)^{q+1}[p-q-1]!N'^{,\pm}_{R,p-q-1}(u).$$

Since we can again estimate

$$\frac{R}{\varepsilon}(p+1) \leq \frac{\mathcal{E}}{\varepsilon} \min\{1, R/\mathcal{E}\} \max\{p+1, R/\mathcal{E}\}^2,$$
$$\frac{R}{\varepsilon} \leq \frac{\mathcal{E}}{\varepsilon} \min\{1, R/\mathcal{E}\} \max\{1, R/\mathcal{E}\},$$

we can combine the above bounds to arrive at

$$H'' \leq \frac{1}{[p]!} \frac{C_{G_2}}{2} \frac{\mathcal{E}}{\varepsilon} \min\{1, R/\mathcal{E}\} \times$$
$$\Big\{ \max\{p+1, R/\mathcal{E}\}^2 \sum_{q=0}^{p} \binom{p}{q} q! \left(\frac{R\gamma_{G_2}}{2}\right)^{q+1} [p-q-2]! N'^{,\pm}_{R,p-q-2}(u)$$
$$+ \max\{p+1, R/\mathcal{E}\} \sum_{q=0}^{p} \binom{p}{q} q! \left(\frac{R\gamma_{G_2}}{2}\right)^{q} [p-q-1]! N'^{,\pm}_{R,p-q-1}(u) \Big\}.$$

□

Proposition 5.5.25 (control of tangential derivatives). *Let $R \in (0,1]$. Let the coefficients A, b, c satisfy (5.5.16). For $\varepsilon \in (0,1]$ define $\mathcal{E}$ by (5.5.3) and assume that f satisfies (5.5.17) and that G_1, G_2 satisfy (5.5.18), (5.5.19). Finally let u be a solution of the transmission problem*

$$-\varepsilon^2 \nabla \cdot (A\nabla u) + b \cdot \nabla u + cu = f \quad \text{on } B_R, \qquad [\varepsilon^2 \partial_{n_A} u] = \varepsilon (G_1 + G_2 u) \quad \text{on } \Gamma_R.$$

Then for some $K_1 > 0$ depending only on the constants of (5.5.16) and γ_f, γ_{G_1}, γ_{G_2} of (5.5.17)–(5.5.19), there holds

$$N'^{,\pm}_{R,p}(u) \leq C_u K_1^{p+2} \frac{\max\{(p+3), R/\mathcal{E}\}^{p+2}}{[p]!}, \qquad p \geq -1, \tag{5.5.33}$$
$$\begin{aligned} C_u := {} & \min\{1, R/\mathcal{E}\} \mathcal{E} \|\nabla u\|_{L^2(B_R)} \\ & + (\mathcal{E}/\varepsilon)^2 \min\{1, R/\mathcal{E}\}^2 \left[C_c \|u\|_{L^2(B_R)} + C_f\right] \\ & + C_{G_1} \min\{1, R/\mathcal{E}\} (\mathcal{E}/\varepsilon) \\ & + C_{G_2} (\mathcal{E}/\varepsilon) \min\{1, R/\mathcal{E}\} \left[\min\{1, R/\mathcal{E}\} \mathcal{E} \|\nabla u\|_{L^2(B_R)} + \|u\|_{L^2(B_R)}\right]. \end{aligned} \tag{5.5.34}$$

Proof: The proof is very similar to the induction argument given in the proof of Proposition 5.5.16. We therefore merely outline the important points. First, we calculate

$$N'^{,\pm}_{R,-1}(u) \leq \frac{R}{2} \|\nabla u\|_{L^2(B_R)} \leq \frac{1}{2} \min\{1, R/\mathcal{E}\} \max\{1, R/\mathcal{E}\} \mathcal{E} \|\nabla u\|_{L^2(B_R)},$$

which shows that the claim is true for $p = -1$, if $K_1 \geq 1$. Next, for $p = 0$, we write the differential equation as $-\nabla \cdot (A\nabla u) = \varepsilon^{-2}[f - b \cdot \nabla u - cu]$ with transmission conditions $[\partial_{n_A} u] = G := \varepsilon^{-1}(G_1 + G_2 u)$ and employ Lemma 5.5.23 to get

$$N_{R,0}^{\prime,\pm}(u) \leq C\Big\{\left(\frac{R}{\varepsilon}\right)^2 \|f\|_{L^2(B_R)} + \frac{C_b R^2}{\varepsilon^2}\|\nabla u\|_{L^2(B_R)} + \frac{C_c R^2}{\varepsilon^2}\|u\|_{L^2(B_R)}$$
$$+ \frac{R}{\varepsilon}\|G_1\|_{L^2(B_R)} + \frac{R^2}{\varepsilon}\|\nabla G_1\|_{L^2(B_R)} + \frac{R}{\varepsilon}\|G_2\|_{L^\infty(B_R)}\|u\|_{L^2(B_R)}$$
$$+ \|\nabla G_2\|_{L^\infty(B_R)}\frac{R^2}{\varepsilon}\|u\|_{L^2(B_R)} + \|G_2\|_{L^\infty(B_R)}\frac{R^2}{\varepsilon}\|\nabla u\|_{L^2(B_R)}\Big\}.$$

Exploiting $R \leq 1$, $R/\mathcal{E} = \min\{1, R/\mathcal{E}\}\max\{1, R/\mathcal{E}\}$, $C_b/\varepsilon^{-2} \leq \mathcal{E}^{-1}$, and the assumptions on f, G_1, G_2 in (5.5.17)–(5.5.19), we obtain for a constant $C > 0$ with C_u as defined in (5.5.34)

$$N_{R,0}^{\prime,\pm}(u) \leq C \max\{1, R/\mathcal{E}\}^2 C_u.$$

Hence, for suitable K_1, we see that (5.5.33) is indeed valid.
The induction argument then proceeds along similar lines as in the proof of Proposition 5.5.16 and is based on Lemma 5.5.23 with $G = \varepsilon^{-1}(G_1 + G_2 u)$. In addition to the terms that arise in the proof of Proposition 5.5.16 (where the induction is based on Lemma 5.5.15), we have to control the extra term

$$H_{R,p}(\varepsilon^{-1}(G_1 + G_2 u)) \leq \varepsilon^{-1} H_{R,p}(G_1) + \varepsilon^{-1} H_{R,p}(G_2 u).$$

The contribution $\varepsilon^{-1} H_{R,p}(G_2 u)$ is estimated in Lemma 5.5.24; the estimate of Lemma 5.5.24 has a form that is amenable to the induction argument of the proof of Proposition 5.5.16 with the exception of one point: The first sum in the estimate of Lemma 5.5.24 contains the term (corresponding to $q = p$ there)

$$T := \frac{C_{G_2}}{\lfloor p \rfloor !}\frac{\mathcal{E}}{\varepsilon} \min\{1, R/\mathcal{E}\}\max\{1, R/\mathcal{E}\}^2 \left[\left(\frac{\gamma_{G_2} R}{2}\right)^p + \left(\frac{\gamma_{G_2} R}{2}\right)^{p+1}\right] p! N_{R,-2}^{\prime,\pm}(u),$$

which has to be treated separately as the induction hypothesis (5.5.33) does not cover the case $p = -2$. However, since $N_{R,-2}^{\prime,\pm}(u) = \|u\|_{L^2(B_R)}$ and we can bound $p! \max\{1, R/\mathcal{E}\}^2 \leq \max\{p+3, R/\mathcal{E}\}^{p+2}$, we get, for any $K_1 > 0$,

$$T \leq K_1^{p+2} \frac{\max\{p+3, R/\mathcal{E}\}^{p+2}}{\lfloor p \rfloor !} C_{G_2}(\mathcal{E}/\varepsilon) \min\{1, R/\mathcal{E}\} \times$$
$$\left[\left(\frac{\gamma_{G_2} R}{2K_1}\right)^p K_1^{-2} + \left(\frac{\gamma_{G_2} R}{2K_1}\right)^{p+1} K_1^{-1}\right].$$

This is of a form suitable for the induction argument if we choose K_1 sufficiently large so that the bracketed term is small than 1/2, say.
For the other contribution, $\varepsilon^{-1} H_{R,p}(G_1)$, we estimate

$$\varepsilon^{-1} H_{R,p}(G_1)$$
$$= \varepsilon^{-1} \frac{1}{[p-1]!} \sup_{R/2 \le r < R} (R-r)^{p+1} \left[\|\nabla_x^p G_1\|_{L^2(B_r^+)} + \frac{R-r}{[p]} \|\nabla_x^p \nabla G_1\|_{L^2(B_r^+)} \right]$$
$$\le C_{G_1} \frac{\varepsilon^{-1}}{[p-1]!} \left\{ \left(\frac{R}{2}\right)^{p+1} \gamma_{G_1}^p \max\{(p+1)/R, 1/\mathcal{E}\}^p \right.$$
$$\left. + \left(\frac{R}{2}\right)^{p+2} \frac{\gamma_{G_1}^{p+1}}{[p]} \max\{(p+2)/R, 1/\mathcal{E}\}^{p+1} \right\}$$
$$\le \frac{1}{2} C_{G_1} (\mathcal{E}/\varepsilon) \frac{R/\mathcal{E}}{p!} \left[\left(\frac{\gamma_{G_1}}{2}\right)^p + \left(\frac{\gamma_{G_1}}{2}\right)^{p+1} \right] \max\{p+2, R/\mathcal{E}\}^{p+1}.$$

Bounding now $R/\mathcal{E} \le \min\{1, R/\mathcal{E}\} \max\{p+3, R/\mathcal{E}\}$, we arrive at

$$\varepsilon^{-1} H_{R,p}(G_1) \le$$
$$C_{G_1} \min\{1, R/\mathcal{E}\} \frac{\max\{p+3, R/\mathcal{E}\}^{p+2}}{p!} \left[\left(\frac{\gamma_{G_1}}{2}\right)^p + \left(\frac{\gamma_{G_1}}{2}\right)^{p+1} \right]$$

and can then proceeds just as in the proof of Proposition 5.5.16. □

Proof of Proposition 5.5.4: Proposition 5.5.4 is now proved just as Proposition 5.5.2 with Proposition 5.5.25 as the induction hypothesis rather than Proposition 5.5.16. □

5.5.5 Regularity of Neumann problems: Proof of Prop. 5.5.3

The proof of Proposition 5.5.3 is very similar to that of the transmission problem, Proposition 5.5.4, and therefore omitted. One merely has to check that each occurrence of $B_R^+ \cup B_R^-$ or $B_r^+ \cup B_r^-$ can be replaced with B_R^+, B_r^+, respectively. The operator $[\partial_{n_A} u]$ is simply replaced with the co-normal derivative operator ∂_{n_A}. For easy of citation, however, we repeat here the analog of Lemma 5.5.23:

Lemma 5.5.26. *Let $A \in C^1(\overline{B_R^+}, \mathbb{S}_>^n)$ with $0 < \lambda_{min} \le A$. Assume $f \in L^2(B_R^+)$. Let $G \in H^1(B_R^+)$. Then there exists $C'_B > 0$ depending only on λ_{min}, $\|A\|_{L^\infty(B_R^+)}$, $R\|A'\|_{L^\infty(B_R^+)}$, and n such that any solution $u \in H^1(B_R)$ of the Neumann problem*

$$-\nabla \cdot (A \nabla u) = f \quad \text{on } B_R^+, \qquad \partial_{n_A} u = G \quad \text{on } \Gamma_R$$

satisfies

$$R^2 \|\nabla^2 u\|_{L^2(B_{R/2}^+)} \le \tag{5.5.35}$$
$$C'_B \left[R^2 \|f\|_{L^2(B_R^+)} + R\|G\|_{L^2(B_R^+)} + R^2 \|\nabla G\|_{L^2(B_R^+)} + R\|\nabla u\|_{L^2(B_R^+)} \right].$$

Part III

Regularity in Terms of Asymptotic Expansions

6. Exponentially Weighted Countably Normed Spaces

6.1 Motivation and outline

6.1.1 Motivation

The present Chapter 6 is closely connected with the ensuing Chapter 7 as the regularity theory in exponentially weighted countably normed spaces developed here will be employed in Chapter 7 for the definition of the corner layers in the asymptotic expansions.
The corner layers for problem (1.2.11) are introduced in Chapter 7 to remove an incompatibility between two boundary layer expansions that meet at a vertex. Specifically, near each vertex A_j, the corner layer is defined as the solution of a transmission problem in a sector near A_j where the jump across an interface and the jump of the co-normal derivative across this interface is chosen so as to match the jumps of the boundary layer expansions. One essential feature of these jumps is that they decay exponentially away from A_j. Thus, we expect the solution to decay exponentially as well. This is indeed the case and in fact, all derivatives decay exponentially away from A_j. In order to make this more precise, the present chapter is devoted to a study of the analytic regularity properties of solutions of transmission problems with exponentially decaying input data. To illustrate the key ideas, let us again consider a sector $S_R(\omega)$. This sector is divided into two subsectors S^+, S^- by the curve $\Gamma = \{(r\cos\omega', r\cos\omega') \,|\, 0 < r < R\}$ for some fixed $\omega' \in (0, \omega)$. We then consider the transmission problem

$$-\varepsilon^2 \Delta u_\varepsilon + u_\varepsilon = f \qquad \text{on } S^+ \cup S^-, \tag{6.1.1a}$$

$$[u_\varepsilon] = h_1 \qquad \text{on } \Gamma, \tag{6.1.1b}$$

$$\varepsilon^2 [\partial_n u_\varepsilon] = \varepsilon h_2 \qquad \text{on } \Gamma, \tag{6.1.1c}$$

$$u_\varepsilon = 0 \qquad \text{on } \partial S_R(\omega). \tag{6.1.1d}$$

Here, the bracket $[\,\cdot\,]$ represents the jump across Γ; h_1, h_2 are assumed to be smooth on Γ and h_1 vanishes at the endpoints of Γ. One essential assumption on the data f, h_1, and h_2 is that they decay exponentially away from the origin. This is formalized using the weight functions $\hat{\Psi}_{p,\beta,\varepsilon,\alpha}$ defined in (6.2.1) as:

$$\hat{\Psi}_{p,\beta,\varepsilon,\alpha}(x) = \hat{\Phi}_{p,\beta,\varepsilon}(x) e^{\alpha |x|/\varepsilon}.$$

Countably normed spaces $\mathcal{B}^l_{\beta,\varepsilon,\alpha}$ are then defined in complete analogy to the spaces $\mathcal{B}^l_{\beta,\varepsilon}$ of Chapter 4. In fact, for $\alpha = 0$, we recover the spaces $\mathcal{B}^l_{\beta,\varepsilon}$ of

Chapter 4. As the results of this chapter are mostly of auxiliary nature for the definition of the corner layers in Chapter 7, we restrict our regularity theory for (6.1.1) with data h_i, f for which we have pointwise control of the growth of the derivatives:

$$\|\hat{\Psi}_{0,0,\varepsilon,\alpha} D^p h_i\|_{L^\infty(\Gamma)} \leq C_h K^p \max\{p+1, \varepsilon^{-1}\}^p \qquad \forall p \in \mathbb{N}_0, \tag{6.1.2a}$$

$$\|\hat{\Psi}_{p,1,\varepsilon,\alpha} \nabla^p f\|_{L^\infty(S^+ \cup S^-)} \leq C_f K^p \max\{p+1, \varepsilon^{-1}\}^p \qquad \forall p \in \mathbb{N}_0. \tag{6.1.2b}$$

Here, C, K, $\alpha > 0$ and the differentiation operator D represents differentiation in the tangential direction on Γ. The solution u_ε of the transmission problem (6.1.1) with these input data is then analytic on $S^+ \cup S^-$, it is exponentially decaying away from the origin, and it has a singularity at the origin. More precisely, Theorem 6.4.13 asserts that for fixed $R' < R$ the solution u_ε is in the space $\mathcal{B}^2_{\beta,\varepsilon,\alpha'}(S', C\varepsilon, \gamma)$ for some C, $K > 0$ and $\alpha' \in (0, \alpha)$ independent of ε, where

$$S' = (S^+ \cup S^-) \cap B_{R'}(0).$$

The regularity assertion $u_\varepsilon \in \mathcal{B}^2_{\beta,\varepsilon,\alpha'}(S', C\varepsilon, \gamma)$ can be written more explicitly as

$$\varepsilon\|\hat{\Psi}_{0,0,\varepsilon,\alpha'} \nabla u_\varepsilon\|_{L^2(S')} + \|\hat{\Psi}_{0,0,\varepsilon,\alpha'} u_\varepsilon\|_{L^2(S')} \leq C\varepsilon, \tag{6.1.3}$$

$$\|\hat{\Psi}_{p,\beta,\varepsilon,\alpha'} \nabla^{p+2} u_\varepsilon\|_{L^2(S')} \leq C\varepsilon\gamma^p \max\{p+2, \varepsilon^{-1}\}^{p+2} \quad \forall p \in \mathbb{N}_0. \tag{6.1.4}$$

These estimates, in particular (6.1.3), show that the solution u_ε is small: (6.1.3) implies that the energy norm of u_ε is $O(\varepsilon)$. A similar observation will be made about the corner layers in Chapter 7, which in turn has ramifications for the design of FE-meshes in Chapter 3.
The proof of this regularity result is similar to that of Proposition 5.3.4. The key observation is that the weight functions $\hat{\Psi}_{p,\beta,\varepsilon,\alpha}$ satisfy (6.2.7), i.e., on balls $B_r(x)$ with $r = c|x|$ for some $c \in (0, 1)$ we have for some C, $K > 0$ independent of ε, x, and $\alpha \geq 0$

$$\min_{z \in B_{c|x|}(x)} \hat{\Psi}_{p,\beta,\varepsilon,\alpha}(z) \geq C K^p e^{-c\alpha|x|/\varepsilon} \hat{\Psi}_{p,\beta,\varepsilon,\alpha}(x) = C K^p \hat{\Psi}_{p,\beta,\varepsilon,(1-c)\alpha}(x), \tag{6.1.5a}$$

$$\max_{z \in B_{c|x|}(x)} \hat{\Psi}_{p,\beta,\varepsilon,\alpha}(z) \leq C K^p e^{c\alpha|x|/\varepsilon} \hat{\Psi}_{p,\beta,\varepsilon,\alpha}(x) = C K^p \hat{\Psi}_{p,\beta,\varepsilon,(1+c)\alpha}(x). \tag{6.1.5b}$$

These properties of the weight functions $\hat{\Psi}_{p,\beta,\varepsilon,\alpha}$ allow us to proceed as in the proof of Proposition 5.3.4 by combing local regularity results into a global one; we proceed in several steps:

1. As a first step, a suitable lifting H_1 to $S^+ \cup S^-$ is constructed for the datum h_1 so that one may consider (6.1.1) with $h_1 = 0$. Owing to the fact that a straight sector is considered, this lifting is most conveniently constructed in polar coordinates. To show that the lifting are indeed in the spaces $\mathcal{B}^2_{\beta,\varepsilon,\alpha'}$, the change of variables from polar coordinates to Cartesian has to be analyzed very carefully. This is done in Section 6.3.

2. After the first step, we may assume that $h_1 = 0$. The "energy" estimate (6.1.3) is obtained by showing that the bilinear form associated with (6.1.1) satisfies an inf-sup condition on pairs of exponentially weighted H^1-spaces, i.e., on pairs $H^1_{\varepsilon,\alpha} \times H^1_{\varepsilon,-\alpha}$. These spaces are the usual Sobolev spaces H^1 equipped with the norms $\|\cdot\|_{\varepsilon,\alpha}$, $\|\cdot\|_{\varepsilon,-\alpha}$ given by

$$\|u\|_{\varepsilon,\alpha} = \|\hat{\Psi}_{0,0,\varepsilon,\alpha} u\|_{L^2(S')} + \varepsilon \|\hat{\Psi}_{0,0,\varepsilon,\alpha} \nabla u\|_{L^2(S')}.$$

Proposition 6.4.6 shows that the bilinear form associated with (6.1.1) satisfies an inf-sup conditions on the pairs of spaces $H^1_{\varepsilon,\alpha'} \times H^1_{\varepsilon,-\alpha'}$ for every $0 < \alpha' < \alpha$ sufficiently small. This allows us to obtain in Proposition 6.4.8 for $\alpha' < \alpha'' < \alpha$

$$\|u\|_{\varepsilon,\alpha'} \le C\Big[\|\hat{\Psi}_{0,1,\varepsilon,\alpha'} f\|_{L^2(S')} + \|\hat{\Psi}_{0,0,\varepsilon,\alpha''} H_2\|_{L^2(S')} + \varepsilon\|\hat{\Psi}_{1,0,\varepsilon,\alpha''} \nabla H_2\|_{L^2(S')}\Big],$$

where H_2 is an extension of h_2 into the sector $S_R(\omega)$. We note that the right-hand side involves L^2-based bounds on f whereas we stipulate L^∞-based bounds on f on (6.1.2). We bound further using $\alpha' < \alpha$

$$\begin{aligned} \|\hat{\Psi}_{0,1,\varepsilon,\alpha'} f\|_{L^2(S')} &\le C \|\hat{\Psi}_{0,1,\varepsilon,\alpha} f\|_{L^\infty(S')} \|\hat{\Psi}_{0,0,\varepsilon,-(\alpha-\alpha')}\|_{L^2(S')} \\ &\le C\varepsilon\, \|\hat{\Psi}_{0,1,\varepsilon,\alpha} f\|_{L^\infty(S')}, \end{aligned} \tag{6.1.6}$$

where we used $\hat{\Psi}_{0,0,\varepsilon,-(\alpha-\alpha')}(x) = e^{-(\alpha-\alpha')|x|/\varepsilon}$. Similar reasoning allows us to show in Lemma 6.4.11 for the extension H_2

$$\|\hat{\Psi}_{0,0,\varepsilon,\alpha''} H_2\|_{L^2(S')} + \varepsilon \|\hat{\Psi}_{1,0,\varepsilon,\alpha''} \nabla H_2\|_{L^2(S')} \le C\varepsilon. \tag{6.1.7}$$

Combining (6.1.6), (6.1.7) therefore yields

$$\|u_\varepsilon\|_{\varepsilon,\alpha'} \le C\varepsilon,$$

for a constant $C > 0$ independent of ε. The reasoning for the presence of the factor ε can be best seen in our derivation of (6.1.6): We assume L^∞-based bounds on the data instead of L^2-based ones. A similar mechanism is responsible for the factor ε in (6.1.7).
3. The next step consists in obtaining bounds in weighted L^2 spaces for $\nabla^2 u_\varepsilon$. For suitable $\beta \in [0,1)$ and $\alpha' < \alpha$ we have by Proposition 6.4.9 for $R'' \in (R', R)$ upon setting $S'' = (S^+ \cup S^-) \cap B_{R''(0)}$

$$\begin{aligned} \|\hat{\Psi}_{0,\beta,\varepsilon,\alpha'} \nabla^2 u_\varepsilon\|_{L^2(S'')} \le C\varepsilon^{-1}\Big\{&\|\hat{\Psi}_{0,1,\varepsilon,\alpha} f\|_{L^\infty(S_R(\omega))} \\ &+ \|\hat{\Psi}_{0,0,\varepsilon,\alpha} H_2\|_{L^\infty(S_R(\omega))} + \varepsilon\|\hat{\Psi}_{0,1,\varepsilon,\alpha} H_2\|_{L^\infty(S_R(\omega))}\Big\}. \end{aligned}$$

Comparing this result with the corresponding one for $\alpha = 0$ in Proposition 5.3.2, we see that we gained again one power of ε. Just as above in our "energy" estimates, this is due to our assumption that L^∞-based bounds on the data f, h_2 are available whereas merely L^2-based ones are required.

4. The bounds on higher derivatives are now obtained from local regularity akin to our procedure in Proposition 5.3.4. We cover S' by balls $B_{r_i}(x_i)$ with the key property that $r_i = c|x_i|$ for some fixed $c \in (0,1)$. Next, we use local regularity results, Proposition 5.5.1 for balls in the interior of $S^+ \cup S^-$, Proposition 5.5.2 for half-balls near the outer boundary, and Proposition 5.5.4 for balls with center on Γ, and then sum over all balls to get

$$\|\hat{\Psi}_{p,\beta,\varepsilon,\alpha'} \nabla^{p+2} u_\varepsilon\|_{L^2(S')} \leq C K^p \max\{p+2, \varepsilon^{-1}\}^{p+2} \times \\ \left[\varepsilon \|\hat{\Psi}_{-1,0,\varepsilon,(1+\delta)\alpha'} \nabla u_\varepsilon\|_{L^2(S'')} + \|\hat{\Psi}_{0,0,\varepsilon,(1+\delta)\alpha'} u_\varepsilon\|_{L^2(S'')} + \varepsilon C_f + \varepsilon C_h\right],$$

where $S'' = (S^+ \cup S^-) \cap B_{R''}(0)$ for some suitable $R'' \in (R', R)$ and $\delta > 0$ is essentially determined by the factor $c \in (0,1)$ as in the relations (6.1.5). Just as in the proof of Proposition 5.3.4, the properties (6.1.5) of the weight function $\hat{\Psi}_{p,\beta,\varepsilon,\alpha}$ are instrumental for obtaining the above global result by summing local ones. Note that the contributions εC_f and εC_h from the right-hand side f and the data h_2 are of size $O(\varepsilon)$. The reason for the presence of the factor ε is the same as in the "energy" estimate and the bound on $\nabla^2 u_\varepsilon$: For example, the L^2-bounds $\|\hat{\Psi}_{p,\beta,\varepsilon,\alpha'} \nabla^p f\|_{L^2(S')}$ are required whereas we stipulate pointwise control of the derivatives of f in (6.1.2).
5. The terms $\|\hat{\Psi}_{-1,0,\varepsilon,(1+\delta)\alpha'} \nabla u_\varepsilon\|_{L^2(S'')}$, $\|\hat{\Psi}_{0,0,\varepsilon,(1+\delta)\alpha'} u_\varepsilon\|_{L^2(S'')}$ are treated as in Chapter 5 using an embedding theorem, Lemma 6.2.3, to give

$$\varepsilon \|\hat{\Psi}_{-1,0,\varepsilon,(1+\delta)\alpha'} \nabla u_\varepsilon\|_{L^2(S'')} + \|\hat{\Psi}_{0,0,\varepsilon,(1+\delta)\alpha'} u_\varepsilon\|_{L^2(S'')} \leq C \|u_\varepsilon\|_{H^{2,2}_{\beta,\varepsilon,(1+\delta)\alpha'}(S'')}.$$

Finally, this last term can be controlled in terms of f by the bounds above if we choose α' so small that $(1+\delta)\alpha'$ is sufficiently small.

6.1.2 Outline of Chapter 6

Section 6.2.1 starts with the definition of the spaces $H^{m,l}_{\beta,\varepsilon,\alpha}$ and the countably normed spaces $\mathcal{B}^l_{\beta,\varepsilon,\alpha}$. Their main properties are also collected in Section 6.2.1. These spaces reduce to the spaces $H^{m,l}_{\beta,\varepsilon}$ and $\mathcal{B}^l_{\beta,\varepsilon}$ for the special choice $\alpha = 0$. They are defined completely analogously to those introduced in Chapter 4 and share many of their properties. As the proofs of these properties are very similar to those of Chapter 4, most of the arguments are merely outlined. As before, the key results are Hardy-type embedding results (Lemma 6.2.3) and the fact that the countably normed spaces $\mathcal{B}^l_{\beta,\varepsilon,\alpha}$ are (up to a modification of α) invariant under analytic changes of variables (Theorem 6.2.6). Results for curvilinear sectors will be inferred from straight sectors via a mapping argument (by appealing to Theorem 6.2.6). As in Chapter 4, pointwise results for the growth of the derivatives of functions from $\mathcal{B}^l_{\beta,\varepsilon,\alpha}$ are made available in Theorem 6.2.7 and Corollary 6.2.8.
For the analysis in straight sectors, it is often convenient to introduce polar coordinates. Section 6.3 therefore collects some results concerning the change of variables from polar to Cartesian coordinates.

In Section 6.4.1 we formulate the transmission problem that we analyze and prove unique solvability of the corresponding variational formulation in exponentially weighted spaces.
Section 6.4.4 finally contains the main result of this chapter, Theorem 6.4.13. This result is again obtained by proving the corresponding result on straight sectors and using a mapping argument by appealing to Theorem 6.2.6. Inhomogeneous boundary and jump conditions are treated using special lifting results in Lemma 6.4.15.

6.2 The exponentially weighted spaces $H^{m,l}_{\beta,\varepsilon,\alpha}$ and $\mathcal{B}^l_{\beta,\varepsilon,\alpha}$ in sectors

6.2.1 Properties of the exponentially weighted spaces

On a sectors S we define for $p \in \mathbb{Z}$, $\alpha \in \mathbb{R}$, $\beta \in [0,1]$, and $\varepsilon > 0$ the weight functions $\hat{\Psi}_{p,\beta,\varepsilon,\alpha}$ by

$$\hat{\Psi}_{p,\beta,\varepsilon,\alpha}(x) := e^{\alpha|x|/\varepsilon}\hat{\Phi}_{p,\beta,\varepsilon}(x), \tag{6.2.1}$$

where the weight functions $\hat{\Phi}_{\beta,\varepsilon}$ are defined in (4.2.2). We note that

$$\hat{\Psi}_{p,\beta,\varepsilon,0} = \hat{\Phi}_{p,\beta,\varepsilon}, \qquad \hat{\Psi}_{0,0,\varepsilon,\alpha}(x) = e^{\alpha|x|/\varepsilon}, \qquad \frac{1}{\hat{\Psi}_{0,0,\varepsilon,\alpha}(x)} = \hat{\Psi}_{0,0,\varepsilon,-\alpha}(x).$$

The definition of the weight function $\hat{\Psi}_{p,\beta,\varepsilon,\alpha}$ as a product of $\hat{\Phi}_{p,\beta,\varepsilon}$ and an exponential function allows us to infer properties of the weights $\hat{\Psi}_{p,\beta,\varepsilon,\alpha}$ from those of the functions $\hat{\Phi}_{p,\beta,\varepsilon}$. The analog of Lemma 4.2.2 reads as follows:

Lemma 6.2.1. *Let $S \in \mathbb{R}^2$ be a sector. Then there holds for all $p \in \mathbb{N}_0$, $\varepsilon \in (0,1]$, $\beta \in [0,1]$, $\alpha \geq 0$, $l \in \mathbb{N}$, and $x \in S$:*

$$\hat{\Psi}_{p,\beta,\varepsilon,\alpha}(x) \sim \hat{\Phi}_{p,0,\varepsilon}(x)\hat{\Psi}_{0,\beta,\varepsilon,\alpha}(x) \sim \hat{\Psi}_{p,0,\varepsilon,\alpha}(x)\hat{\Phi}_{0,\beta,\varepsilon}(x), \tag{6.2.2}$$

$$\hat{\Psi}_{p-l,\beta,\varepsilon,\alpha}(x) \sim \hat{\Psi}_{p,0,\varepsilon,\alpha}(x)\hat{\Phi}_{-l,\beta,\varepsilon}(x) \sim \hat{\Phi}_{p,0,\varepsilon}(x)\hat{\Psi}_{-l,\beta,\varepsilon,\alpha}(x), \tag{6.2.3}$$

$$\hat{\Psi}_{p-l,\beta,\varepsilon,\alpha}(x) \sim \hat{\Psi}_{p,\beta-l,\varepsilon,\alpha}(x), \tag{6.2.4}$$

$$\frac{1}{\hat{\Psi}_{p,\beta,\varepsilon,\alpha}(x)} \sim e^{-\alpha|x|/\varepsilon}\left[1 + \left(\frac{\min\{1, \varepsilon(p+1)\}}{|x|}\right)^{p+\beta}\right], \tag{6.2.5}$$

$$\frac{\max\{p+1, \varepsilon^{-1}\}^p}{\hat{\Psi}_{p,0,\varepsilon,\alpha}(x)} \sim e^{-\alpha|x|/\varepsilon}\max\{(p+1)/|x|, \varepsilon^{-1}\}^p. \tag{6.2.6}$$

Here, the relationship $a \sim b$ means that there exist C, $K > 0$ independent of $p \in \mathbb{N}_0$, $\varepsilon \in (0,1]$, and $x \in S$ such that $C^{-1}K^{-p}a \leq b \leq CK^p a$.
Furthermore, let $c \in (0,1)$ be given. Then there holds for all balls $B_{c|x|}(x)$ with $x \in S$

$$\min_{z\in B_{c|x|}(x)} \hat{\Psi}_{p,\beta,\varepsilon,\alpha}(z) \geq CK^p e^{-c|x|\alpha/\varepsilon}\hat{\Psi}_{p,\beta,\varepsilon,\alpha}(x) = CK^p\hat{\Psi}_{p,\beta,\varepsilon,(1-c)\alpha}(x), \quad (6.2.7)$$

$$\max_{z\in B_{c|x|}(x)} \hat{\Psi}_{p,\beta,\varepsilon,\alpha}(z) \leq CK^p e^{c|x|\alpha/\varepsilon}\hat{\Psi}_{p,\beta,\varepsilon,\alpha}(x) = CK^p\hat{\Psi}_{p,\beta,\varepsilon,(1+c)\alpha}(x), \quad (6.2.8)$$

and also

$$C^{-1}K^{-p}e^{(1-c)\alpha|x|/\varepsilon}\hat{\Phi}_{p,\beta,\varepsilon}(x) \leq \min_{z\in B_{c|x|}(x)} \hat{\Psi}_{p,\beta,\varepsilon,\alpha}(z), \quad (6.2.9)$$

$$\max_{z\in B_{c|x|}(x)} \hat{\Psi}_{p,\beta,\varepsilon,\alpha}(z) \leq CK^p e^{(1+c)\alpha|x|/\varepsilon}\hat{\Phi}_{p,\beta,\varepsilon}(x). \quad (6.2.10)$$

The spaces $H^{l,m}_{\beta,\varepsilon,\alpha}(S)$ are defined as in Section 4.2: For m, $l \in \mathbb{N}_0$, $m \geq l$, $\beta \in [0,1)$, $\varepsilon \in (0,1]$, $\alpha \geq 0$, and a sector S, the spaces $H^{m,l}_{\beta,\varepsilon,\alpha}(S)$ are defined as the completion of the space $C^\infty(S)$ under the norm

$$\|u\|^2_{H^{m,l}_{\beta,\varepsilon,\alpha}(\Omega)} := \quad (6.2.11)$$
$$\sum_{k=0}^{l-1} \varepsilon^{2k}\|\hat{\Psi}_{0,0,\varepsilon,\alpha}\nabla^k u\|^2_{L^2(\Omega)} + \varepsilon^{2l}\sum_{k=l}^{m}\|\hat{\Psi}_{k-l,\beta,\varepsilon,\alpha}\nabla^k u\|^2_{L^2(\Omega)},$$

where for $l = 0$ we used the implicit assumption that an empty sum takes the value 0; i.e., for $l = 0$, the spaces $H^{m,0}_{\beta,\varepsilon,\alpha}$ are the completion under the norm

$$\|u\|^2_{H^{m,0}_{\beta,\varepsilon,\alpha}(\Omega)} := \sum_{k=0}^{m}\|\hat{\Psi}_{k,\beta,\varepsilon,\alpha}\nabla^k u\|^2_{L^2(\Omega)}.$$

For a given sector S and constants C_u, $\gamma_u > 0$, the spaces $\mathcal{B}^l_{\beta,\varepsilon,\alpha}(S, C_u, \gamma_u)$ are defined as

$$\mathcal{B}^l_{\beta,\varepsilon,\alpha}(S, C_u, \gamma_u) = \{u \in H^{l,l}_{\beta,\varepsilon,\alpha}(S) \,|\, \|u\|_{H^{l,l}_{\beta,\varepsilon,\alpha}(S)} \leq C_u \text{ and} \quad (6.2.12)$$
$$\|\hat{\Psi}_{k,\beta,\varepsilon,\alpha}\nabla^{k+l}u\|_{L^2(S)} \leq C_u\gamma_u^k \max\{k+1, \varepsilon^{-1}\}^{k+l}\ \forall k \in \mathbb{N}_0\}.$$

The spaces $H^{l,l}_{\beta,\varepsilon,\alpha}$ and $\mathcal{B}^l_{\beta,\varepsilon,\alpha}$ share some invariance properties under smooth (analytic) changes of variables:

Lemma 6.2.2. *Let $\hat{S}$, S be two sectors. Let $g \in C^2(\hat{S}, S)$ be a C^2-diffeomorphism satisfying additionally $g(0) = 0$. Then for $\varepsilon \in (0,1]$ and $\alpha \in \mathbb{R}$ there exist C, $c > 0$ depending only on $\hat{S}$, g, and the sign of α such that*

$$C^{-1}\|u\|_{H^{l,l}_{\beta,\varepsilon,c\alpha}(S)} \leq \|u\circ g\|_{H^{l,l}_{\beta,\varepsilon,\alpha}(\hat{S})} \leq C\|u\|_{H^{l,l}_{\beta,\varepsilon,\alpha/c}(S)}, \qquad l \in \{0,1,2\}.$$

Proof: The proof is similar to that of Lemma 4.2.6. The key observation is that there are c_1, $c_2 > 0$ such that $c_1|x| \leq |g(x)| \leq c_2|x|$ for all $x \in \hat{S}$. □

In order to prove the analog of Theorem 4.2.20, we start by noting that Lemma 4.2.10 holds with $\hat{\Phi}_{0,\beta,\varepsilon}$ replaced by $\hat{\Psi}_{0,\beta,\varepsilon,\alpha}$:

Lemma 6.2.3. *Let S be a C^2-curvilinear sector, $\beta \in (0,1)$, $\alpha \geq 0$, $l \in \{1,2\}$. Then there exists $C > 0$ depending only on S, β, α, and l such that for every $\varepsilon \in (0,1]$ and every $u \in H^{l,l}_{\beta,\varepsilon,\alpha}(S)$ there exists $\overline{u} \in \mathbb{R}$ such that:*

(i) if $l = 2$ the constant $\overline{u}$ may be taken as $\overline{u} = u(0)$ and there holds

$$\|\hat{\Psi}_{0,\beta-2,\varepsilon,\alpha}(u-\overline{u})\|_{L^2(S\cap B_{2\varepsilon}(0))} \leq \varepsilon\|\hat{\Psi}_{0,\beta-1,\varepsilon,\alpha}\nabla u\|_{L^2(S\cap B_{2\varepsilon}(0))},$$
$$\varepsilon\|\hat{\Psi}_{0,\beta-1,\varepsilon,\alpha}\nabla u\|_{L^2(S\cap B_{2\varepsilon}(0))} \leq C\|u\|_{H^{2,2}_{\beta,\varepsilon,\alpha}(S\cap B_{2\varepsilon}(0))},$$
$$\|\hat{\Psi}_{0,\beta-2,\varepsilon,\alpha}u\|_{L^2(S\setminus B_\varepsilon(0))} + \varepsilon\|\hat{\Psi}_{0,\beta-1,\varepsilon,\alpha}\nabla u\|_{L^2(S\setminus B_\varepsilon(0))} \leq C\|u\|_{H^{2,2}_{\beta,\varepsilon,\alpha}(S)};$$

(ii) if $l = 1$:

$$\|\hat{\Psi}_{0,\beta-1,\varepsilon,\alpha}(u-\overline{u})\|_{L^2(S\cap B_{2\varepsilon}(0))} \leq C\varepsilon\|\hat{\Psi}_{0,\beta,\varepsilon,\alpha}\nabla u\|_{L^2(S\cap B_{2\varepsilon}(0))} \leq C\|u\|_{H^{1,1}_{\beta,\varepsilon,\alpha}(S)},$$
$$\|\hat{\Psi}_{0,\beta-1,\varepsilon,\alpha}u\|_{L^2(S\setminus B_\varepsilon(0))} \leq C\|u\|_{H^{1,1}_{\beta,\varepsilon,\alpha}(S)}.$$

Proof: The proof consists in noting that on $S \cap B_{2\varepsilon}(0)$, the weights $\hat{\Psi}_{0,\beta,\varepsilon,\alpha}$ and $\hat{\Phi}_{0,\beta,\varepsilon}$ are equivalent and then appealing to Lemma 4.2.10. The assertions on the sets $S \setminus B_\varepsilon(0)$ are obvious. □

Lemma 6.2.4 (local characterization of $\mathcal{B}^l_{\beta,\varepsilon,\alpha}$). *Let S be a sector, $l \in \mathbb{N}_0$, $\beta \in (0,1)$, $\varepsilon > 0$, $\alpha \geq 0$. Let $\mathcal{B} = \{B_i \mid i \in \mathbb{N}\}$ be a collection of balls $B_i = B_{r_i}(x_i)$ with the following properties:*

1. *there exists $c \in (0,1)$ with $r_i = c|x_i|$ for all $i \in \mathbb{N}$;*
2. *there exists $N \in \mathbb{N}$ such that $\operatorname{card}\{i \in \mathbb{N} \mid x \in B_i\} \leq N$ for all $x \in S$.*

Let $f \in \mathcal{B}^l_{\beta,\varepsilon,\alpha}(S, C_f, \gamma_f)$. Then there exist C, $\gamma > 0$ depending only on γ_f such that for all $p \in \mathbb{N}_0$, $i \in \mathbb{N}$ there holds

$$\hat{\Psi}_{p,\beta,\varepsilon,(1-c)\alpha}(x_i)\|\nabla^{p+l}f\|_{L^2(S\cap B_i)} \leq CC(i)\gamma^p \max\{p+1,\varepsilon^{-1}\}^{p+l}, \tag{6.2.13}$$
$$\sum_{i=1}^{\infty} C^2(i) \leq C_f^2 N \frac{4}{3} < \infty, \tag{6.2.14}$$

where the constants $C(i)$ are given by

$$C^2(i) := \sum_{p=0}^{\infty} \frac{\|\hat{\Psi}_{p,\beta,\varepsilon,\alpha}\nabla^{p+l}f\|^2_{L^2(S\cap B_i)}}{\max\{p+1,\varepsilon^{-1}\}^{2(p+l)}} \frac{1}{(2\gamma_f)^{2p}} \leq C_f^2 \frac{4}{3} < \infty. \tag{6.2.15}$$

Proof: The proof is very similar to that of Lemma 4.2.17. □

Next, we formulate the analogue of Lemma 4.2.19.

Lemma 6.2.5. *Let S be a sector, $c_0 > 0$ be given. Let $u \in \mathcal{B}^l_{\beta,\varepsilon,\alpha}(S, C_u, \gamma_u)$ for some $l \in \{0,1,2\}$, $\beta \in (0,1)$, $\alpha \geq 0$, $\varepsilon > 0$, and C_u, $\gamma_u > 0$. Then there exist constants C, $\gamma > 0$ independent of ε and a constant $\overline{u} \in \mathbb{R}$ such that*

$$\|\hat{\Psi}_{p-l,\beta,\varepsilon,\alpha}\nabla^p(u-\overline{u})\|_{L^2(S\cap B_{c_0\varepsilon}(0))} \le CC_u \max\{p+1,\varepsilon^{-1}\}^p \qquad \forall p\in\mathbb{N}_0,$$
$$\|\hat{\Psi}_{p-l,\beta,\varepsilon,\alpha}\nabla^p u\|_{L^2(S\setminus B_{c_0\varepsilon}(0))} \le CC_u \max\{p+1,\varepsilon^{-1}\}^p \qquad \forall p\in\mathbb{N}_0.$$

Moreover, in the case $l=2$, *the constant* $\overline{u}$ *may be taken as* $\overline{u}=u(0)$ *and for* $l=0$ *we may take* $\overline{u}=0$.

Proof: The proof is analogous to that of Lemma 4.2.17. □

For functions from the countably normed spaces $\mathcal{B}^l_{\beta,\varepsilon,\alpha}$, we have a result analogous to Theorem 4.2.20:

Theorem 6.2.6 (invariance of $\mathcal{B}^l_{\beta,\varepsilon,\alpha}$ under changes of variables). *Let* S *be a* C^2*-curvilinear sector and* $g: S\to g(S)\subset\mathbb{R}^2$ *be analytic on* $\overline{S}$, $g(0)=0$, *and assume that* g^{-1} *is analytic on* $\overline{g(S)}$. *Let* C_u, $\gamma_u>0$, $\beta\in(0,1)$, $\alpha\ge 0$. *Then there exist constants* C, $\gamma>0$, $c>0$ *depending only on* g, S, β, *and* γ_u *(in particular, they are independent of* ε*) such that for* $l\in\{0,1,2\}$

$$u\in\mathcal{B}^l_{\beta,\varepsilon,\alpha}(g(S),C_u,\gamma_u) \Longrightarrow u\circ g\in\mathcal{B}^l_{\beta,\varepsilon,c\alpha}(S,CC_u,\gamma).$$

Proof: The proof is similar to that of Theorem 4.2.20. □

Finally, we have the analogue of Theorem 4.2.23:

Theorem 6.2.7 (pointwise control of $\mathcal{B}^l_{\beta,\varepsilon,\alpha}$-functions). *Let* S *be an analytic sector,* $l\in\{0,1,2\}$, $u\in\mathcal{B}^l_{\beta,\varepsilon,\alpha}(S,C_u,\gamma_u)$ *for some* C_u, $\gamma_u>0$ *and* $\beta\in(0,1)$, $\varepsilon>0$, $\alpha\ge 0$. *Then for every neighborhood* $\mathcal{U}$ *of* Γ_3 *there exist* C, $\gamma>0$, $c\in(0,1)$ *independent of* ε *and* C_u *such that on* $S':=S\setminus\mathcal{U}$

$$\left\|\frac{\hat{\Psi}_{p-l+1,\beta,\varepsilon,c\alpha}}{\max\{1,|x|/\varepsilon\}}\nabla^p u\right\|_{L^\infty(S')} \le CC_u\gamma^p\max\{(p+1),\varepsilon^{-1}\}^{p+1} \qquad \forall p\in\mathbb{N}.$$

This estimate is also valid for $p=0$ *if either* $l=0$ *or* $l=2$ *together with* $u(0)=0$.

A more convenient form of Theorem 6.2.7 is the following corollary.

Corollary 6.2.8 (pointwise control of $\mathcal{B}^l_{\beta,\varepsilon,\alpha}$-functions). *Assume the hypotheses of Theorem 6.2.7 and additionally* $\alpha\in(0,1]$. *Then there exist* C, $\gamma>0$, $c\in(0,1)$ *independent of* ε *such that for all* $p\in\mathbb{N}$ *and* $x\in S'$ *there holds*

$$|\nabla^p u(x)| \le Cp!\gamma^p\alpha^{-(p+2)}|x|^{-(p+1)}\left(\min\left\{1,\left(\frac{|x|}{\varepsilon}\right)\right\}\right)^{l-\beta} e^{-c\alpha|x|/\varepsilon}. \quad (6.2.16)$$

For $l=0$, *this estimate is also valid for* $p=0$. *For the special case* $l=2$ *and* $u(0)=0$, *we have*

$$|u(x)| \le C\alpha^{-2}\varepsilon^{-1}\min\{1,|x|/\varepsilon\}^{1-\beta}e^{-c\alpha|x|/\varepsilon}. \quad (6.2.17)$$

Proof: From Lemma 6.2.1, we have that

$$\max\{1,|x|/\varepsilon\}\frac{\max\{p+1,\varepsilon^{-1}\}^{p+1}}{\hat{\Psi}_{p-l+1,\beta,\varepsilon,\alpha}(x)} \sim \frac{\max\{1,|x|/\varepsilon\}}{\hat{\Phi}_{-l,\beta,\varepsilon}(x)}e^{-\alpha|x|/\varepsilon}\max\{(p+1)/|x|,\varepsilon^{-1}\}^{p+1}.$$

Next, we bound

$$\max\{1,|x|/\varepsilon\}\max\{(p+1)/|x|,\varepsilon^{-1}\}^{p+1} \leq |x|^{-(p+1)}\max\{1,|x|/\varepsilon\}\max\{p+1,|x|/\varepsilon\}^{p+1} \leq |x|^{-(p+1)}\max\{p+1,|x|/\varepsilon\}^{p+2}$$

and

$$e^{-\alpha|x|/(2\varepsilon)}\max\{p+1,|x|/\varepsilon\}^{p+2} \leq C\gamma^p\alpha^{-(p+2)}\,p!$$

for some appropriate C, $\gamma > 0$ independent of x, ε, α. Combining the above considerations, we obtain

$$|\nabla^p u(x)| \leq C\gamma^p\alpha^{-(p+2)}e^{-\alpha|x|/(2\varepsilon)}|x|^{-p-1}\frac{1}{\hat{\Phi}_{-l,\beta,\varepsilon}(x)}.$$

The bound (6.2.16) now follows for $p \geq 1$. It also holds for $p = 0$ together with $l = 0$ and $p = 0$ together with $u(0) = 0$ for the case $l = 2$ by Theorem 6.2.7. In the latter case, we simplify the estimate as follows:

$$\begin{aligned}|u(x)| &\leq C\gamma\alpha^{-2}|x|^{-1}\min\{1,|x|/\varepsilon\}^{2-\beta}e^{-c\alpha|x|/\varepsilon}\\ &\leq C\gamma\alpha^{-2}\varepsilon^{-1}(\varepsilon/|x|)\min\{1,|x|/\varepsilon\}^{2-\beta}e^{-c\alpha|x|/\varepsilon}\\ &\leq C\gamma\alpha^{-2}\varepsilon^{-1}\min\{1,|x|/\varepsilon\}^{1-\beta}e^{-c\alpha|x|/\varepsilon}.\end{aligned}$$

□

Finally, the following lemma is useful.

Lemma 6.2.9. *Let S be a sector, $\beta \in (0,1)$, $\delta > 0$. Then there exists $C > 0$ depending only on β, δ, and S such that*

$$\|\hat{\Psi}_{0,-\beta,\varepsilon,-\delta}\|_{L^2(S)} \leq C\varepsilon.$$

Proof: The proof follows by a direct calculation. □

6.3 Change of variables: from polar to Cartesian coordinates

Lemma 6.3.1. *Let S be a sector and let $g: S \to P_S$ be the change of variables from Cartesian to polar coordinates, $(r,\varphi) = g(x,y)$. Let f_1, f_2 be two functions, analytic on P_S, that satisfy for all $(p,s) \in \mathbb{N}_0$ and $(r,\varphi) \in P_S$ the estimates:*

$$\left|\partial_r^p \partial_\varphi^s f_i(r,\varphi)\right| \leq CK^{p+s} p! s! r^{-p} \max\{1, r/(\varepsilon(p+1))\}^p r^{t_i} e^{-\alpha_i r/\varepsilon}, \qquad i = 1,2,$$

where C, $K > 0$, and α_i, $t_i \in \mathbb{R}$ are independent of ε. Then there exist C', $\gamma > 0$ depending only on C, K, and S such that the product $(f_1 \cdot f_2) \circ g$ satisfies for all $p \in \mathbb{N}_0$ and all $(x,y) \in S$:

$$|\nabla^p_{(x,y)}((f_1 \cdot f_2) \circ g)(x,y)| \leq C' r^{-p} \gamma^p p! \max\{1, r/(\varepsilon(p+1))\}^p \, r^{t_1+t_2} e^{-(\alpha_1+\alpha_2)r/\varepsilon}$$

where, as usual, $r = r(x,y)$.

Proof: The proof is based the Cauchy integral theorem for derivatives. We introduce the polydisc

$$D(\kappa r, \kappa) := \{(z_r, z_\varphi) \in \mathbb{C}^2 \mid |z_r| < \kappa r, |z_\varphi| < \kappa\}, \qquad \text{where } \kappa := \frac{1}{2K},$$

and claim that the f_i are in fact holomorphic on the polydisc $(r,\varphi) + D(\kappa r, \kappa)$ for $(r,\varphi) \in P_S$. In order to see this, we first note that we have for $|z_r| \leq \kappa r$, $|z_\varphi| \leq \kappa$ the bounds

$$\sum_{p \in \mathbb{N}_0} \frac{1}{p!} K^p p! r^{-p} \max\{1, r/(\varepsilon(p+1))\}^p |z_r|^p \leq 3e^{K|z_r|/(r\varepsilon)},$$

$$\sum_{s \in \mathbb{N}_0} \frac{1}{s!} K^s s! |z_\varphi|^s \leq \frac{1}{1 - K\kappa} \leq 2.$$

(The first bound is obtained by splitting the sum into a part from 0 to r/ε and one from r/ε to ∞—the first sum can be majorized by an exponential series, the second one by a geometric series). From these two estimates it follows immediately that the Taylor series of the functions f_i at the point (r,φ) converge on the polydisc $(r,\varphi) + D(\kappa r, \kappa)$; furthermore, we get the bounds

$$|f_i(r + z_r, \varphi + z_\varphi)| \leq C r^{t_i} e^{-\alpha_i r/\varepsilon} 6 e^{K|z_r|/\varepsilon} \qquad \forall (z_r, z_\varphi) \in D(\kappa r, \kappa), \qquad i = 1,2.$$

Next, the map g is an analytic map and there exists a $C_g > 0$ sufficiently large and a $\tilde{C}_g > 0$ sufficiently small such that for $(x,y) \in S$ and all z_x, $z_y \in \mathbb{C}$ with $|z_x| + |z_y| \leq \tilde{C}_g r(x,y)$:

$$|r(x + z_x, y + z_y) - r(x,y)| \leq C_g \left(|z_x| + |z_y|\right),$$

$$|\varphi(x + z_x, y + z_y) - \varphi(x,y)| \leq C_g \frac{|z_x| + |z_y|}{r(x,y)}.$$

We are now in position to apply Cauchy's integral theorem for derivatives of $(f_1 \cdot f_2) \circ g$: Let $(x,y) \in S$ and let $\delta > 0$ (to be chosen below) be such that

$$0 < 2\delta < \kappa r / C_g, \qquad \text{where } r = r(x,y). \tag{6.3.18}$$

This choice of δ guarantees that the function $(f_1 \cdot f_2) \circ g$ is holomorphic on the polydisc $\{(z_x, z_y) \in \mathbb{C}^2 \mid |z_x - x| < \delta, \quad |z_y - y| < \delta\}$. Cauchy's integral formula for derivatives gives for $(s,t) \in \mathbb{N}_0^2$

$$\partial_x^s \partial_y^t \left((f_1 \cdot f_2) \circ g\right)(x,y) = -\frac{s!t!}{4\pi^2} \int_{|z_1|=\delta} \int_{|z_2|=\delta} \frac{((f_1 \cdot f_2) \circ g)(x+z_1, y+z_2)}{(-z_1)^{s+1}(-z_2)^{t+1}} \, dz_1 dz_2$$

and therefore

$$\left|\partial_x^s \partial_y^t \left((f_1 \cdot f_2) \circ g\right)(x,y)\right| \le C \frac{s!t!}{\delta^{s+t}} e^{-(\alpha_1+\alpha_2)r/\varepsilon} r^{t_1+t_2} e^{2K\delta/\varepsilon}.$$

Setting $p := s+t$ and choosing δ such that

$$2\delta = \frac{1}{2} \min\{(p+1)\varepsilon, \kappa r/C_g\},$$

we obtain for some C, $\gamma > 0$ independent of r, ε, and p:

$$\left|\partial_x^s \partial_y^t \left((f_1 \cdot f_2) \circ g\right)(x,y)\right| \le C r^{t_1+t_2} e^{-(\alpha_1+\alpha_2)r/\varepsilon} \gamma^p \left[\frac{p!}{r^p} + \max(p, \varepsilon^{-1})^p\right].$$

This last estimate can readily be brought to the desired form. □

Lemma 6.3.2. *Let $U \subset \mathbb{R}$ be a bounded neighborhood of 0 and $G \subset \mathbb{R}$ be a bounded open set. Let g_i, $i \in \{1,2\}$, be analytic on $U \times G$ and satisfy for some $\varepsilon \in (0,1]$, C, $K > 0$, $\alpha_i \in \mathbb{R}$, $i \in \{1,2\}$:*

$$\left|(\partial_r^p \partial_\varphi^s g_i)(r,\varphi)\right| \le C K^{p+s} s! \max\{p, \varepsilon^{-1}\}^p e^{-\alpha_i r/\varepsilon} \quad \forall (p,s) \in \mathbb{N}_0^2 \ \forall (r,\varphi) \in U \times G.$$

Then the following holds:

(i) The product $g_1 \cdot g_2$ is analytic on $U \times G$ and there exist C', $\gamma > 0$ depending only on C, K such that for all $(p,s) \in \mathbb{N}_0^2$ and all $(r,\varphi) \in U \times G$

$$\left|(\partial_r^p \partial_\varphi^s (g_1 \cdot g_2))(r,\varphi)\right| \le C' \gamma^{p+s} s! \max\{p, \varepsilon^{-1}\}^p e^{-(\alpha_1+\alpha_2)r/\varepsilon}.$$

(ii) If $g_1(0,\varphi) = 0$ for all $\varphi \in G$, then the function $h(r,\varphi) := \frac{1}{r} g_1(r,\varphi) g_2(r,\varphi)$ is analytic on $U \times G$ and there exist C', $\gamma > 0$ depending only on C, K, and α_1, such that for all $(p,s) \in \mathbb{N}_0^2$ and all $(r,\varphi) \in U \times G$

$$\left|\partial_r^p \partial_\varphi^s h(r,\varphi)\right| \le \varepsilon^{-1} C' \gamma^{p+s} s! \max\{p, \varepsilon^{-1}\}^p e^{-(\alpha_1+\alpha_2)r/\varepsilon}.$$

(iii) If $\alpha_1 + \alpha_2 > 0$ then for every $0 < \alpha' < \alpha_1 + \alpha_2$ there exist constants C, $K > 0$ such that the function $h(r,\varphi) := \varepsilon^{-1} r g_1(r,\varphi) g_2(r,\varphi)$ satisfies for all $(p,s) \in \mathbb{N}_0^2$

$$\left|\partial_r^p \partial_\varphi^s h(r,\varphi)\right| \le C' \gamma^{p+s} s! \max\{p, \varepsilon^{-1}\}^p e^{-\alpha' r/\varepsilon} \qquad \forall (r,\varphi) \in U \times G.$$

Proof: We will only show assertion (ii); the other two are proved similarly. Define the polydisc $D := \{(z_r, z_\varphi) \in \mathbb{C}^2 \mid |z_r| < 1/(2K), |z_\varphi| < 1/(2K)\}$. As in the proof of the Lemma 6.3.1, we conclude by Taylor expansions that the functions g_i satisfy for $(r, \varphi) \in U \times G$ and $(z_r, z_\varphi) \in D$

$$|g_i(r + z_r, \varphi + z_\varphi)| \leq |g_i(r, \varphi)| + 6C|z_r|e^{-\alpha_i r/\varepsilon} e^{K|z_r|/\varepsilon}, \qquad i \in \{1, 2\}. \tag{6.3.19}$$

Without loss of generality, assume that $B_\varepsilon(0) \subset U$. Then, as $g_1(0, \varphi) = 0$ for all $\varphi \in G$, we can bound

$$|\frac{1}{r} g_1(r, \varphi)| \leq C\varepsilon^{-1} \leq C\varepsilon^{-1} e^{-\alpha_1 r/\varepsilon} \qquad \text{for } (r, \varphi) \in B_\varepsilon(0) \times G,$$
$$|\frac{1}{r} g_1(r, \varphi)| \leq C\varepsilon^{-1} e^{-\alpha_1 r/\varepsilon} \qquad \text{for } (r, \varphi) \in (U \setminus B_\varepsilon(0)) \times G.$$

Combining this with (6.3.19), we obtain for all $(r, \varphi) \in U \times G$ and $(z_r, z_\varphi) \in D$:

$$|h(r + z_r, \varphi + z_\varphi)| \leq C e^{-(\alpha_1 + \alpha_2) r/\varepsilon} e^{2K|z_r|/\varepsilon} \left(\frac{|r|/\varepsilon}{|r + z_r|} + \frac{|z_r|}{|r + z_r|} \right).$$

Hence, from Cauchy's integral formula for derivatives we obtain for δ_1, $\delta_2 > 0$ to be chosen sufficiently small below:

$$\partial_r^p \partial_\varphi^s h(r, \varphi) = -\frac{p!s!}{4\pi^2} \int_{|z_r|=\delta_1} \int_{|z_\varphi|=\delta_2} \frac{h(r + z_r, \varphi + z_\varphi)}{(-z_1)^{s+1}(-z_2)^{t+1}} \, dz_r dz_\varphi,$$
$$|\partial_r^p \partial_\varphi^s h(r, \varphi)| \leq C \frac{p!s!}{\delta_1^p \delta_2^s} e^{-(\alpha_1 + \alpha_2) r/\varepsilon} e^{2K\delta_1/\varepsilon} \sup_{|z_r|=\delta_1} \left(\frac{|r|/\varepsilon}{|r + z_r|} + \frac{|z_r|}{|r + z_r|} \right).$$

We now choose δ_1 and δ_2. For fixed $r \in U$, we set

$$\delta_2 := \frac{1}{4K},$$
$$\delta' := \min\{\varepsilon(p + 1), 1/(12K)\},$$
$$\delta_1 := \begin{cases} \delta' & \text{if } 0 < \delta' < r/2 \\ 3\delta' & \text{if } \delta' \geq r/2. \end{cases}$$

It easy to see that then for $|z_r| = \delta_1$, we have the bounds

$$\frac{|r|/\varepsilon}{|r + z_r|} + \frac{|z_r|}{|r + z_r|} \leq C\varepsilon^{-1}$$

for some $C > 0$ independent of r, ε, and p. The remainder of the proof is the same as in Lemma 6.3.1. □

6.4 Analytic regularity in exponentially weighted spaces

6.4.1 Transmission problem: problem formulation

Let S be a sector (cf. Definition 4.2.1) and Γ be a (smooth) curve passing through the origin such that Γ divides S into two Lipschitz domains S^+, S^-. In the present section, we are interested in (analytic) regularity results for transmission problems of the following type:

$$-\varepsilon^2\nabla\cdot(A(x)\nabla u)+c(x)u=f \qquad \text{on } S^+\cup S^-, \tag{6.4.1a}$$
$$u=g \qquad \text{on } \partial S, \tag{6.4.1b}$$
$$[u]=h_1 \qquad \text{on } \Gamma, \tag{6.4.1c}$$
$$\varepsilon[\partial_{n_A} u]=h_2 \qquad \text{on } \Gamma, \tag{6.4.1d}$$

where $f\in H^{-1}(S)$, $g\in H^{1/2}(\partial S)$, $h_1\in H^{1/2}_{00}(\Gamma)$, $h_2\in H^{-1/2}(\Gamma)$. We will make additional regularity assumptions on these data shortly. The bracket operator $[\cdot]$ stands for the jump across the curve Γ, and the co-normal operator $\partial_{n_A} u$ is shorthand for $\mathbf{n}^T A\nabla u$ where $\mathbf{n}$ stands for the outer normal vector of S^+.
In view of the analytic regularity results that we seek later, we assume that the coefficients $A\in\mathcal{A}(S^+\cup S^-,\mathbb{S}^2_>)$, $c\in\mathcal{A}(S^+\cup S^-)$ satisfy

$$\|\nabla^p A\|_{L^\infty(S^+\cup S^-)}\le C_A\gamma_A^p p! \qquad \forall p\in\mathbb{N}_0, \tag{6.4.2a}$$
$$0<\lambda_{min}\le A \qquad \text{on } S^+\cup S^-, \tag{6.4.2b}$$
$$\|\nabla^p c\|_{L^\infty(S^+\cup S^-)}\le C_c\gamma_c^p p! \qquad \forall p\in\mathbb{N}_0, \tag{6.4.2c}$$
$$0<\lambda\le c \qquad \text{on } S^+\cup S^-. \tag{6.4.2d}$$

The weak formulation of (6.4.1) reads: Find $u\in H^1(S^+\cup S^-)$ such that

$$b_\varepsilon(u,v)=\varepsilon\int_\Gamma h_2 v\,ds+\int_S fv\,dx \qquad \forall v\in H^1_0(S) \tag{6.4.3a}$$
$$u=g \quad \text{in } H^{1/2}(\partial S^+\setminus\Gamma), \qquad u=g \quad \text{in } H^{1/2}(\partial S^-\setminus\Gamma), \tag{6.4.3b}$$
$$[u]=h_1 \quad \text{in } H^{1/2}(\Gamma), \tag{6.4.3c}$$

where the bilinear form b_ε is given by

$$b_\varepsilon(u,v):=\int_{S^+\cup S^-}\varepsilon^2\nabla u\cdot(A(x)\nabla v)+c(x)uv\,dx. \tag{6.4.4}$$

In order to see that solutions of (6.4.1) exist, we note that the assumptions on g and h_1 imply the existence of $u_0\in H^1(S)$ with $u_0=g$ on ∂S and $u_1\in H^1(S^+)$ with $u_1=h_1$ on Γ, $u_1=0$ on $\partial S^+\setminus\Gamma$. Hence, the solution u of (6.4.1) can be sought in the form

$$u=\tilde u+u_0+\begin{cases}u_1 & \text{on } S^+\\ 0 & \text{on } S^-,\end{cases}$$

where $\tilde u\in H^1_0(S)$ satisfies for all $v\in H^1_0(S)$

$$b_\varepsilon(\tilde u,v)=\int_S fv\,dx+\varepsilon\int_\Gamma h_2 v\,ds-b_\varepsilon(u_0,v)-\int_{S^+}\varepsilon^2\nabla u_1\cdot(A(x)\nabla v)+c(x)u_1 v\,dx.$$

It is now easy to see that the Lax-Milgram Lemma can be applied to infer the existence of a unique solution $\tilde u\in H^1_0(S)$ to this last variational problem. We now show that a $H^{2,2}_{\beta,1}$ regularity result holds as well. To do so, we consider (6.4.1) with $g = 0$, $h_1 = 0$, $f \in H^{0,0}_{\beta,1}(S)$, and $h_2 = H|_\Gamma$, where $H \in H^{1,1}_{\beta,1}(S^+)$. f generates a bounded linear functional on $H^1(S)$ by Lemma 5.3.7. From [15], we have that also $h_2 \in H^{-1/2}(\Gamma)$:

Lemma 6.4.1. *Let S be a straight sector, $\beta \in (0,1)$, and let be Γ_1 one of the sides of S passing through the origin. Then there exists $C > 0$ such that for all $H \in H^{1,1}_{\beta,1}(S)$ there holds for the trace $h := H|_{\Gamma_1} \in H^{1/2}_{loc}(\Gamma_1)$:*

$$\|\hat{\Phi}_{0,\beta/2,1} h\|_{L^2(\Gamma_1)} \leq C \|H\|_{H^{1,1}_{\beta,1}(S)},$$

$$\left| \int_{\Gamma_1} hv \, ds \right| \leq C \|H\|_{H^{1,1}_{\beta,1}(S)} \|v\|_{H^1(S)} \qquad \forall v \in H^1(S).$$

Proof: The lemma is taken from [15, Lemmata 2.9, 2.11]. □

In this situation, we have the following regularity assertion.

Lemma 6.4.2. *Let $S := S_R(\omega)$ be a straight sector, $\Gamma = \{(r\cos\omega', r\sin\omega') \,|\, 0 < r < R\}$ for some $\omega' \in (0,\omega)$. Γ splits $S_R(\omega)$ into two sectors S^+, S^-. Let A^+, $A^- \in \mathbb{S}^2_>$ and define A by $A|_{S^+} = A^+$, $A|_{S^-} = A^-$. Let $R' < R$. Then there exist $\beta \in [0,1)$ and $C > 0$ such that for every $f \in H^{0,0}_{\beta,1}(S)$ and every $H \in H^{1,1}_{\beta,1}(S^+)$, the problem (6.4.1) with $c = 0$, $g = 0$, $h_1 = 0$, and $h_2 = H|_\Gamma$ has a unique solution $u \in H^1_0(S)$, which satisfies*

$$\|u\|_{H^1(S)} + \|u\|_{H^{2,2}_{\beta,1}((S^+ \cup S^-) \cap B_{R'}(0))} \leq C \left[\|f\|_{H^{0,0}_{\beta,1}(S)} + \|H\|_{H^{1,1}_{\beta,1}(S^+)} \right]. \tag{6.4.5}$$

Proof: We ascertained above that the weak solution $u \in H^1_0(S)$ exists. From the Lax-Milgram theorem and Lemma 6.4.1 we get for every $\beta \in [0,1)$ the existence of a constant $C_\beta > 0$ such that

$$\|u\|_{H^1(S)} \leq C_\beta \left[\|f\|_{H^{0,0}_{\beta,1}(S)} + \|H\|_{H^{1,1}_{\beta,1}(S^+)} \right]. \tag{6.4.6}$$

By our local regularity results, (cf. Propositions 5.5.1, 5.5.2, 5.5.4), it suffices to show the weighted H^2-estimate in a neighborhood of the origin that may be chosen sufficiently small. To that end, let χ be a smooth cut-off function supported by $B_{2\delta}(0)$ with $\chi \equiv 1$ on $B_\delta(0)$ for $\delta > 0$ sufficiently small. A calculation then shows that $\tilde{u} := u\chi$ satisfies the transmission problem

$$-\nabla \cdot (A \nabla \tilde{u}) = \tilde{f} := \chi f - 2\nabla\chi \cdot A\nabla u - u \nabla \cdot (A \nabla \chi) \quad \text{on } S,$$
$$\tilde{u} = 0 \quad \text{on } \partial S,$$
$$[\partial_{n_A} \tilde{u}] = \tilde{h}_2 := \chi h_2 + u[\partial_{n_A} \chi] \quad \text{on } \Gamma.$$

Clearly, $\|\tilde{f}\|_{H^{0,0}_{\beta,1}(S)} \leq C\|f\|_{H^{0,0}_{\beta,1}(S)} + C\|H\|_{H^{1,1}_{\beta,1}(S^+)}$. In view of a) $[\partial_{n_A}\chi] = 0$ near the vertex, b) u is in $H^2((S^+ \cup S^-) \setminus B_\delta(0))$, and c) the bound (6.4.6), we see that $\tilde{h}_2$ can be written in the form $\tilde{h}_2 = \tilde{H}|_\Gamma$, where

$$\|\tilde{H}\|_{H^{1,1}_{\beta,1}(S^+)} \le C\left[\|f\|_{H^{0,0}_{\beta,1}(S)} + \|H\|_{H^{1,1}_{\beta,1}(S^+)}\right].$$

It suffices to establish the bound (6.4.5) for $\tilde{u}$ instead of u. We achieve this by transforming the problem to the particular form considered in Proposition A.2.1. We refer to Lemma A.1.1 for how differential equations change under bi-Lipschitz changes of variables.

As the matrices A^+, A^- are both elements of $\mathbb{S}^2_>$, they can be simultaneously diagonalized by similarity transformations; i.e., there is an affine transformation $F : x \mapsto F(x) := Fx$ such that $F^T A^+ F = \mathrm{Id}$ and $F^T A^- F = D$ for a diagonal matrix $D = \mathrm{diag}\,(d_1, d_2)$ (cf., e.g., [57]). The images of S^+, S^-, Γ under this transformation may again be denoted by S^+, S^-, Γ; also the transforms of the data $\tilde{f}$, $\tilde{h}_2$ are denoted again $\tilde{f}$, $\tilde{h}_2$. The original problem problem has been transformed to one of the form

$$\begin{aligned} -p_1 \Delta \tilde{u}^+ &= \tilde{f} \qquad \text{on } S^+, \qquad -\nabla \cdot (D\nabla \tilde{u}^-) = \tilde{f} \qquad \text{on } S^-, \\ p_1 \partial_n u^+ - \mathbf{n}^T D \nabla \tilde{u}^- &= \tilde{h}_2 \qquad \text{on } \Gamma. \end{aligned}$$

In order to obtain the form of Proposition A.2.1, we use the affine change of variables T^+, T^- on the sectors S^+ and S^- with the additional property that $T^+|_\Gamma = T^-|_\Gamma$. They are chosen as follows: T^- is defined by $(x,y) \mapsto (x, (d_1/d_2)^{1/2} y)$ on S^-. This transforms the differential operator on S^- to one of the form $p_2 \Delta u$, which is the desired form. For S^+, it is easy to find $d > 0$ and an orthogonal matrix $\mathcal{O}$ such that the map $T^+(\mathbf{x}) = d\mathcal{O}\mathbf{x}$ satisfies $T^+|_\Gamma = T^-|_\Gamma$. Thus, the differential equation on S^+ preserves its form, and we get the desired form for an application of Proposition A.2.1. The existence of $\beta \in [0,1)$ and the bound (6.4.5) then follows for the transformed function $\tilde{u}$. Since all changes of variables were piecewise affine, the desired weighted H^2-bound for the original $\tilde{u}$ follows. □

Before extending the regularity assertion of Lemma 6.4.2 to variable coefficients, we make the following observation:

Remark 6.4.3 Assume the hypotheses of Lemma 6.4.2. Let for some $\delta > 0$ the two sectors

$$\begin{aligned} S^{+,\delta} &:= S^+ \cap \{(r\cos\varphi, r\sin\varphi) \,|\, 0 < r < R,\ \omega' - \delta < \varphi < \omega' + \delta\}, \\ S^{-,\delta} &:= S^- \cap \{(r\cos\varphi, r\sin\varphi) \,|\, 0 < r < R,\ \omega' - \delta < \varphi < \omega' + \delta\} \end{aligned}$$

be given. Let β be given by Lemma 6.4.2. Then for h_2 of the form $h_2 = H^+|_\Gamma + H^-|_\Gamma$ for $H^+ \in H^{1,1}_{\beta,1}(S^{+,\delta})$, $H^- \in H^{1,1}_{\beta,1}(S^{-,\delta})$, we have the bound (for a constant C depending additionally on δ)

$$\begin{aligned} &\|u\|_{H^1(S)} + \|u\|_{H^{2,2}_{\beta,1}((S^+ \cup S^-)\cap B_{R'}(0))} \\ &\qquad \le C\left[\|f\|_{H^{0,0}_{\beta,1}(S)} + \|H^+\|_{H^{1,1}_{\beta,1}(S^{+,\delta})} + \|H^-\|_{H^{1,1}_{\beta,1}(S^{-,\delta})}\right]. \end{aligned}$$

To see this, we observe that "reflecting" the data H^- at the line Γ yields a function $\tilde{H}^-$ with $\tilde{H}^-|_\Gamma = H^-|_\Gamma$ and $\|\tilde{H}^-\|_{H^{1,1}_{\beta,1}(S^{+,\delta})} = \|H^-\|_{H^{1,1}_{\beta,1}(S^{-,\delta})}$. If

$\delta < \omega - \omega'$, we can extend the function $H^+ + \tilde{H}^-$ to an $H^{1,1}_{\beta,1}(S^+)$-function with the aid of Lemma 5.4.1. The stated bound now follows from Lemma 6.4.2. ∎

We can now extend this result to the case of variable coefficients A, c:

Proposition 6.4.4 (weighted H^2-regularity of transmission problems). *Assume that (6.4.2) holds and let* $0 < R < R'$, $\omega \in (0, 2\pi)$, $\omega' \in (0, \omega)$. *Set* $S := S_R(\omega)$ *and* $\Gamma = \{(r\cos\omega', r\sin\omega') \,|\, 0 < r < R\}$. *Then there exist* $\beta \in [0,1)$ *and* $C > 0$ *such that for every* $f \in H^{0,0}_{\beta,1}(S)$, $H \in H^{1,1}_{\beta,1}(S^+)$ *the problem (6.4.1) with* $g = 0$, $h_1 = 0$, $h_2 = H|_\Gamma$ *has a unique solution* $u \in H^1_0(S)$, *which satisfies*

$$\|u\|_{H^1(S)} + \|u\|_{H^{2,2}_{\beta,1}((S^+ \cup S^-) \cap B_{R'}(0))} \leq C \left[\|f\|_{H^{0,0}_{\beta,1}(S)} + \|H\|_{H^{1,1}_{\beta,1}(S^+)} \right].$$

Proof: The proof consists in a perturbation argument akin to the one in the proofs of Proposition 5.2.2 and Lemma 5.4.2.
First, we note that the solution $u \in H^1_0(S)$ exists by Lax-Milgram. Next, as in the proof of Proposition 5.2.2, we have for an H^1_0-solution u of the transmission problem

$$-\nabla \cdot (A \nabla u) = f \quad \text{on } S, \qquad [\partial_{n_A} u] = h_2 = H^+|_\Gamma + H^-|_\Gamma \quad \text{on } \Gamma$$

the following weighted H^2-bound, if $R'' \in (R, R')$ is fixed and $H^+ \in H^{1,1}_{1,1}(S^+)$, $H^- \in H^{1,1}_{1,1}(S^-)$ (see also Proposition 5.5.4 for suitable local estimates):

$$\begin{aligned} \| \, |x| \nabla^2 u \|_{L^2(S^+_{R'} \cup S^-_{R'})} \leq C \Big[& \| \, |x| f \|_{L^2(S_{R''})} + \|\nabla u\|_{L^2(S_{R''})} \\ & + \|H^+\|_{L^2(S^+_{R''})} + \| \, |x| \nabla H^+ \|_{L^2(S^+_{R''})} + \|H^-\|_{L^2(S^-_{R''})} + \| \, |x| \nabla H^- \|_{L^2(S^-_{R''})} \Big]. \end{aligned} \tag{6.4.7}$$

We introduce the piecewise constant matrix

$$\tilde{A} := \begin{cases} \tilde{A}^+ := \lim\limits_{\substack{x \to 0 \\ x \in S^+}} A(x) & \text{for } x \in S^+, \\ \tilde{A}^- := \lim\limits_{\substack{x \to 0 \\ x \in S^-}} A(x) & \text{for } x \in S^-. \end{cases}$$

Since by assumption $A^+ := A|_{S^+} \in C^1(\overline{S^+}, \mathbb{S}^2_>)$, $A^- := A|_{S^+} \in C^1(\overline{S^+}, \mathbb{S}^2_>)$, there exists $C > 0$ such that

$$\begin{aligned} & |A(x) - \tilde{A}(x)| \leq C|x| \quad \forall x \in S^+ \cup S^-, \\ & |\mathbf{n}^+ \cdot (A^+(x) - \tilde{A}^+)| \leq C|x|, \quad |\mathbf{n}^- \cdot (A^-(x) - \tilde{A}^-)| \leq C|x| \qquad \forall x \in \Gamma. \end{aligned}$$

where $\mathbf{n}^+$, $\mathbf{n}^-$ denote the outer normal vectors for S^+, S^-, respectively. Next, we observe that the solution $u \in H^1_0(S)$ satisfies the equation

$$\begin{aligned} & -\nabla \cdot (\tilde{A} \nabla u) = \tilde{f} := f - cu + \left(\nabla \cdot (A - \tilde{A}) \right) \nabla u + (A - \tilde{A}) : \nabla^2 u \qquad \text{on } S^+ \cup S^- \\ & [\partial_{n_{\tilde{A}}} u] = \tilde{h}_2 := \tilde{H}^+|_\Gamma + \tilde{H}^-|_\Gamma \quad \text{on } \Gamma, \end{aligned}$$

where $\tilde{H}^+$, defined on S^+, and $\tilde{H}^-$, defined on S^-, are given by

$$\tilde{H}^+ := H - \mathbf{n}^+ \cdot (A^+ - \tilde{A}^+)\nabla u^+, \qquad \tilde{H}^- := -\mathbf{n}^- \cdot (A^- - \tilde{A}^-)\nabla u^-;$$

as in Lemma 5.4.2, the functions $\mathbf{n}^+$, $\mathbf{n}^-$ are suitably defined on the sectors S^+, S^-. The argument can be completed using Remark 6.4.3 and by bounding $\|\tilde{f}\|_{H^{0,0}_{\beta,1}(S_{R''})}$, $\|\tilde{H}^+\|_{H^{0,0}_{\beta,1}(S^+_{R''})}$, $\|\tilde{H}^-\|_{H^{0,0}_{\beta,1}(S^-_{R''})}$ with the aid of (6.4.7). □

6.4.2 Transmission problem in exponentially weighted spaces

We are interested in solutions u of (6.4.1) that decay exponentially away from the origin. In order to treat such solutions in a variational framework, it is convenient to introduce the exponentially weighted spaces $H^1_{\varepsilon,\alpha}$. For $\varepsilon > 0$, $\alpha \in \mathbb{R}$, and a sector $S \subset \mathbb{R}^2$ we define the spaces $H^1_{\varepsilon,\alpha}(S)$ as the usual space $H^1(S)$ equipped with the norm $\|\cdot\|_{\varepsilon,\alpha}$ given by

$$\|u\|^2_{\varepsilon,\alpha} := \varepsilon^2\|\hat{\Psi}_{0,0,\varepsilon,\alpha}\nabla u\|^2_{L^2(S)} + \|\hat{\Psi}_{0,0,\varepsilon,\alpha}u\|^2_{L^2(S)}.$$

Similarly, the space $H^1_0(S)$ equipped with the norm $\|\cdot\|_{\varepsilon,\alpha}$ is denoted by $H^1_{0,\varepsilon,\alpha}(S)$. The following lemma gives a convenient characterization of the spaces $H^1_{\varepsilon,\alpha}(S)$, $H^1_{0,\varepsilon,\alpha}(S)$:

Lemma 6.4.5. *Let S be a sector, $\alpha \in \mathbb{R}$, $\varepsilon \in (0,1]$. Then there exists $C > 0$ independent of ε such that for all $u \in H^1(S)$ there holds*

$$C^{-1}\|u\|_{\varepsilon,\alpha} \leq \varepsilon\|\hat{\Psi}_{0,0,\varepsilon,\alpha}u\|_{H^1(S)} + \|\hat{\Psi}_{0,0,\varepsilon,\alpha}u\|_{L^2(S)} \leq C\|u\|_{\varepsilon,\alpha}.$$

Proof: Denoting $r = |x|$, we have $\hat{\Psi}_{0,0,\varepsilon,\alpha}u = e^{\alpha r/\varepsilon}u$. Hence, $|\nabla(e^{\alpha r/\varepsilon}u)| \leq |\alpha|/\varepsilon|e^{\alpha r/\varepsilon}u| + |e^{\alpha r/\varepsilon}\nabla u|$. This gives the upper bound. The lower bound is proved similarly. □

The key tool for our regularity theory in exponentially weighted spaces is the assertion that the bilinear form b_ε of (6.4.4) satisfies an inf-sup condition on appropriate spaces $H^1_{0,\varepsilon,\alpha} \times H^1_{0,\varepsilon,-\alpha}$:

Proposition 6.4.6 (inf-sup condition on $H^1_{0,\varepsilon,\alpha}\times H^1_{0,\varepsilon,-\alpha}$). *Let S be a sector, b_ε be the bilinear form defined in (6.4.4) with coefficients A, c satisfying (6.4.2). Then for $\alpha \geq 0$ with*

$$0 \leq \alpha^2 C_A^2 < \lambda_{min}\lambda \tag{6.4.8}$$

there holds for $u \in H^1_{0,\varepsilon,\alpha}(S)$ and $v \in H^1_{0,\varepsilon,-\alpha}(S)$:

$$\inf_{0\neq u}\sup_{0\neq v}\frac{b_\varepsilon(u,v)}{\|u\|_{\varepsilon,\alpha}\|v\|_{\varepsilon,-\alpha}} \geq \frac{2}{\sqrt{3+4\alpha^2}}\frac{\lambda_{min}\lambda - C_A^2\alpha^2}{\lambda_{min}+\lambda+\sqrt{4C_A^2\alpha^2+(\lambda_{min}-\lambda)^2}},$$
$$|b_\varepsilon(u,v)| \leq \max\{C_A, C_c\}\|u\|_{\varepsilon,\alpha}\|v\|_{\varepsilon,-\alpha}.$$

Furthermore, for every $f \in H^{0,0}_{1,\varepsilon,\alpha}(S)$ *and every* $h \in H^{1,1}_{\varepsilon,\alpha+\delta}(S)$ *with* $\delta > 0$, *we have*

$$\left|\int_S fv\,dx\right| \leq C\|\hat{\Psi}_{0,1,\varepsilon,\alpha} f\|_{L^2(S)}\|v\|_{\varepsilon,-\alpha} \qquad \forall v \in H^1_{0,\varepsilon,-\alpha}(S),$$

$$\left|\varepsilon\int_\Gamma hv\,ds\right| \leq C_\delta\|h\|_{H^{1,1}_{1,\varepsilon,\alpha+\delta}(S)}\|v\|_{\varepsilon,-\alpha} \qquad \forall v \in H^1_{0,\varepsilon,-\alpha}(S),$$

where $C > 0$ *depends only on* S *and* C_δ *depends only on* S, α, *and* $\delta > 0$.

Proof: We start with the proof of the inf-sup condition. Given $u \in H^1_{\varepsilon,\alpha}(S)$, we set $v := \hat{\Psi}_{0,0,\varepsilon,2\alpha} u = e^{2\alpha r/\varepsilon} u$, where $r = r(x) = |x|$. We note that $\nabla v = e^{2\alpha r/\varepsilon}\nabla u + 2\alpha/\varepsilon \nabla r e^{2\alpha r/\varepsilon} u$. From this, it is easy to get

$$\|v\|_{\varepsilon,-\alpha} \leq \sqrt{3+4\alpha^2}\|u\|_{\varepsilon,\alpha}. \tag{6.4.9}$$

Next, we compute

$$\begin{aligned}
&b_\varepsilon(u,v) = \varepsilon^2\int_S (A(x)\nabla u)\nabla v + c(x)uv\,dx = \\
&\varepsilon^2\int_S e^{2\alpha r/\varepsilon}(A(x)\nabla u)\nabla u\,dx + 2\alpha\varepsilon\int_S e^{2\alpha r/\varepsilon}(A(x)\nabla u)(\nabla r)u\,dx + \int_S c\,e^{2\alpha r/\varepsilon}u^2\,dx \\
&\geq \lambda_{min}\varepsilon^2\|\hat{\Psi}_{0,0,\varepsilon,\alpha}\nabla u\|^2_{L^2(S)} \\
&\qquad - 2\alpha C_A\varepsilon\|\hat{\Psi}_{0,0,\varepsilon,\alpha}\nabla u\|_{L^2(S)}\|\hat{\Psi}_{0,0,\varepsilon,\alpha}u\|_{L^2(S)} + \lambda\|\hat{\Psi}_{0,0,\varepsilon,\alpha}u\|^2_{L^2(S)} \\
&\geq (\lambda_{min} - \alpha C_A\delta)\,\varepsilon^2\|\hat{\Psi}_{0,0,\varepsilon,\alpha}\nabla u\|^2_{L^2(S)} + (\lambda - \alpha C_A/\delta)\,\varepsilon^2\|\hat{\Psi}_{0,0,\varepsilon,\alpha}u\|^2_{L^2(S)} \\
&\geq \min\{(\lambda_{min} - \alpha C_A\delta), (\lambda - \alpha C_A/\delta)\}\|u\|^2_{\varepsilon,\alpha}
\end{aligned}$$

for all $\delta > 0$. It remains to choose δ appropriately. It is not hard to see that α needs to satisfy the constraint (6.4.8) in order for the minimum to be non-negative. Next, given α satisfying this constraint, we choose δ such that the minimum is as large as possible, i.e., we choose δ such that $\lambda_{min} - \alpha C_A\delta = \lambda - \alpha C_A/\delta$. Elementary calculations then show that this choice of δ leads to

$$\min\{(\lambda_{min} - \alpha C_A\delta), (\lambda - \alpha C_A/\delta)\} = \frac{2(\lambda_{min}\lambda - C_A^2\alpha^2)}{\lambda_{min} + \lambda + \sqrt{4C_A^2\alpha^2 + (\lambda_{min} - \lambda)^2}},$$

which, combined with (6.4.9), gives the desired inf-sup condition. The continuity of the bilinear form b_ε on the spaces $H^1_{\varepsilon,\alpha} \times H^1_{\varepsilon,-\alpha}$ follows readily from the Cauchy-Schwarz inequality.

For the last two estimates, we start by noting that Lemma 4.2.2 implies

$$\frac{1}{\hat{\Psi}_{0,1,\varepsilon,\alpha}(x)} = \frac{1}{\hat{\Phi}_{0,1,\varepsilon}(x)}\hat{\Psi}_{0,0,\varepsilon,-\alpha}(x) \leq C\left(1 + \frac{\varepsilon}{d(x)}\right)\hat{\Psi}_{0,0,\varepsilon,-\alpha}(x),$$

where $d(x) = \operatorname{dist}(x, \partial S)$. Hence, we get with Lemma 4.2.12 for $v \in H^1_{0,\varepsilon,-\alpha}(S)$:

$$\|\hat{\Psi}^{-1}_{0,1,\varepsilon,\alpha} v\|_{L^2(S)} \le C\left[\|\hat{\Psi}_{0,0,\varepsilon,-\alpha} v\|_{L^2(S)} + \varepsilon\|\frac{1}{d}(\hat{\Psi}_{0,0,\varepsilon,-\alpha} v)\|_{L^2(S)}\right]$$
$$\le C\left[\|\hat{\Psi}_{0,0,\varepsilon,-\alpha} v\|_{L^2(S)} + \varepsilon\|\hat{\Psi}_{0,0,\varepsilon,-\alpha} v\|_{H^1(S)}\right] \le C\|v\|_{\varepsilon,-\alpha},$$

where we employed Lemma 6.4.5 in the last step. We can now conclude that

$$\left|\int_S f v\, dx\right| \le \|\hat{\Psi}_{0,1,\varepsilon,\alpha} f\|_{L^2(S)} \|\hat{\Psi}^{-1}_{0,1,\varepsilon,\alpha} v\|_{L^2(S)} \le \|\hat{\Psi}_{0,1,\varepsilon,\alpha} f\|_{L^2(S)} \|v\|_{\varepsilon,-\alpha}.$$

For the last estimate, we proceed similarly: In view of Lemma A.1.8, we have

$$\varepsilon\left|\int_\Gamma h v\, ds\right| \le \varepsilon\|\hat{\Psi}_{0,0,\varepsilon,\alpha} h\|_{H^{1,1}_{1,1}(S)} \|\hat{\Psi}^{-1}_{0,0,\varepsilon,\alpha} v\|_{H^1(S)}.$$

As above, Lemma 6.4.5 gives $\varepsilon\|\hat{\Psi}^{-1}_{0,0,\varepsilon\alpha} v\|_{H^1(S)} \le C\|v\|_{\varepsilon,-\alpha}$. Using

$$\sup_{x\in S} \max\{1, |x|/\varepsilon\} e^{-\delta|x|/\varepsilon} \le \max\{1, 1/(e\delta)\} < \infty,$$

we compute for the other factor

$$\|\hat{\Psi}_{0,0,\varepsilon,\alpha} h\|_{L^2(S)} \le \|h\|_{H^{1,1}_{1,\varepsilon,\alpha}(S)} \le \|h\|_{H^{1,1}_{1,\varepsilon,\alpha+\delta}(S)},$$
$$\|r\nabla(\hat{\Psi}_{0,0,\varepsilon,\alpha} h)\|_{L^2(S)} \le C\left[\|\hat{\Psi}_{0,0,\varepsilon,\alpha+\delta} h\|_{L^2(S)} + \varepsilon\|\hat{\Psi}_{0,1,\varepsilon,\alpha+\delta}\nabla h\|_{L^2(S)}\right]$$
$$\le C\|h\|_{H^{1,1}_{1,\varepsilon,\alpha+\delta}(S)},$$

where the constant $C > 0$ is independent of ε but depends on $\delta > 0$. □

Remark 6.4.7 We note that the proof of Proposition 6.4.6 actually shows that the inf-sup condition also holds for the pair $H^1_{\varepsilon,\alpha}(S) \times H^1_{\varepsilon,-\alpha}(S)$–the homogeneous boundary conditions are not essential. ▪

Proposition 6.4.6 is the basis for the solution theory in exponentially weighted spaces:

Proposition 6.4.8 (existence and uniqueness in $H^1_{0,\varepsilon,\alpha}$). *Let S be a sector and Γ be a curve as given at the outset of Section 6.4.1. Let A, c satisfy (6.4.2) and α satisfy (6.4.8). Let $\delta > 0$. Then, for every $f \in H^{0,0}_{1,\varepsilon,\alpha}(S)$, $h \in H^{1,1}_{1,\varepsilon,\alpha+\delta}(S)$ the problem:*

$$\textit{Find } u \in H^1_0(S) \textit{ s.t. } b_\varepsilon(u,v) = \varepsilon\int_\Gamma h v\, ds + \int_S f v\, dx \qquad \forall v \in H^1_0(S)$$

has a unique solution u, and for a $C > 0$ independent of ε, f, h, we have

$$\|u\|_{\varepsilon,\alpha} \le C\left[\|f\|_{H^{0,0}_{1,\varepsilon,\alpha}(S)} + \|h\|_{H^{1,1}_{1,\varepsilon,\alpha+\delta}(S)}\right].$$

Proof: Proposition 6.4.6 shows that the bilinear form b_ε satisfies an inf-sup condition on the pair of spaces $H^1_{0,\varepsilon,\alpha}(S) \times H^1_{0,\varepsilon,-\alpha}(S)$, and that the functions f, h generate bounded linear functionals on $H^1_{0,\varepsilon,-\alpha}(S)$; the proof now follows from a well-known result, see, e.g., [11, Thm. 5.2.1], [31, Sec. II.1.1]. □

6.4.3 Analytic regularity in exponentially weighted spaces

We start with an $H^{2,2}_{\beta,\varepsilon,\alpha}$ estimate for solutions to transmission problems.

Proposition 6.4.9. *Let $S = S_R(\omega)$ and let Γ be a straight curve passing through the origin that splits S into S^+, S^-. Let A, c satisfy (6.4.2), and assume that $\alpha > 0$ satisfies (6.4.8). Let furthermore $g_i = 0$, $i \in \{1,2\}$, $h_1 = 0$. Let $\beta \in (0,1)$ be given by the statement of Proposition 6.4.4. Then there exist $C > 0$, $\alpha' \in (0,\alpha)$ independent of ε such that for every $f \in H^{0,0}_{\beta,\varepsilon,\alpha}(S_R(\omega))$ and every $h_2 \in H^{1,1}_{\beta,\varepsilon,\alpha+\delta}(S_R(\omega))$, $\delta > 0$, the solution u of the transmission problem (6.4.3) is in $H^2_{loc}(S^+ \cup S^-)$ and satisfies*

$$\|\hat{\Psi}_{0,\beta,\varepsilon,\alpha'}\nabla^2 u\|_{L^2((S^+\cup S^-)\cap B_{R'}(0))} \leq C\varepsilon^{-2}\left[\|f\|_{H^{0,0}_{\beta,\varepsilon,\alpha}(S)} + \|H\|_{H^{1,1}_{\beta,\varepsilon,\alpha}(S)}\right].$$

Proof: The bound is obtained with the aid of Proposition 6.4.4; the actual proof is very similar to the analogous $H^{2,2}_{\beta,\varepsilon}$ regularity result of Propositions 5.3.2, 5.4.4. We will therefore merely outline the major differences. We start by introducing the shorthand $S^\pm := (S^+\cup S^-)\cap B_{R''}(0)$. Furthermore, we use $\alpha' < \alpha$ repeatedly in the proof but will specify it toward the end of the proof. In order to account for the fact that a transmission problem is considered, the collection of balls $\mathcal{B} = \{B_i \,|\, i \in \mathbb{N}\}$ of the proof of Proposition 5.3.2 is now taken such that additionally the balls B_i satisfy the dichotomy that either $B_i \subset S^+ \cup S^-$ (implying in fact that either $B_i \subset S^+$ or $B_i \subset S^-$) or that the center of the ball B_i is located on Γ, Γ_1, or Γ_2 (implying that $B_i \cap S^+$ or $B_i \cap S^-$ is a half disc). In this way, the local regularity results Lemmata 5.5.5, 5.5.7, 5.5.23 are applicable. Following the proof of Propositions 5.3.2, 5.4.4 we then obtain bounds on $S_h \cap S^\pm$ and $S^\pm \setminus S_h$ separately (see the proof of Proposition 5.3.2 for the precise choice of h—the essential requirement is that $h \sim \varepsilon$). We note that on S_h the weight functions $\hat{\Phi}_{0,\beta,\varepsilon}$ and $\hat{\Psi}_{0,\beta,\varepsilon,\alpha'}$ are equivalent, i.e., there is $c > 0$ independent of ε such that

$$c^{-1}\hat{\Phi}_{0,\beta,\varepsilon}(x) \leq \hat{\Psi}_{0,\beta,\varepsilon,\alpha'}(x) \leq c\hat{\Phi}_{0,\beta,\varepsilon}(x) \qquad \forall x \in S_h.$$

Thus, we can conclude in the same fashion as in the proof of Proposition 5.4.4 (merely replacing the appeal to Lemma 5.4.2 with that to Proposition 6.4.4)

$$\begin{aligned}\|\hat{\Psi}_{0,\beta,\varepsilon,\alpha'}\nabla^2 u\|_{L^2(S^\pm\cap S_h)} \leq C\Big[&\varepsilon^{-2}\|f\|_{H^{0,0}_{\beta,\varepsilon,\alpha'}(S^\pm\cap S_H)} + \varepsilon^{-2}\|h_2\|_{H^{1,1}_{\beta,\varepsilon,\alpha'}(S^\pm\cap S_H)}\\ &+ \varepsilon^{-1}\|\nabla u\|_{L^2(S^\pm\cap S_H)} + \varepsilon^{-1}\|\frac{1}{|x|}u\|_{L^2(S^\pm\cap S_H)}\Big].\end{aligned}$$

From Lemma 4.2.12, we see that, using the homogeneous boundary conditions, we may bound $\|\frac{1}{|x|}u\|_{L^2(S\cap S_H)} \leq C\|\nabla u\|_{L^2(S\cap S_H)}$. Since $\hat{\Psi}_{0,0,\varepsilon,\alpha'}(x) \sim 1$ for all $x \in S_H$, we obtain

$$\begin{aligned}\|\hat{\Psi}_{0,\beta,\varepsilon,\alpha'}\nabla^2 u\|_{L^2(S^\pm\cap S_h)} \leq C\Big[&\varepsilon^{-2}\|f\|_{H^{0,0}_{\beta,\varepsilon,\alpha'}(S^\pm\cap S_H)}\\ &+ \varepsilon^{-2}\|h_2\|_{H^{1,1}_{\beta,\varepsilon,\alpha'}(S^\pm\cap S_H)} + \varepsilon^{-1}\|\hat{\Psi}_{0,0,\varepsilon,\alpha'}\nabla u\|_{L^2(S^\pm\cap S_H)}\Big].\end{aligned} \tag{6.4.10}$$

Next, we consider $S^{\pm} \setminus S_h$. We observe that for all balls $B_i = B_{c|x_i|}(x_i)$ with $i \in \mathbb{N} \setminus I_\varepsilon$ we have for some C, $\kappa \geq 1$ independent of ε and $\delta' > 0$

$$|x_i| \geq C^{-1}\varepsilon, \tag{6.4.11}$$

$$\begin{aligned} C^{-1}\hat{\Psi}_{0,\beta,\varepsilon,\kappa^{-1}\delta'}(x_i) &\leq \min_{z\in B_i} \hat{\Psi}_{0,\beta,\varepsilon,\delta'}(z) \\ &\leq \max_{z\in B_i} \hat{\Psi}_{0,\beta,\varepsilon,\delta'}(z) \leq C\hat{\Psi}_{0,\beta,\varepsilon,\kappa\delta'}(x_i). \end{aligned} \tag{6.4.12}$$

We now employ the local regularity results Lemmata 5.5.5, 5.5.7, 5.5.23 to conclude as in the proofs of Propositions 5.3.2, 5.4.4 that for $i \in \mathbb{N} \setminus I_\varepsilon$

$$\begin{aligned} \|\nabla^2 u\|_{L^2(\hat{B}_i \cap S^\pm)} \leq C\Big[&\varepsilon^{-2}\|f\|_{L^2(B_i\cap S')} + \varepsilon^{-2}\|h_2\|_{L^2(B_i\cap S')} \\ &+ \varepsilon^{-1}\|\nabla h_2\|_{L^2(B_i\cap S')} + \varepsilon^{-1}\|\nabla u\|_{L^2(B_i\cap S')} + \varepsilon^{-2}\|u\|_{L^2(B_i\cap S')}\Big], \end{aligned}$$

where we abbreviated $S' := S_{R'}(\omega) \cap (S^+ \cup S^-)$. Multiplying this last estimate with $\hat{\Psi}_{0,\beta,\varepsilon,\kappa\alpha'}(x_i)$ we obtain with (6.4.12) (taking $\delta' = \alpha'$ and $\delta' = \kappa^2\alpha'$)

$$\begin{aligned} \|\hat{\Psi}_{0,\beta,\varepsilon,\alpha'}\nabla^2 u\|_{L^2(\hat{B}_i\cap S^\pm)} \leq C\Big[&\varepsilon^{-2}\|\hat{\Psi}_{0,\beta,\varepsilon,\kappa^2\alpha'} f\|_{L^2(B_i\cap S')} \\ &+ \varepsilon^{-2}\|\hat{\Psi}_{0,\beta,\varepsilon,\kappa^2\alpha'} h_2\|_{L^2(B_i\cap S')} + \varepsilon^{-1}\|\hat{\Psi}_{0,\beta,\varepsilon,\kappa^2\alpha'}\nabla h_2\|_{L^2(B_i\cap S')} \\ &+ \varepsilon^{-1}\|\hat{\Psi}_{0,0,\varepsilon,\kappa^2\alpha'}\nabla u\|_{L^2(B_i\cap S')} + \varepsilon^{-2}\|\hat{\Psi}_{0,0,\varepsilon,\kappa^2\alpha'} u\|_{L^2(B_i\cap S')}\Big]. \end{aligned}$$

We choose now $\alpha' < \alpha/\kappa^2$ and set $\alpha'' := \alpha'\kappa^2 < \alpha$. We then get

$$\begin{aligned} \|\hat{\Psi}_{0,\beta,\varepsilon,\alpha'}\nabla^2 u\|_{L^2(\hat{B}_i\cap S^\pm)} \leq C\varepsilon^{-2}\Big\{&\|\hat{\Psi}_{0,\beta,\varepsilon,\alpha''} f\|_{L^2(B_i\cap S')} \\ &+ \|\hat{\Psi}_{0,\beta,\varepsilon,\alpha''} h_2\|_{L^2(B_i\cap S')} + \varepsilon\|\hat{\Psi}_{0,\beta,\varepsilon,\alpha''}\nabla h_2\|_{L^2(B_i\cap S')} \\ &+ \varepsilon\|\hat{\Psi}_{0,0,\varepsilon,\alpha''}\nabla u\|_{L^2(B_i\cap S')} + \|\hat{\Psi}_{0,0,\varepsilon,\alpha''} u\|_{L^2(B_i\cap S')}\Big\}. \end{aligned}$$

Squaring and summing on i we get, after including the contribution from S_h and observing that $\alpha' \leq \alpha''$

$$\begin{aligned} \|\hat{\Psi}_{0,\beta,\varepsilon,\alpha'}\nabla^2 u\|_{L^2(S^\pm)} \leq C\varepsilon^{-2}\Big[&\|f\|_{H^{0,0}_{\beta,\varepsilon,\alpha''}(S')} + \|h_2\|_{H^{1,1}_{\beta,\varepsilon,\alpha''}(S')} \\ &+ \varepsilon\|\hat{\Psi}_{0,0,\varepsilon,\alpha''}\nabla u\|_{L^2(S')} + \|\hat{\Psi}_{0,0,\varepsilon,\alpha''} u\|_{L^2(S')}\Big]. \end{aligned} \tag{6.4.13}$$

Since $\alpha'' < \alpha$, there exists $\delta > 0$ such that $\alpha'' + \delta \leq \alpha$, so that we can use the *a priori* estimate of Proposition 6.4.8 to bound

$$\begin{aligned} &\varepsilon\|\hat{\Psi}_{0,\beta,\varepsilon,\alpha''}\nabla u\|_{L^2(S')} + \|\hat{\Psi}_{0,\beta,\varepsilon,\alpha''} u\|_{L^2(S')} \leq \\ &\quad C\left[\|f\|_{H^{0,0}_{1,\varepsilon,\alpha''}(S)} + \|H\|_{H^{1,1}_{1,\varepsilon,\alpha''+\delta}(S)}\right] \leq C\left[\|f\|_{H^{0,0}_{\beta,\varepsilon,\alpha}(S)} + \|H\|_{H^{1,1}_{1,\varepsilon,\alpha}(S)}\right]. \end{aligned}$$

□

A bootstrapping argument then allows us to control in a similar fashion all derivatives of the solution:

Theorem 6.4.10. *Let $S = S_R(\omega)$ and let Γ be a straight curve passing through the origin that splits S into S^+, S^-. Let $R' \in (0, R)$, let A, c satisfy (6.4.2), and assume that $\alpha > 0$ satisfies (6.4.8). Let furthermore $g_i = 0$, $i \in \{1,2\}$, $h_1 = 0$. Let $\beta \in (0,1)$ be given by the statement of Proposition 6.4.4. Then for every $f \in \mathcal{B}^0_{\beta,\varepsilon,\alpha}(S^+ \cup S^-, C_f, \gamma_f)$ and every $h_2 \in \mathcal{B}^1_{\beta,\varepsilon,\alpha}(S^+, C_h, \gamma_h)$ the solution u of the transmission problem (6.4.3) is analytic on $S^+ \cup S^-$ and satisfies for some C, K, $\alpha' \in (0, \alpha)$ independent of ε, C_f, C_h*

$$\|\hat{\Psi}_{p,\beta,\varepsilon,\alpha'} \nabla^{p+2} u\|_{L^2((S^+ \cup S^-) \cap B_{R'}(0))} \le C K^{p+2} \max\{p+1, \varepsilon^{-1}\}^{p+2} [C_f + C_h].$$

If additionally, the data A, c, are analytic on $S_R(\omega)$, $h_2 = 0$, and $f \in \mathcal{B}^0(S, C_f, \gamma_f)$, then u is analytic on $S_{R'}(\omega)$ and $\|\hat{\Psi}_{p,\beta,\varepsilon,\alpha'} \nabla^{p+2} u\|_{L^2(S_{R'}(0))} \le C C_f K^{p+2} \max\{p+1, \varepsilon^{-1}\}^{p+2}$ for all $p \in \mathbb{N}_0$.

Proof: The proof follows the arguments given in the proof of Proposition 5.4.5. The key observation for handling the exponential decay is (6.4.12). □

6.4.4 Analytic regularity for a special transmission problem

We now consider the regularity of (6.4.1) with analytic data where—contrasting Theorem 6.4.10 in which L^2-based bounds on the data h_2, f were required—pointwise estimates on the data g_i, h_i, f are available. This will lead to an improvement in the estimate by a factor ε.

Straight sectors. For simplicity, we will first consider the case of a straight sector $S = S_R(\omega)$. The case of a general analytic sector will then be obtained by a mapping argument. Let Γ be a straight line of the form $\Gamma = \{(r\cos\omega', r\sin\omega') \,|\, r \in (0,R)\}$ for some ω' with $0 < \omega' < \omega$. Next, we denote by $\Gamma_1 = \{(r\cos\omega, 0) \,|\, r \in (0,R)\}$, $\Gamma_2 = \{(r\cos\omega, r\sin\omega) \,|\, r \in (0,R)\}$ the two straight sides of $\partial S_R(\omega)$ and by Γ_3 the curved side $\Gamma_3 = \{(R\cos\varphi, R\sin\varphi) \,|\, 0 < \varphi < \omega\}$. The curve Γ divides $S_R(\omega)$ into two components S^+, S^-.
We introduce four functions g_1, g_2, h_1, $h_2 : [0,R] \to \mathbb{R}$ and $f : S^+ \cup S^- \to \mathbb{R}$ satisfying for some C, K, $\alpha > 0$, $R' \in (0,R)$:

$$g_1(R) = g_2(R) = 0, \tag{6.4.14a}$$
$$g_1(0) = g_2(0), \tag{6.4.14b}$$
$$h_1(0) = h_1(R) = 0 \tag{6.4.14c}$$
$$\|e^{\alpha r/\varepsilon} D^p g_i\|_{L^\infty((0,R))} \le C\varepsilon^{-p}, \quad p \in \{0,1,2\}, \quad i \in \{1,2\}, \tag{6.4.14d}$$
$$\|e^{\alpha r/\varepsilon} D^p h_i\|_{L^\infty((0,R))} \le C\varepsilon^{-p}, \quad p \in \{0,1,2\}, \quad i \in \{1,2\}, \tag{6.4.14e}$$
$$\|\hat{\Psi}_{0,1,\varepsilon,\alpha} f\|_{L^\infty(S^+ \cup S^-)} \le C. \tag{6.4.14f}$$

Additionally, we stipulate for all $p \in \mathbb{N}_0$:

$$\|e^{\alpha r/\varepsilon} D^p g_i\|_{L^\infty((0,R'))} \le C K^p \max\{p, \varepsilon^{-1}\}^p, \quad i \in \{1,2\}, \tag{6.4.14g}$$
$$\|e^{\alpha r/\varepsilon} D^p h_i\|_{L^\infty((0,R'))} \le C K^p \max\{1, \varepsilon^{-1}\}^p, \quad i \in \{1,2\}, \tag{6.4.14h}$$
$$\|\hat{\Psi}_{p,1,\varepsilon,\alpha} \nabla^p f\|_{L^\infty((S^+ \cup S^-) \cap B_{R'}(0))} \le C K^p \max\{1, \varepsilon^{-1}\}^p. \tag{6.4.14i}$$

We are now in position to formulate the transmission problem that we consider: Find $u \in H^1(S^+ \cup S^-)$ such that

$$b_\varepsilon(u,v) = \varepsilon \int_\Gamma h_2 v \, ds + \int_{S^+ \cup S^-} f v \, dx \qquad \forall v \in H_0^1(S_R(\omega)), \tag{6.4.15a}$$

$$u = g_i \text{ on } \Gamma_i,\ i \in \{1,2\}, \quad u = 0 \text{ on } \Gamma_3, \quad [u] = h_1 \text{ on } \Gamma, \tag{6.4.15b}$$

where, by a slight abuse of notation, the functions g_i, h_i depending on one variable only, the "radial variable r", are defined on the straight lines Γ_i, Γ as, e.g., $g_i|_{\Gamma_i}(x) = g_i(|x|)$ for $x \in \Gamma_i$. We note that the assumption (6.4.14) guarantees that the boundary data g are indeed an element of $H^{1/2}(\partial S)$. Similarly, the conditions on the h_i imply that $h_1 \in H_{00}^{1/2}(\Gamma)$ and $h_2 \in H^{-1/2}(\Gamma)$. Thus, existence and uniqueness of the solution of (6.4.15) is ensured.
We note the following lemma:

Lemma 6.4.11. *Let $\alpha > 0$, $\beta \in (0,1)$, $\alpha' \in (0,\alpha)$, and f, h_2 satisfy for some C_f, C_h, K_f, $K_h > 0$*

$$\|\hat{\Psi}_{p,1,\varepsilon,\alpha} \nabla^p f\|_{L^\infty(S^+ \cup S^-)} \le C_f K_f^p \max\{1, \varepsilon^{-1}\}^p \quad \forall p \in \mathbb{N}_0,$$
$$\|\hat{\Psi}_{p,0,\varepsilon,\alpha} D^p h_2\|_{L^\infty((0,R))} \le C_h K_h^p \max\{1, \varepsilon^{-1}\}^p \quad \forall p \in \mathbb{N}_0.$$

Then there exists a function H_2 such that $H_2|_\Gamma = h_2$ and there exist constants C, $\gamma > 0$ independent of ε, C_f, C_h such that $f \in \mathcal{B}^0_{\beta,\varepsilon,\alpha'}(S^+ \cup S^-, CC_f\varepsilon, \gamma)$, $H_2 \in \mathcal{B}^1_{\beta,\varepsilon,\alpha'}(S, CC_h\varepsilon, \gamma)$.

Proof: The assertion for f follows by a direct computation; the factor ε in the expression $CC_f\varepsilon$ stems from the fact that $\alpha' < \alpha$.
The function H_2 is constructed in polar coordinates as

$$H_2(r,\varphi) := h_2(r);$$

clearly, $H_2|_\Gamma = h_2$. Writing $r = |x|$, we have by Lemma 6.2.1 for some C, $K > 0$

$$C^{-1} K^{-p} \max\{(p+1)/r, \varepsilon^{-1}\}^p \le \frac{e^{-\alpha r/\varepsilon} \hat{\Psi}_{p,0,\varepsilon,\alpha}(x)}{\max\{p+1, \varepsilon^{-1}\}^p} \le C K^p \max\{(p+1)/r, \varepsilon^{-1}\}^p. \tag{6.4.16}$$

Lemma 6.3.1 and the assumptions on h_2 imply the existence of K, $C > 0$ such that

$$|\nabla^p H_2(x)| \le C C_h K^p e^{-\alpha r/\varepsilon} \max\{(p+1)/r, \varepsilon^{-1}\}^p \quad \forall p \in \mathbb{N}_0.$$

The bound (6.4.16) therefore implies again with Lemma 6.2.1

$$\|\hat{\Psi}_{p,0,\varepsilon,\alpha} \nabla^p H_2\|_{L^\infty(S)} \le C C_h K^p \max\{(p+1), \varepsilon^{-1}\}^p \quad \forall p \in \mathbb{N}_0.$$

Since $0 < \alpha' < \alpha$, this last estimate can be used to show the existence of C, $\gamma > 0$ such that $H_2 \in \mathcal{B}^1_{\beta,\varepsilon,\alpha'}(S, CC_h\varepsilon, \gamma)$. □

Our main result of this section are the following Theorems 6.4.12, 6.4.13.

Theorem 6.4.12 (analytic regularity, transmission problem). *Let $S = S_R(\omega)$ be a straight sector, $\Gamma = \{(r\cos\omega', r\sin\omega') \,|\, r \in (0,R)\}$, $\omega' \in (0,\omega)$, be a straight line dividing S into S^+ and S^-. Assume that A, c satisfy (6.4.2) and that $\alpha > 0$ satisfies (6.4.8). Let the data g_i, h_i ($i \in \{1,2\}$), f satisfy (6.4.14). Finally, let $R'' \in (0,R')$. Then the solution $u \in H^1(S^+ \cup S^-)$ of (6.4.15) exists and is analytic on $(S^+ \cup S^-) \cap B_{R'}(0)$. Furthermore, there exist C, $\gamma > 0$, $\alpha' \in (0,\alpha)$, and $\beta \in [0,1)$ independent of ε such that for all $p \in \mathbb{N}_0$*

$$\|\hat{\Psi}_{0,0,\varepsilon,\alpha'} u\|_{L^2(S^+\cup S^-)} + \varepsilon\|\hat{\Psi}_{0,0,\varepsilon,\alpha'}\nabla u\|_{L^2(S^+\cup S^-)} \le C\varepsilon, \tag{6.4.17}$$

$$\|\hat{\Psi}_{p,\beta,\varepsilon,\alpha'}\nabla^{p+2} u\|_{L^2((S^+\cup S^-)\cap B_{R''}(0))} \le C\varepsilon\gamma^p \max\{p+1,\varepsilon^{-1}\}^{p+2}. \tag{6.4.18}$$

If additionally $h_1 = h_2 = 0$ and the coefficients A, c, and the right-hand side f are analytic on $S_{R'}(\omega)$, then the solution u of (6.4.15) is analytic on $S_{R'}(\omega)$ and the set $(S^+ \cup S^-) \cap B_{R''}(0)$ may be replaced with $S_{R''}(\omega)$ in (6.4.17), (6.4.18).

Proof: The proof of Theorem 6.4.12 is based on Theorem 6.4.10, Lemma 6.4.11, and Lemmata 6.4.15 below: The lifting result Lemma 6.4.15 is used to reduce the problem to one with $g_1 = g_2 = h_1 = 0$; the right-hand side f and the jump h_2 still satisfy the bounds (6.4.14f), (6.4.14i), (6.4.14e), (6.4.14h). We conclude with Lemma 6.4.11 that $f \in \mathcal{B}^0_{\beta,\varepsilon,\alpha'}((S^+ \cup S^-) \cap B_{R'}(0), C\varepsilon, \gamma)$, $h_2 \in \mathcal{B}^1_{\beta,\varepsilon,\alpha'}(S^+ \cap B_{R'}(0), C\varepsilon, \gamma)$ for $\alpha' \in (0,\alpha)$. An appeal to Theorem 6.4.10 finally allows us to conclude the argument. □

Curved sectors. With a mapping argument, Theorem 6.4.12 can be extended to curvilinear sectors. To fix ideas, let S be a curvilinear sector (with sides Γ_i, $i \in \{1,2,3\}$) and let Γ be an analytic arc passing through the origin and dividing S into two subsectors S^+, S^-. The arcs Γ_i, Γ may be parametrized by analytic maps $\Lambda_i : (0,R) \to \Gamma_i$ and $\Lambda : (0,R) \to \Gamma$ for some $R > 0$. We assume that the maps Λ_i, Λ satisfy

$$C^{-1}r \le |\Lambda_i(r)| \le Cr, \qquad C^{-1}r \le |\Lambda(r)| \le Cr \quad \forall r \in (0,R). \tag{6.4.19}$$

Concerning the data $g \in H^{1/2}(\partial S) \cap C(\partial S)$, $h_1 \in H^{1/2}_{00}(\Gamma)$, $h_2 \in H^{1/2}(\Gamma)$, f, we make the following assumptions: There are constants C, K, $\alpha > 0$ such that upon setting $g_i := g|_{\Gamma_i}$ we have for $p \in \{0,1,2\}$

$$g \in C(\partial S), \quad g_3 = 0, \tag{6.4.20a}$$

$$\|e^{\alpha|\Lambda_i(r)|/\varepsilon} D^p(g_i \circ \Lambda_i)(r)\|_{L^\infty((0,R))} \le C\varepsilon^{-p}, \quad i \in \{1,2\}, \tag{6.4.20b}$$

$$\|e^{\alpha|\Lambda(r)|/\varepsilon} D^p(h_i \circ \Lambda)(r)\|_{L^\infty((0,R))} \le C\varepsilon^{-p}, \quad i \in \{1,2\}, \tag{6.4.20c}$$

$$\|\hat{\Psi}_{0,1,\varepsilon,\alpha} f\|_{L^\infty(S^+\cup S^-)} \le C, \tag{6.4.20d}$$

and additionally the analyticity properties for all $p \in \mathbb{N}_0$:

$$\|e^{\alpha|\Lambda_i(r)|/\varepsilon} D^p(g_i \circ \Lambda_i)(r)\|_{L^\infty((0,R'))} \le C\varepsilon^{-p}, \quad i \in \{1,2\}, \tag{6.4.20e}$$

$$\|e^{\alpha|\Lambda(r)|/\varepsilon} D^p(h_i \circ \Lambda)(r)\|_{L^\infty((0,R'))} \le C\varepsilon^{-p}, \quad i \in \{1,2\}, \tag{6.4.20f}$$

$$\|\hat{\Psi}_{p,1,\varepsilon,\alpha}\nabla^p f\|_{L^\infty((S^+\cup S^-)\cap B_{R'}(0))} \le C\gamma^p \max\{p+1,\varepsilon^{-1}\}^p. \tag{6.4.20g}$$

We then consider the transmission problem: Find $u \in H^1(S^+ \cup S^-)$ such that

$$b_\varepsilon(u,v) = \varepsilon \int_\Gamma h_2 v\, ds + \int_{S^+\cup S^-} fv\, dx \qquad \forall v \in H_0^1(S), \tag{6.4.21a}$$
$$u = g \quad \text{on } \partial S, \qquad [u] = h_1 \quad \text{on } \Gamma. \tag{6.4.21b}$$

Theorem 6.4.13 (analytic regularity, transmission problem). *Let S be a curvilinear sector and let Γ be an analytic curve passing through the origin that divides S into two curvilinear sectors S^+, S^-. Assume that A, c satisfy (6.4.2). Let the data g_i, h_i, ($i \in \{1,2\}$), f satisfy (6.4.20). Then the transmission problem (6.4.21) has a unique solution $u \in H^1(S^+ \cup S^-)$, and there exist constants C, $\gamma > 0$, $\alpha' \in (0,\alpha)$, $\beta \in [0,1)$, $R'' > 0$ independent of ε such that for all $p \in \mathbb{N}_0$*

$$\|\hat\Psi_{0,0,\varepsilon,\alpha'} u\|_{L^2(S^+\cup S^-)} + \varepsilon \|\hat\Psi_{0,0,\varepsilon,\alpha'} \nabla u\|_{L^2(S^+\cup S^-)} \le C\varepsilon, \tag{6.4.22}$$
$$\|\hat\Psi_{p,\beta,\varepsilon,\alpha'} \nabla^{p+2} u\|_{L^2((S^+\cup S^-)\cap B_{R''}(0))} \le C\varepsilon\gamma^p \max\{p+1, \varepsilon^{-1}\}^{p+2}. \tag{6.4.23}$$

If additionally $h_1 = h_2 = 0$ and the coefficients A, c, and the right-hand side f are analytic on $S \cap B_{R'}(0)$, then the solution u is analytic on $S \cap B_{R'}(0)$; the set $(S^+ \cup S^-) \cap B_{R''}(0)$ may then be replaced with $S \cap B_{R''}(0)$ in (6.4.22), (6.4.23).

Proof: The proof follows from Theorem 6.4.12 by a mapping argument. Let ω_1, ω_2 be the angles between the lines Γ_1, Γ_2 and Γ at the origin. It is then easy to construct an invertible Lipschitz mapping $F : \underline{S_R(\omega_1 + \omega_2)}$ (for some appropriate R) such that F is analytic on $\overline{S_R(\omega_1)}$ and $\overline{S_R(\omega_1+\omega_2) \setminus S_R(\omega_1)}$ (e.g., by the "blending method," [58–60]). This mapping allows us to apply Theorem 6.4.12 in a neighborhood of the origin. Mapping back to the original variables yields the desired result with Theorem 6.2.6. □

A different way of putting the result of Theorem 6.4.12 is to state that the restrictions to the two subsectors S^+, S^- are in exponentially weighted spaces $\mathcal{B}^2_{\beta,\varepsilon,\alpha'}$, that is, $u|_{S^+} \in \mathcal{B}^2_{\beta,\varepsilon,\alpha'}(S^+ \cap B_{R''}(0), C\varepsilon, \gamma)$, $u|_{S^-} \in \mathcal{B}^2_{\beta,\varepsilon,\alpha'}(S^- \cap B_{R''}(0), C\varepsilon, \gamma)$. Hence, applying Corollary 6.2.8 allows us to state the following corollary.

Corollary 6.4.14. *Under the assumptions of Theorem 6.4.13, the solution u of the transmission problem (6.4.21) is analytic on the two sectors $S^+ \cap B_{R''}(0)$, $S^- \cap B_{R''}(0)$. Furthermore, for $\beta \in [0,1)$ of Theorem 6.4.12 and some C, γ, $\alpha' > 0$ independent of ε there holds for $x \in (S^+ \cup S^-) \cap B_{R''}(0)$*

$$|u(x) - g(0)| \le C \left(\frac{|x|}{\varepsilon}\right)^{1-\beta} e^{-\alpha'|x|/\varepsilon},$$
$$|\nabla^p u(x)| \le C p! |x|^{-p} \left(\frac{|x|}{\varepsilon}\right)^{1-\beta} e^{-\alpha'|x|/\varepsilon} \quad \forall p \in \mathbb{N}.$$

Proof: We will only show the statements on $S' := S^+ \cap B_{R''}(0)$. The claim for the other sector is proved analogously. As $u|_{S^+} \in \mathcal{B}^2_{\beta,\varepsilon,\alpha'}(S', C\varepsilon, \gamma)$, we can apply Corollary 6.2.8 with $l = 2$, to get that u satisfies on S' for some C, $\gamma > 0$ independent of ε:

$$|\nabla^p u(x)| \leq C\varepsilon\gamma^p p! \, |x|^{-1-p} \min\{1, |x|/\varepsilon\}^{2-\beta} e^{-\alpha'|x|/\varepsilon} \qquad \forall p \in \mathbb{N}.$$

After properly adjusting the constants C, γ, α', this estimate can be brought to the desired form. For the case $p = 0$, we first note that $u|_{S'} \in \mathcal{B}^2_{\beta,\varepsilon,\alpha'}$ implies $u|_{S'} \in C(\overline{S'})$ by Lemma 4.2.9. Next, as the boundary data g_1 is continuous up to the origin, we conclude that $u(0) = g_1(0) = g_2(0)$. Thus, appealing again to Corollary 6.2.8 gives the desired result. □

Lifting results for inhomogeneous boundary conditions. We prove a lifting result to deal with inhomogeneous boundary conditions:

Lemma 6.4.15 (lifting). *Let $S_R(\omega)$ be a straight sector that is divided into two sectors S^+, S^- by the straight line Γ. Let $R' \in (0, R)$ and set*

$$S^{\pm}_R := (S^+ \cup S^-) \cap B_R(0), \qquad S^{\pm}_{R'} := (S^+ \cup S^-) \cap B_{R'}(0).$$

Let g_i ($i \in \{1,2\}$), h_1 satisfy (6.4.14a)–(6.4.14e) and the analyticity requirements (6.4.14g), (6.4.14h). Let $\beta \in (0,1)$. Then there exist $u \in H^1(S^{\pm}_R)$, u analytic on $S^{\pm}_{R'}$, and constants C, $\gamma > 0$, $\alpha' \in (0, \alpha)$ independent of ε such that

$$u|_{\Gamma_i} = g_i, \quad i \in \{1,2\}, \qquad u|_{\Gamma_3} = 0, \tag{6.4.24a}$$

$$[u] = h_1 \quad \text{on } \Gamma, \tag{6.4.24b}$$

$$\varepsilon \|\hat{\Psi}_{0,0,\varepsilon,\alpha'} \nabla u\|_{L^2(S^{\pm}_R)} \leq C\varepsilon, \tag{6.4.24c}$$

$$\|\hat{\Psi}_{0,0,\varepsilon,\alpha'} u\|_{L^2(S^{\pm}_R)} \leq C\varepsilon, \tag{6.4.24d}$$

$$\|\hat{\Psi}_{p,\beta,\varepsilon,\alpha'} \nabla^{p+2} u\|_{L^2(S^{\pm}_{R'})} \leq C\varepsilon\gamma^p \max\{p+1, \varepsilon^{-1}\}^{p+2} \tag{6.4.24e}$$

for all $p \in \mathbb{N}_0$. Furthermore, for A, c satisfying (6.4.2) the function $f := -\varepsilon^2 \nabla \cdot (A\nabla u) + cu$ and the jump $h_2 := \varepsilon[\partial_{n_A} u]$ across Γ satisfy

$$\|\hat{\Psi}_{0,1,\varepsilon,\alpha} f\|_{L^\infty(S^{\pm}_R)} \leq C, \tag{6.4.25a}$$

$$\|\hat{\Psi}_{p,1,\varepsilon,\alpha} \nabla^p f\|_{L^\infty(S^{\pm}_{R'})} \leq C\gamma^p \max\{p+1, \varepsilon^{-1}\}^p \quad \forall p \in \mathbb{N}_0, \tag{6.4.25b}$$

$$\|\hat{\Psi}_{0,0,\varepsilon,\alpha} D^p h_2\|_{L^\infty((0,R))} \leq CK^p \max\{p+1, \varepsilon^{-1}\}^p \quad \forall p \in \{0,1\}, \tag{6.4.25c}$$

$$\|\hat{\Psi}_{0,0,\varepsilon,\alpha} D^p h_2\|_{L^\infty((0,R'))} \leq CK^p \max\{p+1, \varepsilon^{-1}\}^p \quad \forall p \in \mathbb{N}_0. \tag{6.4.25d}$$

Proof: The proof is lengthy and therefore broken up into three pieces: First, we construct u satisfying (6.4.24). Next, we ascertain the bounds (6.4.25a), (6.4.25b) concerning f. In the final step, we check the bounds on $\partial_{n_A} u$ on Γ.

1. step: For the first assertions concerning u, we construct u satisfying (6.4.24a), (6.4.24b) and the following conditions:

$$\|\hat{\Psi}_{0,0,\varepsilon,\alpha} u\|_{L^\infty(S^{\pm}_R)} + \varepsilon \|\hat{\Psi}_{0,0,\varepsilon,\alpha} \nabla u\|_{L^\infty(S^{\pm}_R)} \leq C, \tag{6.4.26a}$$

$$\|\hat{\Psi}_{1,0,\varepsilon,\alpha} \nabla^2 u\|_{L^\infty(S^{\pm}_R)} \leq C\varepsilon^{-2}, \tag{6.4.26b}$$

$$\|\hat{\Psi}_{p,0,\varepsilon,\alpha} \nabla^{p+1} u\|_{L^\infty(S^{\pm}_{R'})} \leq C\gamma^p \max\{p+1, \varepsilon^{-1}\}^{p+1}. \tag{6.4.26c}$$

The bounds (6.4.24c)–(6.4.24e) then follow easily from Lemma 6.2.9: For example, (6.4.24e) follows from

$$\begin{aligned}
&\|\hat{\Psi}_{p,\beta,\varepsilon,\alpha'}\nabla^{p+2}u\|_{L^2(S^{\pm}_{R'})} \\
&\quad\le CK^p\|\hat{\Psi}_{p+1,0,\varepsilon,\alpha}\nabla^{p+2}u\|_{L^\infty(S^{\pm}_{R'})}\|\hat{\Psi}_{0,-(1-\beta),-(\alpha-\alpha')}\|_{L^2(S^{\pm}_{R'})} \\
&\quad\le CK^p\varepsilon\|\hat{\Psi}_{p+1,0,\varepsilon,\alpha}\nabla^{p+2}u\|_{L^\infty(S^{\pm}_{R'})}.
\end{aligned}$$

Recalling the notation $\Gamma_1 = \{(r,0)\,|\,r\in(0,R)\}$, $\Gamma_2 = \{(r\cos\omega, r\sin\omega)\,|\,r\in(0,R)\}$, $\Gamma = \{(r\cos\omega', r\sin\omega')\,|\,r\in(0,R)\}$, we claim that the following function (defined in polar coordinates) has the desired properties (6.4.26):

$$u := u_g + u_h, \tag{6.4.27}$$

$$u_g(r,\varphi) := g_1(r) + (g_2(r) - g_1(r))\,\frac{\varphi}{\omega}, \tag{6.4.28}$$

$$u_h(r,\varphi) := \begin{cases} h_1(r)\dfrac{\varphi}{\omega'} & \text{if } \varphi\in(0,\omega'], \\ 0 & \text{else.}\end{cases} \tag{6.4.29}$$

By construction, u satisfies the desired boundary conditions (6.4.24a) and the jump condition (6.4.24b). The assumptions on the functions g_i, $i\in\{1,2\}$, and h_1 readily imply that u satisfies $\|\hat{\Psi}_{0,0,\alpha,\varepsilon}u\|_{L^\infty(S_R(\omega))}\le C$ for some $C>0$ independent of ε, giving the first part of (6.4.26a). It suffices therefore to consider higher derivatives of u to prove (6.4.26).
We restrict our attention to the sector $S^+ = \{(r\cos\varphi, r\sin\varphi)\,|\,r\in(0,R),\ \varphi\in(0,\omega')\}$, the other one being handled analogously. We write u in the form $u(r,\varphi) = u_1(r) + l(\varphi)u_2(r)$ where the analytic functions u_1, u_2 depend on r only and l is a linear function. We note that $u_2(0)=0$ in view of $g_1(0)=g_2(0)$ and $h_1(0)=0$. Using the formulae

$$u_x = u_r\cos\varphi - u_\varphi\frac{\sin\varphi}{r}, \qquad u_y = u_r\sin\varphi + u_\varphi\frac{\cos\varphi}{r}, \tag{6.4.30}$$

we infer from Lemma 6.3.2 that $\nabla_{(x,y)}u_1$ satisfies for $\varphi\in(0,\omega')$, $r\in(0,R')$

$$|\partial_\varphi^s\partial_r^p\nabla_{(x,y)}u_1(r,\varphi)| \le C\varepsilon^{-1}\gamma^{p+s}s!\max\{p+1,\varepsilon^{-1}\}^p e^{-\alpha r/\varepsilon}\ \forall (p,s)\in\mathbb{N}_0^2. \tag{6.4.31}$$

Hence, applying Lemma 6.3.1, we get that u_1 satisfies in Cartesian coordinates on $S^{\pm}_{R'}$ for some C, $\gamma>0$ independent of ε:

$$|\nabla^p\nabla u_1(x)| \le C\varepsilon^{-1}\gamma^p\max\{(p+1)/r,\varepsilon^{-1}\}^p e^{-\alpha r/\varepsilon}, \qquad r=|x|.$$

Using Lemma 6.2.1, we see

$$\|\hat{\Psi}_{p,0,\varepsilon,\alpha}\nabla^p\nabla u\|_{L^\infty(S^+)} \le C\varepsilon^{-1}\gamma^p\max\{p+1,\varepsilon^{-1}\}^p \quad \forall p\in\mathbb{N}_0,$$

from which (6.4.26c) follows. For u_2, we note that $u_2(0,\varphi)=0$ and that thus (6.4.30) and (ii) of Lemma 6.3.2 imply that $\nabla_{(x,y)}u_2$ satisfy for $\varphi\in(0,\omega)$, $r\in(0,R')$

$$|\partial_\varphi^s \partial_r^p \nabla_{(x,y)} u_2(r,\varphi)| \le C\varepsilon^{-1}\gamma^{p+s} s! \max\{p+1, \varepsilon^{-1}\}^p e^{-\alpha r/\varepsilon} \qquad \forall (p,s) \in \mathbb{N}_0^2,$$

which is a bound that has the same structure as (6.4.31). Thus, we may reason as in our treatment of u_1 to see that u_2 also satisfies (6.4.26c). The bound (6.4.26a), (6.4.26b) follow by inspection.

2. step: The bound (6.4.26c) implies with Lemma 6.2.1 for ∇u:

$$|\nabla^p \nabla u(x)| \le C\varepsilon^{-1}\gamma^p \max\{(p+1)/r, \varepsilon^{-1}\}^p e^{-\alpha r/\varepsilon} \quad \forall p \in \mathbb{N}_0.$$

Applying Lemma 4.3.3 (with the change of variables taken as the identity) to the product $A\nabla u$, we get the existence of C, $\gamma > 0$ such that

$$|\nabla^p (A\nabla u)(x)| \le C\varepsilon^{-1}\gamma^p \max\{(p+1)/r, \varepsilon\}^p e^{-\alpha r/\varepsilon} \quad \forall p \in \mathbb{N}_0.$$

Thus, we conclude for $\nabla^p(\nabla \cdot (A\nabla u))$

$$|\nabla^p (\nabla \cdot (A\nabla u))(x)| \le C\varepsilon^{-1}\gamma^p \max\{(p+1)/r, \varepsilon\}^{p+1} e^{-\alpha r/\varepsilon} \quad \forall p \in \mathbb{N}_0.$$

Hence, with Lemma 6.2.1

$$\|\hat{\Psi}_{p,1,\varepsilon,\alpha} \nabla^p (\nabla \cdot (A\nabla u))\|_{L^\infty(S^\pm_{R'})} \le C\varepsilon^{-1}\gamma^p \max\{p+1, \varepsilon^{-1}\}^{p+1}.$$

After adjusting the constants C, γ, we arrive at

$$\|\hat{\Psi}_{p,1,\varepsilon,\alpha} \nabla^p \left(\varepsilon^2 \nabla \cdot (A\nabla u)\right)\|_{L^\infty(S^\pm_{R'})} \le C\gamma^p \max\{p+1, \varepsilon^{-1}\}^p \quad \forall p \in \mathbb{N}_0,$$

which is part of the desired bound for f. Controlling $\nabla^p(cu)$ is done analogously. Thus (6.4.25b) holds. In order to prove (6.4.25a), one has to verify $\|\hat{\Psi}_{0,1,\varepsilon,\alpha} f\|_{L^\infty(S^\pm_R \setminus S^\pm_{R'})} \le C$, which follows from (6.4.26b).

3. step: We finally turn to the bound for the jump of the co-normal derivatives. We start by noting that the co-normal derivatives operator ∂_{n_A} on the curve Γ for the subsectors S^+, S^- can be written as

$$\partial_{n_A} = \frac{a^\pm(r)}{r}\partial_\varphi + b^\pm(r)\partial_r \tag{6.4.32}$$

for some analytic functions a^+, a^-, b^+, b^-. Here, the $+$ and $-$ sign indicate whether the co-normal derivative corresponds to S^+ or S^-. We fix one of the sectors, S^+, say, and consider the co-normal derivative $\partial_{n_A} u|_\Gamma$. With the functions u_1, u_2, $l(\varphi)$ of the first step, we then get

$$\partial_{n_A} u|_\Gamma = \frac{a^+}{r} u_2(r,\omega')l'(\omega') + b^+(r)\partial_r u_2(r,\omega')l(\omega') + b^+(r)u_1'(r).$$

Lemma 6.3.2 then allows us to conclude

$$\|\hat{\Psi}_{0,0,\varepsilon,\alpha} D^p(\partial_{n_A} u|_\Gamma)\|_{L^\infty((0,R'))} \le C\varepsilon^{-1} \max\{p+1, \varepsilon^{-1}\}^p \quad \forall p \in \mathbb{N}_0,$$

from which we get (6.4.25d) in view of $h_2 = \varepsilon[\partial_{n_A} u]$. The bound (6.4.25c) is obtained by inspection. □

7. Regularity through Asymptotic Expansions

7.1 Motivation and outline

7.1.1 Motivation

Preliminaries. In Chapter 5, we expressed regularity of the solution u_ε of (1.2.11) through the countably normed spaces $\mathcal{B}^l_{\beta,\varepsilon}$ and obtained a shift theorem, Corollary 5.3.12. This shift theorem states that for right-hand sides $f \in \mathcal{B}^0_{\beta,\varepsilon}$, the solution u_ε of (1.2.11) is in $\mathcal{B}^2_{\beta,\varepsilon}$. The condition for f to be in $\mathcal{B}^0_{\beta,\varepsilon}$ is not restrictive if ε is small as the derivatives of f may be very large everywhere in Ω. For example, a function f such as $\sin(x/\varepsilon)$ or, more generally, a function satisfying

$$\|\nabla^p f\|_{L^\infty(\Omega)} \leq C\gamma^p \max\{p+1, \varepsilon^{-1}\}^p \qquad \forall p \in \mathbb{N}_0$$

is an element of $\mathcal{B}^0_{\beta,\varepsilon}$. For highly oscillatory right-hand sides f such as $f(x) = \sin(x/\varepsilon)$, one has to expect that the solution u_ε is highly oscillatory on Ω as well; i.e., one has to expect that the derivatives of the solution u_ε are large everywhere in Ω. For right-hand sides $f \in \mathcal{B}^0_{\beta,\varepsilon}$ the statement $u_\varepsilon \in \mathcal{B}^2_{\beta,\varepsilon}$ is therefore the best one can expect. In practice, however, the right-hand side f may be more "regular" in the sense bounds on its derivatives are available that are independent of ε:

$$\|\nabla^p f\|_{L^\infty(\Omega)} \leq C_f \gamma_f^p p! \qquad \forall p \in \mathbb{N}_0 . \tag{7.1.1}$$

Clearly, such a function is still in $\mathcal{B}^0_{\beta,\varepsilon}$ and thus Corollary 5.3.12 is applicable. However, Corollary 5.3.12 is no longer sharp. In this situation, the typical behavior of the solution u_ε of (1.2.11) is that it is smooth (with bounds on the derivatives independent of ε) in the interior of Ω and that u_ε has boundary layer character with sharp gradients near the boundary only. A precise characterization of this behavior is achieved through asymptotic expansions. This is the approach taken in the present chapter. In order to illustrate our claim that for smooth right-hand sides f the typical solution behavior is indeed to be smooth in the interior and to have boundary layer character near the boundary, we consider a one-dimensional model problem in the following lemma. As in the introductions to the preceding chapters, it is mostly the construction employed in the proof that is of interest here. We mention that closely related analysis can be found in [92].

Lemma 7.1.1 (regularity of asymptotic expansions). *Let $\Omega = (-1,1)$ and assume that f is analytic on Ω with*

$$\|f^{(p)}\|_{L^\infty(\Omega)} \le C_f \gamma_f^p p! \qquad \forall p \in \mathbb{N}_0. \tag{7.1.2}$$

Let u_ε be the solution to

$$L_\varepsilon u_\varepsilon := -\varepsilon^2 u_\varepsilon'' + u_\varepsilon = f \quad \text{on } \Omega, \qquad u_\varepsilon(\pm 1) = 0. \tag{7.1.3}$$

Then there exist C, γ, $\alpha > 0$ independent of ε such that for each $\varepsilon \in (0,1]$ the solution u_ε can be decomposed as

$$u_\varepsilon = w_\varepsilon + u_\varepsilon^{BL} + r_\varepsilon$$

with the following properties (we set $\rho(x) = \operatorname{dist}(x, \partial\Omega)$):

$$\begin{aligned}
\|w_\varepsilon^{(p)}\|_{L^\infty(\Omega)} &\le C\gamma^p p! \qquad \forall p \in \mathbb{N}_0, \\
|\left(u_\varepsilon^{BL}\right)^{(p)}(x)| &\le C\gamma^p \max\{p!, \varepsilon^{-p}\} e^{-\rho(x)/\varepsilon} \qquad \forall p \in \mathbb{N}_0 \quad \forall x \in \Omega, \\
\|r_\varepsilon\|_{L^2(\Omega)} + \varepsilon \|r_\varepsilon'\|_{L^2(\Omega)} &\le C e^{-\alpha/\varepsilon}, \\
r_\varepsilon(\pm 1) &= 0.
\end{aligned}$$

Proof: The functions w_ε, u_ε^{BL}, and r_ε are constructed with the aid of the classical asymptotic expansions for (7.1.3). In a first step, we construct for each $M \in \mathbb{N}_0$ functions w_M, u_M^{BL}, and r_M such that

$$u_\varepsilon = w_M + u_M^{BL} + r_M.$$

In a second step, we choose the expansion order M appropriately in dependence on ε.

Let $M \in \mathbb{N}_0$ be given. We define the outer expansion w_M by

$$w_M(x) := \sum_{i=0}^{M} \varepsilon^{2i} f^{(2i)}(x).$$

We note that the functions w_M are good approximations to particular solutions of (7.1.3) as the defect $f - L_\varepsilon w_M$ is small for small ε:

$$f(x) - L_\varepsilon w_M(x) = \varepsilon^{2M+2} f^{(2M+2)}(x).$$

The outer expansion w_M, however, does not conform to the boundary conditions. In order to correct this, we introduce the boundary layer function u^{BL} as the solution of

$$L_\varepsilon u_M^{BL} = 0 \quad \text{on } \Omega, \qquad u_M^{BL}(\pm 1) = -w_M(\pm 1).$$

u_M^{BL} has the form

$$u_M^{BL} = A_M^- e^{-(1+x)/\varepsilon} + A_M^+ e^{-(1-x)/\varepsilon},$$

where the constants A_M^-, A_M^+ are bounded by

$$\left|A_M^-\right| + \left|A_M^+\right| \leq C\left(|w_M(-1)| + |w_M(1)|\right) \leq C\|w_M\|_{L^\infty(\Omega)}$$

for some constant C independent of ε and M. Using the definition $\rho(x) = \operatorname{dist}(x, \partial\Omega)$, we see that

$$\left|\left(u_M^{BL}\right)^{(p)}(x)\right| \leq C\varepsilon^{-p}e^{-\rho(x)/\varepsilon}\|w_M\|_{L^\infty(\Omega)},$$

where again $C > 0$ is independent of ε and M. The remainder r_M is simply defined by $r_M := u_\varepsilon - w_M - u_M^{BL}$. By construction we have $r_M(\pm 1) = 0$. For bounds on r_M, we observe that it solves the following equation

$$L_\varepsilon r_M = L_\varepsilon u_\varepsilon - L_\varepsilon w_M = f - L_\varepsilon w_M = \varepsilon^{(2M+2)} f^{(2M+2)} \quad \text{in } \Omega, \quad r_\varepsilon(\pm 1) = 0.$$

Appealing to the Lax-Milgram theorem, we obtain in the energy norm $\|\cdot\|_\varepsilon$ defined as in (1.2.7)

$$\|r_M\|_\varepsilon \leq \|\varepsilon^{(2M+2)} f^{(2M+2)}\|_{L^2(\Omega)} \leq C_f\left(\gamma_f \varepsilon(2M+2)\right)^{2M+2}.$$

Using the elementary fact $(2i+p)^{2i+p} \leq (2i)^{2i}p^p e^{2i+p}$, we can bound $w_M^{(p)}$ by

$$\|w_M^{(p)}\|_{L^\infty(\Omega)} \leq C_f p^p e^p S(M) \qquad \forall p \in \mathbb{N}_0,$$
$$S(M) := \sum_{i=0}^{M} \left(e\gamma_f\varepsilon(2i)\right)^{2i}.$$

We have thus defined for each M a decomposition $u_\varepsilon = w_M + u_M^{BL} + r_M$ with the following properties

$$\|w_M^{(p)}\|_{L^\infty(\Omega)} \leq C_f e^p p^p S(M) \qquad \forall p \in \mathbb{N}_0, \tag{7.1.4a}$$
$$\left|\left(u_M^{BL}\right)^{(p)}(x)\right| \leq C\varepsilon^{-p}e^{-\rho(x)/\varepsilon}S(M) \qquad \forall p \in \mathbb{N}_0 \quad \forall x \in \Omega, \tag{7.1.4b}$$
$$\|r_M\|_{L^2(\Omega)} + \varepsilon\|r_M'\|_{L^2(\Omega)} \leq C_f\left(\gamma_f\varepsilon(2M+2)\right)^{2M+2}. \tag{7.1.4c}$$

We now choose M so as to minimize $\left(\gamma_f\varepsilon(2M+2)\right)^{2M+2}$. Specifically, upon choosing M such that

$$2M+2 = \left\lfloor \frac{\alpha}{\varepsilon} \right\rfloor \quad \text{with} \quad \alpha := \frac{1}{e^2\gamma_f},$$

we get $e\gamma_f\varepsilon(2M+2) \leq e^{-1}$ and $(2M+2) \geq \alpha/\varepsilon - 1$. Thus,

$$\left(e\gamma_f\varepsilon(2M+2)\right)^{2M+2} \leq e^{-(2M+2)} \leq ee^{-\alpha/\varepsilon},$$
$$S(M) \leq C_f \sum_{i=0}^{M} \left(\gamma_f\varepsilon(2M)\right)^{2i} \leq C_f \sum_{i=0}^{\infty} e^{-2(2i)} \leq C_f \frac{1}{1-e^{-4}}.$$

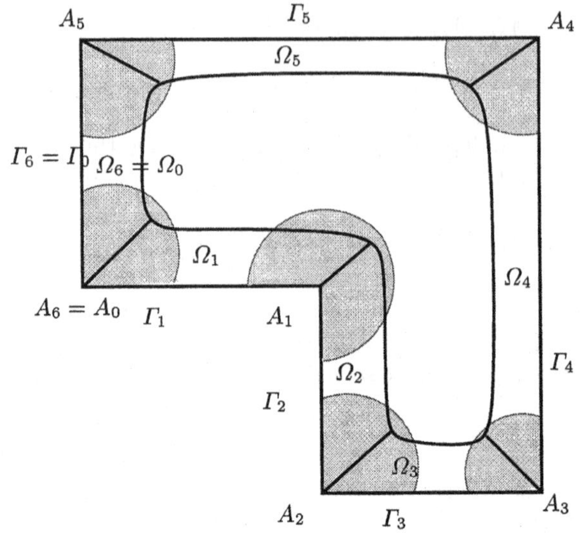

Fig. 7.1.1. Scheme of the supports of χu_ε^{BL} and $\hat{\chi} u_\varepsilon^{CL}$.

Inserting these bounds in (7.1.4) shows that this specific choice of the expansion order M implies the statement of the lemma. □

Lemma 7.1.1 illustrates several things. First of all, neglecting the exponentially small contribution r_ε, the exact solution u_ε has indeed two components: a smooth (analytic) component w_ε and a boundary layer component u_ε^{BL} that decays exponentially away from $\partial\Omega$. We will meet these two solution components in the two-dimensional context again. Next, when using asymptotic expansions one often does not have bounds for the remainder that are explicit in both the perturbation parameter ε and the expansion order. However, this is necessary if one wants to choose the expansion order for a given ε so as to minimize the remainder. In our case, the analyticity of the datum f permits such explicit control and thus allows us to get an "optimal" expansion order $M \sim \varepsilon^{-1}$. This relationship will pervade much of our two-dimensional analysis. A further observation to be made about Lemma 7.1.1 is that, while it provides precise analytic regularity assertions for the terms w_ε and u_ε^{BL}, the remainder r_ε is merely asserted to be small in some norm. This is typical for techniques from asymptotic expansions. It implies in particular that in the context of numerical analysis, decompositions of the form given in Lemma 7.1.1 can only be used up to a certain "error level", namely, the bound on r_ε. Beyond that, additional information about the remainder r_ε is required.

Outline of the construction of the decomposition. Lemma 7.1.1 provides a decomposition of the solution u_ε of (7.1.3) that captures the main features of u_ε, namely, the smooth behavior in the interior and the boundary layer behavior near the boundary. The goal of the present chapter is to provide an analogous decomposition for (1.2.11) in the two-dimensional case. Specifically, solutions u_ε

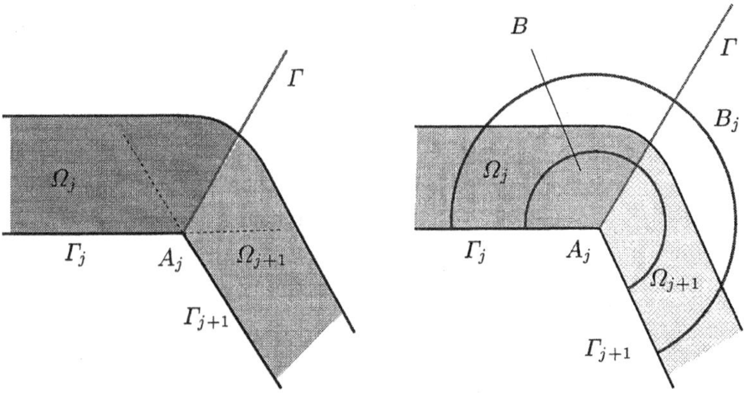

Fig. 7.1.2. Scheme near vertex A_j for the definition of u_M^{CL}.

of (1.2.11) on curvilinear polygons Ω are decomposed as

$$u_\varepsilon = w_\varepsilon + \chi u_\varepsilon^{BL} + \hat{\chi} u_\varepsilon^{CL} + r_\varepsilon. \tag{7.1.5}$$

The terms w_ε, u_ε^{BL}, and r_ε are defined in a similar way as in Lemma 7.1.1 and are introduced to capture the smooth part and the boundary layer behavior near the boundary. However, neither the smooth part w_ε nor the boundary layer part u_ε^{BL} capture the corner singularities that the solution u_ε must have. These effects are captured by the corner layer part u_ε^{CL}. As the boundary layer part u_ε^{BL} and the corner layer part u_ε^{CL} can be defined in a meaningful way in a neighborhood of $\partial\Omega$ and the vertices only, we employ cut-off functions χ, $\hat{\chi}$ for suitable localizations.

The actual construction of the terms in (7.1.5) is rather lengthy and technical; the whole of Chapter 7 is devoted to this undertaking. The main results are collected in Theorem 7.4.5 and Corollary 7.4.6. In order to give the reader a guideline for the proceedings of this chapter, we outline the main steps (which parallel those of Lemma 7.1.1) for the following model problem:

$$L_\varepsilon u_\varepsilon := -\varepsilon^2 \Delta u_\varepsilon + u_\varepsilon = f \quad \text{on } \Omega, \qquad u_\varepsilon = g \quad \text{on } \partial\Omega. \tag{7.1.6}$$

Here, Ω is assumed to be the polygon shown in Fig. 7.1.1. The right-hand side f is assumed analytic on $\overline{\Omega}$ and satisfies (7.1.1). The Dirichlet data g are analytic on each portion Γ_i, $i \in \{1, \dots, 6\}$, of $\partial\Omega$.

1. As in the proof of Lemma 7.1.1, we first seek classical asymptotic-expansions-based decompositions for the solution u_ε of (7.1.6)

$$u_\varepsilon = w_M + \chi u_M^{BL} + \hat{\chi} u_M^{CL} + r_M \tag{7.1.7}$$

   for each expansion order M. At the end, we will choose $M \sim \varepsilon^{-1}$ to obtain the desired decomposition (7.1.5).
2. **Definition of the outer expansion w_M in Section 7.2.** In the present constant-coefficient case, we define w_M as

$$w_M = \sum_{i=0}^{M} \varepsilon^{2i} \Delta^i f.$$

w_M can be viewed as a particular solution to the equation that satisfies the differential equation (up to a small defect) but does not conform to the boundary conditions. A simple calculation shows that the defect $f - L_\varepsilon w_M$ is given by $f - L_\varepsilon w_M = \varepsilon^{2M+2} \Delta^{M+1} f$. Hence, assuming that f satisfies (7.1.1), one can show that the defect satisfies

$$\|f - L_\varepsilon w_M\|_{L^\infty(\Omega)} \leq C \left(\gamma\varepsilon(2M+2)\right)^{2M+2} \tag{7.1.8}$$

for some C, $\gamma > 0$ independent of ε and M. It can also be shown that there are C, γ, $K > 0$ independent of ε and M such that under the assumption

$$\varepsilon(2M+2)K \leq 1, \tag{7.1.9}$$

the outer expansion w_M is analytic on $\overline{\Omega}$ and satisfies

$$\|\nabla^p w_M\|_{L^\infty(\Omega)} \leq C\gamma^p p! \qquad \forall p \in \mathbb{N}_0 .$$

3. **Definition of the boundary layer expansion u_M^{BL} in Section 7.3.** We assume that ε and M satisfy (7.1.9). Then the outer expansion w_M yields (up to the small defect) a particular solution to (7.1.6). However, it fails to conform to the boundary conditions. This is corrected in a second step by means of boundary layer expansions u_M^{BL}, which are defined in a neighborhood of the boundary $\partial\Omega$. They are most conveniently defined in boundary-fitted coordinates. However, as $\partial\Omega$ is only piecewise smooth, the use of boundary-fitted coordinates implies that we can define the boundary layer expansion u_M^{BL} only in a piecewise fashion. Specifically (cf. Fig. 7.1.1), on each subdomain Ω_j, $j \in \{1, \ldots, 6\}$, we introduce boundary-fitted coordinates (ρ_j, θ_j) where ρ_j measures the distance to the curve Γ_j and θ_j is the "tangential" coordinate in this boundary-fitted system. The boundary layer expansion u_M^{BL} is then defined on each subdomain Ω_j separately in terms of boundary-fitted coordinates (ρ_j, θ_j). It is an (approximate) solution to

$$L_\varepsilon u^{BL} = 0 \quad \text{on } \Omega_j, \qquad u^{BL} = g - w_M|_{\Gamma_j}. \tag{7.1.10}$$

The boundary layer expansions u_M^{BL} then have the following key properties: There are C, γ, α, $K > 0$ independent of ε, M such that under the assumption

$$\varepsilon(2M+2)K \leq 1 \tag{7.1.11}$$

(which is the same as (7.1.9) after properly adjusting the constant K) there holds

$$u_M^{BL}|_{\Gamma_j} = g - w_M, \tag{7.1.12}$$

$$\sup_{\theta_j} \left|\partial_{\rho_j}^s \partial_{\theta_j}^t u_M^{BL}(\rho_i, \theta_j)\right| \leq C\gamma^{s+t} t! \varepsilon^{-s} e^{-\alpha\rho_j/\varepsilon}, \tag{7.1.13}$$

$$|L_\varepsilon u_M^{BL}(\rho_i, \theta_j)| \leq \left(\gamma\varepsilon(2M+2)\right)^{2M+2} e^{-\alpha\rho_j/\varepsilon}. \tag{7.1.14}$$

For those combinations of ε and M for which $\gamma\varepsilon(2M+2)$ is small, (7.1.12), (7.1.14) expresses the fact that the function u_M^{BL} is indeed an approximate solution of (7.1.10). (7.1.12) shows that the functions u_M^{BL} have indeed the typical boundary layer behavior: They decay exponentially away from the boundary curves Γ_j, and they behave smoothly (in fact, analytically) in the tangential variable θ_j. As the boundary layer functions u_M^{BL} are defined in a meaningful way only near the boundary $\partial\Omega$, their effect is confined to $\cup_{j=1}^{6}\overline{\Omega_j}$ with the cut-off function χ.

4. **Definition of the corner layer u_M^{CL} in Section 7.4.3.** We assume that ε and M satisfy (7.1.9) and additionally (7.1.11). The boundary layer expansions u_M^{BL} are defined on each subdomain Ω_j separately. In general, they do not match where two subdomains Ω_j, Ω_{j+1} meet. Their common boundary is denoted Γ in Fig. 7.1.2. This incompatibility is removed by corner layers u_M^{CL} that are defined in the shaded regions of Fig. 7.1.1 near the vertices A_j (see also Fig. 7.1.2 for a detailed view near A_j). Specifically, each such shaded region is interpreted as a sector S_j that is divided into two subsectors S_j^+, S_j^- by Γ. The corner layer is then defined as the solution of a transmission problem of the following type:
$$L_\varepsilon u_M^{CL} = 0 \qquad \text{on } S_j^+ \cup S_j^-,$$
$$[u_M^{CL}] = -[\chi u_M^{BL}] \qquad \text{on } \Gamma,$$
$$[\partial_n u_M^{CL}] = -[\partial_n (\chi u_M^{BL})] \qquad \text{on } \Gamma,$$
$$u_M^{CL} = 0 \qquad \text{on } \partial S_j.$$
As the boundary layer expansions u_M^{BL} decay exponentially away from A_j on the common interface Γ, this transmission problem is of the form analyzed in Chapter 6. One therefore obtains regularity assertions of the following type: For a suitable ball $B = B_\kappa(A_j)$ (see Fig. 7.1.2), u_M^{CL} satisfies on $(S_j^+ \cup S_j^-)\cap B$
$$|\nabla^p u_M^{CL}(x)| \le C\gamma^p \varepsilon^{\beta_j - 1} r_j^{1-\beta_j-p} e^{-\alpha r_j/\varepsilon} \qquad \forall p \in \mathbb{N}_0,$$
where $r_j = \operatorname{dist}(x, A_j)$, and the constants C, $\gamma > 0$, $\beta_j \in [0,1)$ are independent of ε and M. We note that the solution u_M^{CL} of this transmission problem decays exponentially away from A_j and has the typical corner singularity behavior near A_j. It should be stressed, however, that it is only piecewise smooth and that its jump across Γ matches that of χu_M^{BL}. As u_M^{CL} is only defined as the solution of a local problem near the vertices A_j, we confine it to a neighborhood of the vertices A_j with the aid of a cut-off function $\hat{\chi}$.

5. **Definition of the remainder r_M in Section 7.4.3.** We still assume that ε and M satisfy (7.1.9), (7.1.11). Once w_M, u_M^{BL}, and u_M^{CL} are defined, the remainder r_M is determined by (7.1.5). As in Lemma 7.1.1, one obtains bounds on r_M by observing that it solves a differential equation. From the bounds on the residuals $f - L_\varepsilon w_M$ and $L_\varepsilon u_M^{BL}$ (and some technicalities involving the cut-off functions χ, $\hat{\chi}$), we obtain from the shift theorem Theorem 5.3.8
$$\|r_M\|_{H^{2,2}_{\beta',\varepsilon}(\Omega)} \le C\left\{(\gamma\varepsilon(2M+2))^{2M+2} + e^{-\alpha/\varepsilon}\right\}$$

for some α, C, $\gamma > 0$ and $\beta' \in [0,1)^J$ independent of ε, M. Here, the term $e^{-\alpha/\varepsilon}$ stems from our treatment of the cut-off functions and the term $(\gamma\varepsilon(2M+2))^{2M+2}$ reflects the residuals $f - L_\varepsilon w_M$ and $L_\varepsilon u_M^{BL}$.

6. In the final step, we choose
$$M = \lfloor \lambda \varepsilon^{-1} \rfloor$$
for λ sufficiently small depending only on the input data. For this choice of M we write $w_\varepsilon = w_M$, $u_\varepsilon^{BL} = u_M^{BL}$, $u_\varepsilon^{CL} = u_M^{CL}$, $r_\varepsilon = r_M$ and get a decomposition of the form (7.1.5) with the following properties:
$$\|\nabla^p w_\varepsilon\|_{L^\infty(\Omega)} \leq C\gamma^p p!,$$
$$\sup_{\theta_j} \left|\partial_{\rho_j}^s \partial_{\theta_j}^t u_\varepsilon^{BL}(\rho_j,\theta_j)\right| \leq C\gamma^{s+t} t! \varepsilon^{-s} e^{-\alpha\rho_j/\varepsilon},$$
$$|\nabla^p u_\varepsilon^{CL}(x)| \leq C\gamma^p \varepsilon^{\beta_j - 1} r_j^{1-\beta_j-p} e^{-\alpha r_j/\varepsilon}, \quad x \in (S_j^+ \cup S_j^-) \cap B_\kappa(A_j),$$
$$\|r_\varepsilon\|_{H^{2,2}_{\beta',\varepsilon}(\Omega)} \leq e^{-\alpha/\varepsilon},$$
for all $p \in \mathbb{N}_0$; all constants are independent of ε.

A few general comments about asymptotic expansions and our decomposition (7.1.5) are in order. First, asymptotic expansions do not converge in general, i.e., including more terms (in our notation: increasing M) does not necessarily increase the accuracy for a fixed ε. Realizing this, it is natural to seek the expansion order M that minimizes the remainder r_M for a given ε. This is achieved in our construction by choosing $M \sim \varepsilon^{-1}$. Instrumental for this procedure, however, is the piecewise analyticity of the input data. Next, while we define outer expansion w_M and the boundary layer expansion u_M^{BL} in essentially the classical way, we introduce the corner layer u_M^{CL} with our application in mind, namely, the convergence analysis of the *hp*-FEM in Chapter 3. One of the aims of classical asymptotic expansions is to obtain smooth (differentiable) expansions. For the convergence analysis of the FEM, however, it is sufficient that the function to be approximated be smooth on each element. Hence, we may use piecewise smooth functions in our decomposition of the solution u provided that the curves of discontinuity coincide with mesh-lines. We exploit this observation in our definition of the corner layer u_M^{CL} as solutions of transmission problems of the form considered in Chapter 6. The restriction that the corner layers are only piecewise smooth is more of an aesthetic restriction than of practical importance as the line of discontinuity, the curve Γ in Fig. 7.1.2, can be chosen somewhat arbitrarily.

7.1.2 Outline of Chapter 7

The outline of this chapter is as follows. We start in Section 7.2 by defining the outer expansion w_M. These functions are (up to a small defect) particular solutions of the equation but ignore the boundary conditions. Theorem 7.2.2 in Section 7.2 makes precise analytic regularity statements for the outer expansion w_M. As we consider (1.2.11) with variable coefficients, the outer expansion w_M

cannot be defined in as explicit a way as was done above. Rather, it is defined in a recursive way and regularity results as stated above have to be proved by induction.
In a second step, the boundary conditions are corrected. This is done in Section 7.3 by means of the boundary layer expansion u_M^{BL}. Technically, this is achieved by introducing boundary-fitted coordinates (ρ_j, θ_j) for each boundary curve Γ_j and then seeking a solution of the homogeneous differential equation with inhomogeneous boundary data on the subdomain Ω_j. More precisely, the solution u^{BL} is expanded formally as

$$u^{BL}(\rho_j, \theta_j) \sim \sum_{i=0}^{\infty} \varepsilon^i U_i(\rho_j/\varepsilon, \theta_j),$$

and then inserted into the differential equation. Equating like powers of ε ultimately yields a recurrence relation for the functions U_i, each being the solution of an ordinary differential equation. The proper treatment of these recursively defined functions U_i is again achieved by induction arguments. The boundary layer expansion u_M^{BL} is then obtained by truncating the formal sum for u^{BL}. The main results concerning the boundary layer expansions u_M^{BL} are collected in Theorem 7.3.3.
As Ω is a curvilinear polygon, this boundary layer expansions lead to incompatibilities near the vertices A_j. These incompatibilities are removed with the aid of corner layers in Section 7.4.3. Analysis of the remainder r_M is also done in Section 7.4.3.
We leave some freedom in the choice of the curve Γ: For concave corners A_j, this smooth curve can be chosen arbitrarily as long as the angle between Γ and each of the two boundary curves Γ_j, Γ_{j+1} is less than π. At convex corners A_j, an alternate corner layer may even be defined without taking recourse to an auxiliary curve Γ. A corner layer defined in this way in convex corners is then analytic instead of being merely piecewise analytic. Section 7.4.3 includes the construction in this case as well.
The construction of this alternate corner layer in convex corners is closely related to the choice of the subdomains Ω_j. In the example of this introductory section, the subdomains Ω_j in Fig. 7.1.1 do not overlap, and the corner layers u_M^{CL} are only piecewise smooth. These corner layers are defined near each vertex A_j and depend on the particular choice of the auxiliary line Γ passing through A_j; this line is $\partial\Omega_j \cap \partial\Omega_{j+1}$. If a vertex A_j is a convex corner, then an alternative definition of a corner layer is possible that is analytic and not merely piecewise analytic. For such a construction, it is more convenient to allow sudomains Ω_j, Ω_{j+1} to overlap at convex corners. Our technical assumptions on the subdomains Ω_j and the cut-off functions χ, $\hat{\chi}$ at the outset of Section 7.4.1 reflect this.

7.2 Regularity of the outer expansion

The *outer* expansion for a problem of the form (1.2.11a) is obtained by seeking a (formal) particular solution of the differential equation in the form

$u \sim \sum_{j=0}^{\infty} \varepsilon^j u_j$. This (formal) power series in ε is inserted into (1.2.11a) and like powers of ε are equated, yielding the following recurrence relation for the unknown functions u_j:

$$u_0(x) := \frac{1}{c(x)} f(x), \quad u_{2j+2}(x) := \frac{1}{c(x)} \nabla \cdot (A(x) \nabla u_j(x)), \quad u_{2j+1} = 0, \quad j \in \mathbb{N}_0. \tag{7.2.1}$$

For each $M \in \mathbb{N}_0$ we can then define the *outer expansion* as

$$w_M := \sum_{j=0}^{2M+1} \varepsilon^j u_j(x) = \sum_{j=0}^{M} \varepsilon^{2j} u_{2j}(x). \tag{7.2.2}$$

From this definition of the outer expansion, we compute

$$L_\varepsilon w_M = f - \varepsilon^{2M+2} \nabla \cdot (A \nabla u_{2M}). \tag{7.2.3}$$

We remark the particularly simple structure of w_M for the special case $A = \mathrm{Id}$:

$$w_M = \sum_{j=0}^{M} \varepsilon^{2j} \Delta^j f. \tag{7.2.4}$$

The following lemma (cf. also [92,96]) is useful for controlling the functions u_j defined by (7.2.1):

Lemma 7.2.1. *Let $\Omega \subset \mathbb{R}^2$ be a domain, A, c be analytic on $\overline{\Omega}$ and assume that $c \geq c_0$ on $\overline{\Omega}$ for some $c_0 > 0$. Assume that u_0 is analytic on Ω and satisfies there*

$$\|\nabla^p u_0\|_{L^\infty(\Omega)} \leq C_u \gamma_u^p p! \qquad \forall p \in \mathbb{N}_0.$$

Define the functions u_{2j}, $j \in \mathbb{N}_0$, recursively by

$$u_{j+2}(x) := \frac{1}{c(x)} \nabla \cdot (A(x) \nabla u_j(x)).$$

Then there exist C, K, $\gamma > 0$ depending only on A, c such that there holds

$$\|\nabla^p u_j\|_{L^\infty(\Omega)} \leq C C_u K^j j! p! (\gamma + 2\gamma_u)^{j+p} \qquad \forall j \in \mathbb{N}_0, \quad p \in \mathbb{N}_0.$$

Proof: By our assumptions, the functions A, $\frac{1}{c}$, u_0 have holomorphic extensions (again denoted A, $1/c$, u_0) to a set $\tilde{G} := \{(x + z_1, y + z_2) \in \mathbb{C}^2 \mid (x, y) \in \Omega, z_i, \in \mathbb{C}, |z_i| < \delta_0\}$ with $\delta_0 < \min\{1, 1/\gamma_u\}$ appropriately chosen. Moreover, by replacing δ_0 with $\delta_0/2$, we may assume that

$$\|u_0\|_{L^\infty(\tilde{G})} \leq 2C_u,$$

$$\left\|\frac{1}{c} A_{ij}\right\|_{L^\infty(\tilde{G})} + \left\|\frac{1}{c} \partial_x A_{ij}\right\|_{L^\infty(\tilde{G})} + \left\|\frac{1}{c} \partial_y A_{ij}\right\|_{L^\infty(\tilde{G})} \leq C_A, \qquad i, j \in \{1, 2\}$$

for some C_A depending only on the data A, c. For $\delta \in (0, \delta_0)$, we define the sets

$$\tilde{G}_\delta := \{(x + z_1, y + z_2) \in \tilde{G} \,|\, (x, y) \in \Omega,\ |z_i| < \delta_0 - \delta\}.$$

We now prove a stronger statement than required by the statement of the lemma, namely, that for $K^2 > 10eC_A$

$$\|u_j\|_{L^\infty(\tilde{G}_\delta)} \le \delta^{-j} K^j j! \|u_0\|_{L^\infty(\tilde{G})} \qquad \forall j \in 2\mathbb{N}_0 \quad \forall \delta \in (0, \delta_0). \tag{7.2.5}$$

The claim of the lemma then follows easily from (7.2.5) with Cauchy's integral formula for derivatives (with the path of integration being a circle with radius $\sim \delta_0/2$) and the choice $\delta \sim \delta_0/2$. For the proof of (7.2.5), we note that it holds for $j = 0$ and proceed by induction on the even j. To that end, assume that for some $j \in 2\mathbb{N}_0$ (7.2.5) is already proven for all even $0 \le j' \le j$. From Cauchy's integral theorem for derivatives, we have for all $\delta \in (0, \delta_0)$ and all $\kappa \in (0, 1)$:

$$\|D^\alpha u_j\|_{L^\infty(\tilde{G}_\delta)} \le \frac{|\alpha|!}{(\kappa\delta)^{|\alpha|}} \|u_j\|_{L^\infty(\tilde{G}_{(1-\kappa)\delta})}, \qquad \forall \alpha \in \mathbb{N}_0^2.$$

Thus, we get from the definition of the recurrence relation, the fact that $\delta < 1$, $\kappa < 1$, and the induction hypothesis:

$$\begin{aligned}
\|u_{j+2}\|_{L^\infty(\tilde{G}_\delta)} &\le C_A \left[\frac{2}{\kappa\delta} + \frac{4 \cdot 2}{(\kappa\delta)^2}\right] \|u_j\|_{L^\infty(\tilde{G}_{(1-\kappa)\delta})} \\
&\le 10C_A \frac{1}{(\kappa\delta)^2} \|u_j\|_{L^\infty(\tilde{G}_{(1-\kappa)\delta})} \\
&\le 10C_A \delta^{-(j+2)} j! K^j (1-\kappa)^{-j} \kappa^{-2} \|u_0\|_{L^\infty(\tilde{G})} \\
&\le \delta^{-(j+2)} K^{j+2} (j+2)! \|u_0\|_{L^\infty(\tilde{G})} \left[10C_A K^{-2} \frac{1}{(j+1)(j+2)\kappa^2} (1-\kappa)^{-j}\right].
\end{aligned} \tag{7.2.6}$$

Next, choosing $\kappa = 1/(j+2)$ and observing that this choice implies

$$\frac{(1-\kappa)^j}{\kappa^2 (j+1)(j+2)} = \frac{(j+2)^{j+1}}{(j+1)^{j+1}} = \left(1 + \frac{1}{j+1}\right)^{j+1} \le e,$$

we can bound the expression in brackets $[\,\cdot\,]$ in (7.2.6) by 1 and thus conclude the induction argument. □

Lemma 7.2.1 enables us to formulate the following for the outer expansion w_M:

Theorem 7.2.2 (regularity of the outer expansion). *Let $G \subset \mathbb{R}^2$ be an open set and let $f \in \mathcal{A}(G)$ satisfy*

$$\|\nabla^p f\|_{L^\infty(G)} \le C_f \gamma_f^p p! \qquad \forall p \in \mathbb{N}_0.$$

Assume that A, c are analytic on $\overline{G}$ and that $c > 0$ on $\overline{G}$. Then, for each $M \in \mathbb{N}_0$ the function w_M given by (7.2.4) is analytic on G and there exist C, K, $\gamma > 0$ depending only on the coefficients A, c such that under the constraint $0 < 2M\varepsilon(\gamma + 2\gamma_f)K \le 1$ there holds for all $p \in \mathbb{N}_0$

$$\begin{aligned}
\|\nabla^p w_M\|_{L^\infty(G)} &\le C_f C (\gamma + 2\gamma_f)^p p!, \\
\|\nabla^p (L_\varepsilon w_M - f)\|_{L^\infty(G)} &\le C_f C \left\{(2M+2)\varepsilon(\gamma + 2\gamma_f)K\right\}^{2M+2} (\gamma + 2\gamma_f)^p p!.
\end{aligned}$$

Proof: We employ Lemma 7.2.1. From the definition $u_0 = f/c$, we readily ascertain that there exists C, $\gamma > 0$ depending only on c such that

$$\|\nabla u_0\|_{L^\infty(G)} = \|\nabla^p(f/c)\|_{L^\infty(G)} \le CC_f(\gamma + 2\gamma_f)^p p! \qquad \forall p \in \mathbb{N}_0 .$$

Thus, Lemma 7.2.1 yields the existence of C, γ, $\tilde{K} > 0$ depending only on A, c such that the functions u_{2j} defined in (7.2.1) satisfy

$$\|\nabla^p u_{2j}\|_{L^\infty(G)} \le CC_f(\gamma + 2\gamma_f)^{p+2j}(2j)!p!\tilde{K}^{2j} \quad \forall p \in \mathbb{N}_0,\ j \in \mathbb{N}_0. \tag{7.2.7}$$

Hence, the definition of w_M implies

$$\begin{aligned}\|\nabla^p w_M\|_{L^\infty(G)} &\le CC_f(\gamma + 2\gamma_f)^p p! \sum_{j=0}^{M} \varepsilon^{2j}(\gamma + 2\gamma_f)^{2j}(2j)!\tilde{K}^{2j} \\ &\le CC_f(\gamma + 2\gamma_f)^p p! \sum_{j=0}^{M} \varepsilon^{2j}\left\{2M(\gamma + 2\gamma_f)\tilde{K}\right\}^{2j} \\ &\le CC_f(\gamma + 2\gamma_f)^p p!;\end{aligned}$$

in the last step, we took $K := 2\tilde{K}$ for the constant K of the statement of this theorem in order to bound the sum by a geometric series converging to 2. This proves the first claim of the theorem. For the second one, we observe that (7.2.3) implies by (7.2.7)

$$\begin{aligned}\|\nabla^p(L_\varepsilon w_M - f)\|_{L^\infty(G)} &= \|\nabla^p \varepsilon^{2M+2}\nabla \cdot (A\nabla u_{2M})\|_{L^\infty(G)} \\ &\le CC_f\left((2M+2)\varepsilon(\gamma + 2\gamma_f)K\right)^{2M+2}(\gamma + 2\gamma_f)^p p!.\end{aligned}$$

□

Remark 7.2.3 Theorem 7.2.2 gives bounds on the derivatives of the outer expansion w_M that are explicit in their dependence on the domain of analyticity of the datum f (measured in terms of γ_f). This could be of interest if singular right-hand sides f are considered. ▪

In what follows, we will use the outer expansion only as a means to obtain one particular solution to (1.2.11a) that is analytic and whose residue is small. It is therefore convenient to choose the expansion order M in dependence on ε such that the residue is small. This is achieved in the ensuing corollary.

Corollary 7.2.4. *Assume the hypotheses of Theorem 7.2.2. Then there exist C, K, $\gamma > 0$ independent of ε such that for every $\varepsilon \in (0, 1]$ there exists $w_\varepsilon \in \mathcal{A}(G)$ satisfying*

$$\begin{aligned}\|\nabla^p w_\varepsilon\|_{L^\infty(G)} &\le CK^p p! \qquad \forall p \in \mathbb{N}_0, \\ \|\nabla^p(L_\varepsilon w_\varepsilon - f)\|_{L^\infty(G)} &\le CK^p p! e^{-\gamma/\varepsilon} \qquad \forall p \in \mathbb{N}_0.\end{aligned}$$

Proof: Let γ, K be given by Theorem 7.2.2. We then choose

$$M := \left\lfloor \frac{1}{2K(\gamma + 2\gamma_f)} \right\rfloor$$

and set $w_\varepsilon := w_M$. The desired result then follows from Theorem 7.2.2. We remark that this proof also gives explicit bounds on the constants C, K, γ appearing in the statement of Corollary 7.2.4. □

7.3 Regularity of the boundary layer expansion

7.3.1 Definition and properties of the boundary layer expansion

The purpose of the present subsection is the definition of boundary layer expansions. For our purposes, let I_x, $I_y \subset \mathbb{R}$ be bounded intervals with the additional assumption that the left endpoint of I_y is the origin. Define $R = I_x \times I_y$ and let $A \in \mathcal{A}(\overline{R}, \mathbb{S}^2_>)$, $c \in \mathcal{A}(\overline{R})$ be given satisfying

$$0 < \lambda_{min} \leq A \quad \text{on } \overline{R}, \qquad c > 0 \quad \text{on } \overline{R} \tag{7.3.1}$$

for some $\lambda_{min} > 0$. On the rectangle R, we consider the differential operator

$$L_\varepsilon u = -\varepsilon^2 \nabla \cdot (A(x,y)\nabla u) + c(x,y)u$$

and construct (approximate) solutions to

$$L_\varepsilon u = 0 \qquad \text{on } R, \qquad u|_{y=0} = g,$$

where g is assumed to be analytic on $\overline{I_x}$ and satisfies

$$\|\nabla^p g\|_{L^\infty(I_x)} \leq C_g \gamma_g^p p! \qquad \forall p \in \mathbb{N}_0. \tag{7.3.2}$$

The solution u that we construct furthermore satisfies a decay condition in y. Let us denote the components of the matrix A by a_{ij} ($i, j \in \{1,2\}$) with $a_{12} = a_{21}$. The expression $\nabla \cdot (A\nabla u)$ then takes the form

$$\nabla\cdot(A\nabla u) = \left(a_{11}\partial_x^2 + 2a_{12}\partial_{xy} + a_{22}\partial_y^2 + (a_{11,x} + a_{12,y})\partial_x + (a_{12,x} + a_{22,y})\partial_y\right) u$$

where we abbreviated $a_{kj,x} = \partial_x a_{kl}$, $a_{kj,y} = \partial_y a_{kl}$. Next, we introduce the *stretched variable* $\hat{y} := y/\varepsilon$. In the new coordinates $(x, \hat{y})$, the differential operator L_ε reads

$$\begin{aligned} L_\varepsilon u = -\Big\{&\varepsilon^2 a_{11}\partial_x^2 + 2\varepsilon a_{12}\partial_{x\hat{y}} + a_{22}\partial_{\hat{y}}^2 \\ &+ \varepsilon^2(a_{11,x} + a_{12,y})\partial_x + \varepsilon(a_{12,x} + a_{22,y})\partial_{\hat{y}}\Big\}u + cu. \end{aligned} \tag{7.3.3}$$

The coefficients a_{kl}, $a_{kl,x}$ and c are expanded as power series in y as follows:

$$a_{kl}(x,y) = \sum_{i=0}^{\infty} a_{kl}^i(x) y^i, \quad a_{kl,x}(x,y) = \sum_{i=0}^{\infty} a_{kl,x}^i(x) y^i, \quad c(x,y) = \sum_{i=0}^{\infty} c^i(x) y^i.$$

Our assumptions on the data imply convergence of these power series on the ball $\{y \in \mathbb{C} \,|\, |y| < y_0\}$ for some $y_0 > 0$. Inserting these Taylor expansions into (7.3.3) and equating like powers of ε, we obtain

$$L_\varepsilon = \sum_{i=0}^{\infty} \varepsilon^i L_i, \tag{7.3.4}$$

where the differential operators L_i are given by

$$L_i = -b_{22}^i \hat{y}^i \partial_{\hat{y}}^2 - b_{12}^{i-1} \hat{y}^{i-1} \partial_{x\hat{y}} - b_{11}^{i-2} \hat{y}^{i-2} \partial_x^2 - b_1^{i-2} \hat{y}^{i-2} \partial_x - b_2^{i-1} \hat{y}^{i-1} \partial_{\hat{y}} + c^i \hat{y}^i, \tag{7.3.5}$$

with coefficients b_{kl}^i, b_k^i defined by

$$b_{22}^i(x) = a_{22}^i(x), \quad b_{12}^i(x) = 2a_{12}^i(x), \quad b_{11}^i(x) = a_{11}^i(x), \tag{7.3.6a}$$
$$b_1^i(x) = \left[a_{11,x}^i(x) + a_{12}^{i+1}(x)(i+1)\right], \tag{7.3.6b}$$
$$b_2^i(x) = \left[a_{12,x}^i(x) + a_{22}^{i+1}(x)(i+1)\right], \tag{7.3.6c}$$
$$b_{11}^i(x) = b_{12}^i(x) = b_{22}^i(x) = b_1^i(x) = b_2^i(x) = 0 \qquad \text{for } i < 0. \tag{7.3.6d}$$

In particular, the operator L_0 has the form

$$L_0 = -a_{22}^0(x) \partial_{\hat{y}}^2 + c^0(x). \tag{7.3.7}$$

Remark 7.3.1 The formal series in (7.3.4) in fact converges for $|\varepsilon \hat{y}| < y_0$ because the power series defining a_{kl}, $a_{kl,x}$, c converge. We will make use of this observation below. The constant $y_0 > 0$ depends only on the data A, c. ▪

In order to define the boundary layer expansion, we make the formal ansatz $u(x,y) \sim \sum_{j=0}^{\infty} \varepsilon^j \widehat{U}_j(x,\hat{y})$. Upon inserting this ansatz into (7.3.4), we obtain $\sum_{j=0}^{\infty} \varepsilon^j \sum_{j=0}^{i} L_j \widehat{U}_{i-j} = 0$. Next, setting the coefficients of this formal power series in ε to zero, yields a recurrence relation of ordinary differential equations (in $\hat{y}$) for the unknown functions $\widehat{U}_j$:

$$\left\{-a_{22}^0(x)\partial_{\hat{y}}^2 + c^0(x)\right\} \widehat{U}_i = L_0 \widehat{U}_i = -\sum_{j=1}^{i} L_j \widehat{U}_{i-j} := \widehat{F}_i = \sum_{k=1}^{6} \widehat{F}_i^k, \tag{7.3.8a}$$

$$\widehat{F}_i^1 = \sum_{j=0}^{i-1} b_{22}^{j+1} \hat{y}^{j+1} \partial_{\hat{y}}^2 \widehat{U}_{i-1-j}, \quad \widehat{F}_i^2 = \sum_{j=0}^{i-1} b_{12}^j \hat{y}^j \partial_{x\hat{y}} \widehat{U}_{i-1-j}, \tag{7.3.8b}$$

$$\widehat{F}_i^3 = \sum_{j=0}^{i-2} b_{11}^j \hat{y}^j \partial_x^2 \widehat{U}_{i-2-j}, \quad \widehat{F}_i^4 = \sum_{j=0}^{i-2} b_1^j \hat{y}^j \partial_x \widehat{U}_{i-2-j}, \tag{7.3.8c}$$

$$\widehat{F}_i^5 = \sum_{j=0}^{i-1} b_2^j \hat{y}^j \partial_{\hat{y}} \widehat{U}_{i-1-j}, \quad \widehat{F}_i^6 = -\sum_{j=0}^{i-1} c^{j+1} \hat{y}^{j+1} \widehat{U}_{i-1-j}, \tag{7.3.8d}$$

where we used the tacit convention that empty sums take the value zero. In order to complete this system of ordinary differential equation, we have to prescribe two boundary conditions for each $\widehat{U}_i$. As we want u to decay for $\hat{y} \to \infty$ and as we want to satisfy the boundary condition $u(x,0) = g(x)$, we prescribe

$$\widehat{U}_i \to 0 \qquad \text{for } \hat{y} \to \infty, \tag{7.3.9a}$$

$$\widehat{U}_i(x) = \begin{cases} g(x) & \text{for } i = 0 \\ 0 & \text{for } i > 0. \end{cases} \tag{7.3.9b}$$

(7.3.8), (7.3.9) define a unique sequence $\widehat{U}_i$, $i = 0, 1, \ldots$ of functions. The boundary layer expansion u_M^{BL} is taken as

$$u_M^{BL}(x,y) := \sum_{i=0}^{2M+1} \varepsilon^i \widehat{U}_i(x,\hat{y}) = \sum_{i=0}^{2M+1} \varepsilon^i \widehat{U}_i(x, y/\varepsilon). \tag{7.3.10}$$

Remark 7.3.2 The series $\sum_{i=0}^{\infty} \varepsilon^i \widehat{U}_i$ is a formal series, which cannot be expected to converge. We truncated this formal series after the term $\varepsilon^{2M+1}\widehat{U}_{2M+1}$ so that the first neglected term is of order ε^{2M+2}. This is the same (formal) error as the one introduced by truncating the outer expansion. ▪

We are now in position to state our main result concerning the regularity of the boundary layer expansion (7.3.10).

Theorem 7.3.3 (regularity of the inner expansion). *Let $A \in \mathcal{A}(\overline{R}, \mathbb{S}^2_>)$, $c \in \mathcal{A}(\overline{R})$ satisfy (7.3.1), and let g satisfy (7.3.2). Then the function u_M^{BL} of (7.3.10) is analytic on $\overline{R}$ and satisfies $u_M^{BL}(\cdot,0) = g$ on I_x. Moreover, there exist C, γ, K, K', y_0, $\lambda > 0$ depending only on the data A, c such that for all $\varepsilon \in (0,1]$, $M \in \mathbb{N}_0$ with $0 \le \varepsilon(2M+2)(\gamma + 2\gamma_g)K' \le 1$ the following holds: for all $(p,q) \in \mathbb{N}_0^2$*

$$|\partial_x^q \partial_y^p u_M^{BL}(x,y)| \le C C_g q! K^{p+q} \varepsilon^{-p} (\gamma + 2\gamma_g)^q e^{-\lambda y/\varepsilon} \qquad \forall (x,y) \in I_x \times \mathbb{R}_0^+$$

and for the residual, we have for all $(x,y) \in I_x \times (0, y_0)$

$$|\partial_x^q \partial_y^p L_\varepsilon u_M^{BL}(x,y)|$$
$$\le C C_g K^{p+q} p! q! \varepsilon^{-p} (\gamma + 2\gamma_g)^q \left(\varepsilon(2M+2)(\gamma+2\gamma_g)K'\right)^{2M+2} e^{-\lambda y/\varepsilon}.$$

The proof is lengthy and therefore relegated to the next subsection.
As in the case of the outer expansion, we can extract from Theorem 7.3.3 a corollary by choosing the expansion order M proportional to ε^{-1}.

Corollary 7.3.4. *Under the hypothesis of Theorem 7.3.3 there exist C, K, γ, $\lambda > 0$, and $y_0 > 0$ independent of ε such that for every $\varepsilon \in (0,1]$ there exists $u_\varepsilon^{BL} \in \mathcal{A}(\overline{R})$ with $u_\varepsilon^{BL}(\cdot, 0) = g$ on I_x and there holds for all $(p,q) \in \mathbb{N}_0^2$*

$$|\partial_x^q \partial_y^p u_\varepsilon^{BL}(x,y)| \le C K^{p+q} q! \varepsilon^{-p} e^{-\lambda y/\varepsilon}, \qquad \forall (x,y) \in I_x \times \mathbb{R}^+,$$
$$|\partial_x^q \partial_y^p L_\varepsilon u_\varepsilon^{BL}(x,y)| \le C K^{p+q} p! q! e^{-\gamma/\varepsilon} e^{-\lambda y/\varepsilon} \qquad \forall (x,y) \in I_x \times (0, y_0).$$

Proof: The proof follows by the same reasoning as in the proof of Corollary 7.2.4 by choosing the expansion order M proportional to ε^{-1}. □

7.3.2 Proof of Theorem 7.3.3

Let I_x be the interval defined at the outset of Section 7.3.1. For $X > 0$, we define complex neighborhoods of I_x as

$$S_X := \{z \in \mathbb{C} \mid \operatorname{dist}(z, I_x) < X\},$$
$$S_X(\delta) := \{z \in S_X \mid \operatorname{dist}(z, \partial S_X) > \delta\}, \qquad \delta > 0. \tag{7.3.11}$$

By geometric considerations, it is easy to see that $S_X(\delta) = S_{X-\delta}$. In view of the form of the operator L_0, it is convenient to introduce the function

$$\lambda(x) := \sqrt{\frac{c^0(x)}{a_{22}^0(x)}}, \tag{7.3.12}$$

which is positive on I_x and which can be extended holomorphically to a (complex) neighborhood of I_x.

Lemma 7.3.5. *Let the coefficients A, c be analytic on $\overline{R}$ and let $g \in \mathcal{A}(\overline{I_x})$ satisfy (7.3.2). Then there exist constants C_A, γ_A, $\lambda_0 > 0$ depending only on A, c such that for $X := (\gamma_A + 2\gamma_g)^{-1} > 0$ the functions g, λ, and b_{11}^i, b_{12}^i, b_{22}^i, b_1^i, b_2^i, c^i of (7.3.5) have holomorphic extensions to S_X and satisfy for all $z \in S_X$*

$$|g(z)| \le 2C_g,$$
$$\left|\frac{1}{a_{22}^0(z)}\right| \le C_A, \qquad |\lambda(z)| \le C_A, \qquad \operatorname{Re}\lambda(z) \ge \lambda_0, \qquad \operatorname{Re}\lambda^2(z) \ge \lambda_0^2,$$
$$|b_{11}^i(z)| + |b_{12}^i(z)| + |b_{22}^i(z)| + |b_1^i(z)| + |b_2^i(z)| + |c^i(z)| \le C_A\gamma_A^i \qquad \forall i \in \mathbb{N}_0.$$

Proof: For $z \in S_{1/(2\gamma_g)}$, the bound on g follows from power series expansions around points of I_x. The second bounds follow easily from the assumptions on A, c. For the last bound, we note that there exists a complex neighborhood $\tilde{R} \subset \mathbb{C} \times \mathbb{C}$ of $\overline{R}$ on which A and c are holomorphic. The result follows from Cauchy's integral formula for derivatives; for example,

$$c^i(x) = \frac{1}{2\pi\mathbf{i}} \int_C \frac{c(x,y)}{(-y)^{i+1}}\,dy$$

for some closed loop C around the origin (in the complex plane). □

Lemma 7.3.6. *Let $\lambda \in \mathbb{C}$ with $\operatorname{Re}\lambda > 0$, $\operatorname{Re}\lambda^2 > 0$. Let f be an entire function satisfying for some $C_f > 0$, $j \in \mathbb{N}_0$, $q \ge (j + 1/2)/|\lambda| > 0$*

$$|f(z)| \le C_f e^{-\operatorname{Re}(\lambda z)}(q + |z|)^j \qquad \forall z \in \mathbb{C}.$$

Let $g \in \mathbb{C}$ and let $u : (0,\infty) \to \mathbb{C}$ be the solution to

$$-u'' + \lambda^2 u = f \quad \text{on } (0,\infty), \qquad u(0) = g, \qquad \lim_{x\to\infty} u(x) = 0.$$

Then u can be extended to an entire function (again denoted u), which satisfies

$$|u(z)| \le \left[C_f \frac{1}{|\lambda|}(q + |z|)^{j+1}(j+1)^{-1} + |g|\right] e^{-\operatorname{Re}(\lambda z)} \qquad \forall z \in \mathbb{C}.$$

Proof: For $z \in (0,\infty)$, the use of a Green's function gives the following representation of the solution $u(z)$:

$$u(z) = \frac{1}{2\lambda^2}\Big(e^{-\lambda z}\int_0^{\lambda z} e^{y} f(y/\lambda)\,dy + e^{\lambda z}\int_{\lambda z}^{\infty} e^{-y} f(y/\lambda)\,dy \\ - e^{-\lambda z}\int_0^{\infty} e^{-y} f(y/\lambda)\,dy\Big) + g e^{-\lambda z}.$$

Analytic continuation then removes the restriction to $(0,\infty)$. In order to get the desired bound, we estimate each of these four terms separately. For the first integral, we use as the path of integration the straight line connecting 0 and λz to get

$$\left|e^{-\lambda z}\int_0^{\lambda z} e^{y} f(y/\lambda)\,dy\right| \le e^{-\operatorname{Re}\lambda z}\int_0^1 C_f (q+t|z|)^j |\lambda z|\, e^{-\operatorname{Re} t\lambda z} e^{\operatorname{Re} t\lambda z}\,dt \\ \le C_f e^{-\operatorname{Re}\lambda z}\frac{|\lambda|}{j+1}\left\{(q+|z|)^{j+1} - q^{j+1}\right\}.$$

For the second term, we calculate

$$\left|e^{\lambda z}\int_{\lambda z}^{\infty} e^{-y} f(y/\lambda)\,dy\right| = \left|\int_0^{\infty} e^{-y} f(z+y/\lambda)\,dy\right| \\ \le e^{-\operatorname{Re}\lambda z} C_f |\lambda|^{-j}\int_0^{\infty} e^{-2y}(|\lambda| q + |\lambda z| + y)^j\,dy \\ = C_f e^{-\operatorname{Re}\lambda z}|\lambda|^{-j} 2^{-(j+1)} e^{2|\lambda|(q+|z|)}\Gamma(j+1, 2|\lambda|(q+|z|)),$$

where $\Gamma(\cdot,\cdot)$ denotes the incomplete Gamma function, and we used [61, eq. 8.353.5] in the last step. We observe that $2|\lambda| q \ge 2j+1 \ge j$. Thus, we may employ the estimate

$$|\Gamma(\alpha,\xi)| \le \frac{|e^{-\xi}\xi^{\alpha}|}{|\xi| - \alpha_0}, \qquad \alpha_0 = \max\{\alpha - 1, 0\}, \quad \operatorname{Re}\xi \ge 0, \quad |\xi| > \alpha_0$$

(see, e.g., [103, Chap. 4, Sec. 10]) to finally arrive at

$$\left|e^{\lambda z}\int_{\lambda z}^{\infty} e^{-y} f(y/\lambda)\,dy\right| \le C_f e^{-\operatorname{Re}\lambda z}|\lambda|\frac{(q+|z|)^{j+1}}{2|\lambda|(q+|z|) - j} \\ \le C_f e^{-\operatorname{Re}\lambda z}|\lambda|\frac{(q+|z|)^{j+1}}{j+1}.$$

For the third term, we observe that the integral $\int_0^\infty f(y) e^{-y}\,dy$ is precisely the second term with $z = 0$. We conclude therefore that for the third term

$$\left|e^{-\lambda z}\int_0^{\infty} f(y/\lambda) e^{-y}\,dy\right| \le C_f e^{-\operatorname{Re}\lambda z}|\lambda|\frac{q^{j+1}}{j+1}.$$

Hence, we arrive at

$$\left| e^{-\lambda z} \int_0^{\lambda z} e^y f(y/\lambda)\, dy + e^{\lambda z} \int_{\lambda z}^{\infty} e^{-y} f(y/\lambda)\, dy - e^{-\lambda z} \int_0^{\infty} e^{-y} f(y/\lambda)\, dy \right| \leq$$
$$2C_f e^{-\operatorname{Re}\lambda z}(q+|z|)^{j+1}\frac{|\lambda|}{j+1}.$$

Combining this estimate with the obvious one for the fourth term, we arrive at the desired bound. □

Lemma 7.3.7. *Let $X > 0$ and let λ be a function holomorphic on S_X satisfying $|\lambda(x)| \leq C_A$ for all $x \in S_X$. Let U be holomorphic on $S_X \times \mathbb{C}$ and assume that there exist $C_U > 0$, λ_0, $i \in \mathbb{N}_0$ such that for all $\delta \in (0, X)$ and all $(x, z) \in S_X(\delta) \times \mathbb{C}$*

$$|U(x,z)| \leq C_U\{(2i+1)/\lambda_0 + |z|\}^{2i} e^{-\operatorname{Re}(\lambda(x)z)}\delta^{-i}.$$

Then for all $\delta \in (0, X)$ and for all $(x, z) \in S(X - \delta) \times \mathbb{C}$ there holds

$$|\partial_z U(x,z)| \leq \frac{\lambda_0}{2} e^{2C_A/\lambda_0} C_U \left(\frac{2(i+1)+1}{\lambda_0} + |z|\right)^{2i} e^{-\operatorname{Re}(\lambda(x)z)}\delta^{-i},$$
$$|\partial_z^2 U(x,z)| \leq \frac{\lambda_0^2}{2} e^{2C_A/\lambda_0} C_U \left(\frac{2(i+1)+1}{\lambda_0} + |z|\right)^{2i} e^{-\operatorname{Re}(\lambda(x)z)}\delta^{-i},$$
$$|\partial_x U(x,z)| \leq 4\lambda_0 e^{2C_A/\lambda_0} C_U \left(\frac{2i+2}{\lambda_0} + |z|\right)^{2i+1} e^{-\operatorname{Re}(\lambda(x)z)}\delta^{-(i+1)},$$
$$|\partial_x^2 U(x,z)| \leq 4\lambda_0^2 e^{2C_A/\lambda_0} C_U \left(\frac{2i+2}{\lambda_0} + |z|\right)^{2i+2} e^{-\operatorname{Re}(\lambda(x)z)}\delta^{-(i+2)},$$
$$|\partial_{xz} U(x,z)| \leq 2\lambda_0^2 e^{4C_A/\lambda_0} C_U \left(\frac{2(i+1)+1}{\lambda_0} + |z|\right)^{2i+1} e^{-\operatorname{Re}(\lambda(x)z)}\delta^{-(i+1)}.$$

Proof: For the first estimate, we use Cauchy's integral theorem for derivatives:

$$|\partial_z U(x,z)| = \left| \frac{1}{2\pi \mathbf{i}} \int_{|t|=2/\lambda_0} \frac{U(x,z+t)}{t^2}\, dt \right|$$
$$\leq \lambda_0/2 C_U\{(2i+1)/\lambda_0 + |z| + 2/\lambda_0\}^{2i} e^{-\operatorname{Re}\lambda z + |\lambda| 2/\lambda_0}\delta^{-i}.$$

The second estimate is proved similarly. For the derivatives with respect to the x-variable, we first note that for all $\delta \in (0, X)$, there holds

$$|\lambda'(z)| \leq C_A \delta^{-1} \qquad \forall z \in S_X(\delta), \tag{7.3.13}$$

which can easily be ascertained with Cauchy's integral theorem for derivatives, taking as the path of integration the (complex) circle of radius $\delta' < \delta$ around x and then letting $\delta' \to \delta$. In order to get the third estimate, we use Cauchy's integral theorem for derivatives but choose the path of integration as $\partial B_{\kappa\delta}(x)$

with $\kappa \in (0,1)$ to be chosen below. Noting that this path is completely contained in $S_X((1-\kappa)\delta)$, we have for $t \in \mathbb{C}$ with $|t| = \kappa\delta$ that $\mathrm{Re}(\lambda(x+t)z) = \mathrm{Re}(\lambda(x)z) + \mathrm{Re}\left((\lambda(x+t) - \lambda(x))z\right)$ and arrive

$$\begin{aligned}
|\partial_x U(x,z)| &= \left| \frac{1}{2\pi \mathbf{i}} \int_{|t|=\kappa\delta} \frac{U(x+t,z)}{t^2}\, dt \right| \\
&\leq C_U \frac{1}{(\kappa\delta)((1-\kappa)\delta)^i} \left(\frac{2i+1}{\lambda_0} + |z| \right)^{2i} e^{-\mathrm{Re}(\lambda(x)z)} e^{\kappa\delta C_A((1-\kappa)\delta)^{-1}} \\
&\leq C_U e^{C_A\kappa/(1-\kappa)} \frac{1}{\kappa(1-\kappa)^i} \delta^{-(i+1)} \left(\frac{2i+1}{\lambda_0} + |z| \right)^{2i} e^{-\mathrm{Re}(\lambda(x)z)}.
\end{aligned}$$

Choosing $\kappa = 1/(2i+2+\lambda_0|z|)$ and observing that this choice implies

$$\frac{1}{2} \leq 1-\kappa, \qquad \frac{1}{4} \leq (1-\kappa)^{2i+2+\lambda_0|z|} \leq (1-\kappa)^i \leq 1 \qquad \forall i \in \mathbb{N}_0,$$

we obtain

$$\begin{aligned}
|\partial_x U(x,z)| &\leq C_U e^{2C_A/\lambda_0} \delta^{-(i+1)} \frac{4}{\kappa} \left(\frac{2i+1}{\lambda_0} + |z| \right)^{2i} e^{-\mathrm{Re}(\lambda(x)z)} \\
&\leq C_U e^{2C_A/\lambda_0} 4\delta^{-(i+1)} \lambda_0 \left(\frac{2i+2}{\lambda_0} + |z| \right)^{2i+1} e^{-\mathrm{Re}(\lambda(x)z)}.
\end{aligned}$$

Finally, the fourth and fifth estimate are proved completely analogously. □

We now turn to bounding the terms $\widehat{U}_i$ defined by (7.3.8), (7.3.9). We note that with λ defined in (7.3.12), the functions $\widehat{U}_i$ satisfy the equation

$$-\partial_{\hat{y}}^2 \widehat{U}_i + \lambda^2(x)\widehat{U}_i = \frac{1}{a_{22}^0(x)} \widehat{F}_i$$

with the boundary conditions (7.3.9). From $\widehat{F}_0 = 0$ and (7.3.9), we obtain

$$\widehat{U}_0(x,\hat{y}) = g(x) e^{-\lambda(x)\hat{y}}. \tag{7.3.14}$$

We compute

$$\begin{aligned}
\widehat{F}_1(x,\hat{y}) &= \{a(x) + b(x)\hat{y}\}\, e^{-\lambda(x)\hat{y}}, \\
a(x) &:= \left[b_2^0(x)\lambda(x)g(x) - b_{12}^0(x)(\lambda'(x)g(x) + \lambda(x)g'(x) - \lambda(x)\lambda'(x)g(x)) \right], \\
b(x) &:= \left[\lambda^2(x) b_{22}^1(x) - c^1(x) \right] g(x),
\end{aligned}$$

and find that the function $\widehat{U}_1$ is then given by

$$\widehat{U}_1(x,\hat{y}) = \frac{1}{a_{22}^0(x)} \left\{ \left[\frac{a(x)}{2\lambda(x)} - \frac{b(x)}{4\lambda^2(x)} \right] \hat{y} - \frac{b(x)}{4\lambda(x)} \hat{y}^2 \right\} e^{-\lambda(x)\hat{y}}. \tag{7.3.15}$$

It can be shown inductively (in fact, this is shown in Corollary 7.3.9 below) that for all $i \in \mathbb{N}_0$ the function $\widehat{U}_i(x, \hat{y})$ is of the form $\widehat{U}_i(x, \hat{y}) = P_{2i}(x, \hat{y})e^{-\lambda(x)\hat{y}}$ where the functions P_{2i} are polynomials of degree $2i$ in the variable $\hat{y}$ with coefficients analytic in x. This observation also motivates the induction hypothesis for the following proposition.

Proposition 7.3.8. *Let the coefficients $A \in \mathcal{A}(\overline{R}, \mathbb{S}^2_>)$, $c \in \mathcal{A}(\overline{R})$ satisfy (7.3.1) and let $g \in \mathcal{A}(I_x)$ satisfy (7.3.2). Let the functions $\widehat{U}_i$ be defined by (7.3.8), (7.3.9). Then there are C_U, K, $\gamma_A > 0$ depending only on the coefficients A, c such that with $X = (\gamma_A + 2\gamma_g)^{-1}$ there holds for all $i \in \mathbb{N}_0$ and all $(x, \hat{y}) \in S_X(\delta) \times \mathbb{C}$*

$$|\widehat{U}_i(x, \hat{y})| \leq C_U C_g K^i \delta^{-i} \frac{1}{i!} \left(\frac{2i+1}{\lambda_0} + |\hat{y}| \right)^{2i} e^{-\operatorname{Re}(\lambda(x)\hat{y})}. \tag{7.3.16}$$

Proof: Let X and the constants C_A, γ_A be given by Lemma 7.3.5. The choice of X in Lemma 7.3.5 and Cauchy's integral theorem for derivatives imply for the holomorphic extension of g to S_X:

$$|g(z)| \leq 2C_g \quad \text{on } S_X, \qquad |g'(z)| \leq 2\delta^{-1} C_g \quad \text{on } S_X(\delta), \quad \delta > 0.$$

These bounds allow us to find a constant $C > 0$ depending only on the coefficients A and c such that for all $(x, \hat{y}) \in S_X(\delta) \times \mathbb{C}$ the functions $\widehat{U}_0$, $\widehat{U}_1$ given in (7.3.14), (7.3.15) satisfy

$$|\widehat{U}_0(x, \hat{y})| \leq C C_g e^{-\operatorname{Re}(\lambda(x)z)}, \qquad |\widehat{U}_1(x, \hat{y})| \leq C C_g \delta^{-1} \left(|\hat{y}| + |\hat{y}|^2\right) e^{-\operatorname{Re}(\lambda(x)z)}.$$

We note that for some $C > 0$ depending only on the coefficients A, c we can bound: $|\hat{y}| + |\hat{y}|^2 \leq C(2/\lambda_0 + |\hat{y}|)^2$ for all $(x, \hat{y}) \in S_X \times \mathbb{C}$. We conclude that (7.3.16) holds true for $i = 0$ and $i = 1$. In order to proceed by induction on i, we now define the constants C_U, K, whose existence is asserted in the statement of the proposition. C_U is chosen such that (7.3.16) holds true for $i = 0$ and $i = 1$. For the definition of K, we introduce

$$C_1 := \frac{\lambda_0^2}{2} e^{2C_A/\lambda_0}, \qquad C_2 := 2\lambda_0^2 e^{4C_A/\lambda_0}, \qquad C_3 := 4\lambda_0^2 e^{2C_A/\lambda_0}, \tag{7.3.17a}$$

$$C_4 := 4\lambda_0 e^{2C_A/\lambda_0}, \qquad C_5 := \frac{\lambda_0}{2} e^{2C_A/\lambda_0} \tag{7.3.17b}$$

and then choose $K > 1$ such that

$$\left[C_A^2 \frac{2K}{2K - \lambda_0} \left(\frac{C_1 \lambda_0}{2K} + \frac{\lambda_0}{2K} + \frac{C_2 \lambda_0}{2K} + \frac{C_3 \lambda_0^2}{4K^2} + \frac{C_4 \lambda_0^3}{8\gamma_A K^2} + \frac{C_5 \lambda_0^2}{4\gamma_A K} \right) \right] \leq 1. \tag{7.3.18}$$

Next, we have for $i \geq 2$

$$\frac{1}{(i-j-1)!} \leq \frac{i^{j+1}}{i!} \leq \frac{1}{i!}\left(\frac{\lambda_0}{2}\right)^{j+1}\left(\frac{2i}{\lambda_0}+|\hat{y}|\right)^{j+1}, \qquad j \leq i-1,$$
$$\frac{1}{(i-j-2)!} \leq \frac{i^{j+2}}{i!} \leq \frac{1}{i!}\left(\frac{\lambda_0}{2}\right)^{j+2}\left(\frac{2i}{\lambda_0}+|\hat{y}|\right)^{j+2}, \qquad j \leq i-2,$$
$$\left(\frac{2i}{\lambda_0}+|\hat{y}|\right)^{-1} \leq \frac{\lambda_0}{4} \leq \frac{\lambda_0}{2}.$$

This allows us to bound for $b \in \{0,1\}$

$$|\hat{y}|^{j+1}\frac{1}{(i-1-j)!}\left(\frac{2i}{\lambda_0}+|\hat{y}|\right)^{2(i-j-1)} \leq \left(\frac{\lambda_0}{2}\right)^{j+1}\frac{1}{i!}\left(\frac{2i}{\lambda_0}+|\hat{y}|\right)^{2i}, \quad (7.3.19)$$
$$|\hat{y}|^{j}\frac{1}{(i-1-j)!}\left(\frac{2i}{\lambda_0}+|\hat{y}|\right)^{2(i-j-1)+1-b} \leq \left(\frac{\lambda_0}{2}\right)^{j+1+b}\frac{1}{i!}\left(\frac{2i}{\lambda_0}+|\hat{y}|\right)^{2i}, (7.3.20)$$
$$|\hat{y}|^{j}\frac{1}{(i-2-j)!}\left(\frac{2i}{\lambda_0}+|\hat{y}|\right)^{2(i-j-2)+2-b} \leq \left(\frac{\lambda_0}{2}\right)^{j+2+b}\frac{1}{i!}\left(\frac{2i}{\lambda_0}+|\hat{y}|\right)^{2i}. (7.3.21)$$

We can now prove the statement of the proposition by induction on i. To that end, we assume that (7.3.16) holds true for all $0 \leq i' < i$ for $i \geq 2$ and show that it holds for i as well. This is achieved with the aid of Lemma 7.3.6. By the linearity of the operator L_0, we can obtain a bound on $\widehat{U}_i$ by considering 6 (slightly) different types of subproblems. First, let us consider solutions u_j^1 of

$$-\partial_{\hat{y}}^2 u_j^1 + \lambda^2 u_j^1 = \frac{b_{22}^{j+1}}{a_{22}^0}\hat{y}^{j+1}\partial_{\hat{y}}^2 \widehat{U}_{i-1-j}, \qquad u_j^1(x,0)=0, \qquad \lim_{\hat{y}\to\infty} u_j^1(x,\hat{y}) = 0.$$

Noting that $\frac{\lambda_0^2}{2}e^{2|\lambda(x)|/\lambda_0} \leq C_1$ on S_X, we obtain from Lemma 7.3.7, the induction hypothesis, and (7.3.19) that there holds on $S_X(\delta)\times\mathbb{C}$

$$\begin{aligned}&\left|\hat{y}^{j+1}\partial_{\hat{y}}^2 \widehat{U}_{i-1-j}\right| \\ &\leq |\hat{y}|^{j+1}C_1 C_U C_g \delta^{-(i-1-j)}K^{i-1-j}\frac{1}{(i-1-j)!}\left(\frac{2i}{\lambda_0}+|\hat{y}|\right)^{2(i-j-1)} e^{-\operatorname{Re}(\lambda\hat{y})} \\ &\leq C_1 C_U C_g \delta^{-(i-1-j)}K^{i-1-j}(\lambda_0/2)^{j+1}\frac{1}{i!}\left(\frac{2i}{\lambda_0}+|\hat{y}|\right)^{2i} e^{-\operatorname{Re}(\lambda\hat{y})}.\end{aligned}$$

Applying now Lemma 7.3.6 and noting the bounds on b_{22}^{j+1}, a_{22}^0 from Lemma 7.3.5, we get on $S_X(\delta)\times\mathbb{C}$

$$|u_1^j(x,\hat{y})| \leq K^i\delta^{-i}C_U C_g\frac{1}{i!}\left(\frac{2i}{\lambda_0}+|\hat{y}|\right)^{2i} e^{-\operatorname{Re}(\lambda\hat{y})}\left\{C_A^2 C_1\left(\frac{\lambda_0\gamma_A\delta}{2K}\right)^{j+1}\right\}.$$

We immediately note that a similar reasoning applies to the functions u_6^j defined as the solutions to $-\partial_{\hat{y}}^2 u_6^j + \lambda^2 u_6^j = -c^{j+1}/a_{22}^0\hat{y}^{j+1}\widehat{U}_{i-1-j}$ with the corresponding boundary conditions:

$$|u_6^j(x,\hat{y})| \le K^i\delta^{-i}C_U C_g \frac{1}{i!}\left(\frac{2i}{\lambda_0}+|\hat{y}|\right)^{2i} e^{-\operatorname{Re}(\lambda\hat{y})}\left\{C_A^2\left(\frac{\lambda_0\gamma_A\delta}{2K}\right)^{j+1}\right\}.$$

The remaining 4 cases are treated similarly. We bound with C_2 of (7.3.17) and (7.3.20) for the solution u_2^j of $-\partial_{\hat{y}}^2 u_2^j + \lambda^2 u_2^j = b_{12}^j/a_{22}^0 \hat{y}^j \partial_{x\hat{y}} \widehat{U}_{i-1-j}$ as follows:

$$\begin{aligned}
&\left|\hat{y}^j \partial_{x\hat{y}} \widehat{U}_{i-1-j}\right| \\
&\quad \le |\hat{y}|^j C_2 C_U C_g K^{i-1-j}\delta^{-(i-j)} \frac{1}{(i-1-j)!}\left(\frac{2i}{\lambda_0}+|\hat{y}|\right)^{2(i-j-1)+1} e^{-\operatorname{Re}(\lambda\hat{y})} \\
&\quad \le C_2 C_U C_g K^{i-1-j}\delta^{-(i-j)}(\lambda_0/2)^{j+1}\frac{1}{i!}\left(\frac{2i}{\lambda_0}+|\hat{y}|\right)^{2i} e^{-\operatorname{Re}(\lambda\hat{y})}.
\end{aligned}$$

Hence, reasoning as before, we conclude that u_2^j satisfies

$$|u_2^j(x,\hat{y})| \le C_U C_g K^i\delta^{-i}\frac{1}{i!}\left(\frac{2i}{\lambda_0}+|\hat{y}|\right)^{2i} e^{-\operatorname{Re}(\lambda\hat{y})}\left\{\frac{C_A^2 C_2\lambda_0}{2K}\left(\frac{\gamma_A\delta\lambda_0}{2K}\right)^j\right\}.$$

For the solution u_3^j of $-\partial_{\hat{y}}^2 u_3^j + \lambda^2 u_3^j = b_{11}^j/a_{22}^0 \hat{y}^j \partial_x^2 \widehat{U}_{i-2-j}$, we bound with C_3 of (7.3.17) and (7.3.21)

$$\begin{aligned}
&\left|\hat{y}^j \partial_x^2 \widehat{U}_{i-2-j}\right| \\
&\quad \le |\hat{y}|^j C_3 C_U C_g K^{i-2-j}\delta^{-(i-j)} \frac{1}{(i-2-j)!}\left(\frac{2i}{\lambda_0}+|\hat{y}|\right)^{2(i-j-2)+2} e^{-\operatorname{Re}(\lambda\hat{y})} \\
&\quad \le C_3 C_U C_g K^{i-2-j}\delta^{-(i-j)}(\lambda_0/2)^{j+2}\frac{1}{i!}\left(\frac{2i}{\lambda_0}+|\hat{y}|\right)^{2i} e^{-\operatorname{Re}(\lambda\hat{y})}.
\end{aligned}$$

Thus, we conclude that u_3^j satisfies

$$|u_3^j(x,\hat{y})| \le C_U C_g K^i\delta^{-i}\frac{1}{i!}\left(\frac{2i}{\lambda_0}+|\hat{y}|\right)^{2i} e^{-\operatorname{Re}(\lambda\hat{y})}\left\{\frac{C_A^2 C_3\lambda_0^2}{4K^2}\left(\frac{\gamma_A\delta\lambda_0}{2K}\right)^j\right\}.$$

For the solution u_4^j of $-\partial_{\hat{y}}^2 u_4^j + \lambda^2 u_4^j = b_1^j/a_{22}^0 \hat{y}^j \partial_x \widehat{U}_{i-2-j}$, we bound with C_4 of (7.3.17) and (7.3.21)

$$\begin{aligned}
&\left|\hat{y}^j \partial_x \widehat{U}_{i-2-j}\right| \\
&\quad \le |\hat{y}|^j C_4 C_U C_g K^{i-2-j}\delta^{-(i-1-j)} \frac{1}{(i-2-j)!}\left(\frac{2i}{\lambda_0}+|\hat{y}|\right)^{2(i-j-2)+1} e^{-\operatorname{Re}(\lambda\hat{y})} \\
&\quad \le C_4 C_U C_g K^{i-2-j}\delta^{-(i-1-j)}(\lambda_0/2)^{j+3}\frac{1}{i!}\left(\frac{2i}{\lambda_0}+|\hat{y}|\right)^{2i} e^{-\operatorname{Re}(\lambda\hat{y})}.
\end{aligned}$$

This leads to a bound for u_j^4 of the form

$$|u_4^j(x,\hat{y})| \le C_U C_g K^i \delta^{-i} \frac{1}{i!} \left(\frac{2i}{\lambda_0} + |\hat{y}|\right)^{2i} e^{-\operatorname{Re}(\lambda \hat{y})} \left\{ \frac{C_A^2 C_4 \lambda_0^3 \delta}{8K^2} \left(\frac{\gamma_A \delta \lambda_0}{2K}\right)^j \right\}.$$

Finally, for the solution u_5^j of $-\partial_{\hat{y}}^2 u_5^j + \lambda^2 u_5^j = b_2^j / a_{22}^0 \hat{y}^j \partial_{\hat{y}} \widehat{U}_{i-1-j}$, we bound with C_5 of (7.3.17) and (7.3.20)

$$\begin{aligned} &\left|\hat{y}^j \partial_{\hat{y}} \widehat{U}_{i-1-j}\right| \\ &\quad \le |\hat{y}|^j C_5 C_U C_g K^{i-1-j} \delta^{-(i-1-j)} \frac{1}{(i-1-j)!} \left(\frac{2i}{\lambda_0} + |\hat{y}|\right)^{2(i-j-1)} e^{-\operatorname{Re}(\lambda \hat{y})} \\ &\quad \le C_U C_g K^{i-1-j} \delta^{-(i-1-j)} (\lambda_0/2)^{j+2} \frac{1}{i!} \left(\frac{2i}{\lambda_0} + |\hat{y}|\right)^{2i} e^{-\operatorname{Re}(\lambda \hat{y})}. \end{aligned}$$

Thus, reasoning as before, we conclude that u_5^j satisfies

$$|u_5^j(x,\hat{y})| \le C_U C_g K^i \delta^{-i} \frac{1}{i!} \left(\frac{2i}{\lambda_0} + |\hat{y}|\right)^{2i} e^{-\operatorname{Re}(\lambda \hat{y})} \left\{ \frac{C_A^2 C_5 \lambda_0^2 \delta}{4K} \left(\frac{\gamma_A \delta}{K}\right)^j \right\}.$$

We notice that $\delta \le X \le \gamma_A^{-1}$. Thus, for $2K > \lambda_0$ we can bound

$$\sum_{j=0}^{\infty} \left(\frac{\gamma_A \lambda_0 \delta}{2K}\right)^j \le \sum_{j=0}^{\infty} \left(\frac{\lambda_0}{2K}\right)^{j+1} \le \frac{2K}{2K - \lambda_0},$$

$$\sum_{j=0}^{\infty} \left(\frac{\gamma_A \lambda_0 \delta}{2K}\right)^{j+1} \le \frac{\lambda_0}{2K} \frac{2K}{2K - \lambda_0},$$

and we obtain for the function $\widehat{U}_i$ on $S_X(\delta) \times \mathbb{C}$ by combining the above six estimates:

$$\begin{aligned} |\widehat{U}_i(x,\hat{y})| &\le C_U C_g K^i \delta^{-i} \frac{1}{i!} \left(\frac{2i}{\lambda_0} + |\hat{y}|\right)^{2i} \times \\ &\left[C_A^2 \frac{2K}{2K - \lambda_0} \left(\frac{C_1 \lambda_0}{2K} + \frac{\lambda_0}{2K} + \frac{C_2 \lambda_0}{2K} + \frac{C_3 \lambda_0^2}{4K^2} + \frac{C_4 \lambda_0^3 \delta}{8K^2} + \frac{C_5 \lambda_0^2 \delta}{4K} \right) \right]. \end{aligned}$$

Since $\delta \le \gamma_A^{-1}$, Our choice of K in (7.3.18) implies that the expression in brackets $[\cdot]$ is bounded by 1; this concludes the induction argument. □

Corollary 7.3.9. *Under the hypotheses of Proposition 7.3.8, the functions $\widehat{U}_i$ defined by the recursion (7.3.8), (7.3.9) are of the form*

$$\widehat{U}_i(x,\hat{y}) = \left\{ \sum_{j=0}^{2i} \alpha_{i,j}(x) \hat{y}^j \right\} e^{-\lambda(x)\hat{y}}$$

for some functions $\alpha_{i,j}$ holomorphic on S_X.

Proof: The proof follows from a variation of Liouville's theorem. To that end, we consider for fixed i, x the function $v := \widehat{U}_i e^{\lambda \hat{y}}$. From Proposition 7.3.8, v is an entire function that is $O(|\hat{y}|^{2i})$ at ∞. With the aid of Liouville's theorem that entire bounded functions are constant, it is now easy to see that v is a polynomial of degree $2i$. The holomorphy of the coefficients $\alpha_{i,j}$ now follows from the holomorphy of $\widehat{U}_i$ in the first variable x. We remark that for $i \geq 1$, the condition $\widehat{U}_i(x,0) = 0$ implies that $\alpha_{i,0}(x) \equiv 0$ for $i \geq 1$. □

Lemma 7.3.10. *Let a, $b \geq 0$ with $a + b \geq 1$. Then the function*

$$j \mapsto \frac{1}{\Gamma(j+1)}(2j+a+b)^{2j} b^{-j}$$

is monotonically increasing on $\mathbb{R}_0^+$.

Proof: We define the function

$$f(j) := \ln\left(\frac{1}{j!}(2j+a+b)^{2j} b^{-j}\right) = 2j \ln(2j+a+b) - \ln \Gamma(j+1) - j \ln b,$$

$$f'(j) = 2\ln(2j+a+b) + \frac{2j}{2j+a+b} - \ln b - \psi(j+1),$$

where $\psi(x) = \frac{d}{dx}\Gamma(x)$. Using the bound $\psi(x) \leq \ln x$ (cf., e.g., [61, eq. 8.361.3]), we can estimate

$$f'(j) \geq \ln \frac{2j+a+b}{b} + \ln \frac{2j+a+b}{j+1} + \frac{2j}{2j+a+b} \geq 0$$

by our assumptions $b > 0$ and $a + b \geq 1$. □

Lemma 7.3.11. *Assume the hypotheses of Proposition 7.3.8 and let $X > 0$ be as given there. Then there exist constants C, K, $\gamma > 0$ depending only on the data A, c such that for all $i \in \mathbb{N}_0$ and $(x, \hat{y}) \in S_X(\delta) \times \mathbb{C}$ there holds for k, $l \in \mathbb{N}_0$ with $k + l \leq 2$*

$$\left|\partial_x^k \partial_{\hat{y}}^l \widehat{U}_i(x,\hat{y})\right| \leq C C_g K^i \delta^{-(i+k)} \frac{1}{i!}\left(\frac{2i+3}{\lambda_0} + |\hat{y}|\right)^{2i+k} e^{-\operatorname{Re}(\lambda(x)\hat{y})}.$$

Proof: The proof follows immediately from Proposition 7.3.8, Lemma 7.3.7, slightly enlarging the constant K of Proposition 7.3.8, and appropriately choosing C. □

Lemma 7.3.12. *Let u_M^{BL} be given by (7.3.10). Then there exist constants C, K, $\lambda_0 > 0$, $y_0 > 0$ depending only the coefficients A, c such that on $S_X(\delta) \times B_{y_0}(0) \subset \mathbb{C} \times \mathbb{C}$ there holds*

$$\left|L_\varepsilon u_M^{BL}(x,y)\right| \leq C C_g K^{2M+2} \varepsilon^{2M+2} \delta^{-(2M+2)} \times$$
$$\frac{(4M+3+\lambda_0|y/\varepsilon|)^{4M+3}}{(2M+1)!} e^{-\operatorname{Re}(\lambda(x)y/\varepsilon)} \left\{1 + \varepsilon\delta^{-1}(4M+3+\lambda_0|y/\varepsilon|)\right\}.$$

Proof: In the proof we will write $\hat{y}$ for $\hat{y} = y/\varepsilon$ whenever notationally convenient. From Remark 7.3.1 there is $y_0 > 0$ such that $L_\varepsilon = \sum_{i=0}^{\infty} \varepsilon^i L_i$ uniformly on compact subsets of $S_X \times B_{y_0}(0) \subset \mathbb{C} \times \mathbb{C}$. Next, combining this with the defining property of the functions $\widehat{U}_i$, namely, $\sum_{j=0}^{i} L_j \widehat{U}_{i-j} = \sum_{j=0}^{i} L_{i-j} \widehat{U}_j = 0$ for all $i \in \mathbb{N}_0$, we get on compact subsets of $S_X \times B_{y_0}(0)$

$$L_\varepsilon u_M^{BL}(x,y) = \sum_{i=0}^{\infty} \varepsilon^i L_i \sum_{j=0}^{2M+1} \varepsilon^j \widehat{U}_j(x, y/\varepsilon) = \sum_{i=0}^{\infty} \varepsilon^i \sum_{j=0}^{\min\{i,2M+1\}} L_{i-j} \widehat{U}_j(x, y/\varepsilon)$$
$$= \sum_{i=2M+2}^{\infty} \varepsilon^i \sum_{j=0}^{2M+1} L_{i-j} \widehat{U}_j(x, y/\varepsilon).$$

We now have to bound $L_{i-j}\widehat{U}_j$. In view of the definition of the operator L_i in (7.3.5), (7.3.6), we can write $L_{i-j}\widehat{U}_j = T_1 + T_2 + \cdots + T_6$. In order to bound these six terms, we introduce the abbreviation

$$F_j := \frac{1}{\Gamma(j+1)} \left(\frac{2j+3}{\lambda_0} + |\hat{y}| \right)^{2j} e^{-\operatorname{Re}(\lambda(x)\hat{y})}.$$

For the constant γ_A of Lemma 7.3.5 and with the aid of Lemma 7.3.11, we get the existence of constants C, $K > 0$ (K suitably larger than K given in Lemma 7.3.11) depending only on the data A, c such that

$$\begin{aligned}
|T_1| &\le C C_g \gamma_A^{i-j} |\hat{y}|^{i-j} K^j \delta^{-j} F_j, \\
|T_2| &\le C C_g \gamma_A^{i-1-j} |\hat{y}|^{i-1-j} K^j \delta^{-j-1} F_{j+1/2}, \\
|T_3| &\le C C_g \gamma_A^{i-2-j} |\hat{y}|^{i-2-j} K^j \delta^{-j-2} F_{j+1}, \\
|T_4| &\le C C_g \gamma_A^{i-2-j} |\hat{y}|^{i-2-j} K^j \delta^{-j-1} F_{j+1/2}, \\
|T_5| &\le C C_g \gamma_A^{i-1-j} |\hat{y}|^{i-1-j} K^j \delta^{-j} F_j, \\
|T_6| &\le C C_g \gamma_A^{i-j} |\hat{y}|^{i-j} K^j \delta^{-j} F_j;
\end{aligned}$$

furthermore, $T_3 = T_4 = 0$ for $i - j - 2 < 0$.
Let us first consider the contribution of the terms T_1 to $L_\varepsilon u_M^{BL}$. We have

$$\sum_{j=0}^{2M+1} \sum_{i=2M+2}^{\infty} \varepsilon^i |T_1| \le C C_g \left(\sum_{j=0}^{2M+1} \gamma_A^{-j} |\hat{y}|^{-j} K^j \delta^{-j} F_j \right) \left(\sum_{i=2M+2}^{\infty} (\varepsilon \gamma_A |\hat{y}|)^i \right).$$

Assuming that y_0 in the statement of the lemma is so small that $\gamma_A y_0 < 1/2$, we can bound $\sum_{i=2M+2}^{\infty} (\varepsilon\gamma_A|\hat{y}|)^i \le 2(\varepsilon\gamma_A|\hat{y}|)^{2M+2}$ and therefore arrive at

$$\sum_{j=0}^{2M+1} \sum_{i=2M+2}^{\infty} \varepsilon^i T_1 \le C C_g \left(\varepsilon\gamma_A|\hat{y}|\right)^{2M+2} \sum_{j=0}^{2M+1} \gamma_A^{-j} |\hat{y}|^{-j} K^j \delta^{-j} F_j.$$

Next, from Lemma 7.3.10, we obtain for $0 \le j \le 2M+1$

$$|\hat{y}|^{-j} F_j \le \lambda_0^{-j} (\lambda_0 |\hat{y}|)^{-j} \frac{1}{\Gamma(j+1)} (2j+3+\lambda_0|\hat{y}|)^{2j} \, e^{-\operatorname{Re}(\lambda\hat{y})}$$
$$\le \lambda_0^{-(2M+1)-j} |\hat{y}|^{-(2M+1)} \frac{1}{(2M+1)!} (4M+5+\lambda_0|\hat{y}|)^{2(2M+1)} \, e^{-\operatorname{Re}(\lambda\hat{y})}.$$

Thus, by properly adjusting the constant K, and assuming that $\delta \le 1$ (as we may since $\delta \le X = (\gamma + 2\gamma_g)^{-1}$ and γ is at our disposal), we get

$$\sum_{j=0}^{2M+1} \sum_{i=2M+2}^{\infty} \varepsilon^i |T_1|$$
$$\le C C_g K^{2M+2} \delta^{-(2M+1)} \varepsilon^{2M+2} \frac{|\hat{y}|}{(2M+1)!} (4M+5+\lambda_0|\hat{y}|)^{2(2M+1)} \, e^{-\operatorname{Re}(\lambda\hat{y})}$$
$$\le C C_g K^{2M+2} \delta^{-(2M+1)} \varepsilon^{2M+2} \frac{1}{(2M+1)!} (4M+5+\lambda_0|\hat{y}|)^{4M+3} \, e^{-\operatorname{Re}(\lambda\hat{y})}.$$

The contributions due to T_2, T_5, T_6 are treated similarly. We have

$$\sum_{j=0}^{2M+1} \sum_{i=2M+2}^{\infty} \varepsilon^i |T_2|$$
$$\le C C_g K^{2M+2} \delta^{-(2M+2)} \varepsilon^{2M+2} \frac{1}{(2M+1)!} (4M+5+\lambda_0|\hat{y}|)^{4M+3} \, e^{-\operatorname{Re}(\lambda\hat{y})},$$

$$\sum_{j=0}^{2M+1} \sum_{i=2M+2}^{\infty} \varepsilon^i |T_5|$$
$$\le C C_g K^{2M+2} \delta^{-(2M+1)} \varepsilon^{2M+2} \frac{1}{(2M+1)!} (4M+5+\lambda_0|\hat{y}|)^{4M+2} \, e^{-\operatorname{Re}(\lambda\hat{y})},$$

$$\sum_{j=0}^{2M+1} \sum_{i=2M+2}^{\infty} \varepsilon^i |T_6|$$
$$\le C C_g K^{2M+2} \delta^{-(2M+1)} \varepsilon^{2M+2} \frac{1}{(2M+1)!} (4M+5+\lambda_0|\hat{y}|)^{4M+3} \, e^{-\operatorname{Re}(\lambda\hat{y})}.$$

For the last two contributions, due to T_3, T_4, we observe that they vanish for $i-2-j<0$, i.e., for $(j,i) = (2M+1, 2M+2)$. Hence, we bound

$$\sum_{j=0}^{2M+1} \sum_{i=2M+2}^{\infty} \varepsilon^i |T_3| \le \sum_{j=0}^{2M} \sum_{i=2M+2}^{\infty} \varepsilon^i |T_3| + \sum_{j=2M+1}^{2M+1} \sum_{i=2M+3}^{\infty} \varepsilon^i |T_3|.$$

We compute

$$\sum_{j=0}^{2M} \sum_{i=2M+2}^{\infty} \varepsilon^i |T_3|$$
$$\leq CC_g K^{2M+2} \delta^{-(2M+2)} \varepsilon^{2M+2} \frac{1}{(2M+1)!} (4M+5+\lambda_0|\hat{y}|)^{4M+2} e^{-\operatorname{Re}(\lambda \hat{y})},$$
$$\sum_{j=2M+1}^{2M+1} \sum_{i=2M+3}^{\infty} \varepsilon^i |T_3|$$
$$\leq CC_g K^{2M+2} \delta^{-(2M+3)} \varepsilon^{2M+3} \frac{1}{(2M+1)!} (4M+5+\lambda_0|\hat{y}|)^{4M+4} e^{-\operatorname{Re}(\lambda \hat{y})}.$$

The contribution from T_4 is bounded analogously. We conclude that (again, after appropriately adjusting the constants C, K) there exist C, $K > 0$ depending only on A, c such that for $(x,y) \in S_X(\delta) \times B_{y_0}(0) \subset \mathbb{C} \times \mathbb{C}$ there holds

$$\left|L_\varepsilon u_M^{BL}(x,y)\right| \leq CC_g K^{2M+2} \delta^{-(2M+2)} \varepsilon^{2M+2} \times$$
$$\frac{1}{(2M+1)!} (4M+3+\lambda_0|\hat{y}|)^{4M+3} e^{-\operatorname{Re}(\lambda \hat{y})} \left\{1 + \varepsilon\delta^{-1}(2M+5+\lambda_0|\hat{y}|)\right\}.$$

□

Lemma 7.3.13. *For every M, $a \geq 0$, $\alpha \in (0,1)$ there holds*

$$\sup_{r>0} (M+a+r)^M e^{-\alpha r} \leq \left(\frac{M}{\alpha}\right)^M e^{(1-\alpha)M} e^{\alpha a}.$$

Proof: We note that

$$\sup_{r>0} (M+a+r)^M e^{-\alpha r} = \sup_{x>a} (M+x)^M e^{-\alpha x} e^{\alpha a} \leq \sup_{x>0} (M+x)^M e^{-\alpha x} e^{\alpha a}.$$

The claim now follows from elementary considerations and $\alpha < 1$. □

Proof of Theorem 7.3.3: We start by considering the derivatives of the functions $\widehat{U}_i$. Let λ_0, C_A, K, $X = (\gamma_A + 2\gamma_g)^{-1}$ be given by Proposition 7.3.8. From Lemma 7.3.5, we have that $\lambda_0 < \operatorname{Re}\lambda$ and $|\lambda| \leq C_A$ on S_X. In order to calculate $\partial_{\hat{y}}^p \partial_x^q \widehat{U}_i(x,\hat{y})$, we choose $\kappa \in (0,1)$ such that

$$\frac{\kappa}{1-\kappa} C_A \leq \frac{\lambda_0}{2}.$$

This choice of κ guarantees that

$$e^{\kappa/(1-\kappa) C_A} e^{-\lambda(x)\hat{y}} \leq e^{-(\lambda_0/2)\hat{y}} \qquad \forall (x,\hat{y}) \in I_x \times \mathbb{R}_0^+. \tag{7.3.22}$$

Let $(x,y) \in I_x \times \mathbb{R}_0^+$ and set $\hat{y} = y/\varepsilon$. Cauchy's integral theorem for derivatives and Proposition 7.3.8 give the existence of C, $K > 0$ depending only on the coefficients A, c such that for $0 < \delta < X$

$$\left|\partial_{\hat{y}}^p \partial_x^q \widehat{U}_i(x,\hat{y})\right| = \frac{p!q!}{4\pi^2}\left|\int_{|t|=\kappa\delta}\int_{|s|=R} \frac{\widehat{U}_i(x+t,\hat{y}+s)}{(-t)^{q+1}(-s)^{p+1}}\,ds\,dt\right|$$

$$\leq CC_g p!q!(\kappa\delta)^{-q}R^{-p}\left((1-\kappa)\delta\right)^{-i}K^i\frac{1}{i!}\left(\frac{2i+1}{\lambda_0}+\hat{y}+R\right)^{2i} e^{\kappa/(1-\kappa)C_A\hat{y}}e^{-\lambda(x)\hat{y}}.$$

Choosing now $R = p+1$ and adjusting the constants C, K, we get

$$\left|\partial_{\hat{y}}^p \partial_x^q \widehat{U}_i(x,\hat{y})\right| \leq CC_g q!\tilde{\gamma}^{p+q}\delta^{-(i+q)}K^i\frac{1}{i!}\left(2i+1+\lambda_0\hat{y}+\lambda_0(p+1)\right)^{2i}e^{-(\lambda_0/2)\hat{y}}$$

for some $\tilde{\gamma} > 0$ depending only on the coefficients A, c. Splitting $e^{-\lambda_0/2\hat{y}} = e^{-\lambda_0/4\hat{y}}e^{-\lambda_0/4\hat{y}}$ we get from Lemma 7.3.13 that

$$(2i+1+\lambda_0\hat{y}+\lambda_0(p+1))^{2i}e^{-(\lambda_0/4)\hat{y}} \leq C(8i)^{2i}e^{(\lambda_0/4)p}$$

and thus

$$\left|\partial_{\hat{y}}^p \partial_x^q \widehat{U}_i(x,\hat{y})\right| \leq CC_g q!\tilde{\gamma}^{p+q}\delta^{-(i+q)}K^i\frac{1}{i!}i^{2i}e^{-(\lambda_0/4)\hat{y}}.$$

Readjusting again the various constants and letting $\delta \to X = (\gamma_A+2\gamma_g)^{-1}$ (with γ_A, γ_g as in the statement of Proposition 7.3.8) gives

$$\left|\partial_{\hat{y}}^p \partial_x^q \widehat{U}_i(x,\hat{y})\right| \leq CC_g q!\tilde{\gamma}^{p+q}(\gamma_A+2\gamma_g)^q(\gamma_A+2\gamma_g)^iK^ii^ie^{-\lambda_0/4\hat{y}}.$$

where the constants C, K, $\tilde{\gamma}$, γ_A depend only on the coefficients A, c. With these constants γ_A, K, we now impose the assumption

$$\varepsilon(2M+2)(\gamma_A+2\gamma_g)K \leq \frac{1}{2} \tag{7.3.23}$$

to conclude that under this assumption on ε, M there holds

$$\left|\sum_{i=0}^{2M+1}\varepsilon^i\partial_{\hat{y}}^p \partial_x^q \widehat{U}_i(x,\hat{y})\right| \leq CC_g q!\tilde{\gamma}^{p+q}(\gamma_A+2\gamma_g)^q e^{-(\lambda_0/4)\hat{y}} \qquad \forall (x,\hat{y}) \in I_x \times \mathbb{R}_0^+.$$

The first bound in Theorem 7.3.3 now follows. For the second bound, we proceed similarly, but base the proof on Lemma 7.3.12 rather than Proposition 7.3.8. For $\delta < X$ and $R < y_0/2$ (y_0 as given by Lemma 7.3.12) we want to apply Cauchy's integral theorem for derivatives. To that end, we observe that Lemma 7.3.12 gives the existence of C, $K > 0$ depending only on A, c such that for all $(t,s) \in B_{\kappa\delta}(0) \times B_{R\varepsilon}(0) \subset \mathbb{C}\times\mathbb{C}$ there holds (cf. also (7.3.13))

$$\begin{aligned}\left|L_\varepsilon u_M^{BL}(x+t,y+s)\right| &\leq CC_gK^{2M+2}\varepsilon^{2M+2}\left((1-\kappa)\delta\right)^{-(2M+2)} \times \\ &\frac{1}{(2M+1)!}\left(4M+3+\lambda_0(\hat{y}+R)\right)^{4M+3}e^{-\lambda(x)\hat{y}}e^{\kappa/(1-\kappa)C_A\hat{y}} \times \\ &\left\{1+\varepsilon\left((1-\kappa)\delta\right)^{-1}\left(4M+3+\lambda_0(\hat{y}+R)\right)\right\}.\end{aligned}$$

We reason now as above in order to simplify this expression. First, we note that the assumption $x \in I_x$ and the choice of κ guarantee (7.3.22). Next, Lemma 7.3.13 allows us to conclude

$$\begin{aligned}\frac{1}{(2M+1)!}\left(4M+3+\lambda_0(\hat{y}+R)\right)^{4M+4} e^{-(\lambda_0/4)\hat{y}} &\le CK^M \frac{1}{(2M+1)!}(16M)^4 M \\ &\le CK^M M^{2M}\end{aligned}$$

for some appropriate constants C, K. Thus, we arrive at

$$\begin{aligned}\left|L_\varepsilon u_M^{BL}(x+t,y+s)\right| &\le CC_g K^{2M+2}\varepsilon^{2M+2}\delta^{-(2M+2)}M^{2M}e^{-(\lambda_0/4)\hat{y}}\left\{1+\varepsilon\delta^{-1}\right\} \\ &\le CC_g\left(\varepsilon(2M+2)\delta^{-1}K\right)^{2M+2}e^{-(\lambda_0/4)\hat{y}}\left\{1+\varepsilon\delta^{-1}\right\}.\end{aligned}$$

Cauchy's integral theorem for derivatives therefore yields

$$\begin{aligned}&\left|\partial_y^p\,\partial_x^q\,L_\varepsilon u_M^{BL}(x,y)\right| = \frac{p!q!}{4\pi^2}\left|\int_{|t|=\kappa\delta}\int_{|s|=R\varepsilon}\frac{L_\varepsilon u_M^{BL}(x+t,\hat{y}+s)}{(-t)^{q+1}(-s)^{p+1}}\,ds\,dt\right| \\ &\le CC_g p!q!(\kappa\delta)^{-q}(\varepsilon R)^{-p}\left(\varepsilon(2M+2)\delta^{-1}K\right)^{2M+2}e^{-(\lambda_0/4)\hat{y}}\left\{1+\varepsilon\delta^{-1}\right\}.\end{aligned}$$

Next, observing that we may let $\delta \to (\gamma_A+2\gamma_g)^{-1}$, we obtain for some suitable constants

$$\begin{aligned}&\left|\partial_y^p\,\partial_x^q\,L_\varepsilon u_M^{BL}(x,y)\right| \le CC_g p!q!\varepsilon^{-p}\tilde{\gamma}^{p+q}(\gamma_A+2\gamma_g)^q\times \\ &\left(\varepsilon(2M+2)(\gamma_A+2\gamma_g)K\right)^{2M+2}e^{-(\lambda_0/4)\hat{y}}\left\{1+\varepsilon(\gamma_A+2\gamma_g)\right\}.\end{aligned}$$

As by assumption (7.3.23), $\varepsilon(\gamma_A+2\gamma_g)$ is bounded, the desired result now follows. □

7.4 Regularity through asymptotic expansions

7.4.1 Notation and main result

Let Ω be a curvilinear polygon satisfying the assumptions set out in Section 1.2. The boundary $\partial\Omega$ consists of J analytic arcs Γ_j, $j=1,\ldots,J$, whose endpoints are the vertices A_{j-1}, A_j $j=1,\ldots,J$ (we set $A_0 := A_J$).
As described in Section 1.2, each analytic arc Γ_j is parametrized as $\Gamma_j = \{(x_j(\theta),y_j(\theta))\,|\,\theta\in(0,1)\}$ for some analytic functions x_j, y_j. Our assumptions imply furthermore the existence of $\Theta>0$ such that the functions x_j, y_j are analytic on $(-2\Theta,1+2\Theta)$. We write $\widetilde{\Gamma}_j := \{(x_j(\theta),y_j(\theta)\,|\,\theta\in(-\Theta,1+\Theta)\}$. Without loss of generality, we may furthermore assume that the parametrization of Γ_j is done such that the normal vector $(-y_j'(\theta),x'(\theta)) \neq 0$ points into Ω. Next, we define *boundary-fitted* coordinates (ρ_j,θ_j) through the mapping

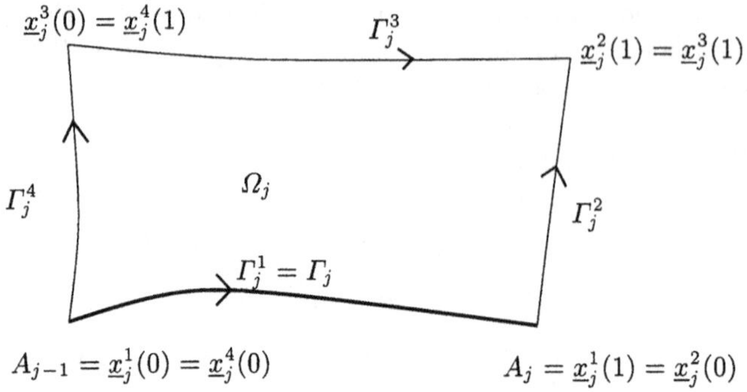

Fig. 7.4.1. Orientation of the curves Γ_j^k comprising the boundary of Ω_j.

$$\psi_j : [0,\rho_0] \times [-\Theta, 1+\Theta] \to \overline{\Omega}$$
$$(\rho,\theta) \mapsto (x_j(\theta), y_j(\theta)) + \frac{\rho}{\sqrt{|x_j'(\theta)|^2 + |y_j'(\theta)|^2}}(-y_j'(\theta), x_j'(\theta)).$$

The maps ψ_j are real analytic and in fact invertible in a neighborhood of $\{0\} \times (0,1)$ (with inverse ψ_j^{-1} being again real analytic). Without loss of generality, we may assume that ρ_0 and Θ are chosen such that ψ_j is real analytic and invertible on $[-\rho_0, \rho_0] \times [-\Theta, 1+\Theta]$. The inverse ψ_j^{-1} thus defines boundary-fitted coordinates (ρ_j, θ_j) in a neighborhood of $\widetilde{\Gamma}_j$ via $(\rho_j, \theta_j) = (\rho_j(\underline{x}), \theta_j(\underline{x})) = \psi_j^{-1}(\underline{x})$.

Remark 7.4.1 ρ_j in the boundary-fitted coordinates (ρ_j, θ_j) has a geometric interpretation: for $\underline{x} \in \Omega$ in a neighborhood of $\widetilde{\Gamma}_j$, we have $\rho_j(\underline{x}) = \operatorname{dist}(\underline{x}, \widetilde{\Gamma}_j)$. ▪

Boundary layers and corner layers are phenomena that are restricted to a (small) neighborhood of the boundary $\partial\Omega$. We therefore cover the half-tubular neighborhood $\mathcal{U} := \{(x,y) \in \Omega \,|\, \operatorname{dist}(x, \partial\Omega) < \rho_0\}$ by sets Ω_j with the following properties:

1. The domains $\Omega_j \subset \Omega$ are curvilinear rectangles, i.e., $\partial\Omega_j$ consists of four analytic arcs $\Gamma_j^k = \{\underline{x}_j^k(t) \,|\, t \in (0,1)\}$, $k \in \{1,\dots,4\}$, where the functions $\underline{x}_j^k$ are analytic on $[0,1]$.
2. The four angles of Ω_j are strictly between 0 and π and there holds $A_{j-1} = \underline{x}_j^1(0) = \underline{x}_j^4(0)$, $A_j = \underline{x}_j^1(1) = \underline{x}_j^2(0)$, $\underline{x}_j^2(1) = \underline{x}_j^3(1)$, $\underline{x}_j^4(1) = \underline{x}_j^3(0)$. In particular, therefore, we have $\Gamma_j^1 = \Gamma_j$ (cf. Fig. 7.4.1).
3. The arcs Γ_j^4, Γ_j^2 meet Γ_j *only* at the points A_{j-1}, A_j, respectively: There exists $C > 0$ such that $\rho_j(\underline{x}_j^4(t)) \geq Ct$, $\rho_j(\underline{x}_j^2(t)) \geq Ct$.
4. $\cup_{j=1}^J \overline{\Omega_j}$ covers the half-tubular neighborhood $\mathcal{U}$ of $\partial\Omega$.
5. $\Omega_j \subset \psi_j((0,\rho_0) \times (-\Theta, 1+\Theta))$ for all j.
6. ρ_0 is so small that $\overline{\Omega_j} \cap \overline{\Omega_k} \neq \emptyset$ if and only $k \in \{j-1, j, j+1\}$.
7. If the angle ω_j at the vertex A_j satisfies $\omega_j \geq \pi$, then the intersection $\Gamma_j' := \partial\Omega_j \cap \partial\Omega_{j+1}$ satisfies $\Gamma_j' = \Gamma_j^2 = \Gamma_{j+1}^4$ (i.e., at re-entrant corners A_j the sets Ω_j, Ω_{j+1} do not overlap).

8. If the angle ω_j at the vertex A_j satisfies $\omega_j < \pi$, then either the intersection $\Gamma_j' := \partial\Omega_j \cap \partial\Omega_{j+1}$ is an analytic arc and satisfies $\Gamma_j' = \Gamma_j^2 = \Gamma_{j+1}^4$ or $\Gamma_j^4 \subset \Gamma_{j-1}$, $\Gamma_j^2 \subset \Gamma_j$.

Remark 7.4.2 These definitions are rather technical. They formalize the following idea (cf. Fig. 7.4.2):

1. At a re-entrant corner A_j (cf. Fig. 7.4.2), the two subdomains Ω_j, Ω_{j+1} abutting on A_j share only a common analytic arc Γ_j'; the crucial features is that Γ_j' divides the angle at A_j into two angles, each being smaller than π.
2. At the convex corners A_{j-1}, the two subdomains Ω_{j-1}, Ω_j meet in one of two ways. Either, the two subdomains meet in the same fashion as in the case of re-entrant corners (left panel of Fig. 7.4.2) or they overlap (right panel of Fig. 7.4.2). In the latter case there is the additional condition that *two* boundary curves of the subdomains Ω_{j-1}, Ω_j lie on the boundary $\partial\Omega$: For $\partial\Omega_{j-1}$ we have $\Gamma_{j-1}^1 = \Gamma_{j-1}$ and $\Gamma_{j-1}^2 \subset \Gamma_j$ and for $\partial\Omega_j$ we have $\Gamma_j^1 = \Gamma_j$ together with $\Gamma_j^4 \subset \Gamma_{j-1}$.

It is noteworthy that we leave choices for the selection of the domains Ω_j at convex corners. This freedom will later on correspond to different choices in the decomposition of u_ε. ∎

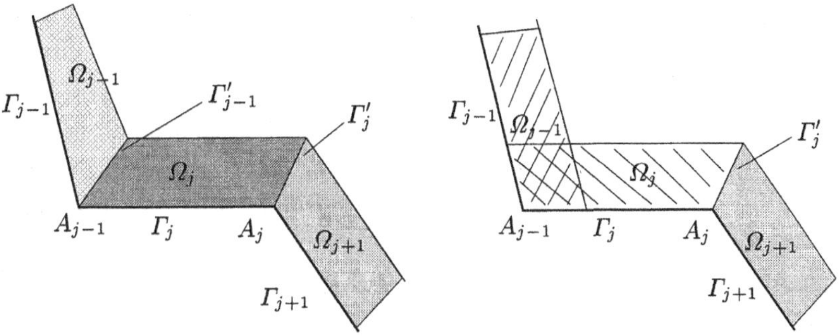

Fig. 7.4.2. Different choices of subdomains Ω_j at convex corners.

Next, we choose cut-off functions χ_j^{BL}, χ_j^{CL}. We start with the cut-off functions χ_j^{CL}, supported by neighborhoods of the vertices A_j. To do so, let $R > 0$ be so small that

$$\overline{B_{2R}(A_j)} \subset \overline{\Omega_j} \cup \overline{\Omega_{j+1}}, \qquad j = 1, \ldots, J,$$
$$\overline{B_{2R}(A_j)} \cap \overline{B_{2R}(A_k)} = \emptyset \qquad \text{if } j \neq k$$

and that the sets

$$S_j := \Omega \cap B_{2R}(A_j) \tag{7.4.2}$$

form curvilinear sectors (with apex A_j) in the sense of Definition 4.2.1 whose two sides emanating from A_j lie on the arcs Γ_j, Γ_{j+1}. Furthermore, we set

$$S_j^+ := S_j \cap \Omega_{j+1}, \qquad S_j^- := S_j \cap \Omega_j. \tag{7.4.3}$$

Remark 7.4.3 In the case that Ω_j, Ω_{j+1} do overlap, we have $S_j = S_j^+ = S_j^-$. In the case that Ω_j, Ω_{j+1} do not overlap, the sector S_j is divided by the analytic arc Γ_j' into two subsectors that are precisely S_j^+, S_j^-. ▪

We choose cut-off functions $\chi_j^{CL} : \mathbb{R}^2 \to \mathbb{R}_0^+$ with

$$\operatorname{supp} \chi_j^{CL} \subset B_R(A_j), \qquad j = 1, \dots, J,$$
$$\chi_j^{CL} \equiv 1 \qquad \text{on } B_{R/2}(A_j).$$

Note that $\operatorname{supp} \chi_j^{CL} \cap \Omega \subset S_j$.

We now turn to the definition of the cut-off functions χ_j^{BL}, which is most conveniently done in boundary-fitted coordinates. To that end, we introduce a smooth cut-off functions $\chi = \chi(\rho, \theta)$ satisfying $\chi(\rho, \theta) \equiv 1$ for $|\rho| \leq R'/2$ and $\chi(\rho, \theta) \equiv 0$ for $\rho \geq R'$ for some $0 < R' < \rho_0$ to be determined shortly. We then set (the function $\mathbb{E}_{\overline{\Omega_j}}$ denotes the characteristic function for the set $\overline{\Omega_j}$)

$$\chi_j^{BL} := (\chi \circ \psi_j^{-1}) \cdot \mathbb{E}_{\overline{\Omega_j}}. \tag{7.4.4}$$

$R' > 0$ is now chosen so small that

$$\chi_j^{CL} \equiv 1 \quad \text{on the set } (\operatorname{supp} \chi_j^{BL} \cap \operatorname{supp} \chi_{j+1}^{BL}). \tag{7.4.5}$$

Remark 7.4.4 The precise choice of the cut-off functions χ_j^{BL} is not very important. The essential feature is that a) $\chi_j^{BL} \equiv 1$ in a neighborhood of Γ_j and b) χ_j^{BL} is supported by $\overline{\Omega_j}$. Condition a) is enforced by defining χ_j^{BL} in boundary-fitted coordinates and condition b) is enforced by multiplying by the characteristic function $\mathbb{E}_{\overline{\Omega_j}}$. ▪

By means of the method of matched asymptotic expansions, the solutions u of (1.2.11a) with boundary conditions (1.2.11b) can be written as follows: For every $M \in \mathbb{N}_0$, there are functions w_M (the *smooth part*), $u_{j,M}^{BL}$, the *boundary layer parts*, $u_{j,M}^{CL}$, the *corner layer parts*, and r_M, the remainder, such that the exact solution u admits the following decomposition:

$$u_\varepsilon = w_M + \sum_{j=1}^{J} \chi_j^{BL} \left(u_{j,M}^{BL} \circ \psi_j^{-1}\right) + \sum_{j=1}^{J} \chi_j^{CL} u_{j,M}^{CL} + r_M. \tag{7.4.6}$$

In order to formulate the main result of this section, Theorem 7.4.5, which asserts regularity properties of the components of (7.4.6), we introduce, analogous to the definition of $\Phi_{p,\beta,\varepsilon}$ at the outset of Section 5.3.2, the weight function

$$\Psi_{p,\beta,\varepsilon,\alpha}(x) := \Pi_{j=1}^{J} \hat{\Psi}_{p,\beta_j,\varepsilon,\alpha}(x - A_j). \tag{7.4.7}$$

Theorem 7.4.5 (asymptotic expansion, regularity). *Let Ω be a curvilinear polygon, f be analytic on $\overline{\Omega}$ satisfying*

$$\|\nabla^p f\|_{L^\infty(\Omega)} \le C_f \gamma_f^p p! \qquad \forall p \in \mathbb{N}_0,$$

and let the piecewise analytic boundary data g satisfy (1.2.4). Then there exist C, γ, K, $\alpha > 0$, $R' > 0$, $\beta \in (0,1)^J$, $\beta' \in (0,1)^J$ independent of ε and γ_f, and there exist constants C', $\gamma' > 0$ independent of ε such that the terms appearing in (7.4.6) satisfy the following: If $0 < \varepsilon(2M+2)(1+\gamma_f)K \le 1$ then there holds for all $p \in \mathbb{N}_0$ and all $q \in \mathbb{N}_0$

$$\|\nabla^p w_M\|_{L^\infty(\Omega)} \le C\left(\gamma(1+\gamma_f)\right)^p p!,$$

$$\sup_{\theta_j \in [-\Theta, 1+\Theta]} |\partial^p_{\rho_j} \partial^q_{\theta_j} u^{BL}_{j,M}(\rho_j, \theta_j)| \le C\left(\gamma(1+\gamma_f)\right)^{p+q} q! \varepsilon^{-p} e^{-\alpha\rho_j/\varepsilon}, \quad \rho_j \ge 0,$$

$$\|r_M\|_{H^{2,2}_{\beta',\varepsilon}(\Omega)} \le C'\left\{e^{-\alpha/\varepsilon} + \left(\varepsilon(2M+2)\gamma(1+\gamma_f)\right)^{2M+2}\right\},$$

$$(u^{BL}_{j,M} \circ \psi_j^{-1} + w_M)|_{\Gamma_j} = g|_{\Gamma_j},$$

$$r_M = 0 \qquad \text{on } \partial\Omega.$$

The corner layers $u^{CL}_{j,M}$ are analytic on $(S_j^+ \cup S_j^-) \cap B_{R'}(A_j)$ and satisfy for all $p \in \mathbb{N}_0$

$$\|\Psi_{0,0,\varepsilon,\alpha} u^{CL}_{j,M}\|_{L^2(S_j^+ \cup S_j^-)} + \varepsilon \|\Psi_{0,0,\varepsilon,\alpha} \nabla u^{CL}_{j,M}\|_{L^2(S_j^+ \cup S_j^-)} \le C'\varepsilon, \tag{7.4.8a}$$

$$\|\Psi_{p,\beta,\varepsilon,\alpha} \nabla^{p+2} u^{CL}_{j,M}\|_{L^2((S_j^+ \cup S_j^-) \cap B_{R'}(A_j))} \le C'\varepsilon(\gamma')^p \max\{p, \varepsilon^{-1}\}^{p+2}. \tag{7.4.8b}$$

In particular, if $S_j^+ = S_j^- = S_j$, then the corner layer function $u^{CL}_{j,M}$ is analytic on $S_j \cap B_{R'}(A_j)$. Furthermore, the corner layers u_j^{CL} satisfy the following pointwise estimates for $p \ge 0$ and $x \in (S_j^+ \cup S_j^-) \cap B_{R'}(A_j)$

$$\left|\nabla^p \left(u^{CL}_{j,M}(x) - u^{CL}_{j,M}(A_j)\right)\right| \le C' p! r_j^{-p} (\gamma')^p \left(\frac{r_j}{\varepsilon}\right)^{1-\beta_j} e^{-\alpha r_j/\varepsilon}, \tag{7.4.9}$$

where $r_j = \operatorname{dist}(x, A_j)$. In particular, if $S_j^+ = S_j^- = S_j$, then $u^{CL}_{j,M}(A_j) = -g(A_j)$ whereas in the case $S_j^+ \cap S_j^- = \emptyset$ we have $u^{CL}_{j,M}(A_j) = 0$.

The proof of Theorem 7.4.5 is lengthy and therefore broken up into several steps handled in the subsequent subsections. Using the same arguments as in Corollaries 7.2.4, 7.3.4, we can extract from Theorem 7.4.5 the following result.

Corollary 7.4.6 (asymptotic expansion, regularity). *Under the assumptions of Theorem 7.4.5, there exist C, γ, R', $\alpha > 0$ and $\beta \in (0,1)^J$, $\beta' \in (0,1)^J$ independent of ε such that for every ε the solution u_ε of (1.2.11) can be decomposed as*

$$u_\varepsilon = w_\varepsilon + \sum_{j=1}^J \chi_j^{BL}(u^{BL}_{j,\varepsilon} \circ \psi_j^{-1}) + \sum_{j=1}^J \chi_j^{CL} u^{CL}_{j,\varepsilon} + r_\varepsilon,$$

where the terms w_ε, $u^{BL}_{j,\varepsilon}$, $u^{CL}_{j,\varepsilon}$, r_ε satisfy the boundary conditions

$$(u^{BL}_{j,\varepsilon} \circ \psi_j^{-1} + w_\varepsilon)|_{\Gamma_j} = g|_{\Gamma_j},$$
$$r_\varepsilon = 0 \qquad on\ \partial\Omega,$$

and have the following regularity properties: For all $p,\ q \in \mathbb{N}_0$

$$\|\nabla^p w_\varepsilon\|_{L^\infty(\Omega)} \leq C\gamma^p p!,$$
$$\sup_{\theta_j \in [-\Theta, 1+\Theta]} |\partial^p_{\rho_j}\partial^q_{\theta_j} u^{BL}_{j,\varepsilon}(\rho_j,\theta_j)| \leq C\gamma^{p+q} q! \varepsilon^{-p} e^{-\alpha\rho_j/\varepsilon}, \quad \rho_j \geq 0,$$
$$\|r_\varepsilon\|_{H^{2,2}_{\beta',\varepsilon}(\Omega)} \leq C e^{-\gamma'/\varepsilon},$$
$$\|\Psi_{0,0,\varepsilon,\alpha} u^{CL}_{j,\varepsilon}\|_{L^2(S_j^+ \cup S_j^-)} + \varepsilon \|\Psi_{0,0,\varepsilon,\alpha}\nabla u^{CL}_{j,\varepsilon}\|_{L^2(S_j^+\cup S_j^-)} \leq C\varepsilon,$$
$$\|\Psi_{p,\beta,\varepsilon,\alpha}\nabla^{p+2} u^{CL}_{j,\varepsilon}\|_{L^2((S_j^+\cup S_j^-)\cap B_{R'}(A_j))} \leq C'\varepsilon\gamma^p \max\{p, \varepsilon^{-1}\}^{p+2},$$
$$\left|\nabla^p \left(u^{CL}_{j,M}(x) - u^{CL}_{j,M}(A_j)\right)\right| \leq Cp! r_j^{-p}\gamma^p \left(\frac{r_j}{\varepsilon}\right)^{1-\beta_j} e^{-\alpha r_j/\varepsilon},$$

for all $x \in (S_j^+ \cup S_j^-) \cap B_{R'}(A_j)$*; here* $r_j = \operatorname{dist}(x, A_j)$*. In particular, if* $S_j^+ = S_j^- = S_j$*, then* $u^{CL}_{j,M}(A_j) = -g(A_j)$ *whereas in the case* $S_j^+ \cap S_j^- = \emptyset$ *we have* $u^{CL}_{j,M}(A_j) = 0$*.*

Proof: As in the proof of Corollaries 7.2.4, 7.3.4, we merely choose $M = \lambda/\varepsilon$ in Theorem 7.4.5 for some suitable $\lambda > 0$. □

7.4.2 Proof of Theorem 7.4.5: smooth and boundary layer parts

This section is devoted to the definition and the proof of the bounds on the smooth part, w_M and the boundary layer components, $u^{BL}_{j,M}$. Theorem 7.2.2 gives the existence of C, K, $\gamma > 0$ depending only on the coefficients A, c such that under the assumption

$$0 < \varepsilon(2M+2)(1+\gamma_f)\gamma K \leq 1 \tag{7.4.10}$$

there holds for w_M as defined in (7.2.2)

$$\|\nabla^p w_M\|_{L^\infty(\Omega)} \leq CC_f\left(\gamma(1+\gamma_f)\right)^p p! \qquad \forall p \in \mathbb{N}_0, \tag{7.4.11}$$
$$\|L_\varepsilon w_M - f\|_{L^\infty(\Omega)} \leq CC_f\left(\varepsilon(2M+2)(1+\gamma_f)\gamma K\right)^{2M+2}. \tag{7.4.12}$$

(7.4.11) gives the desired bound on w_M. Next, we proceed with defining and analyzing the terms $u^{BL}_{j,M}$. This is done as in the classical asymptotic analysis of the problem under consideration; the technical tools for its execution have been provided in Section 7.3.1.

We note that the outer expansion w_M represents, up to a small defect, a particular solution to (1.2.11a). However, w_M does not satisfy the correct boundary conditions on $\partial\Omega$. In order to remedy that, one would like to introduce a boundary layer function u^{BL} satisfying

$$L_\varepsilon u^{BL} = 0 \quad \text{on } \Omega, \qquad u^{BL} = g - w_M \quad \text{on } \partial\Omega,$$

for $w_M + u^{BL}$ then satisfies the correct boundary conditions and, up to a small defect, also the differential equation. The boundary layer functions $u^{BL}_{j,M}$ are now approximations to u^{BL} in neighborhoods of the arcs Γ_j. They are constructed as follows. The function $u_j := u^{BL} \circ \psi_j$ satisfies on the rectangle $R = (0, \rho_0) \times (-\Theta, 1+\Theta)$

$$\widehat{L}_\varepsilon u_j := -\nabla_{(\rho_j,\theta_j)} \cdot \left(\hat{A}\nabla_{(\rho_j,\theta_j)} u_j\right) + \hat{c} u_j = 0 \quad \text{on } R, \tag{7.4.13a}$$

$$u_j|_{\rho_j=0} = g_j := (g \circ \psi_j - w_M \circ \psi)|_{\rho_j=0}, \tag{7.4.13b}$$

where the subscript in the notation $\nabla_{(\rho_j,\theta_j)}$ emphasizes that differentiation takes place with respect to the boundary-fitted coordinates (ρ_j, θ_j). The functions $\hat{A}$, $\hat{c}$ are given by (cf. Lemma A.1.1)

$$\hat{A} = (\det \psi_j')(\psi_j')^{-T}(A \circ \psi_j)(\psi_j')^{-1}, \qquad \hat{c} = (c \circ \psi_j)\det \psi_j'.$$

By the analyticity of ψ_j, there holds $\hat{A} \in \mathcal{A}(\overline{R}, \mathbb{S}^2_>)$, $\hat{c} \in \mathcal{A}(\overline{R})$ with $\hat{c} > 0$ on $\overline{R}$. Furthermore, from Lemma 4.3.4, there are C, $\gamma > 0$ independent of ε and M such that for g_j of (7.4.13)

$$\|D^p g_j\|_{L^\infty((-\Theta,1+\Theta))} \le C\,(\gamma(1+\gamma_f))^p\, p! \qquad \forall p \in \mathbb{N}_0. \tag{7.4.14}$$

The situation of (7.4.13) is therefore the one considered in Section 7.3.1. Defining $u^{BL}_{j,M}$ as in (7.3.10) we get from Theorem 7.3.3 that under assumption (7.4.10) (with appropriately modified constants γ, K, which are still independent of ε, M, γ_f) there holds for some $\alpha > 0$ independent of ε, M, γ_f and all $(p, q) \in \mathbb{N}_0^2$ and $\rho_j \ge 0$:

$$u^{BL}_{j,M} = (g \circ \psi_j - w_M \circ \psi_j) \qquad \text{on } \rho_j = 0, \tag{7.4.15}$$

$$\sup_{\theta_j \in (-\Theta,1+\Theta)} |\partial^p_{\rho_j}\partial^q_{\theta_j} u^{BL}_{j,M}(\rho_j,\theta_j)| \le C\,(\gamma(1+\gamma_f))^{p+q}\, q!\varepsilon^{-p} e^{-\alpha\rho_j/\varepsilon}, \tag{7.4.16}$$

$$|\widehat{L}_\varepsilon u^{BL}_{j,M}(\rho_j,\theta_j)| \le C\,(\varepsilon(2M+2)\gamma(1+\gamma_f)K)^{2M+2}\, e^{-\alpha\rho_j/\varepsilon}. \tag{7.4.17}$$

Lemma A.1.1 implies that

$$\widehat{L}_\varepsilon(v \circ \psi_j) = ((L_\varepsilon v) \circ \psi_j) \cdot (\det \psi_j')^{-1} \qquad \text{for all smooth functions } v. \tag{7.4.18}$$

Thus, we obtain on the "physical" domain $\psi_j(R)$ with $\rho_j = \operatorname{dist}(x, \widetilde{\Gamma}_j)$

$$|L_\varepsilon(u^{BL}_{j,M} \circ \psi_j^{-1})| \le C\,(\varepsilon(2M+2)\gamma(1+\gamma_f)K)^{2M+2}\, e^{-\alpha\rho_j/\varepsilon} \quad \text{on } \psi_j(R) \supset \Omega_j. \tag{7.4.19}$$

These estimates imply the desired bounds on the functions $u^{BL}_{j,M}$. In fact, the transition from (7.4.17) to (7.4.19) requires only adjusting the constant C—the constants γ and K are the same in both instances.

The properties of the functions $u^{BL}_{j,M}$ obtained so far lead to the following lemma.

Lemma 7.4.7. *Let $u^{BL}_{j,M}$ satisfy (7.4.16), (7.4.19). Then on* $\operatorname{supp}\chi^{BL}_j$ *there holds for some C, γ, $\alpha > 0$ independent of*

$$\|L_\varepsilon\left(\chi^{BL}_j u^{BL}_{j,M}\circ\psi_j^{-1}\right)\|_{L^\infty(\operatorname{supp}\chi^{BL}_j)} \le C\left\{e^{-\alpha'/\varepsilon} + (\varepsilon(2M+2)\gamma(1+\gamma_f))^{2M+2}\right\}.$$

Proof: In view of (7.4.18) and the definition of $\chi^{BL}_j = \chi\circ\psi_j^{-1}\cdot \mathbb{E}_{\overline{\Omega_j}}$, it suffices to show

$$\|\widehat{L}_\varepsilon\left(\chi u^{BL}_{j,M}\right)\|_{L^\infty(\operatorname{supp}\chi)} \le C\left\{e^{-\alpha'/\varepsilon} + (\varepsilon(2M+2)\gamma(1+\gamma_f))^{2M+2}\right\}.$$

We observe that for smooth functions χ, u there holds

$$\begin{aligned}\widehat{L}_\varepsilon(\chi u) &= -\varepsilon^2\nabla\cdot\left(\hat{A}\nabla(\chi u)\right) + \hat{c}\chi u\\ &= \chi\widehat{L}_\varepsilon u - \varepsilon^2(\nabla\chi)\cdot(\hat{A}\nabla u) - \varepsilon^2(\nabla u)\cdot(\hat{A}\nabla\chi).\end{aligned}$$

For $u = u^{BL}_{j,M}$, the bound (7.4.22) gives

$$\|\chi\widehat{L}_\varepsilon u^{BL}_{j,M}\|_{L^\infty(\operatorname{supp}\chi)} \le C\left(\varepsilon(2M+2)\gamma(1+\gamma_f)\right)^{2M+2}.$$

Next, in order to treat the remaining terms, we observe that in a neighborhood of $\rho = 0$, the cut-off function $\chi \equiv 1$, i.e., $\nabla\chi \equiv 0$. Furthermore, for $\rho > R'/2$, the function $u^{BL}_{j,M}$ and all its derivatives are exponentially small by (7.4.16), implying

$$\|-\varepsilon^2\nabla\chi\cdot(\hat{A}\nabla u^{BL}_{j,M}) - \varepsilon^2(\nabla u^{BL}_{j,M})\cdot\hat{A}\nabla\chi\|_{L^\infty(\operatorname{supp}\chi)} \le C\varepsilon e^{-\alpha'/\varepsilon}$$

for some $\alpha' > 0$ depending on α of (7.4.16) and χ^{BL}_j. □

7.4.3 Proof of Theorem 7.4.5: corner layer and remainder

So far, we have defined w_M and the functions $u^{BL}_{j,M}$. Consider now the function

$$u_{IO} := w_M + \sum_{j=1}^{J}\chi^{BL}_j\left(u^{BL}_{j,M}\circ\psi_j^{-1}\right). \tag{7.4.20}$$

The definition (7.4.4) implies that the cut-off functions χ^{BL}_j are only piecewise smooth with possible jumps across $\partial\Omega_j$. Likewise, the function u_{IO} may jump across $\partial\Omega_j$. We note that these regions are contained in the set

$$V := \cup_{j=1}^{J}\operatorname{supp}\chi^{CL}_j.$$

First, we show that

$$u_{IO} = g \qquad \text{on } (\partial\Omega)\setminus V, \tag{7.4.21}$$

$$\|L_\varepsilon u_{IO} - f\|_{L^\infty(\Omega\setminus V)} \le C\left\{e^{-\alpha'/\varepsilon} + (\varepsilon(2M+2)\gamma(1+\gamma_f))^{2M+2}\right\}, \tag{7.4.22}$$

where the constants C, α', $\gamma > 0$ are independent of ε, M, and γ_f. The estimate (7.4.21) follows from the fact that $\chi_j^{BL} \equiv 1$ in a neighborhood of Γ_j, the construction of the functions $u_{j,M}^{BL}$, and the assumptions on the supports of the cut-off functions χ_j^{BL}. (7.4.22) follows directly from Lemma 7.4.7.
Bounds (7.4.21), (7.4.22) imply that the function u_{IO} satisfies the differential equation (up to a small defect) and the boundary conditions except for a neighborhood of the vertices. There, the function u_{IO} is not necessarily continuous and does not satisfy the boundary conditions. This last inconsistency is now removed with the aid of the *corner layers* $u_{j,M}^{CL}$. An additional side effect of the introduction of the corner layers $u_{j,M}^{CL}$ is that the function

$$u_{IOC} := w_M + \sum_{j=1}^{J} \chi_j^{BL}(u_{j,M}^{BL} \circ \psi_j^{-1}) + \sum_{j=1}^{J} \chi_j^{CL} u_{j,M}^{CL} \tag{7.4.23}$$

is an element of $C^1(\Omega)$.
When we defined the subdomains Ω_j in Section 7.4.1, we pointed out that two subdomain Ω_j, Ω_{j+1} meet in one of two ways near a vertex A_j. In the first situation, Ω_j, Ω_{j+1} overlap, in which case we call the vertex A_j a "convex" corner. In the second situation, Ω_j, Ω_{j+1} do not overlap, and the vertex A_j is called a "general" corner. We emphasize that a convex corner may also be treated as a general corner. Each of these two scenarios is dealt with in turn in the following two subsections.

Corner layers at convex corners. We start with the setting at a convex corner because the construction of the corner layer is more intuitive and straight forward. We point out, however, that a convex corner may also be treated as a general corner, elaborated in the ensuing subsection.
Let A_j be the convex corner under consideration and recall the definition of the sector S_j in (7.4.2). Its two sides emanating from A_j are denoted Γ^1, Γ^2 with $\Gamma^1 \subset \Gamma_j$, $\Gamma^2 \subset \Gamma_{j+1}$ (see Fig. 7.4.3). By our assumption that A_j is a convex corner (in the sense that the domains Ω_j, Ω_{j+1} overlap) we see that on S_j both boundary-fitted coordinate systems (ρ_j, θ_j), $(\rho_{j+1}, \theta_{j+1})$ are available; u_{IO} of (7.4.20) is smooth on S_j and can be written as

$$u_{IO} = w_M + \chi_j^{BL}(u_{j,M}^{BL} \circ \psi_j^{-1}) + \chi_{j+1}^{BL}(u_{j+1,M}^{BL} \circ \psi_{j+1}^{-1}) \qquad \text{on } S_j.$$

Lemma 7.4.7 therefore implies that

$$\|L_\varepsilon u_{IO} - f\|_{L^\infty(S_j)} \le C\left\{e^{-\alpha'/\varepsilon} + (\varepsilon(2M+2)\gamma(1+\gamma_f))^{2M+2}\right\}.$$

Furthermore, by the construction of the functions $u_{j,M}^{BL}$, $u_{j+1,M}^{BL}$, it is easy to see that

$$u_{IO} = \chi_{j+1}^{BL}\left(u_{j+1}^{BL} \circ \psi_{j+1}^{-1}\right) \qquad \text{on } \Gamma^1 \subset \Gamma_j, \tag{7.4.24a}$$

$$u_{IO} = \chi_j^{BL}\left(u_j^{BL} \circ \psi_j^{-1}\right) \qquad \text{on } \Gamma^2 \subset \Gamma_{j+1}. \tag{7.4.24b}$$

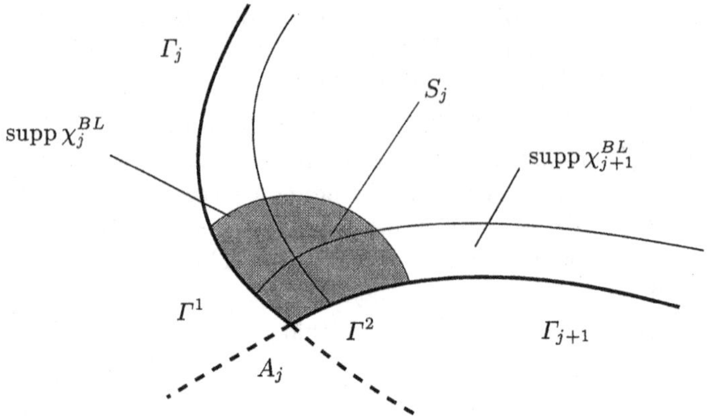

Fig. 7.4.3. Situation at a convex corner.

We are now in position to apply Theorem 6.4.13 to the sector S_j with homogeneous transmission conditions $h_1 = h_2 = 0$ (in fact, we have not specified the arc Γ that subdivides S_j into two subsectors!), homogeneous right-hand side f but inhomogeneous boundary data g. On ∂S_j, the boundary data g are given by

$$g_1 = g|_{\Gamma_1} := -\chi_{j+1}^{BL} u_{j+1,M}^{BL} \circ \psi_{j+1}^{-1}, \qquad g_2 = g|_{\Gamma_2} := -\chi_j^{BL} u_{j,M}^{BL} \circ \psi_j^{-1}.$$

It remains to check that the boundary data g satisfy the assumptions of Theorem 6.4.13, i.e., (6.4.20). These follow by combining the following facts:

1. χ_j^{BL}, χ_{j+1}^{BL} are identically one in a neighborhood of A_j;
2. the assumptions on the arcs Γ_j^4, Γ_{j+1}^2 guarantee that (6.4.19) is satisfied;
3. $u_{j,M}^{BL}$, $u_{j+1,M}^{BL}$ satisfy the desired estimates in the boundary-fitted coordinate systems (ρ_j, θ_j), $(\rho_{j+1}, \theta_{j+1})$.

These three facts together with Lemma 4.3.3 (dealing with changes of variables) give (6.4.20). Thus, Theorem 6.4.13 yields the existence of a function $u_{j,M}^{CL}$, analytic in a neighborhood of A_j (more precisely: on $S_j \cap B_{R'}(A_j)$ for suitable $R' > 0$), with the following properties:

$$L_\varepsilon u_{j,M}^{CL} \equiv 0 \qquad \text{on } S_j, \tag{7.4.25}$$

$$u_{j,M}^{CL} = -\chi_j^{BL} u_{j,M}^{BL} \circ \psi_j^{-1} \qquad \text{on } \Gamma_{j+1} \cap \partial S_j, \tag{7.4.26}$$

$$u_{j,M}^{CL} = -\chi_{j+1}^{BL} u_{j+1,M}^{BL} \circ \psi_{j+1}^{-1} \qquad \text{on } \Gamma_j \cap \partial S_j, \tag{7.4.27}$$

together with *a priori* bounds for all $p \in \mathbb{N}_0$

$$\|\Psi_{0,0,\varepsilon,\alpha'} u_{j,M}^{CL}\|_{L^2(S_j)} + \varepsilon \|\Psi_{0,0,\varepsilon,\alpha'} \nabla u_{j,M}^{CL}\|_{L^2(S_j)} \le C\varepsilon, \tag{7.4.28}$$

$$\|\Psi_{p,\beta,\varepsilon,\alpha'} \nabla^{p+2} u_{j,M}^{CL}\|_{L^2(S_j \cap B_{R'}(A_j))} \le C\varepsilon \max\{p, \varepsilon^{-1}\}^{p+2}. \tag{7.4.29}$$

for some constants C, γ, R', $\alpha' > 0$ independent of ε. We note that (7.4.28), (7.4.29) are the desired estimates. Finally, by the assumptions on the supports of the cut-off functions χ_j^{CL}, χ_j^{BL} we have

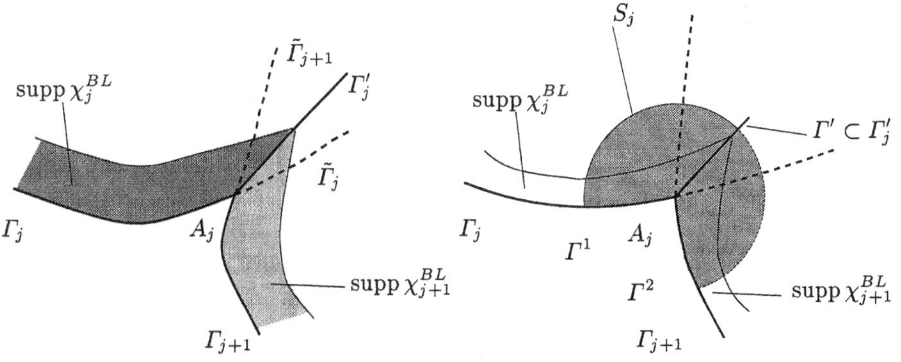

Fig. 7.4.4. Situation at a general corner.

$$\chi_j^{CL}\chi_j^{BL} = \chi_j^{BL} \quad \text{on } \Gamma_{j+1}, \qquad \chi_j^{CL}\chi_{j+1}^{BL} = \chi_{j+1}^{BL} \quad \text{on } \Gamma_j.$$

Thus, the function $\chi_j^{BL}u_{j,M}^{CL}$ also satisfies

$$\begin{aligned} \chi_j^{CL}u_{j,M}^{CL} &= -\chi_j^{BL}u_{j,M}^{BL} \circ \psi_j^{-1} \quad \text{on } \Gamma_{j+1}, \\ \chi_j^{CL}u_{j,M}^{CL} &= -\chi_{j+1}^{BL}u_{j+1,M}^{BL} \circ \psi_{j+1}^{-1} \quad \text{on } \Gamma_j. \end{aligned}$$

We conclude that on S_j the function u_{IOC} of (7.4.23) satisfies

$$\begin{aligned} u_{IOC} &= g \qquad \text{on } \partial\Omega \cap \partial S_j, \\ \|L_\varepsilon u_{IOC} - f\|_{L^2(S_j)} &\le C\left\{e^{-\alpha'/\varepsilon} + (\varepsilon(2M+2)\gamma(1+\gamma_f))^{2M+2}\right\}. \end{aligned} \tag{7.4.30}$$

Remark 7.4.8 Why might it be advantageous to choose to treat even a convex corner as a general one? The above arguments show that, in order for Theorem 6.4.13 to be applicable, we need the functions $u_{j,M}^{BL} \circ \psi_j^{-1}$, $u_{j+1,M}^{BL} \circ \psi_{j+1}^{-1}$ to decay exponentially on the the arcs Γ_{j+1}, Γ_j, respectively. As long as the angle between Γ_j, Γ_{j+1} is strictly less than π, this is guaranteed (in a neighborhood of A_j at least). However, the decay rate deteriorates as the angle tends to π. Treating a convex corner as a general one avoids this problem. ▪

Corner layers at general corners. We now turn to the construction of the corner layer in a general corner.

Let A_j be a general corner in the sense that the subdomains Ω_j, Ω_{j+1} abutting on A_j do not overlap. Recall the definition $\Gamma_j' = \partial\Omega_j \cap \partial\Omega_{j+1}$ and the definition of the sector S_j in (7.4.2). Denote by Γ^1, Γ^2 the two sides of S_j emanating from A_j with $\Gamma^1 \subset \Gamma_j$, $\Gamma^2 \subset \Gamma_{j+1}$. Next, we set $\Gamma' := S_j \cap \Gamma_j'$ (see Fig. 7.4.4). We recognize that by our assumptions on the supports of the cut-off functions χ_j^{BL}, χ_{j+1}^{BL}, the function u_{IO} of (7.4.20) restricted to S_j has the form

$$u_{IO} = w_M + \begin{cases} \chi_j^{BL}u_{j,M}^{BL} \circ \psi_j^{-1} & \text{on } S_j^- \\ \chi_{j+1}^{BL}u_{j+1,M}^{BL} \circ \psi_{j+1}^{-1} & \text{on } S_j^+. \end{cases} \tag{7.4.31}$$

By construction of the functions $u^{BL}_{j,M}$, $u^{BL}_{j+1,M}$, it is then clear that u_{IO} satisfies the correct boundary conditions on $\partial\Omega \cap \partial S_j$, i.e., $u_{IO} = g$ on $\partial\Omega \cap \partial S_j$. From Lemma 7.4.7, we also see that

$$\|L_\varepsilon u_{IO}\|_{L^\infty(S_j^+ \cup S_j^-)} \leq C\left\{e^{-\alpha'/\varepsilon} + (\varepsilon(2M+2)\gamma(1+\gamma_f))^{2M+2}\right\}.$$

The representation (7.4.31) shows that u_{IO} is discontinuous across Γ'. The corner layer function $u^{CL}_{j,M}$ is now chosen such that this discontinuity (and the jump of the co-normal derivative) is corrected. The jumps are

$$h_1 := \chi^{BL}_{j+1} u^{BL}_{j+1,M} \circ \psi^{-1}_{j+1} - \chi^{BL}_j u^{BL}_{j,M} \circ \psi^{-1}_j \qquad \text{on } \Gamma', \tag{7.4.32a}$$

$$\begin{aligned} h_2 := {} & \varepsilon(\mathbf{n}^+)^T A \nabla(\chi^{BL}_{j+1} u^{BL}_{j+1,M} \circ \psi^{-1}_{j+1}) \\ & + \varepsilon(\mathbf{n}^-)^T A \nabla(\chi^{BL}_j u^{BL}_{j,M} \circ \psi^{-1}_j) \quad \text{on } \Gamma', \end{aligned} \tag{7.4.32b}$$

where $\mathbf{n}^+$, $\mathbf{n}^-$ represent the outer normal vector of S_j^+, S_j^- on the curve Γ'. We note that the continuity of g and w_M at the vertex A_j implies that $h_1(A_j) = 0$. Furthermore, because for some $\kappa > 0$ the cut-off functions χ^{BL}_j, χ^{BL}_{j+1} are identically one on $B_\kappa(A_j) \cap \Gamma'$, the data h_1, h_2 are analytic on $B_\kappa(A_j) \cap \Gamma'$. In fact, one can check using the properties of the functions $u^{BL}_{j,M}$, $u^{BL}_{j+1,M}$ and Lemma 4.3.3 that (6.4.20) hold. Thus Theorem 6.4.13 is applicable (with $g_1 = g_2 = 0$ and h_1, h_2 given by (7.4.32)) and yields the existence of $u^{CL}_{j,M}$ such that

$$L_\varepsilon u^{CL}_{j,M} \equiv 0 \qquad \text{on } S_j^+ \cup S_j^-, \tag{7.4.33a}$$

$$u^{CL}_{j,M} = 0 \qquad \text{on } \partial S_j \cap \partial\Omega, \tag{7.4.33b}$$

$$[u^{CL}_{j,M}] = h_1 \qquad \text{on } \Gamma', \tag{7.4.33c}$$

$$\varepsilon[\partial_{n_A} u^{CL}_{j,M}] = h_2 \qquad \text{on } \Gamma', \tag{7.4.33d}$$

together with the following *a priori* estimates for all $p \in \mathbb{N}_0$:

$$\|\Psi_{0,0,\varepsilon,\alpha'} u^{CL}_{j,M}\|_{L^2(S_j^+ \cup S_j^-)} + \varepsilon\|\Psi_{0,0,\varepsilon,\alpha'} \nabla u^{CL}_{j,M}\|_{L^2(S_j^+ \cup S_j^-)} \leq C\varepsilon, \tag{7.4.34}$$

$$\|\Psi_{p,\beta,\varepsilon,\alpha'} \nabla^{p+2} u^{CL}_{j,M}\|_{L^2((S_j^+ \cup S_j^-) \cap B_{R'}(A_j))} \leq C\varepsilon \max\{p, \varepsilon^{-1}\}^{p+2} \tag{7.4.35}$$

for some constants C, γ, R', $\alpha' > 0$ independent of ε. Thus, the function $u^{CL}_{j,M}$ satisfies the desired bounds.

We observe that on S_j, the function u_{IOC} of (7.4.23) takes the form

$$u_{IOC} = u_{IO} + \chi^{CL}_j u^{CL}_{j,M} = w_M + \chi^{BL}_j u^{BL}_{j,M} \circ \psi^{-1}_j + \chi^{BL}_{j+1,M} u^{BL}_{j+1,M} \circ \psi^{-1}_{j+1} + \chi^{CL}_j u^{CL}_{j,M}.$$

As $u^{CL}_{j,M} \equiv 0$ on $\partial S_j \cap \partial\Omega$, we have by construction of the functions w_M, $u^{BL}_{j,M}$, $u^{BL}_{j+1,M}$ that $u_{IOC} = g$ on $\partial S_j \cap \partial\Omega$. Thus, the correct Dirichlet boundary conditions are satisfied. By the assumptions on the cut-off functions χ^{BL}_j, χ^{BL}_{j+1} we have

$$\chi^{CL}_j \chi^{BL}_j = \chi^{BL}_j \quad \text{on } \Gamma', \qquad \chi^{CL}_j \chi^{BL}_{j+1} = \chi^{BL}_{j+1} \quad \text{on } \Gamma'.$$

We therefore obtain that u_{IOC} is smooth on $S_j^+ \cup S_j^-$ and satisfies

$$[u_{IOC}] = 0, \qquad [\partial_{n_A} u_{IOC}] = 0 \quad \text{on } \Gamma'.$$

By the smoothness of the coefficient matrix A, we conclude that $u_{IOC} \in C^1(S_j)$. Because additionally $u_{IOC} \in H^1(S_j^+ \cup S_j^-)$, we get

$$u_{IOC} \in H^1(S_j) \cap C^1(S_j) \cap C^\infty(S_j^+ \cup S_j^-) \subset H^1(S_j) \cap H^2_{loc}(S_j) \cap C^\infty(S_j^+ \cup S_j^-).$$

This allows us to verify that u_{IOC} is a weak solution to

$$L_\varepsilon u_{IOC} = \tilde{f} \in L^\infty(S_j) \qquad \text{on } S_j,$$
$$\tilde{f} := \begin{cases} L_\varepsilon u_{IOC} & \text{on } S_j^+ \\ L_\varepsilon u_{IOC} & \text{on } S_j^-. \end{cases}$$

Using (7.4.35) and the properties of χ_j^{CL}, we infer furthermore

$$\begin{aligned} \|L_\varepsilon u_{IOC} - f\|_{L^\infty(S_j)} &= \|\tilde{f} - f\|_{L^\infty(S_j)} \\ &\leq C\left\{e^{-\alpha'/\varepsilon} + (\varepsilon(2M+2)\gamma(1+\gamma_f))^{2M+2}\right\}. \end{aligned} \tag{7.4.36}$$

Remainder r_M and pointwise estimates. As u_{IOC} has now been defined, the remainder r_M may be defined through (7.4.6). In order to obtain an estimate for r_M, we start by observing that combining (7.4.22) with (7.4.36) implies

$$\|L_\varepsilon u_{IOC} - f\|_{L^\infty(\Omega)} \leq C\left\{e^{-\alpha'/\varepsilon} + (\varepsilon(2M+2)\gamma(1+\gamma_f))^{2M+2}\right\}. \tag{7.4.37}$$

We observe that the function u_{IOC} solves the following (weakly posed) problem:

$$L_\varepsilon r_M = L_\varepsilon(u_\varepsilon - u_{IOC}) = f_{r_M} = f - L_\varepsilon u_{IOC} \qquad \text{on } \Omega,$$
$$r_M = 0 \qquad \text{on } \partial\Omega,$$

where the right hand-side $f_{r_M} \in L^\infty(\Omega)$ satisfies (7.4.37). Thus, appealing to Theorem 5.3.8 gives the existence of $\beta' \in (0,1)^J$ and $C > 0$ such that

$$\|r_M\|_{H^{2,2}_{\beta',\varepsilon}(\Omega)} \leq C\left\{e^{-\alpha'/\varepsilon} + (\varepsilon(2M+2)\gamma(1+\gamma_f))^{2M+2}\right\},$$

allowing us to conclude the proof of Theorem 7.4.5 with the exception of the pointwise estimates for the corner layer functions $u_{j,M}^{CL}$. These are obtained from the L^2-based estimates using Corollary 6.2.8. The proof of Theorem 7.4.5 is now complete.

Appendix

A.1 Some technical lemmata

A.1.1 Transformations of elliptic equations

Lemma A.1.1. *Let $G \subset \mathbb{R}^n$ be open, $A = (a_{ij}(\cdot))_{i,j=1}^n \in L^\infty(G, \mathbb{S}^n)$, and $f \in L^2(G)$. Let $u \in H^1_{loc}(G)$ be a weak solution of*

$$-\nabla \cdot (A\nabla u) = f \qquad \text{on } G, \tag{A.1.1}$$

i.e.,

$$\int_G (A\nabla u) \cdot \nabla v \, dx = \int_G fv \qquad \forall v \in C_0^\infty(G). \tag{A.1.2}$$

Let $F : \hat{G} \to G$ be a bi-Lipschitzian mapping from the open set $\hat{G}$ to G. Then the function $\hat{u} := u \circ F \in H^1_{loc}(\hat{G})$ is a weak solution of

$$-\nabla \cdot \left((\det F')(F')^{-T} \hat{A}\, (F')^{-1} \nabla \hat{u} \right) = \hat{f} \det F' \qquad \text{on } \hat{G},$$

where $\hat{A} = A \circ F$, $\hat{f} = f \circ F$.

Proof: We have to show that

$$\int_{\hat{G}} \nabla \hat{u} \cdot \left((\det F')(F')^{-T} \hat{A}\, (F')^{-1} \right) \nabla \varphi \, dx = \int_{\hat{G}} \hat{f} \, \det F' \varphi \, dx \qquad \forall \varphi \in C_0^\infty(\hat{G}).$$

Equivalently, by transforming to the domain G (using the fact that F is bi-Lipschitzian and [136, Theorem 2.2.2]), we have to show that

$$\int_G (A\nabla u) \cdot \nabla \varphi \, dx = \int_G f\varphi \, dx \qquad \forall \varphi \in \widetilde{C}(G),$$

where $\widetilde{C}(G) = \{\varphi(x) = \hat{\varphi} \circ F^{-1}(x) \text{ for some } \hat{\varphi} \in C_0^\infty(\hat{G})\}$. Clearly, $\widetilde{C}(G) \subset C^1_{comp}(G)$, where $C^1_{comp}(G)$ denotes the set of all C^1-functions on G that have compact support. As $C_0^\infty(G)$ is dense (in the H^1-topology) in $C^1_{comp}(G)$, assumption (A.1.2) readily implies that

$$\int_G \nabla u \cdot A \nabla \varphi \, dx = \int_G f\varphi \, dx \qquad \forall \varphi \in C^1_{comp}(G).$$

This completes the argument. $\square$

Lemma A.1.2. *Let $R > 0$, $\omega \in (0, 2\pi)$, $A \in C^2(\overline{S_R(\omega)}, \mathbb{S}^2_>)$ and $0 < \lambda_{min} \le A$ on $S_R(\omega)$. Then there exists $C > 0$ such that for every $\delta \in (0,1)$ we can find a function $\chi_\delta \in C^2(\overline{S_R(\omega)})$ with the following properties:*

1. $\chi_\delta \equiv 1$ *on* $B_{\delta/2}(0) \cap S_R(\omega)$;
2. $\chi_\delta \equiv 0$ *on* $\cap S_R(\omega) \setminus B_\delta(0)$;
3. $|\nabla^j \chi_\delta| \le C\delta^{-j}$, $j \in \{0,1,2\}$;
4. $\partial_{n_A} \chi_\delta = 0$ *on* $\partial S_R(\omega) \setminus \partial B_R(0)$.

Proof: In polar coordinates (r, φ), the co-normal derivatives operator ∂_{n_A} takes the following form on each of the two lateral parts $\Gamma_1 \subset \{(r,0) \,|\, r > 0\}$, $\Gamma_2 \subset \{(r\cos\omega, r\sin\omega) \,|\, r > 0\}$ of the sector $S_R(\omega)$:

$$\partial_{n_A} v = a_i(r)\frac{1}{r}\partial_\varphi v + b_i \partial_r v,$$

where the functions a_i, b_i, $i \in \{1,2\}$, are C^2. Additionally, we may reason as in the proof of Lemma 5.5.21 to see $|a_i(r)| \ge a_0 > 0$, $i \in \{1,2\}$ for all $r \in (0, R)$. Choose a fixed function $\chi \in C^\infty(\mathbb{R})$ with the following properties

$$\chi \equiv 1 \text{ on } B_{1/2}(0), \qquad \chi \equiv 0 \text{ on } \mathbb{R} \setminus B_1(0).$$

We then define the cut-off function χ_δ in polar coordinates by

$$\chi_\delta(r,\varphi) = \chi(r/\delta) - \frac{r}{\delta}\frac{b_2(r)\chi'(r/\delta)}{a_2(r)}(\varphi/\omega)^2(\varphi - \omega) - \frac{r}{\delta}\frac{b_1(r)\chi'(r/\delta)}{a_1(r)}\varphi(1 - \varphi/\omega)^2.$$

A calculation shows that χ_δ has all the desired properties. □

A.1.2 Leibniz formulas

Lemma A.1.3. *Let U, V be C^∞ tensors. Then there holds*

$$|\nabla^p(U \cdot V)(x)| \le \sum_{q=0}^{p} \binom{p}{q} |\nabla^q U(x)||\nabla^{p-q}V(x)|.$$

Proof: The proof is an extended version of the proof of [98, Lemma 5.7.4]. By arranging the set of common indices and the set of the remaining indices into single sequences, we may assume that

$$U = \{U_{ij}\}, \qquad V = \{V_{ij}\}, \qquad UV = \{\omega_{ij}\}, \quad \omega_{ij} = \sum_k U_{ik}V_{jk}.$$

The claim for $p = 0$ is obvious. For $p = 1$, we calculate

$$\begin{aligned}|\nabla(\sum_j U_{ij}V_{jk})| &= \sqrt{\sum_{i,k,l} |\sum_j \partial_l U_{ij}V_{jk} + \sum_j U_{ij}\partial_l V_{jk}|^2} \\ &\le \sqrt{\sum_{i,k,l} |\sum_j \partial_l U_{ij}V_{jk}|^2} + \sqrt{\sum_{i,k,l} |\sum_j U_{ij}\partial_l V_{jk}|^2}.\end{aligned}$$

Now,

$$\sum_l \sum_{i,k} |\sum_j U_{ij} \partial_l V_{jk}|^2 \leq \sum_{i,k,l} \Big(\sum_j |U_{ij}|^2\Big)\Big(\sum_j |\partial_l V_{jk}|^2\Big)$$
$$= \Big(\sum_{i,j} |U_{ij}|^2\Big)\Big(\sum_{l,k,j} |\partial_l V_{jk}|^2\Big) = |U|^2\,|\nabla V|^2.$$

Completely analogously, we obtain $\sum_{i,k,l} |\sum_j \partial_l U_{ij} V_{jk}|^2 \leq |\nabla U|^2\,|V|^2$. This proves the case $p = 1$. In order to prove the claim for $p > 1$, we first note that

$$|\nabla^{p+1} U| = |\nabla^p (\nabla U)|.$$

We are now in position to prove the claim by induction on p. Suppose, the claim of the lemma holds for p. We then estimate

$$\begin{aligned}
|\nabla^{p+1} UV| &= |\nabla^p(\nabla(UV))| \leq |\nabla^p((\nabla U)V)| + |\nabla^p(U\nabla V)| \\
&\leq \sum_{q=0}^{p} \binom{p}{q} |\nabla^{q+1} U|\,|\nabla^{p-q} V| + \sum_{q=0}^{p} \binom{p}{q} |\nabla^q U|\,|\nabla^{p+1-q} V| \\
&= |\nabla^{p+1} U|\,|V| + \sum_{q=0}^{p-1} \binom{p}{q} |\nabla^{q+1} U|\,|\nabla^{p-q} V| + |U|\,|\nabla^{p+1} V| \\
&\quad + \sum_{q=1}^{p} \binom{p}{q} |\nabla^q U|\,|\nabla^{p+1-q} V| \\
&= |\nabla^{p+1} U|\,|V| + |U|\,|\nabla^{p+1} V| + \sum_{q=1}^{p} \binom{p}{q-1} |\nabla^q U|\,|\nabla^{p+1-q} V| \\
&\quad + \sum_{q=1}^{p} \binom{p}{q} |\nabla^q U|\,|\nabla^{p+1-q} V| \\
&= \sum_{q=0}^{p+1} \binom{p+1}{q} |\nabla^q U|\,|\nabla^{p+1-q} V|.
\end{aligned}$$

□

Lemma A.1.4. *Let U, V be C^∞ tensor. Then for all $p \in \mathbb{N}_0$ there holds*

$$\sum_{|\alpha|=p} \frac{|\alpha|!}{\alpha!} |D^\alpha(UV)(x) - (UD^\alpha V)(x)|^2 \leq \left(\sum_{q=1}^{p} \binom{p}{q} |\nabla^q U(x)|\,|\nabla^{p-q} V(x)|\right)^2.$$

Proof: Without loss of generality, the point x may be taken as $x = 0$. We will also drop the explicit dependence on x for the remainder of the proof. Furthermore, for simplicity of notation, we will also assume that U and V are scalar functions. We start by noting that by Lemma A.1.3 there holds

$$\sum_{|\alpha|=p} \frac{|\alpha|!}{\alpha!} |D^\alpha(UV)|^2 = |\nabla^p(UV)|^2 \le \left(\sum_{q=1}^{p} \binom{p}{q} |\nabla^q U| \, |\nabla^{p-q} V| \right)^2 \tag{A.1.3}$$

provided that $\nabla^p V = 0$ at the fixed point $x = 0$. Next, introduce the auxiliary function

$$H(z) := \sum_{|\alpha|=p} \frac{D^\alpha V(0)}{\alpha!} z^\alpha.$$

It satisfies

$$D^\beta H(0) = \begin{cases} 0 & \text{if } |\beta| < p \\ D^\beta V(0) & \text{if } |\beta| = p. \end{cases}$$

From (A.1.3) we get

$$\begin{aligned} |\nabla^p\Big(U(V-H)\Big)|^2 &\le \left(\sum_{q=1}^{p} \binom{p}{q} |\nabla^q U| \, |\nabla^{p-q}(V-H)| \right)^2 \\ &= \left(\sum_{q=1}^{p} \binom{p}{q} |\nabla^q U| \, |\nabla^{p-q} V| \right)^2. \end{aligned} \tag{A.1.4}$$

Next, for $|\alpha| = p$, we calculate

$$\begin{aligned} D^\alpha\Big(U(V-H)\Big) &= \sum_{0 \le \beta \le \alpha} \binom{\alpha}{\beta} D^\beta U D^{\alpha-\beta}(V-H) = \sum_{0 \ne \beta \le \alpha} \binom{\alpha}{\beta} D^\beta U D^{\alpha-\beta} V \\ &= D^\alpha(UV) - U D^\alpha V. \end{aligned} \tag{A.1.5}$$

Combining (A.1.4), (A.1.5), we arrive at

$$\begin{aligned} \sum_{|\alpha|=p} \frac{|\alpha|!}{\alpha!} |D^\alpha(UV) - U D^\alpha V|^2 &= \sum_{|\alpha|=p} \frac{|\alpha|!}{\alpha!} |D^\alpha\Big(U(V-H)\Big)|^2 \\ &= |\nabla^p\Big(U(V-H)\Big)|^2 \le \left(\sum_{q=1}^{p} \binom{p}{q} |\nabla^q U| \, |\nabla^{p-q} V| \right)^2, \end{aligned}$$

which proves the lemma. □

Lemma A.1.5. *Let U, V be C^∞ tensor fields defined on $G \subset \mathbb{R}^n$. Denote by ∇_x differentiation with respect to the variables $x_1, \ldots, x_{n-1}$. Then there holds for all $(x, x_n) \in G$, p, $q \in \mathbb{N}_0$:*

$$|\partial_{x_n}^q \nabla_x^p \nabla^p (U \cdot V)| \le \sum_{r=0}^{p} \sum_{s=0}^{q} \binom{p}{r} \binom{q}{s} |\partial_{x_n}^s \nabla_x^r U| |\partial_{x_n}^{q-s} \nabla_x^{p-r} V|.$$

Proof: The proof is completely analogous to that of Lemma A.1.3 by induction on q. The case $q = 0$ is handled by Lemma A.1.3. □

A.1.3 Hardy inequalities

We have the following well-known Hardy inequality in one dimension:

Lemma A.1.6. *Let $\beta < 1/2$ and $x_0 \in (0,1)$. Then there exists $C > 0$ depending only on β and x_0 such that*

$$\int_0^1 x^{-2\beta} u^2(x)\, dx \le C \left[\int_0^1 x^{2(1-\beta)} |u'(x)|^2\, dx + \int_{x_0}^1 u^2(x)\, dx \right]$$

for all functions u such that the right-hand side is finite.

Proof: The proof follows from the Hardy inequality [68, Thm. 330]. □

This one-dimensional result can be extended to sectors:

Lemma A.1.7. *Let $\beta \in [0,1)$, $T = \{(x,y)\,|\,0 < x < 1,\ 0 < y < 1-x\}$, and $T' \subset T$. Then there exists $C > 0$ depending only on β and T' such that*

$$\|r^{\beta-1}u\|_{L^2(T)} \le C\left[\|r^\beta \nabla u\|_{L^2(T)} + \|u\|_{L^2(T')}\right], \qquad r(x) := \operatorname{dist}(x,(0,0)),$$

for all functions u such that the right-hand side is finite.

Proof: The result follows easily from Lemma A.1.6 if all integrals are expressed in polar coordinates. □

Lemma A.1.8. *Let $S_R(\omega) = \{(r\cos\varphi, r\sin\varphi)\,|\,r \in (0,R),\ \varphi \in (0,\omega)\}$,$\omega > 0$ be a sector. Let $\omega' \in (0,\omega)$ and let $\Gamma := \{(r\cos\omega', r\sin\omega')\,|\,r \in (0,R)\}$ divide $S_R(\omega)$ into two sectors S^+, S^-. Then there exists $C > 0$ such that*

$$\left|\int_\Gamma Hv\, ds\right| \le C\|H\|_{H^{1,1}_{1,1}(S^+)} \|v\|_{H^1(S_R(\omega))} \quad \forall v \in H^1_0(S_R(\omega)), \quad H \in H^{1,1}_{1,1}(S^+),$$

where the norm $\|\cdot\|_{H^{1,1}_{1,1}(S^+)}$ is given by $\|H\|_{L^2(S^+)} + \|\,|x|\nabla H\|_{L^2(S^+)}$.

Proof: An analogous result for $v \in H^1(S_R(\omega))$ is given in Lemma 6.4.1. It is the Dirichlet boundary conditions imposed on v that allow us to take $\beta = 1$. We may restrict our attention to a neighborhood of the origin. Then, using smooth changes of variables, it is easy to see that it suffices to show the following analogous bound on a square:

$$\left|\int_0^1 H(x,0)v(x,0)\, dx\right| \le C\|H\|_{H^{1,1}_{1,1}(S)}\|v\|_{H^1(S)} \quad \forall v \in H^1_{(0)}(S), \qquad \text{(A.1.6)}$$

where $S = (0,1)^2$, $H^1_{(0)}(S) = \{v \in H^1(S)\,|\,v(x,\cdot) = 0, x = 0 \text{ or } x = 1\}$, and $\|H\|_{H^{1,1}_{1,1}(S)} = \|H\|_{L^2(S)} + \|\,|x|\nabla H\|_{L^2(S)}$. We define on the interval $(0,1)$ the weight function $d(x) := x(1-x)$. The assumption $v \in H^1_{(0)}(S)$ implies

$$\int_0^1 \frac{1}{d(x)} v^2(x,0)\,dx \le C\|v(\cdot,0)\|^2_{H^{1/2}_{00}((0,1))} \le C\|v\|^2_{H^1(S)}. \tag{A.1.7}$$

Next, a calculation shows that the function $\tilde{H} : (x,y) \mapsto d(x)H(x,y)$ is also in $H^1_{(0)}(S)$ and satisfies

$$\int_0^1 \frac{\tilde{H}^2(x,0)}{d(x)}\,dx \le C\|\tilde{H}(\cdot,0)\|^2_{H^{1/2}_{00}((0,1))} \le C\|\tilde{H}\|^2_{H^1(S)} \le C\|H\|_{H^{1,1}_{1,1}(S)}. \tag{A.1.8}$$

Hence, since $\sqrt{d(x)}H(x,0) = 1/\sqrt{d(x)}\tilde{H}(x,0)$, we get

$$\begin{aligned}\left|\int_0^1 H(\cdot,0)v(\cdot,0)\,dx\right| &= \left|\int_0^1 d^{1/2}H(\cdot,0)d^{-1/2}v(\cdot,0)\,dx\right| \\ &\le \|d^{1/2}H(\cdot,0)\|_{L^2((0,1))}\,\|d^{-1/2}v(\cdot,0)\|_{L^2((0,1))} \le C\|H\|_{H^{1,1}_{1,1}(S)}\|v\|_{H^1(S)}.\end{aligned}$$

□

A.2 Kondrat'ev's theory for a special transmission problem

A.2.1 Problem formulation and notation

The present section is devoted to the regularity analysis of a transmission problem in infinite sectors. Let $-2\pi < \omega_1 < 0 < \omega_2 < 2\pi$ be given and define the two infinite sectors

$$\begin{aligned} S_1 &:= \{(r\cos\varphi, r\sin\varphi)\,|\,0<r<\infty, \omega_1<\varphi<0\}, \\ S_2 &:= \{(r\cos\varphi, r\sin\varphi)\,|\,0<r<\infty, 0<\varphi<\omega_2\}. \end{aligned}$$

We denote by Γ_1, Γ_2, Γ the three rays $\Gamma_i = \{(r\cos\omega_i, r\sin\omega_i)\,|\,0<r<\infty\}$, $\Gamma = \{(r,0)\,|\,0<r<\infty\}$. For fixed $p_1, p_2 > 0$ we consider the following transmission problem:

$$-p_i\Delta u_i = f_i \quad \text{on } S_i, \quad i \in \{1,2\}, \tag{A.2.1a}$$
$$u_i = 0 \quad \text{on } \Gamma_i, \quad i \in \{1,2\}, \tag{A.2.1b}$$
$$u_1 = u_2 \quad \text{on } \Gamma, \tag{A.2.1c}$$
$$p_1\partial_n u_1 - p_2\partial_n u_2 = G|_\Gamma \quad \text{on } \Gamma, \tag{A.2.1d}$$

where the data f_i and G are assumed to satisfy $f_i \in H^{0,0}_{\beta,1}(S_i)$, $G \in H^{1,1}_{\beta,1}(S_1)$ for some $\beta \in [0,1)$. We point out that $|\omega_1|+|\omega_2|$ may be larger than 2π. In order to clarify the notion of solutions, we introduce the space V^1_0 of pairs of functions defined on $S_1 \times S_2$ as follows:

$$V^1_0 := \{(u_1,u_2)\,|\,u_i \in H^1(S_i \cap B_R(0))\ \forall R>0,\ u_i = 0 \text{ on } \Gamma_i \text{ and } u_1 = u_2 \text{ on } \Gamma\}.$$

Solutions of (A.2.1) are understood in a weak sense, i.e., a pair $(u_1, u_2) \in V_0^1$ is a solution of (A.2.1) if

$$\int_{S_1} p_1 \nabla u_1 \nabla v_1 \, dx + \int_{S_2} p_2 \nabla u_2 \nabla v_2 \, dx = \int_{S_1} f_1 v_1 \, dx + \int_{S_2} f_2 v_2 \, dx + \int_{\Gamma} G v_1$$

for all $(v_1, v_2) \in V^1_{0,comp}$, where

$$V^1_{0,comp} = \{(v_1, v_2) \in V_0^1 \mid v_i = 0 \text{ on } S_i \setminus B_R(0) \text{ for some } R > 0\}.$$

We can now formulate our main result:

Proposition A.2.1. *There exist $\beta \in [0,1)$ and $C > 0$ depending only on the parameters ω_i and p_i such that the following regularity assertion holds: Let $f_i \in H^{0,0}_{\beta,1}(S_i)$ and $G \in H^{1,1}_{\beta,1}(S_1)$ with $f_i = 0$ on $S_i \setminus B_1(0)$ and $G = 0$ on $S_1 \setminus B_1(0)$. Assume that a pair $(u_1, u_2) \in V_0^1$ is a solution of (A.2.1) and satisfies $u_i = 0$ on $S_i \setminus B_1(0)$. Then*

$$\sum_{i=1}^{2} \|u_i\|_{H^{2,2}_{\beta,1}(S_i)} \le C \left[\sum_{i=1}^{2} \|f_i\|_{H^{0,0}_{\beta,1}(S_i)} + \|G\|_{H^{1,1}_{\beta,1}(S_1)} \right].$$

Similar results have been proved and announced in [64, 101, 102]. In the ensuing subsection, we present an outline of a proof that follows closely the argument given in [15], where such a result was proved for a single sector with Dirichlet, Neumann, or mixed boundary conditions. The actual proof is lengthy and therefore relegated to the ensuing subsection. Essentially, the proof consists in two steps: In a first step, a solution $(w_1, w_2) \in W^2_{\beta,0}$ ($W^2_{\beta,0}$ will be defined shortly) of (A.2.1) is constructed by the classical techniques of Kondrat'ev. In the second step, it is ascertained that (w_1, w_2) in fact coincides with the given solution (u_1, u_2).

A.2.2 Proof of Proposition A.2.1

We will follow closely [15] and use their notation as much as possible. The Kondrat'ev spaces (cf., e.g., [39, 79]) are defined for $k \in \mathbb{N}_0$ and $\beta \ge 0$ by

$$W^k_\beta(S_i) := \{u \in L^2_{loc}(S_i) \mid \sum_{0 \le p \le k} \|r^{p-k+\beta} \nabla^p u\|_{L^2(S_i)} < \infty\}.$$

For $\beta \in [0,1)$ and $k = 2$ we define furthermore

$$W^2_{\beta,0} := \{(w_1, w_2) \in W^2_\beta(S_1) \times W^2_\beta(S_2) \mid w_i = 0 \text{ on } \Gamma_i \text{ and } w_1 = w_2 \text{ on } \Gamma\}.$$

Next, we set

$$\begin{aligned} D_1 &= \{(\tau, \theta) \mid \tau \in \mathbb{R}, \quad \theta \in (\omega_1, 0)\}, \\ D_2 &= \{(\tau, \theta) \mid \tau \in \mathbb{R}, \quad \theta \in (0, \omega_2)\}, \\ D &= \{(\tau, \theta) \mid \tau \in \mathbb{R}, \quad \theta \in (\omega_1, \omega_2)\}, \end{aligned}$$

and for $h \geq 0$ and $k \in \mathbb{N}_0$, we set

$$\mathcal{H}_h^k(D) = \{u \mid \int_D e^{2h\tau} \sum_{p=0}^{k} |\nabla^p u|^2 \, d\tau d\theta =: \|u\|^2_{\mathcal{H}_h^k(D)} < \infty\}$$

with analogous definitions for $\mathcal{H}_h^k(D_i)$, $i = 1, 2$. We use the notation $\mathcal{L}_h(D) := \mathcal{H}_h^0(D)$ for the special case $k = 0$. We write the transmission problem (A.2.1) in polar coordinates (r, φ) and then introduce the new variable

$$\tau = \ln(1/r)$$

to arrive at the following transmission problem on the strip D

$$-p_i \left(\partial_\tau^2 \tilde{u}_i + \partial_\theta^2 \tilde{u}_i\right) = \tilde{f}_i(\tau, \theta), \qquad i \in \{1, 2\}, \tag{A.2.2a}$$

$$\tilde{u}_i|_{\theta=\omega_i} = 0, \qquad i \in \{1, 2\}, \tag{A.2.2b}$$

$$\tilde{u}_1|_{\theta=0} = \tilde{u}_2|_{\theta=0}, \tag{A.2.2c}$$

$$p_1 \partial_\theta \tilde{u}_1|_{\theta=0} - p_2 \partial_\theta \tilde{u}_2|_{\theta=0} = \tilde{G}|_{\theta=0}, \tag{A.2.2d}$$

where we abbreviated (thinking of the functions u_i, f_i, G as given in polar coordinates)

$$\tilde{u}_i(\tau, \theta) = u_i(e^{-\tau}, \theta), \tag{A.2.3a}$$

$$\tilde{f}_i(\tau, \theta) = e^{-2\tau} f_i(e^{-\tau}, \theta), \tag{A.2.3b}$$

$$\tilde{G}(\tau, \theta) = e^{-\tau} G(e^{-\tau}, \theta). \tag{A.2.3c}$$

We now turn to the question of solvability of (A.2.2) and that of uniqueness and regularity of the solutions. It is convenient to combine the two functions $\tilde{u}_1$, $\tilde{u}_2$ into a single function $\tilde{u}$ defined on D:

$$\tilde{u} := \begin{cases} \tilde{u}_1 & \text{on } D_1, \\ \tilde{u}_2 & \text{on } D_2 \end{cases}$$

In complete analogy to [15, Lemma 2.3], we have

Lemma A.2.2. *There exists $h_0 > 0$ depending only on p_i, ω_i such that for each $0 < h < h_0$ there exists $C > 0$ with the following properties: If $\tilde{f}_i \in \mathcal{L}_h(D_i)$ and $\tilde{G} \in \mathcal{H}_h^1(D_1)$, then there exists a unique solution $\tilde{u} \in \{v \in \mathcal{H}_h^1(D) \mid v|_{D_i} \in \mathcal{H}_h^2(D_i)\}$ of (A.2.2), which satisfies*

$$\|\tilde{u}\|^2_{\mathcal{H}_h^1(D)} + \sum_{i=1}^{2} \|\tilde{u}|_{D_i}\|^2_{\mathcal{H}_h^2(D_i)} \leq C \left[\sum_{i=1}^{2} \|\tilde{f}_i\|^2_{\mathcal{L}_h(D_i)} + \|\tilde{G}\|^2_{\mathcal{H}_h^1(D_1)}\right]. \tag{A.2.4}$$

Proof: We follow [15, Lemma 2.3] closely and may therefore be brief. We define on the strip D the piecewise constant function p by $p|_{D_i} = p_i$ and the function $\tilde{f}$ by $\tilde{f}|_{D_i} = f_i$.

By means of partial Fourier transform (in the τ variable), i.e.,

$$\hat{f}(\lambda,\theta) = 1/\sqrt{2\pi}\int_{-\infty}^{\infty} e^{-i\lambda\tau}\tilde{f}(\tau,\theta)\,d\tau,$$

and an analogous formula for $\hat{G}$, equation (A.2.2) is transformed into a system of parameter dependent transmission problems, and we get for $\lambda = \xi + ih$, $\xi \in \mathbb{R}$,

$$\begin{aligned} -p\partial_\theta^2\,\tilde{u} + \lambda^2 p\tilde{u} &= \hat{f}(\lambda,\cdot) \qquad \text{on } (\omega_1,0)\cup(0,\omega_2),\\ \tilde{u}(\lambda,\omega_i) &= 0, \qquad i\in\{1,2\},\\ \tilde{u}_1(\lambda,0) &= \tilde{u}_2(\lambda,0),\\ [p\partial_\theta\,\tilde{u}(\lambda,\cdot)] &= \hat{G}(\lambda,0) \qquad \text{on } \theta = 0. \end{aligned}$$

The next step is to obtain bounds for the solution $\tilde{u}$ that are explicit in the parameter λ. This transmission problem has the form considered in Lemma A.2.5 below, which provides the required bounds and in particular the existence of h_0. Proceeding analogously to the proof of [15, Lemma 2.3], we obtain the bound (A.2.4). The final step of the proof of Lemma 2.3 consist in the uniqueness assertion of the solution $\tilde{u}$. In the present situation of a transmission problem, we therefore have to show that a function $\tilde{u} \in \{v \in \mathcal{H}_h^1(D)\,|\,v|_{D_i} \in \mathcal{H}_h^2(D_i)\}$ satisfying

$$\begin{aligned} -p_i\Delta\tilde{u} &= 0 \qquad \text{on } D_i,\\ \tilde{u}_i|_{\theta=\omega_i} &= 0 \qquad i\in\{1,2\},\\ [p\partial_n\,\tilde{u}] &= 0 \qquad \text{on } \theta = 0 \end{aligned}$$

vanishes identically. We note that $\tilde{u}$ has to be piecewise smooth by the local regularity assertions Propositions 5.5.1, 5.5.2, 5.5.4. Let now $(\varphi_j,\mu_j) \in H_0^1(I)\times\mathbb{C}$, $j\in\mathbb{N}$, be the eigenfunctions and eigenvalues μ_j of (A.2.12) in the proof of Lemma A.2.5. The eigenfunctions φ_j are taken orthonormal with respect to the inner product $(u,v)_p = \int_{\omega_1}^{\omega_2} puv\,dx$. The solution $\tilde{u}$ may be expanded as

$$\tilde{u}(\tau,\theta) = \sum_{j=1}^{\infty} a_j(\tau)\varphi_j(\theta), \tag{A.2.5}$$

$$a_j(\tau) = \int_I p\tilde{u}(\tau,\theta)\varphi_j(\theta)\,d\theta. \tag{A.2.6}$$

Hence,

$$a_j''(\tau) = \int_I p\partial_\tau^2\,\tilde{u}(\tau,\theta)\varphi_j(\theta)\,d\theta = -\mu_j^2 a_j(\tau).$$

Thus, we conclude that each a_j satisfies the differential equation

$$a_j''(\tau) + \mu_j^2 a_j(\tau) = 0.$$

Recalling that the eigenvalues μ_j are purely imaginary, i.e., $\mu_j = \mathbf{i}\,\tilde{\mu}_j$ with $\tilde{\mu}_j > 0$ and that $\tilde{\mu}_j \to \infty$ as $j\to\infty$, we see that the coefficients $a_j(\tau)$ have the form

$$a_j(\tau) = c_j e^{-\tilde{\mu}_j\tau} + d_j e^{\tilde{\mu}_j\tau} \tag{A.2.7}$$

for some constants c_j, d_j. As in [15], we now get for $A > 0$ and $D_A := \{(\tau,\theta) \mid -A < \tau < A, \theta \in (\omega_1, \omega_2)\}$

$$\int_{D_A} e^{2h\tau} |\tilde{u}|^2 \, d\tau d\theta \sim \int_{-A}^{A} \int_{\omega_1}^{\omega_2} p \left\{ \sum_{j=1}^{\infty} a_j(\tau) \varphi_j(\theta) \right\}^2 e^{2h\tau} \, d\tau d\theta \quad \text{(A.2.8)}$$

$$= \sum_{j=1}^{\infty} \int_{-A}^{A} e^{2h\tau} |a_j(\tau)|^2 \, d\tau. \quad \text{(A.2.9)}$$

As $\tilde{u} \in \mathcal{H}_h^1(D)$ the left-hand side of (A.2.8) is finite, and it can be seen from this as in [15] that the finiteness of each term in the sum in (A.2.9) implies $c_j = d_j = 0$ for all j. This proves uniqueness and completes the proof of the lemma. □

[15, Lemma 2.4] also carries over almost verbatim:

Lemma A.2.3. *Let the assumptions of Lemma A.2.2 hold. Let in addition* $\tilde{f}_i(\tau,\theta) = 0$, $\tilde{G}(\tau,\theta) = 0$ *for* $\tau < 0$. *Then for* $\varepsilon > 0$ *and* $0 \leq \gamma = h + h_0 - \varepsilon$, $\tilde{D} := \{(\tau,\theta) \mid \tau < 0, \omega_1 < \theta < \omega_2\}$, $\tilde{D}_1 := \{(\tau,\theta) \mid \tau < 0, \omega_1 < \theta < 0\}$, $\tilde{D}_2 := \{(\tau,\theta) \mid \tau < 0, 0 < \theta < \omega_2\}$ *we have for* $|\alpha| \leq 2$ *with* $\alpha \neq (0,2)$

$$\int_{\tilde{D}} |\partial_\tau^{\alpha_1} \partial_\theta^{\alpha_2} \tilde{u}|^2 e^{2(h-\gamma)\tau} \, d\tau d\theta \leq C(\varepsilon) \int_{\tilde{D}} |\partial_\tau^{\alpha_1} \partial_\theta^{\alpha_2} \tilde{u}|^2 e^{2h\tau} d\tau d\theta,$$

$$\int_{\tilde{D}_1 \cup \tilde{D}_2} |\partial_\theta^2 \tilde{u}|^2 e^{2(h-\gamma)\tau} \, d\tau d\theta \leq C(\varepsilon) \int_{\tilde{D}_1 \cup \tilde{D}_2} |\partial_\theta^2 \tilde{u}|^2 e^{2h\tau} d\tau d\theta.$$

Proof: The proof is almost identical to the proof of [15, Lemma 2.4]. One uses the expansion (A.2.5), (A.2.6) and shows that the assumptions $\tilde{f}_i = 0$, $\tilde{G} = 0$ for $\tau < 0$ imply that the coefficients $c_j = 0$ for all $j \in \mathbb{N}$. The result then follows by elementary considerations. A slight difference to the procedure in [15, Lemma 2.4] is that the eigenfunctions φ_j are only piecewise smooth. This forces us to split $\tilde{D}$ into $\tilde{D}_1$ and $\tilde{D}_2$ when considering $\partial_\theta^2 \tilde{u}$. □

We now proceed with the analog of [15, Lemma 2.6].

Lemma A.2.4. *Let* h_0 *be as in Lemma A.2.2 and let* $\beta \in (0,1)$ *satisfy* $\beta > 1 - h_0$. *Then there exists* $C > 0$ *depending only on* ω_i, p_i, *and* β *such that for all* $f_i \in H_{\beta,1}^{0,0}(S_i)$, $G \in H_{\beta,1}^{1,1}(S_1)$ *with* $f_i = 0$ *on* $S_i \setminus B_1(0)$, $G = 0$ *on* $S_1 \setminus B_1(0)$, *there exists a solution* w *of (A.2.1) with*

(i) $w \in W_{\beta,0}^2$, $w_i \in H_{\beta,1}^{2,2}(S_i \cap B_1(0))$,

(ii) $\sum_{i=1}^2 \|\nabla w_i\|_{L^2(S_i)} < \infty$,

(iii) $$\sum_{i=1}^{2} \|w_i\|_{H_{\beta,1}^{2,2}(S_i \cap B_1(0))} \leq C \left[\sum_{i=1}^{2} \|f_i\|_{H_{\beta,1}^{0,0}(S_i)} + \|G\|_{H_{\beta,1}^{1,1}(S_1)} \right].$$

Proof: Follows by imitating all the steps of [15, Lemma 2.6]. In essence, it is checked that the transforms $\tilde{f}_i$, $\tilde{G}$ as defined in (A.2.3) do satisfy the assumptions of Lemma A.2.2. The pair (w_1, w_2) given by Lemma A.2.2 then satisfies (i) and (iii). In order to see that (ii) also holds, we have to check that $||\nabla w_i||_{L^2(S_i \setminus B_1(0))} < \infty$. This is done with the aid of Lemma A.2.3. □

Proof of Proposition A.2.1: This is essentially [15, Lemma 2.8]. Let $(u_1, u_2) \in V_0^1$ be a solution of (A.2.1) and assume that $u_i = 0$ on $S_i \setminus B_1(0)$. Let $(w_1, w_2) \in W^2_{\beta,0}$ be the solution of (A.2.1) given by Lemma A.2.4. It suffices to show $u_i = w_i$. We start by noting that the difference $(u_1 - w_1, u_2 - w_2) \in V_0^1$ and

$$\sum_{i=1}^{2} \int_{S_i} p_i \nabla(u_i - w_i) \cdot \nabla v_i \, dx = 0 \qquad \forall v_i \in V^1_{0,comp}. \tag{A.2.10}$$

Introduce $\widetilde{H}_0^1(S_i) = \{v \,|\, ||\nabla v||_{L^2(S_i)} < \infty, \quad v = 0 \text{ on } \Gamma_i \text{ and } \Gamma\}$. By our assumptions on u_i we have $\nabla u_i \in L^2(S_i)$. Lemma A.2.4 also implies that $\nabla w_i \in L^2(S_i)$. Hence, as $u_1 = u_2$ on Γ and $w_1 = w_2$ on Γ, we obtain that $u_i - w_i \in \widetilde{H}_0^1(S_i)$. Finally, as

$$\{(v_1, v_2) \in V_0^1 \,|\, v_1 = 0 = v_2 \text{ on } \Gamma\}$$

is dense in $\widetilde{H}_0^1(S_1) \times \widetilde{H}_0^1(S_2)$, we may choose $(v_1, v_2) = (u_1 - w_1, u_2 - w_2)$ in (A.2.10) to conclude that $\nabla(u_i - w_i) = 0$ on S_i. The boundary conditions now imply that $u_i = w_i$. □

Lemma A.2.5. *Let $-\infty < \omega_1 < 0 < \omega_2 < \infty$ (in particular, $|\omega_1|$, $|\omega_2|$ may be bigger than 2π) and set $I_1 := (\omega_1, 0)$, $I_2 := (0, \omega_2)$, $I := (\omega_1, \omega_2)$. Define the piecewise constant function p by $p|_{I_i} := p_i > 0$, $i \in \{1, 2\}$. Then there exist h_0, $C > 0$ such that for all $\lambda \in \Lambda := \{\lambda \in \mathbb{C} |\ |\operatorname{Im} \lambda| < h_0\}$ the transmission problem: Find $u \in H_0^1(I)$ such that*

$$-(pu')' + p\lambda^2 u = f \in L^2(I) \qquad \text{on } I, \qquad [pu'] = g \in \mathbb{C} \qquad \text{at } x = 0,$$

has a unique solution satisfying

$$||u||^2_{H^2(I_1 \cup I_2)} + (1+|\lambda|^2)||u'||^2_{L^2(I)} + (1+|\lambda|^4)||u||^2_{L^2(I)} \leq C\left[||f||^2_{L^2(I)} + (1+|\lambda|)|g|^2\right].$$

Proof: Introduce the weighted L^2-inner product $(u, v)_p := \int_I puv \, dx$. The weak formulation of our problem then reads: Find $u \in H_0^1(I)$ such that

$$(u', v')_p + \lambda^2 (u, v)_p = (f/p, v)_p + gv(0) \qquad \forall v \in H_0^1(I). \tag{A.2.11}$$

Consider the eigenvalue problem

$$-(pu')' + p\mu^2 u = 0 \qquad \text{on } I, \qquad [pu'] = 0 \qquad \text{at } x = 0. \tag{A.2.12}$$

By the spectral theorem, there is a countable sequence $(\varphi_j, \mu_j)_{j\in\mathbb{N}} \subset H_0^1(I) \times \mathbb{C}$ such that the functions φ_j form an orthonormal basis of $L^2(I)$ (with respect to $(\cdot,\cdot)_p$) and an orthogonal basis of $H_0^1(I)$ (equipped with the inner product $(u', v')_p$). Furthermore, the eigenvalues μ_j are purely imaginary, i.e., they are of the form $\mu_j = \mathbf{i}\tilde{\mu}_j$ with $\tilde{\mu}_j \to \infty$. Without loss of generality, we may therefore assume that the $\tilde{\mu}_j$ are sorted in ascending order, that is, $\tilde{\mu}_{j+1} \geq \tilde{\mu}_j$ for all $j \in \mathbb{N}$. Finally, $\tilde{\mu}_1 > 0$ implies

$$(u', u')_p \geq \tilde{\mu}_1^2 (u, u)_p \qquad \forall u \in H_0^1(I).$$

We now choose $h_0 \in (0, \tilde{\mu}_1)$ and set $\Lambda := \{\lambda \in \mathbb{C} \mid |\mathrm{Im}\,\lambda| < h_0\}$. Hence, for all $\lambda \in \Lambda$ there holds $\mathrm{Re}\,\lambda^2 > -h_0^2 > -\tilde{\mu}_1^2$ and thus the bilinear form generated by the left-hand side of (A.2.11) is coercive on $H_0^1(I) \times H_0^1(I)$ for $\lambda \in \Lambda$; in particular the coercivity constant is independent of $\lambda \in \Lambda$. It is now easy to see that the desired bound holds for $\lambda \in \Lambda$ bounded. We will therefore restrict our attention to the case $\lambda \in \Lambda$, $|\lambda| \to \infty$.

We consider the case $g = 0$ first. We write $f/p = \sum_{j\in\mathbb{N}} f_j \varphi_j$ with $f_j = (f/p, \varphi_j)_p$ and note that there exists $C > 0$ such that

$$\sum_{j\in\mathbb{N}} |f_j|^2 \leq C\|f\|^2_{L^2(I)}.$$

The solution of (A.2.11) can be represented in the form $u = \sum_{j\in\mathbb{N}} u_j \varphi_j$ where

$$u_j = -\frac{f_j}{\mu_j^2 - \lambda^2}.$$

Thus, we get

$$|\lambda|^4\, \|u\|^2_{L^2(I)} \leq C|\lambda|^4 (u, u)_p = C|\lambda|^4 \sum_{j\in\mathbb{N}} |u_j|^2 = C \sum_{j\in\mathbb{N}} |f_j|^2 \frac{|\lambda|^4}{|\mu_j^2 - \lambda^2|^2},$$

$$|\lambda|^2\, \|u'\|^2_{L^2(I)} \leq C|\lambda|^2 (u', u')_p = C|\lambda|^2 \sum_{j\in\mathbb{N}} |u_j|^2 |\mu_j|^2 = C \sum_{j\in\mathbb{N}} |f_j|^2 \frac{|\lambda|^2 |\mu_j|^2}{|\mu_j^2 - \lambda^2|^2}$$

$$\leq C \sum_{j\in\mathbb{N}} |f_j|^2 \frac{|\lambda|^4 + |\mu_j|^4}{|\mu_j^2 - \lambda^2|^2}.$$

Elementary considerations show

$$\sup_{j\in\mathbb{N}} \sup_{\lambda\in\Lambda} \frac{|\lambda|^4}{|\mu_j^2 - \lambda^2|^2} < \infty, \qquad \sup_{j\in\mathbb{N}} \sup_{\lambda\in\Lambda} \frac{|\mu_j|^4}{|\mu_j^2 - \lambda^2|^2} < \infty.$$

The desired bounds for $|\lambda|^4 \|u\|^2_{L^2(I)}$ and $|\lambda|^2 \|u'\|^2_{L^2(I)}$ now follow. To see that the bound for $\|u\|_{H^2(I_1 \cup I_2)}$ also holds, we note that the differential equation is satisfied pointwise a.e. on $I_1 \cup I_2$. Hence, we get $\|u''\|_{L^2(I_1\cup I_2)} \leq C\left[\|f\|_{L^2(I)} + |\lambda|^2 \|u\|_{L^2(I)}\right]$.

It remains to consider the case $g \neq 0$ and $f = 0$. This is done by constructing a special function u_g satisfying the homogeneous boundary conditions, the differential equation with homogeneous right-hand side, and the correct jump condition at $x = 0$. We make the ansatz

$$u_g := \begin{cases} \alpha \sinh(\lambda(x - \omega_1)) & \text{for } x \in I_1, \\ \beta \sinh(\lambda(x - \omega_2)) & \text{for } x \in I_2. \end{cases}$$

The requirement that u_g be continuous at $x = 0$ and that the jump condition be satisfied yields two conditions for α, β. This solution is given by

$$\alpha = g\frac{\sinh(\lambda\omega_2)}{\lambda W}, \qquad \beta = g\frac{\sinh(\lambda\omega_1)}{\lambda W},$$
$$W = -\left[p_1 \sinh(\lambda\omega_2)\cosh(\lambda\omega_1) - p_2 \cosh(\lambda\omega_2)\sinh(\lambda\omega_1)\right].$$

Elementary consideration now show that for $|\lambda|$ large (implying that $|\operatorname{Re}\lambda|$ is large), u_g satisfies the desired bound. This concludes the proof. □

A.3 Stability properties of the Gauss-Lobatto interpolant

Theorem A.3.1. *Let $I = [-1, 1]$. Then there exists $C > 0$ such that for every $p \in \mathbb{N}$ the Gauss-Lobatto interpolation operator $i_p : C(I) \to \mathcal{P}_p$ satisfies*

$$\|i_p u\|_{H^1(I)} \leq C\|u\|_{H^1(I)} \qquad \forall u \in H^1(I),$$
$$\|i_p u_q\|_{L^2(I)} \leq C(1 + q/p)\|u_q\|_{L^2(I)} \qquad \forall u \in \mathcal{P}_q,$$
$$\|i_p u\|_{L^2(I)} \leq C\left[\|u\|_{L^2(I)} + p^{-1}\|u\|_{H^1(I)}\right] \qquad \forall u \in H^1(I),$$
$$\|i_p u\|_{H^{1/2}(I)} \leq C\sqrt{p}\|u\|_{L^\infty(I)} \qquad \forall u \in C(\overline{I}),$$
$$\|i_p u\|_{H^{1/2}_{00}(I)} \leq C\sqrt{p}\|u\|_{L^\infty(I)} \qquad \forall u \in C(\overline{I}) \text{ with } u(\pm 1) = 0.$$

Proof: For the first three estimates, see, e.g., [28, Sec. 13]. The last two follow from Theorem A.3.2. For example, for the fourth inequality, we have

$$\|i_p u\|_{H^{1/2}(I)} \leq C\left\{\sum_{i=0}^{p}(i_p u(x_i))^2\right\}^{1/2} \leq C\sqrt{p}\|u\|_{L^\infty(I)}.$$

□

Theorem A.3.2. *Let $I = [-1, 1]$. Then there exists $C > 0$ with the following property. For $p \in \mathbb{N}$ let $(x_i)_{i=0}^p$ be the Gauss-Lobatto points on I, i.e., the zeros of the polynomial $x \mapsto (1 - x^2)L_p'(x)$. Then for all $u \in \mathcal{P}_p$ there holds with $\|\underline{u}\|_2 = \left\{\sum_{i=0}^p |u(x_i)|^2\right\}^{1/2}$:*

1. $\|u\|_{L^\infty(I)} \leq \|\underline{u}\|_2$,

2. $\|u\|_{L^2(I)} \leq Cp^{-1/2}\|\underline{u}\|_2$,
3. $\|u\|_{H^{1/2}(I)} \leq C\|\underline{u}\|_2$,
4. $\|u\|_{H^1(I)} \leq Cp\|\underline{u}\|_2$,
5. $\|u\|_{H_{00}^{1/2}(I)} \leq C\|\underline{u}\|_2$ *if additionally* $u(\pm 1) = 0$,
6. $\|\underline{u}\|_2^2 \leq Cp\int_{-1}^{1} \frac{1}{\sqrt{1-x^2}}u^2(x)\,dx \leq Cp\|u\|^2_{H^{1/2}(0,1)}$.

Furthermore, the results are sharp with respect to the spectral order p.

Proof: See [93, Thm. 4.1]. Closely related results can be found in [34]. □

A.4 L^∞ projectors

The following theorem concerning the existence of projection operators onto finite dimensional subspaces is due to Kadec-Snobar, [72]. We present it in the form given in [130] and [41, Chap. 9, Sec. 7].

Theorem A.4.1. *Let Y be any Banach space and $X_n \subset Y$ be a subspace of dimension $n \in \mathbb{N}$. Then there exists a (bounded linear) projector $\Pi : Y \to X_n$ with*

$$\Pi v = v \qquad \forall v \in X_n,$$
$$\|\Pi v\|_Y \leq \sqrt{n}\|v\|_Y \qquad \forall v \in Y.$$

If Y is an L^p-space with $p \in (1,\infty)$, then the norm estimate of Π can be improved to $\|\Pi\| \leq n^{|\frac{1}{2}-\frac{1}{p}|}$.

An easy application is therefore the following corollary.

Corollary A.4.2. *Let $K \subset \mathbb{R}^2$ be a bounded open set. Then for every $p \geq 1$ and every subspace V of the space of polynomials of degree p there exists a bounded linear operator $\Pi : L^\infty(K) \to V$ with the following properties:*

1. *$\Pi v = v$ for all $v \in V$;*
2. *$\|\Pi v\|_{L^\infty(K)} \leq (p+1)\|v\|_{L^\infty(K)}$.*

Proof: The space of polynomials of degree p is clearly a subset of $\mathcal{Q}_p$, which has dimension $(p+1)^2$. The result now follows from Theorem A.4.1. □

Corollary A.4.3. *Let $K \subset \mathbb{R}^2$ be a bounded open set. Let $|||\cdot|||$ be any norm on $W^{1,\infty}(K)$ (equivalent to the standard $W^{1,\infty}(K)$ norm). Then for every $p \geq 1$ and every subspace V of the space of polynomials of degree p there exists a bounded linear operator $\Pi : W^{1,\infty}(K) \to V$ with the following properties:*

1. *$\Pi v = v$ for all $v \in V$;*
2. *$|||\Pi v||| \leq (p+1)|||v|||$ for all $v \in W^{1,\infty}(K)$.*

References

1. R. A. Adams. *Sobolev Spaces.* Academic Press, 1975.
2. S. Agmon. *Lectures on elliptic boundary value problems.* Van Nostrand-Reinhold, Princton, 1965.
3. S. Agmon, A. Douglis, and L. Nirenberg. Estimates near the boundary for solutions of elliptic partial differential equations satisfying general boundary conditions I. *Comm. Pure Appl. Math.*, 12:623–727, 1959.
4. S. Agmon, A. Douglis, and L. Nirenberg. Estimates near the boundary for solutions of elliptic partial differential equations satisfying general boundary conditions II. *Comm. Pure Appl. Math.*, 17:35–92, 1964.
5. M. Ainsworth and P. Coggins. The stability of mixed *hp*-finite element methods for Stokes flow on high aspect ratio elements. *SIAM J. Numer. Anal.*, 38(5):1721–1761, 2000.
6. M. Ainsworth and D. Kay. The approximation theory for the *p*-version finite element method and application to non-linear elliptic PDEs. *Numer. Math.*, 82:351–388, 1999.
7. T. Apel. *Anisotropic Finite Elements: Local Estimates and Applications.* Advances in Numerical Mathematics. Teubner, 1999.
8. T. Apel and G. Lube. Anisotropic mesh refinement for singularly perturbed reaction-diffusion problems. *Appl. Numer. Math.*, 26:415–433, 1998.
9. D. N. Arnold and R. S. Falk. The boundary layer for the Reissner-Mindlin plate model. *SIAM J. Math. Anal*, 21:281–312, 1990.
10. D. N. Arnold and R. S. Falk. Asymptotic analysis of the boundary layer for the Reissner-Mindlin plate model. *SIAM J. Math. Anal*, 27:486–514, 1996.
11. A.K. Aziz and I.M. Babuška, editors. *Mathematical Foundations of the Finite Element Method with Applications to Partial Differential Equations.* Academic Press, New York, 1972.
12. I. Babuška and A.K. Aziz. On the angle condition in the finite element method. *SIAM J. Numer. Anal.*, 13:214–226, 1976.
13. I. Babuška, A. Craig, J. Mandel, and J. Pitkäranta. Efficient preconditioning for the *p* version finite element method in two dimensions. *SIAM J. Numer. Anal.*, 28(3):624–661, 1991.
14. I. Babuška and B. Guo. The *hp* version of the finite element method for domains with curved boundaries. *SIAM J. Numer. Anal.*, 25:837–861, 1988.
15. I. Babuška and B. Guo. Regularity of the solution of elliptic problems with piecewise analytic data. Part I. Boundary value problems for linear elliptic equations of second order. *SIAM J. Math. Anal*, 19(1):172–203, 1988.
16. I. Babuška and B. Guo. Regularity of the solution of elliptic problems with piecewise analytic data. Part II. The trace spaces and application to the boundary value problems with nonhomogeneous boundary conditions. *SIAM J. Math. Anal*, 20(4):763–781, 1989.
17. I. Babuška and B. Guo. On the regularity of elasticity problems with piecewise analytic data. *Advances in Applied Mathematics*, 14:307–347, 1993.

18. I. Babuška, B. Guo, and M. Suri. Implementation of nonhomogeneous Dirichlet boundary conditions in the p-version of the finite element method. *Impact of Computing in Science and Engineering*, 1:36–63, 1989.
19. I. Babuška and B.Q. Guo. The $h-p$ version of the finite element method. Part 1: The basic approximation results. *Computational Mechanics*, 1:21–41, 1986.
20. I. Babuška and B.Q. Guo. The $h-p$ version of the finite element method. Part 2: General results and applications. *Computational Mechanics*, 1:21–41, 1986.
21. I. Babuška, R.B. Kellogg, and J. Pitkäranta. Direct and inverse error estimates for finite elements with mesh refinements. *Numer. Math.*, 33:447–471, 1979.
22. I. Babuška and M. Suri. The optimal convergence rate of the p-version of the finite element method. *SIAM J. Numer. Anal.*, 24:750–776, 1987.
23. I. Babuška and M. Suri. The treatment of nonhomogeneous Dirichlet boundary conditions in the p-version of the finite element method. *Numer. Math.*, 55:97–121, 1989.
24. I. Babuška and M. Suri. On locking and robustness in the finite element method. *SIAM J. Numer. Anal.*, 29(5):1261–1293, 1992.
25. I. Babuška and M. Suri. The p and h-p versions of the finite element method, basic principles and properties. *SIAM review*, 36(4):578–632, 1994.
26. N.S. Bakhvalov. Optimization of methods for the solution of boundary value problems in the presence of a boundary layer. *Zh. Vychisl. Mat. Mat. Fiz.*, 9:841–859, 1969.
27. C. Bernardi and Y. Maday. *Approximations spectrales de problèmes aux limites elliptiques.* Mathématiques & Applications. Springer Verlag, 1992.
28. C. Bernardi and Y. Maday. Spectral methods. In P.G. Ciarlet and J.L. Lions, editors, *Handbook of Numerical Analysis, Vol. 5.* North Holland, 1997.
29. I.A. Blatov. The Galerkin finite-element method for elliptic quasilinear singularly perturbed boundary value problems. III. *Differ. Uravn.*, 30:467–479, 1994.
30. P. Bolley, J. Camus, and M. Dauge. Régularité Gevrey pour le problème de Dirichlet dans des domains coniques. *Comm. PDE*, 10, 1985.
31. F. Brezzi and M. Fortin. *Mixed and Hybrid Finite Element Methods.* Springer Verlag, 1991.
32. V.F. Butuzov. The asymptotic properties of solutions of the equation $\mu^2 \Delta u - k^2(x,y)u = f(x,y)$ in a rectangle. *Differentsial'nye Uravnenia*, 9, 1973.
33. C. Canuto, M.Y. Hussaini, A. Quarteroni, and T.A. Zhang. *Spectral Methods in Fluid Dynamics.* Springer Verlag, 1986.
34. M.A. Casarin. Diagonal edge preconditioners in p-version and spectral element methods. *SIAM J. Sci. Stat. Comp.*, 18(2):610–620, 1997.
35. J. Céa. Approximation variationelle des problèmes aux limites. *Ann. Inst. Fourier (Grenoble)*, 14:345–444, 1964.
36. P. G. Ciarlet. *The Finite Element Method for Elliptic Problems.* North-Holland Publishing Company, 1976.
37. M. Costabel and M. Dauge. Construction of corner singularities for Agmon-Douglis-Nirenberg elliptic systems. *Math. Nachr.*, 162:209–237, 1993.
38. M. Costabel, M. Dauge, and M. Suri. Numerical approximation of a singularly perturbed contact problem. *Comput. Meth. Appl. Mech. Engrg.*, 157:349–363, 1998.
39. M. Dauge. *Elliptic boundary value problems on corner domains*, volume 1341 of *Lecture Notes in Mathematics.* Springer Verlag, 1988.
40. P.J. Davis. *Interpolation and Approximation.* Dover, 1974.
41. R.A. DeVore and G.G. Lorentz. *Constructive Approximation.* Springer Verlag, 1993.
42. E.P. Doolan, J.J.H. Miller, and W. Schilders. *Uniform numerical methods for problems with initial and boundary layers.* Boole Press, Dublin, 1980.
43. M. Dubiner. Spectral methods on triangles and other domains. *J. Sci. Comp.*, 6:345–390, 1991.

44. I. Duff, A.M. Erisman, and J.K. Reid. *Direct Methods for Sparse Matrices.* Oxford Clarendon Press, 1992.
45. M.G. Duffy. Quadrature over a pyramid or cube of integrands with a singularity at a vertex. *SIAM J. Numer. Anal.*, 19:1260–1262, 1982.
46. W. Eckhaus. Boundary layers in linear elliptic singular perturbations. *SIAM Review*, 14:225–270, 1972.
47. W. Eckhaus. *Asymptotic Analysis of Singular Perturbations.* North-Holland, 1979.
48. ESRD Inc., 10845 Olive Boulevard, Suite 170, St. Louis, MO 63141-7760, `http://www.esrd.com`. *User's manual to STRESSCHECK.*
49. L.C. Evans. *Partial Differential Equations.* American Mathematical Society, 1998.
50. P.J. Frey and P.-L. George. *Mesh generation.* Hermes Science Publishing, Oxford, 2000.
51. E.C. Gartland. Graded-mesh difference schemes for singularly perturbed two-point boundary value problems. *Math. Comput.*, 51:631–657, 1988.
52. A. George, J.R. Gilbert, and J.W.H. Liu. *Graph Theory and Sparse Matrix Computation.* Springer Verlag, 1993.
53. P.-L. George and H. Borouchaki. *Delaunay triangulation and meshing.* Editions Hermès, Paris, 1998.
54. K. Gerdes, A.-M. Matache, and C. Schwab. Analysis of membrane locking in *hp*-fem for a cylindrical shell. *ZAMM*, 78:663–686, 1998.
55. K. Gerdes, J.M. Melenk, D. Schötzau, and C. Schwab. The *hp*-version of the streamline-diffusion finite element method in two space dimensions. *Math. Meths. Appl. Sci.*, 11(2):301–337, 2001.
56. D. Gilbarg and N. S. Trudinger. *Elliptic Partial Differential Equations of Second Order.* Grundlagen der mathematischen Wissenschaften 224. Springer, 1977.
57. G. Golub and C. van Loan. *Matrix Computations.* Johns Hopkins, 3rd edition, 1996.
58. W.J. Gordon. Blending-function methods of bivariate and multivariate interpolation and approximation. *SIAM J. Numer. Anal.*, 8:158–177, 1973.
59. W.J. Gordon and Ch.A. Hall. Construction of curvilinear co-ordinate systems and applications to mesh generation. *Internat. J. Numer. Meths. Engrg.*, 7:461–477, 1973.
60. W.J. Gordon and Ch.A. Hall. Transfinite element methods: Blending function interpolation over arbitrary curved element domains. *Numer. Math.*, 21:109–129, 1973.
61. I.S. Gradshteyn and I.M. Ryzhik. *Table of Integrals, Series, and Products, corrected and enlarged edition.* Academic Press, New York, 1980.
62. P. Grisvard. *Elliptic Problems in Nonsmooth Domains.* Pitman, 1985.
63. P. Grisvard. *Singularities in Boundary Value Problems.* Springer Verlag/Masson, 1992.
64. B. Guo, N. Heuer, and E. Stephan. The *hp*-version of the boundary element method for transmission problems with piecewise analytic data. *SIAM J. Numer. Anal.*, 33(2):789–808, 1996.
65. W. Hackbusch. *Multi-Grid Methods and Applications.* Springer-Verlag, 1985.
66. H. Hakula, Y. Leino, and J. Pitkäranta. Scale resolution, locking, and high-order finite element modelling of shells. *Comput. Meth. Appl. Mech. Engrg.*, 133:157–182, 1996.
67. H. Han and R.B. Kellogg. Differentiability properties of solutions of the equation $-\varepsilon^2 \Delta u + ru = f(x,y)$ in a square. *SIAM J. Math. Anal*, 21:394–408, 1990.
68. G. Hardy, J.E. Littlewood, and G. Pólya. *Inequalities.* Cambridge Mathematical Library. Cambridge University Press, 1991.
69. S. Holzer. Mesh generation for *hp*-type finite element analysis of Reissner-Mindlin plates. Technical Report 99–02, Institut für Informationsverarbeitung im Konstruktiven Ingenieursbau, Universität Stuttgart, `http://www.uni-stuttgart.de/iv-kib`, 1999.

70. L. Hörmander. On the theory of general partial differential operators. *Acta math.*, 94:161–248, 1955.
71. A.M. Il'in. *Matching of asymptotic expansions of solutions of boundary value problems.* American Mathematical Society, 1992.
72. M.I. Kadec and M.G. Snobar. Certain functionals on the Minkowski compactum. *Mat. Zametki*, 10:453–458, 1971.
73. G.E. Karniadakis and S.J. Sherwin. *Spectral/hp Element Methods for CFD.* Oxford University Press, 1999.
74. R.B. Kellogg. Boundary layers and corner singularities for a self-adjoint problem. (unpublished revision of [75]).
75. R.B. Kellogg. Boundary layers and corner singularities for a self-adjoint problem. In M. Costabel, M. Dauge, and S. Nicaise, editors, *Boundary Value Problems and Integral Equations in Non-smooth Domains*, volume 167 of *Lecture Notes in Pure and Applied Mathematics*, pages 121–149. Marcel Dekker, New York, 1995.
76. V. A. Kondratjev. Boundary value problems for elliptic equations in domains with conical or angular points. *Trudy Moskovkogo Mat. Obschetsva*, 16:209–292, 1967.
77. V. A. Kondratjev. Boundary value problems for elliptic equations in domains with conical or angular points. *Trans. Moscow Math. Soc.*, 16:227–313, 1967.
78. V.A. Kozlov, V.G. Maz'ya, and J. Rossmann. *Elliptic Boundary Value Problems in Domains with Point Singularities.* American Mathematical Society, 1997.
79. A. Kufner and A.-M. Sändig. *Some Applications of Weighted Sobolev Spaces.* Teubner, 1987.
80. C. Lage. Concept oriented design of numerical software. Technical Report 98–07, Seminar für Angewandte Mathematik, ETH Zürich, 1998.
81. P.A. Lagerstrom. *Matched Asymptotic Expansions: Ideas and Techniques.* Springer Verlag, 1988.
82. P.D. Lax and A.N. Milgram. *Parabolic Equations*, volume 33 of *Annals of Mathematics Studies*, pages 167–190. Princeton University Press, Princeton, 1954.
83. R.S. Lehmann. Developments at an analytic corner of solutions of elliptic partial differential equations. *J. Math. Mech.*, 8:727–760, 1959.
84. J. Li and I.M. Navon. Uniformly convergent finite element methods for singularly perturbed elliptic boundary value problems I: Reaction-diffusion type. *Computers Math. Applic.*, 33(3):57–70, 1998.
85. J.L. Lions. *Perturbations Singulières dans les Problèmes aux Limites et en Contrôle Optimal*, volume 323 of *Lecture Notes in Mathematics.* Springer Verlag, 1973.
86. J.L. Lions and E. Magenes. *Problèmes aux Limites Non Homogènes et Applications, Vol. III.* Dunod, Paris, 1968.
87. V.D. Liseikin. *Grid Generation Methods.* Springer Verlag, 1999.
88. V.D. Liseikin. *Layer-Resolving Grids and Transformations for Singular Perturbation Problems.* VSP Publisher, 2001.
89. V. G. Mazja and B. A. Plamenevskij. On the coefficients in the asymptotics of solutions of elliptic boundary value problems near conical points. *Sov. Math. Dokl.*, 19:1570–1574, 1974.
90. V. G. Mazja and B. A. Plamenevskij. Estimates in L^p and in Hölder classes and the Miranda-Agmon maximum principle for solutions of elliptic boundary value problems in domains with singular points on the boundary. *Amer. Math. Soc. Transl.(2)*, 123:1–56, 1984.
91. W.G. Mazja, S.A. Nazarow, and B.A. Plamenewski. *Asymptotische Theorie elliptischer Randwertaufgaben in singulär gestörten Gebieten.* Akademie-Verlag Berlin, 1991.
92. J.M. Melenk. On the robust exponential convergence of *hp* finite element methods for problems with boundary layers. *IMA J. Numer. Anal.*, 17(4):577–601, 1997.
93. J.M. Melenk. On condition numbers in *hp*-FEM with Gauss-Lobatto based shape functions. *J. Comput. Appl. Math.*, 139:21–48, 2001.

94. J.M. Melenk and C. Schwab. *hp* FEM for reaction-diffusion equations I: Robust exponentional convergence. *SIAM J. Numer. Anal.*, 35:1520–1557, 1998.
95. J.M. Melenk and C. Schwab. Analytic regularity for a singularly perturbed problem. *SIAM J. Math. Anal*, 30:379–400, 1999.
96. J.M. Melenk and C. Schwab. An *hp* FEM for convection-diffusion problems in one dimension. *IMA J. Numer. Anal.*, 19(3):425–453, 1999.
97. J.J.H. Miller, E. O'Riordan, and G.I. Shishkin. *Fitted Numerical Methods for Singular Perturbation Problems.* World Scientific, 1996.
98. C.B. Morrey. *Multiple Integrals in the Calculus of Variations.* Springer Verlag, 1966.
99. K.W. Morton. *Numerical Solution of Convection-Diffusion Problems*, volume 12 of *Applied Mathematics and Mathematical Computation.* Chapman & Hall, 1996.
100. S.A. Nazarov and B.A. Plamenevsky. *Elliptic problems in domains with piecewise smooth boundaries.* de Gruyter, 1994.
101. S. Nicaise and A.-M. Sändig. General interface problems–I. *Math. Meths. Appl. Sci.*, 17:395–429, 1994.
102. S. Nicaise and A.-M. Sändig. General interface problems–II. *Math. Meths. Appl. Sci.*, 17:431–451, 1994.
103. F.W.J. Olver. *Asymptotics and Special Functions.* Academic Press, 1974.
104. J. Pitkäranta, Y. Leino, O. Ovaskainen, and J. Piila. Shell deformation states and the finite element method: A benchmark study of cylindrical shells. *Comput. Meth. Appl. Mech. Engrg.*, 128:81–121, 1995.
105. A. Quarteroni. Some results of Bernstein and Jackson type for polynomial approximation in L^q spaces. *Japan J. Appl. Math.*, 1:173–181, 1984.
106. G. Raugel. Résolution numérique par une méthode d'élements finis du problème de Dirichlet pour le Laplacien dans un polygone. *C. R. Acad. Sci. Paris*, 286:791–794, 1978.
107. H.-G. Roos. Layer-adapted grids for singular perturbation problems. *ZAMM*, 78:291–310, 1998.
108. H.-G. Roos, M. Stynes, and L. Tobiska. *Numerical Methods for Singularly Perturbed Differential Equations*, volume 24 of *Springer Series in Computational Mathematics.* Springer Verlag, 1996.
109. A. H. Schatz and L. B. Wahlbin. On the finite element method for singularly perturbed reaction-diffusion problems in two and one dimension. *Math. Comput.*, 40:47–89, 1983.
110. D. Schötzau and C. Schwab. Mixed *hp*-FEM on anisotropic meshes. *Math. Meths. Appl. Sci.*, 8(5):787–820, 1998.
111. D. Schötzau, C. Schwab, and R. Stenberg. Mixed *hp*-FEM on anisotropic meshes. II: Hanging nodes and tensor products of boundary layer meshes. *Numer. Math.*, 83(4):667–697, 1999.
112. C. Schwab. *p- and hp-Finite Element Methods.* Oxford University Press, 1998.
113. C. Schwab and M. Suri. The *p* and *hp* versions of the finite element method for problems with boundary layers. *Math. Comput.*, 65(216):1403–1429, 1996.
114. C. Schwab, M. Suri, and C. Xenophontos. Boundary layer approximation by spectral/*hp* methods. In *ICOSAHOM 95: Proceedings of the Third International Conference on Spectral and High Order Methods.* Houston Journal of Mathematics, Dept. of Mathematics, University of Houston, Houston, TX 77204–3476, ISSN 0362–1588, 1996.
115. C. Schwab, M. Suri, and C. Xenophontos. The *hp* finite element method for problems in mechanics with boundary layers. *Comput. Meth. Appl. Mech. Engrg.*, 157:311–333, 1998.
116. S.D. Shih and R.B. Kellogg. Asymptotic analysis of a singular perturbation problem. *SIAM J. Math. Anal*, 18:1467–1511, 1987.

117. G.I. Shishkin. Approximation of solutions of singularly perturbed boundary value problems with a corner boundary layer. *Zh. Vychisl. Mat. Mat. Fiz.*, 27:1360–1374, 1987.
118. G.I. Shishkin. *Grid approximation of singularly perturbed elliptic and parabolic problems.* Second Doctoral Thesis, Keldish Institute, Moscow, 1990.
119. G.I. Shishkin. *Approximation of solutions of singularly perturbed boundary value problems.* Russ. Acad. Sci. Ekaterinburg, 1992.
120. G. Sun and M. Stynes. Finite-element methods for singulary perturbed high-order elliptic two-point boundary value problems I: Reaction-diffusion-type problems. *IMA J. Numer. Anal.*, 15:117–139, 1995.
121. B. Sündermann. Lebesgue constants in Lagrangian interpolation at the Fekete points. Ergebnisberichte der Lehrstühle Mathematik III und VIII (Angewandte Mathematik) 44, Universität Dortmund, 1980.
122. B. Sündermann. Lebesgue constants in Lagrangian interpolation at the Fekete points. *Mitt. Math. Ges. Hamb.*, 11:204–211, 1983.
123. B. Szabó and I. Babuška. *Finite Element Analysis.* Wiley, 1991.
124. B. Szegö. *Orthogonal Polynomials.* American Mathematical Society, fourth edition, 1975.
125. A. Toselli and C. Schwab. Mixed *hp*-finite element approximations on geometric edge and boundary layer meshes in three dimensions. Technical Report 2001-02, Seminar für Angewandte Mathematik, ETH Zürich, 2001.
126. T. von Petersdorff. *Randwertprobleme der Elastizitätstheorie für Polyeder—Singularitäten und Approximation mit Randelementmethoden.* PhD thesis, Technische Hochschule Darmstadt, 1989.
127. R. Vulanović. *Mesh construction for discretization of singularly perturbed boundary value problems.* PhD thesis, University of Novi Sad, 1986.
128. W. Wasow. Asymptotic development of the solution of Dirichlet's problem at analytic corners. *Duke Math. J.*, 24:47–56, 1957.
129. J. Wloka. *Partielle Differentialgleichungen.* Teubner, 1982.
130. P. Wojtaszczyk. *Banach spaces for analysts.* Cambridge University Press, 1991.
131. C. Xenophontos. *The hp finite element method for singularly perturbed problems.* PhD thesis, University of Maryland Baltimore County, 1996.
132. C. Xenophontos. Finite element computations for the Reissner-Mindlin plate model. *Comm. Numer. Meths. Eng.*, 14(12):1119–1131, 1998.
133. C. Xenophontos. The *hp* finite element method for singularly perturbed problems in smooth domains. M^3AS, 8(2):299–326, 1998.
134. C. Xenophontos. The *hp* finite element method for singularly perturbed problems in nonsmooth domains. *Numer. Methods Partial Differ. Equations*, 15(1):63–89, 1999.
135. C. Xenophontos. A note on the application of *p*/*hp* finite element methods to reaction-diffusion problems in polygonal domains. *Comm. Numer. Meths. Eng.*, 16(6):391–400, 2000.
136. W.P. Ziemer. *Weakly Differentiable Functions.* Springer Verlag, 1989.

Index

Lecture Notes in Mathematics

For information about Vols. 1–1619
please contact your bookseller or Springer-Verlag

Vol. 1620: R. Donagi, B. Dubrovin, E. Frenkel, E. Previato, Integrable Systems and Quantum Groups. Montecatini Terme, 1993. Editors:M. Francaviglia, S. Greco. VIII, 488 pages. 1996.

Vol. 1621: H. Bass, M. V. Otero-Espinar, D. N. Rockmore, C. P. L. Tresser, Cyclic Renormalization and Automorphism Groups of Rooted Trees. XXI, 136 pages. 1996.

Vol. 1622: E. D. Farjoun, Cellular Spaces, Null Spaces and Homotopy Localization. XIV, 199 pages. 1996.

Vol. 1623: H.P. Yap, Total Colourings of Graphs. VIII, 131 pages. 1996.

Vol. 1624: V. Brýnzanescu, Holomorphic Vector Bundles over Compact Complex Surfaces. X, 170 pages. 1996.

Vol. 1625: S. Lang, Topics in Cohomology of Groups. VII, 226 pages. 1996.

Vol. 1626: J. Azéma, M. Emery, M. Yor (Eds.), Séminaire de Probabilités XXX. VIII, 382 pages. 1996.

Vol. 1627: C. Graham, Th. G. Kurtz, S. Méléard, Ph. E. Protter, M. Pulvirenti, D. Talay, Probabilistic Models for Nonlinear Partial Differential Equations. Montecatini Terme, 1995. Editors: D. Talay, L. Tubaro. X, 301 pages. 1996.

Vol. 1628: P.-H. Zieschang, An Algebraic Approach to Association Schemes. XII, 189 pages. 1996.

Vol. 1629: J. D. Moore, Lectures on Seiberg-Witten Invariants. VII, 105 pages. 1996.

Vol. 1630: D. Neuenschwander, Probabilities on the Heisenberg Group: Limit Theorems and Brownian Motion. VIII, 139 pages. 1996.

Vol. 1631: K. Nishioka, Mahler Functions and Transcendence. VIII, 185 pages. 1996.

Vol. 1632: A. Kushkuley, Z. Balanov, Geometric Methods in Degree Theory for Equivariant Maps. VII, 136 pages. 1996.

Vol. 1633: H. Aikawa, M. Essén, Potential Theory – Selected Topics. IX, 200 pages. 1996.

Vol. 1634: J. Xu, Flat Covers of Modules. IX, 161 pages. 1996.

Vol. 1635: E. Hebey, Sobolev Spaces on Riemannian Manifolds. X, 116 pages. 1996.

Vol. 1636: M. A. Marshall, Spaces of Orderings and Abstract Real Spectra. VI, 190 pages. 1996.

Vol. 1637: B. Hunt, The Geometry of some special Arithmetic Quotients. XIII, 332 pages. 1996.

Vol. 1638: P. Vanhaecke, Integrable Systems in the realm of Algebraic Geometry. VIII, 218 pages. 1996.

Vol. 1639: K. Dekimpe, Almost-Bieberbach Groups: Affine and Polynomial Structures. X, 259 pages. 1996.

Vol. 1640: G. Boillat, C. M. Dafermos, P. D. Lax, T. P. Liu, Recent Mathematical Methods in Nonlinear Wave Propagation. Montecatini Terme, 1994. Editor: T. Ruggeri. VII, 142 pages. 1996.

Vol. 1641: P. Abramenko, Twin Buildings and Applications to S-Arithmetic Groups. IX, 123 pages. 1996.

Vol. 1642: M. Puschnigg, Asymptotic Cyclic Cohomology. XXII, 138 pages. 1996.

Vol. 1643: J. Richter-Gebert, Realization Spaces of Polytopes. XI, 187 pages. 1996.

Vol. 1644: A. Adler, S. Ramanan, Moduli of Abelian Varieties. VI, 196 pages. 1996.

Vol. 1645: H. W. Broer, G. B. Huitema, M. B. Sevryuk, Quasi-Periodic Motions in Families of Dynamical Systems. XI, 195 pages. 1996.

Vol. 1646: J.-P. Demailly, T. Peternell, G. Tian, A. N. Tyurin, Transcendental Methods in Algebraic Geometry. Cetraro, 1994. Editors: F. Catanese, C. Ciliberto. VII, 257 pages. 1996.

Vol. 1647: D. Dias, P. Le Barz, Configuration Spaces over Hilbert Schemes and Applications. VII. 143 pages. 1996.

Vol. 1648: R. Dobrushin, P. Groeneboom, M. Ledoux, Lectures on Probability Theory and Statistics. Editor: P. Bernard. VIII, 300 pages. 1996.

Vol. 1649: S. Kumar, G. Laumon, U. Stuhler, Vector Bundles on Curves - New Directions. Cetraro, 1995. Editor: M. S. Narasimhan. VII, 193 pages. 1997.

Vol. 1650: J. Wildeshaus, Realizations of Polylogarithms. XI, 343 pages. 1997.

Vol. 1651: M. Drmota, R. F. Tichy, Sequences, Discrepancies and Applications. XIII, 503 pages. 1997.

Vol. 1652: S. Todorcevic, Topics in Topology. VIII, 153 pages. 1997.

Vol. 1653: R. Benedetti, C. Petronio, Branched Standard Spines of 3-manifolds. VIII, 132 pages. 1997.

Vol. 1654: R. W. Ghrist, P. J. Holmes, M. C. Sullivan, Knots and Links in Three-Dimensional Flows. X, 208 pages. 1997.

Vol. 1655: J. Azéma, M. Emery, M. Yor (Eds.), Séminaire de Probabilités XXXI. VIII, 329 pages. 1997.

Vol. 1656: B. Biais, T. Björk, J. Cvitanic, N. El Karoui, E. Jouini, J. C. Rochet, Financial Mathematics. Bressanone, 1996. Editor: W. J. Runggaldier. VII, 316 pages. 1997.

Vol. 1657: H. Reimann, The semi-simple zeta function of quaternionic Shimura varieties. IX, 143 pages. 1997.

Vol. 1658: A. Pumarino, J. A. Rodrýguez, Coexistence and Persistence of Strange Attractors. VIII, 195 pages. 1997.

Vol. 1659: V, Kozlov, V. Maz'ya, Theory of a Higher-Order Sturm-Liouville Equation. XI, 140 pages. 1997.

Vol. 1660: M. Bardi, M. G. Crandall, L. C. Evans, H. M. Soner, P. E. Souganidis, Viscosity Solutions and Applications. Montecatini Terme, 1995. Editors: I. Capuzzo Dolcetta, P. L. Lions. IX, 259 pages. 1997.

Vol. 1661: A. Tralle, J. Oprea, Symplectic Manifolds with no Kähler Structure. VIII, 207 pages. 1997.

Vol. 1662: J. W. Rutter, Spaces of Homotopy Self-Equivalences - A Survey. IX, 170 pages. 1997.

Vol. 1663: Y. E. Karpeshina; Perturbation Theory for the Schrödinger Operator with a Periodic Potential. VII, 352 pages. 1997.

Vol. 1664: M. Väth, Ideal Spaces. V, 146 pages. 1997.

Vol. 1665: E. Giné, G. R. Grimmett, L. Saloff-Coste, Lectures on Probability Theory and Statistics 1996. Editor: P. Bernard. X, 424 pages, 1997.

Vol. 1666: M. van der Put, M. F. Singer, Galois Theory of Difference Equations. VII, 179 pages. 1997.

Vol. 1667: J. M. F. Castillo, M. González, Three-space Problems in Banach Space Theory. XII, 267 pages. 1997.

Vol. 1668: D. B. Dix, Large-Time Behavior of Solutions of Linear Dispersive Equations. XIV, 203 pages. 1997.

Vol. 1669: U. Kaiser, Link Theory in Manifolds. XIV, 167 pages. 1997.

Vol. 1670: J. W. Neuberger, Sobolev Gradients and Differential Equations. VIII, 150 pages. 1997.

Vol. 1671: S. Bouc, Green Functors and G-sets. VII, 342 pages. 1997.

Vol. 1672: S. Mandal, Projective Modules and Complete Intersections. VIII, 114 pages. 1997.

Vol. 1673: F. D. Grosshans, Algebraic Homogeneous Spaces and Invariant Theory. VI, 148 pages. 1997.

Vol. 1674: G. Klaas, C. R. Leedham-Green, W. Plesken, Linear Pro-p-Groups of Finite Width. VIII, 115 pages. 1997.

Vol. 1675: J. E. Yukich, Probability Theory of Classical Euclidean Optimization Problems. X, 152 pages. 1998.

Vol. 1676: P. Cembranos, J. Mendoza, Banach Spaces of Vector-Valued Functions. VIII, 118 pages. 1997.

Vol. 1677: N. Proskurin, Cubic Metaplectic Forms and Theta Functions. VIII, 196 pages. 1998.

Vol. 1678: O. Krupková, The Geometry of Ordinary Variational Equations. X, 251 pages. 1997.

Vol. 1679: K.-G. Grosse-Erdmann, The Blocking Technique. Weighted Mean Operators and Hardy's Inequality. IX, 114 pages. 1998.

Vol. 1680: K.-Z. Li, F. Oort, Moduli of Supersingular Abelian Varieties. V, 116 pages. 1998.

Vol. 1681: G. J. Wirsching, The Dynamical System Generated by the 3n+1 Function. VII, 158 pages. 1998.

Vol. 1682: H.-D. Alber, Materials with Memory. X, 166 pages. 1998.

Vol. 1683: A. Pomp, The Boundary-Domain Integral Method for Elliptic Systems. XVI, 163 pages. 1998.

Vol. 1684: C. A. Berenstein, P. F. Ebenfelt, S. G. Gindikin, S. Helgason, A. E. Tumanov, Integral Geometry, Radon Transforms and Complex Analysis. Firenze, 1996. Editors: E. Casadio Tarabusi, M. A. Picardello, G. Zampieri. VII, 160 pages. 1998.

Vol. 1685: S. König, A. Zimmermann, Derived Equivalences for Group Rings. X, 146 pages. 1998.

Vol. 1686: J. Azéma, M. Émery, M. Ledoux, M. Yor (Eds.), Séminaire de Probabilités XXXII. VI, 440 pages. 1998.

Vol. 1687: F. Bornemann, Homogenization in Time of Singularly Perturbed Mechanical Systems. XII, 156 pages. 1998.

Vol. 1688: S. Assing, W. Schmidt, Continuous Strong Markov Processes in Dimension One. XII, 137 page. 1998.

Vol. 1689: W. Fulton, P. Pragacz, Schubert Varieties and Degeneracy Loci. XI, 148 pages. 1998.

Vol. 1690: M. T. Barlow, D. Nualart, Lectures on Probability Theory and Statistics. Editor: P. Bernard. VIII, 237 pages. 1998.

Vol. 1691: R. Bezrukavnikov, M. Finkelberg, V. Schechtman, Factorizable Sheaves and Quantum Groups. X, 282 pages. 1998.

Vol. 1692: T. M. W. Eyre, Quantum Stochastic Calculus and Representations of Lie Superalgebras. IX, 138 pages. 1998.

Vol. 1694: A. Braides, Approximation of Free-Discontinuity Problems. XI, 149 pages. 1998.

Vol. 1695: D. J. Hartfiel, Markov Set-Chains. VIII, 131 pages. 1998.

Vol. 1696: E. Bouscaren (Ed.): Model Theory and Algebraic Geometry. XV, 211 pages. 1998.

Vol. 1697: B. Cockburn, C. Johnson, C.-W. Shu, E. Tadmor, Advanced Numerical Approximation of Nonlinear Hyperbolic Equations. Cetraro, Italy, 1997. Editor: A. Quarteroni. VII, 390 pages. 1998.

Vol. 1698: M. Bhattacharjee, D. Macpherson, R. G. Möller, P. Neumann, Notes on Infinite Permutation Groups. XI, 202 pages. 1998.

Vol. 1699: A. Inoue,Tomita-Takesaki Theory in Algebras of Unbounded Operators. VIII, 241 pages. 1998.

Vol. 1700: W. A. Woyczyński, Burgers-KPZ Turbulence, XI, 318 pages. 1998.

Vol. 1701: Ti-Jun Xiao, J. Liang, The Cauchy Problem of Higher Order Abstract Differential Equations, XII, 302 pages. 1998.

Vol. 1702: J. Ma, J. Yong, Forward-Backward Stochastic Differential Equations and Their Applications. XIII, 270 pages. 1999.

Vol. 1703: R. M. Dudley, R. Norvaiša, Differentiability of Six Operators on Nonsmooth Functions and p-Variation. VIII, 272 pages. 1999.

Vol. 1704: H. Tamanoi, Elliptic Genera and Vertex Operator Super-Algebras. VI, 390 pages. 1999.

Vol. 1705: I. Nikolaev, E. Zhuzhoma, Flows in 2-dimensional Manifolds. XIX, 294 pages. 1999.

Vol. 1706: S. Yu. Pilyugin, Shadowing in Dynamical Systems. XVII, 271 pages. 1999.

Vol. 1707: R. Pytlak, Numerical Methods for Optimal Control Problems with State Constraints. XV, 215 pages. 1999.

Vol. 1708: K. Zuo, Representations of Fundamental Groups of Algebraic Varieties. VII, 139 pages. 1999.

Vol. 1709: J. Azéma, M. Émery, M. Ledoux, M. Yor (Eds), Séminaire de Probabilités XXXIII. VIII, 418 pages. 1999.

Vol. 1710: M. Koecher, The Minnesota Notes on Jordan Algebras and Their Applications. IX, 173 pages. 1999.

Vol. 1711: W. Ricker, Operator Algebras Generated by Commuting Projections: A Vector Measure Approach. XVII, 159 pages. 1999.

Vol. 1712: N. Schwartz, J. J. Madden, Semi-algebraic Function Rings and Reflectors of Partially Ordered Rings. XI, 279 pages. 1999.

Vol. 1713: F. Bethuel, G. Huisken, S. Müller, K. Steffen, Calculus of Variations and Geometric Evolution Problems. Cetraro, 1996. Editors: S. Hildebrandt, M. Struwe. VII, 293 pages. 1999.

Vol. 1714: O. Diekmann, R. Durrett, K. P. Hadeler, P. K. Maini, H. L. Smith, Mathematics Inspired by Biology. Martina Franca, 1997. Editors: V. Capasso, O. Diekmann. VII, 268 pages. 1999.

Vol. 1715: N. V. Krylov, M. Röckner, J. Zabczyk, Stochastic PDE's and Kolmogorov Equations in Infinite Dimensions. Cetraro, 1998. Editor: G. Da Prato. VIII, 239 pages. 1999.

Vol. 1716: J. Coates, R. Greenberg, K. A. Ribet, K. Rubin, Arithmetic Theory of Elliptic Curves. Cetraro, 1997. Editor: C. Viola. VIII, 260 pages. 1999.
Vol. 1717: J. Bertoin, F. Martinelli, Y. Peres, Lectures on Probability Theory and Statistics. Saint-Flour, 1997. Editor: P. Bernard. IX, 291 pages. 1999.
Vol. 1718: A. Eberle, Uniqueness and Non-Uniqueness of Semigroups Generated by Singular Diffusion Operators. VIII, 262 pages. 1999.
Vol. 1719: K. R. Meyer, Periodic Solutions of the N-Body Problem. IX, 144 pages. 1999.
Vol. 1720: D. Elworthy, Y. Le Jan, X-M. Li, On the Geometry of Diffusion Operators and Stochastic Flows. IV, 118 pages. 1999.
Vol. 1721: A. Iarrobino, V. Kanev, Power Sums, Gorenstein Algebras, and Determinantal Loci. XXVII, 345 pages. 1999.
Vol. 1722: R. McCutcheon, Elemental Methods in Ergodic Ramsey Theory. VI, 160 pages. 1999.
Vol. 1723: J. P. Croisille, C. Lebeau, Diffraction by an Immersed Elastic Wedge. VI, 134 pages. 1999.
Vol. 1724: V. N. Kolokoltsov, Semiclassical Analysis for Diffusions and Stochastic Processes. VIII, 347 pages. 2000.
Vol. 1725: D. A. Wolf-Gladrow, Lattice-Gas Cellular Automata and Lattice Boltzmann Models. IX, 308 pages. 2000.
Vol. 1726: V. Marić, Regular Variation and Differential Equations. X, 127 pages. 2000.
Vol. 1727: P. Kravanja M. Van Barel, Computing the Zeros of Analytic Functions. VII, 111 pages. 2000.
Vol. 1728: K. Gatermann Computer Algebra Methods for Equivariant Dynamical Systems. XV, 153 pages. 2000.
Vol. 1729: J. Azéma, M. Émery, M. Ledoux, M. Yor Séminaire de Probabilités XXXIV. VI, 431 pages. 2000.
Vol. 1730: S. Graf, H. Luschgy, Foundations of Quantization for Probability Distributions. X, 230 pages. 2000.
Vol. 1731: T. Hsu, Quilts: Central Extensions, Braid Actions, and Finite Groups. XII, 185 pages. 2000.
Vol. 1732: K. Keller, Invariant Factors, Julia Equivalences and the (Abstract) Mandelbrot Set. X, 206 pages. 2000.
Vol. 1733: K. Ritter, Average-Case Analysis of Numerical Problems. IX, 254 pages. 2000.
Vol. 1734: M. Espedal, A. Fasano, A. Mikelić, Filtration in Porous Media and Industrial Applications. Cetraro 1998. Editor: A. Fasano. 2000.
Vol. 1735: D. Yafaev, Scattering Theory: Some Old and New Problems. XVI, 169 pages. 2000.
Vol. 1736: B. O. Turesson, Nonlinear Potential Theory and Weighted Sobolev Spaces. XIV, 173 pages. 2000.
Vol. 1737: S. Wakabayashi, Classical Microlocal Analysis in the Space of Hyperfunctions. VIII, 367 pages. 2000.
Vol. 1738: M. Émery, A. Nemirovski, D. Voiculescu, Lectures on Probability Theory and Statistics. XI, 356 pages. 2000.
Vol. 1739: R. Burkard, P. Deuflhard, A. Jameson, J.-L. Lions, G. Strang, Computational Mathematics Driven by Industrial Problems. Martina Franca, 1999. Editors: V. Capasso, H. Engl, J. Periaux. VII, 418 pages. 2000.
Vol. 1740: B. Kawohl, O. Pironneau, L. Tartar, J.-P. Zolesio, Optimal Shape Design. Tróia, Portugal 1999. Editors: A. Cellina, A. Ornelas. IX, 388 pages. 2000.
Vol. 1741: E. Lombardi, Oscillatory Integrals and Phenomena Beyond all Algebraic Orders. XV, 413 pages. 2000.
Vol. 1742: A. Unterberger, Quantization and Non-holomorphic Modular Forms. VIII, 253 pages. 2000.
Vol. 1743: L. Habermann, Riemannian Metrics of Constant Mass and Moduli Spaces of Conformal Structures. XII, 116 pages. 2000.
Vol. 1744: M. Kunze, Non-Smooth Dynamical Systems. X, 228 pages. 2000.
Vol. 1745: V. D. Milman, G. Schechtman, Geometric Aspects of Functional Analysis. VIII, 289 pages. 2000.
Vol. 1746: A. Degtyarev, I. Itenberg, V. Kharlamov, Real Enriques Surfaces. XVI, 259 pages. 2000.
Vol. 1747: L. W. Christensen, Gorenstein Dimensions. VIII, 204 pages. 2000.
Vol. 1748: M. Ruzicka, Electrorheological Fluids: Modeling and Mathematical Theory. XV, 176 pages. 2001.
Vol. 1749: M. Fuchs, G. Seregin, Variational Methods for Problems from Plasticity Theory and for Generalized Newtonian Fluids. VI, 269 pages. 2001.
Vol. 1750: B. Conrad, Grothendieck Duality and Base Change. X, 296 pages. 2001.
Vol. 1751: N. J. Cutland, Loeb Measures in Practice: Recent Advances. XI, 111 pages. 2001.
Vol. 1752: Y. V. Nesterenko, P. Philippon, Introduction to Algebraic Independence Theory. XIII, 256 pages. 2001.
Vol. 1753: A. I. Bobenko, U. Eitner, Painlevé Equations in the Differential Geometry of Surfaces. VI, 120 pages. 2001.
Vol. 1754: W. Bertram, The Geometry of Jordan and Lie Structures. XVI, 269 pages. 2001.
Vol. 1755: J. Azéma, M. Émery, M. Ledoux, M. Yor, Séminaire de Probabilités XXXV. VI, 427 pages. 2001.
Vol. 1756: P. E. Zhidkov, Korteweg de Vries and Nonlinear Schrödinger Equations: Qualitative Theory. VII, 147 pages. 2001.
Vol. 1757: R. R. Phelps, Lectures on Choquet's Theorem. VII, 124 pages. 2001.
Vol. 1758: N. Monod, Continuous Bounded Cohomology of Locally Compact Groups. X, 214 pages. 2001.
Vol. 1759: Y. Abe, K. Kopfermann, Toroidal Groups. VIII, 133 pages. 2001.
Vol. 1760: D. Filipović, Consistency Problems for Heath-Jarrow-Morton Interest Rate Models. VIII, 134 pages. 2001.
Vol. 1761: C. Adelmann, The Decomposition of Primes in Torsion Point Fields. VI, 142 pages. 2001.
Vol. 1762: S. Cerrai, Second Order PDE's in Finite and Infinite Dimension. IX, 330 pages. 2001.
Vol. 1763: J.-L. Loday, A. Frabetti, F. Chapoton, F. Goichot, Dialgebras and Related Operads. IV, 132 pages. 2001.
Vol. 1764: A. Cannas da Silva, Lectures on Symplectic Geometry. XII, 217 pages. 2001.
Vol. 1765: T. Kerler, V. V. Lyubashenko, Non-Semisimple Topological Quantum Field Theories for 3-Manifolds with Corners. VI, 379 pages. 2001.
Vol. 1766: H. Hennion, L. Hervé, Limit Theorems for Markov Chains and Stochastic Properties of Dynamical Systems by Quasi-Compactness. VIII, 145 pages. 2001.
Vol. 1767: J. Xiao, Holomorphic Q Classes. VIII, 112 pages. 2001.
Vol. 1768: M.J. Pflaum, Analytic and Geometric Study of Stratified Spaces. VIII, 230 pages. 2001.
Vol. 1769: M. Alberich-Carramiñana, Geometry of the Plane Cremona Maps. XVI, 257 pages. 2002.

Vol. 1770: H. Gluesing-Luerssen, Linear Delay-Differential Systems with Commensurate Delays: An Algebraic Approach. VIII, 176 pages. 2002.

Vol. 1771: M. Émery, M. Yor, Séminaire de Probabilités 1967-1980. A Selection in Martingale Theory. IX, 553 pages. 2002.

Vol. 1772: F. Burstall, D. Ferus, K. Leschke, F. Pedit, U. Pinkall, Conformal Geometry of Surfaces in S^4. VII, 89 pages. 2002.

Vol. 1773: Z. Arad, M. Muzychuk, Standard Integral Table Algebras Generated by a Non-real Element of Small Degree. X, 126 pages. 2002.

Vol. 1774: V. Runde, Lectures on Amenability. XIV, 296 pages. 2002.

Vol. 1775: W. H. Meeks, A. Ros, H. Rosenberg, The Global Theory of Minimal Surfaces in Flat Spaces. Martina Franca 1999. Editor: G. P. Pirola. X, 117 pages. 2002.

Vol. 1776: K. Behrend, C. Gomez, V. Tarasov, G. Tian, Quantum Comohology. Cetraro 1997. Editors: P. de Bartolomeis, B. Dubrovin, C. Reina. VIII, 319 pages. 2002.

Vol. 1777: E. García-Río, D. N. Kupeli, R. Vázquez-Lorenzo, Osserman Manifolds in Semi-Riemannian Geometry. XII, 166 pages. 2002.

Vol. 1778: H. Kiechle, Theory of K-Loops. X, 186 pages. 2002.

Vol. 1779: I. Chueshov, Monotone Random Systems. VIII, 234 pages. 2002.

Vol. 1780: J. H. Bruinier, Borcherds Products on O(2,1) and Chern Classes of Heegner Divisors. VIII, 152 pages. 2002.

Vol. 1781: E. Bolthausen, E. Perkins, A. van der Vaart, Lectures on Probability Theory and Statistics. Ecole d' Eté de Probabilités de Saint-Flour XXIX-1999. Editor: P. Bernard. VIII, 466 pages. 2002.

Vol. 1782: C.-H. Chu, A. T.-M. Lau, Harmonic Functions on Groups and Fourier Algebras. VII, 100 pages. 2002.

Vol. 1783: L. Grüne, Asymptotic Behavior of Dynamical and Control Systems under Perturbation and Discretization. IX, 231 pages. 2002.

Vol. 1784: L.H. Eliasson, S. B. Kuksin, S. Marmi, J.-C. Yoccoz, Dynamical Systems and Small Divisors. Cetraro, Italy 1998. Editors: S. Marmi, J.-C. Yoccoz. VIII, 199 pages. 2002.

Vol. 1785: J. Arias de Reyna, Pointwise Convergence of Fourier Series. XVIII, 175 pages. 2002.

Vol. 1786: S. D. Cutkosky, Monomialization of Morphisms from 3-Folds to Surfaces. V, 235 pages. 2002.

Vol. 1787: S. Caenepeel, G. Militaru, S. Zhu, Frobenius and Separable Functors for Generalized Module Categories and Nonlinear Equations. XIV, 354 pages. 2002.

Vol. 1788: A. Vasil'ev, Moduli of Families of Curves for Conformal and Quasiconformal Mappings.IX, 211 pages. 2002.

Vol. 1789: Y. Sommerhäuser, Yetter-Drinfel'd Hopf algebras over groups of prime order. V, 157 pages. 2002.

Vol. 1790: X. Zhan, Matrix Inequalities. VII, 116 pages. 2002.

Vol. 1791: M. Knebusch, D. Zhang, Manis Valuations and Prüfer Extensions I: A new Chapter in Commutative Algebra. VI, 267 pages. 2002.

Vol. 1792: D. D. Ang, R. Gorenflo, V. K. Le, D. D. Trong, Moment Theory and Some Inverse Problems in Potential Theory and Heat Conduction. VIII, 183 pages. 2002.

Vol. 1793: J. Cortés Monforte, Geometric, Control and Numerical Aspects of Nonholonomic Systems. XV, 219 pages. 2002.

Vol. 1794: N. Pytheas Fogg, Substitution in Dynamics, Arithmetics and Combinatorics. Editors: V. Berthé, S. Ferenczi, C. Mauduit, A. Siegel. XVII, 402 pages. 2002.

Vol. 1795: H. Li, Filtered-Graded Transfer in Using Noncommutative Gröbner Bases. IX, 197 pages. 2002.

Vol. 1796: J.M. Melenk, hp-Finite Element Methods for Singular Perturbations. XIV, 318 pages. 2002.

Vol. 1797: B. Schmidt, Characters and Cyclotomic Fields in Finite Geometry. VIII, 100 pages. 2002.

Vol. 1798: W.M. Oliva, Geometric Mechanics. XI, 270 pages. 2002.

Recent Reprints and New Editions

Vol. 1200: V. D. Milman, G. Schechtman, Asymptotic Theory of Finite Dimensional Normed Spaces. 1986. – Corrected Second Printing. X, 156 pages. 2001.

Vol. 1618: G. Pisier, Similarity Problems and Completely Bounded Maps. 1995 – Second, Expanded Edition VII, 198 pages. 2001.

Vol. 1629: J. D. Moore, Lectures on Seiberg-Witten Invariants. 1997 – Second Edition. VIII, 121 pages. 2001.

Vol. 1638: P. Vanhaecke, Integrable Systems in the realm of Algebraic Geometry. 1996 – Second Edition. X, 256 pages. 2001.

Vol. 1702: J. Ma, J. Yong, Forward-Backward Stochastic Differential Equations and Their Applications. 1999. – Corrected Second Printing. XIII, 270 pages. 2000.